Lecture Notes in Mathematics

Edited by A. Dold and B. Eckmann

1234

Combinatoire énumérative

Proceedings of the "Colloque de combinatoire énumérative", held at Université du Québec à Montréal, May 28 – June 1, 1985

Edité par G. Labelle et P. Leroux

Springer-Verlag

Berlin Heidelberg New York London Paris Tokyo

Editors

Gilbert Labelle
Pierre Leroux
Département de mathématiques et d'informatique
Université du Québec à Montréal
C.P. 8888, Succ. A
Montréal, Québec, Canada H3C 3P8

Mathematics Subject Classification (1980): Primary: 05XX
Secondary: 15AXX, 18AXX, 33AXX, 34AXX

ISBN 3-540-17207-6 Springer-Verlag Berlin Heidelberg New York
ISBN 0-387-17207-6 Springer-Verlag New York Berlin Heidelberg

Printing and binding: Druckhaus Beltz, Hemsbach/Bergstr.
2146/3140-543210

INTRODUCTION

This is the Proceedings volume of the "Colloque de combinatoire énumérative, UQAM 1985", which was held at "Université du Québec à Montréal" (UQAM) from May 28 to June 1, 1985, and complemented by a Special Session on Combinatorics at the annual meeting of the Canadian Mathematical Society at "Université Laval", June 6-8, 1985.

The subjects covered in this volume include: enumeration and analysis of specific combinatorial structures like planar maps, Young tableaux, bridges or Dyck paths, and latin rectangles; combinatorics on words; applications of enumerative combinatorics to q-series, to orthogonal polynomials, to differential equations, to linear representations of the symmetric group, to the celebrated Macdonald and Jacobian conjectures, to Lie algebras, etc; recent developments in the combinatorial theory of species of structures; survey papers on Young's work, on Pólya theory, and on a new theory of "heaps of pieces"; a problem session.

The rest of this introduction gives a more detailed description of the scientific activities of the colloquium and of the content of the Proceedings. It is written in french to reflect the bilingual nature of the meeting. Note that while many talks were given in french, a majority (80%) of papers in this volume are written in Shakespeare's language.

Depuis quelques années, la recherche en combinatoire a connu un développement considérable. Il s'agit maintenant d'un véritable domaine des mathématiques qui possède ses objectifs propres (le dénombrement, l'analyse, la construction et la classification des structures finies), des méthodes et des outils de plus en plus efficaces (bijections et involutions, séries génératrices et indicatrices diverses, formules d'inversion, théorie de Pólya, théorie des espèces de structures, développements asymptotiques, etc.) et des champs d'applications très vastes, notamment en analyse classique (polynômes orthogonaux, q-séries, équations différentielles, etc.), en algèbre (algèbre linéaire, fonctions symétriques, représentations des groupes symétriques, algèbre commutative, ...), en informatique (structures de données, conception et analyse d'algorithmes, combinatoire des mots, etc.), en théorie des probabilités, groupes et algèbres de Lie, analyse numérique, topologie algébrique, physique statistique, biologie moléculaire, etc.

Dans le but de faire le point sur ces développements récents, le groupe de recherche en combinatoire de l'Université du Québec à Montréal a organisé un colloque international qui a réuni pendant cinq jours (du 28 mai au l^{er} juin 1985) plus de cent participants et donné lieu à 35 conférences et communications. De plus, deux membres de l'équipe ont organisé une Session spéciale de combinatoire dans le cadre de la réunion annuelle d'été de la Société mathématique du Canada, tenue quelques jours plus tard, soit du 6 au 8 juin 1985, à l'Université Laval, à Québec. On trouvera ci-après une liste de participants au colloque de l'UQAM (une photo de groupe est disponible sur demande) ainsi que le programme scientifique du colloque et de la session spéciale à Québec.

Ce volume constitue donc les comptes-rendus du colloque de combinatoire énumérative, UQAM 1985, et de sa continuation à Québec. Les articles qu'il contient recouvrent une grande partie des thèmes abordés à ces occasions. Ils portent plus particulièrement sur les sujets suivants:

1. Articles de synthèse, en particulier sur les travaux de Young, sur la théorie de Pólya, ainsi que sur les empilements de pièces, théorie qui jette un regard géométrique nouveau sur les monoïdes de commutation de Cartier-Foata et qui est susceptible de multiples applications.

2. La théorie des espèces de structures: plusieurs articles font le point sur certains aspects de cette théorie combinatoire globale, à la fois élégante et efficace: règles et méthodes de calcul, décompositions et classifications, généralisations, etc.

3. Applications de la combinatoire énumérative, par exemple à l'étude des représentations linéaires du groupe symétrique, des fonctions hypergéométriques basiques, des polynômes orthogonaux, des équations différentielles, d'algèbre de Lie d'opérateurs différentiels, ou des célèbres conjectures de Macdonald et jacobienne, à la généralisation des fonctions tangente et sécante, etc.

4. Problèmes de dénombrements de structures particulières selon certains paramètres, par exemple, les "cartes et hypercartes", les "rectangles latins", "les tableaux de Young", les "ponts" également nommés "chemins de Dyck", les "partitions" ou "partages d'entiers" et les "partitions planes", les "tournois", etc.

5. Problèmes de la combinatoire des mots: palindromes, algèbres de mélange et algèbres de Lie, etc.

6. Rapport de la séance de problèmes tenue pendant le colloque.

Signalons que quelques auteurs ont initié dans ces comptes-rendus des séries importantes d'articles portant plus particulièrement sur la résolution combinatoire des équations différentielles, sur la combinatoire des polynômes de Jacobi, et sur la théorie des empilements de pièces.

Remerciements

Au nom des organisateurs du Colloque de combinatoire énumérative UQAM 1985, André Joyal, Gilbert Labelle, Jacques Labelle, Pierre Leroux et Volker Strehl et au nom de tous les participants, nous remercions chaleureusement les personnes et organismes suivants:

La Fondation UQAM, le Conseil de recherche en science naturelle et en génie du Canada, le Fonds FCAR du Québec, l'Université du Québec à Montréal, pour leur aide financière généreuse.

La société mathématique du Canada, qui a parrainé le Colloque de Montréal et suscité la tenue de la session spéciale de combinatoire à Québec, ainsi que l'Université Laval pour son hospitalité.

Les conférenciers invités qui ont agi comme éditeurs associés ainsi que les nombreux arbitres qui ont effectué un excellent travail d'examen critique de tous les articles soumis à ces comptes-rendus.

Hélène Décoste, étudiante au doctorat, pour son expertise et son aide constante dans l'organisation du colloque et la préparation des comptes-rendus, notamment au niveau du traitement de texte et d'images sur micro-ordinateur Macintosh.

Dominique Chabot, France Gauthier, Hélène Meunier, secrétaires du Département de mathématiques et d'informatique de l'UQAM qui ont assuré la mise en forme d'un grand nombre d'articles de ces comptes-rendus.

Les différents services de l'UQAM, en particulier le service des relations publiques ainsi que Manon Gauthier et Madeleine Loubert, et les nombreux étudiants qui ont contribué à rendre ce colloque des plus accueillants et à en faire ainsi un franc succès.

Gilbert Labelle, Pierre Leroux

PARTICIPANTS

Jaromir ABRHAM
Dept. Industrial Engineering
University of Toronto
Toronto, Ontario
Canada, M5S 1A4

Ashok K. AGARWAL
Dept. of Mathematics
Pennsylvania State University
University Park, PA 16802
U.S.A.

Georges E. ANDREWS
Dept. of Mathematics
Pennsylvania State University
University Park, PA 16802
U.S.A.

Pierre ANTAYA
91, rue Dollard
Chateauguay (Québec)
Canada
J6K 1W5

Didier ARQUÈS
Institut des Sciences ex. et appl.
4, rue des frères Lumières
F-68093 Mulhouse
France

Richard ASKEY
Dept. of Mathematics
University of Wisconsin
Madison, WI 53706
U.S.A.

Joffroy BEAUQUIER
Labor. Rech. Informatique
Université Paris Sud
F-91405 Orsay
France

François BÉDARD
Dép. mathématiques et informatique
Université du Québec à Montréal
C. P. 8888, Succ. A, Montréal, Qué.
Canada, H3C 3P8

Marie-France BÉLANGER
Dép. mathématiques et informatique
Université de Sherbrooke
Sherbrooke, (Québec)
Canada, J1K 1N7

François BERGERON
Dép. mathématiques et informatique
Université du Québec à Montréal
C. P. 8888, Succ. A, Montréal, Qué.
Canada, H3C 3P8

Nantel BERGERON
Dép. mathématiques et informatique
Université du Québec à Montréal
C. P. 8888, Succ. A, Montréal, Qué.
Canada, H3C 3P8

Terry BISSON
Department of Mathematics
Canisius College
Buffalo, N. Y. 14208
U.S.A.

Anders BJÖRNER
Dept. of Mathematics
Mass. Institute of Technology
Cambridge, MA 02139
U.S.A.

Marie BLAIN
785, Rue Franchère
Laval, (Québec)
Canada
H7E 3R1

Sylvain BOUCHER
Dép. mathématiques et informatique
Université du Québec à Montréal
C. P. 8888, Succ. A, Montréal, Qué.
Canada, H3C 3P8

Pierre BOUCHARD
Dép. mathématiques et informatique
Université du Québec à Montréal
C. P. 8888, Succ. A, Montréal, Qué.
Canada, H3C 3P8

Jacques BOURRET
Dép. mathématiques et informatique
Université du Québec à Montréal
C. P. 8888, Succ. A, Montréal, Qué.
Canada, H3C 3P8

David BRESSOUD
Dept. of Mathematics
Pennsylvania State University
University Park, PA 16802
U.S.A.

Srecko BRLEK
Dép. mathématiques et informatique
Université du Québec à Montréal
C. P. 8888, Succ. A, Montréal, Qué.
Canada, H3C 3P8

Jim BYRNES
Dept. of Math., Harbor Campus
Univ. of Massachussets at Boston
Boston, MA 02125
U.S.A.

N. J. CALKIN
Dept. of Combin. & Optim.
University of Waterloo
Waterloo, Ontario
Canada, N2L 3G1

Luigi CERLIENCO
Instituto Matematico Dell'Univ.
Via Ospedale 72
I-09100 Cagliari
Italie

Phillip J. CHASE
8716, Oxwell Lane
Laurel
Maryland, 20708
U.S.A.

Young-Ming CHEN
Dept. of Math. & Comp. Sci.
S.U.N.Y. College at Brockport
Brockport, New York 14420
U.S.A.

Julien CONSTANTIN
Dép. mathématiques et informatique
Université de Sherbrooke
Sherbrooke, (Québec)
Canada, J1K 1N7

Ivan CONSTANTINEAU
Dép. mathématiques et informatique
Université du Québec à Montréal
C. P. 8888, Succ. A, Montréal, Qué.
Canada, H3C 3P8

Henry CRAPO
Bat. 24, INRIA
B. P. 105
F-78153 Le Chesnay Cedex
France

Pierre DAMPHOUSSE
Dép. de Mathématiques
Univ. de Tours
F-37200 Tours
France

Hélène DÉCOSTE
Dép. mathématiques et informatique
Université du Québec à Montréal
C. P. 8888, Succ. A, Montréal, Qué.
Canada, H3C 3P8

Maylis DELEST
UER Math & Info., Univ. Bordeaux I
351, Cours de la Libération
F-33405 Talence Cedex
France

Myriam DESAINTE-CATHERINE
Math & Info., Univ. Bordeaux I
351, Cours de la Libération
F-33405 Talence Cedex
France

Serge DULUCQ
Math. & Info., Un. Bordeaux I
351, Cours de la Libération
F-33405 Talence Cedex
France

Omer EGECIOGLU
Dept. of Computer Science
Univ. of California, Santa Barbara
Santa Barbara, CA 93106
U.S.A.

Mohsine ELEUDJ
Dép. mathématiques et informatique
Université du Québec à Montréal
C. P. 8888, Succ. A, Montréal, Qué.
Canada, H3C 3P8

Luc FAVREAU
Dép. mathématiques et informatique
Université du Québec à Montréal
C. P. 8888, Succ. A, Montréal, Qué.
Canada, H3C 3P8

Dominique FOATA
Dept. de Math. Univ. Strasbourg
7, rue René Descartes
F-67084 Strasbourg
France

John M. FREEMAN
Dept. of Mathematics
Florida Atlantic University
Boca Raton, FL 33431
U.S.A.

Jean-François GAGNÉ
Dép. mathématiques et informatique
Université du Québec à Montréal
C. P. 8888, Succ. A, Montréal, Qué.
Canada, H3C 3P8

Adriano GARSIA
Dept. of Mathematics
Un. California, San Diego
La Jolla, CA 92093
U.S.A.

Daniel GATIEN
Dept. of Mathematics
Mass. Institute of of Technology
Cambridge, MA 02139
U.S.A.

Gilles GAUTHIER
Dep. Sciences fondamentales
Université du Québec à Chicoutimi
Chicoutimi, Qué.
Canada, G7H 2B1

Ira M. GESSEL
Dept of Mathematics
Brandeis University
Waltham, MA 02254
U.S.A.

Chris GODSIL
Department of Mathematics
Simon Fraser University
Brunaby, British Columbia
Canada, V5A 1S6

Ian GOULDEN
Dept of Comb. & Optim.
University of Waterloo
Waterloo, Ontario
Canada, N2L 3W4

Dominique GOUYOU-BEAUCHAMPS
UER Math & Info. Univ. Bordeaux
351, Cours de la Libération
F-33405 Talence Cedex
France

Curtis GREENE
Dept. of Mathematics
Haverford College
Haverford, PA 19041
U.S.A.

Werner HÄSSELBARTH
Inst. Quantenchemie Tu Berlin
Holbeinstr. 48
D-1000 Berlin
R.F.A.

David J. JACKSON
Dept. Comb. & Optim.
University of Waterloo
Waterloo, Ontario
Canada, N2L 3G1

André JOYAL
Dép. mathématiques et informatique
Université du Québec à Montréal
C. P. 8888, Succ. A, Montréal, Qué.
Canada, H3C 3P8

Gil KALAI
Dept. of Mathematics
Mass. Institute of Technology
Cambridge, MA 02139
U.S.A.

Adalbert KERBER
Math. Inst. Univ. Bayreuth
Postfach 3008
D-8580 Bayreuth
R.F.A.

Germain KREWERAS
Inst. de Stat. Univ. P. et M. Curie
4, Place Jussieu
F-75005, Paris
France

Nicholas KRIER
Dept. of Mathematics
Colorado State University
Fort Collins, CO 80523
U.S.A.

Gilbert LABELLE
Dép. mathématiques et informatique
Université du Québec à Montréal
C. P. 8888, Succ. A, Montréal, Qué.
Canada, H3C 3P8

Jacques LABELLE
Dép. mathématiques et informatique
Université du Québec à Montréal
C. P. 8888, Succ. A, Montréal, Qué.
Canada, H3C 3P8

Martine LABRÈCHE
Dép. mathématiques et informatique
Université du Québec à Montréal
C. P. 8888, Succ. A, Montréal, Qué.
Canada, H3C 3P8

Clément LAM
Computer Sc., Concordia University
1455, Boul. de Maisonneuve Ouest
Montréal, Qué.
Canada, H3G 1M8

Denis LARIVIÈRE
Dép. mathématiques et informatique
Université du Québec à Montréal
C. P. 8888, Succ. A, Montréal, Qué.
Canada, H3C 3P8

Pierre LEROUX
Dép. mathématiques et informatique
Université du Québec à Montréal
C. P. 8888, Succ. A, Montréal, Qué.
Canada, H3C 3P8

Jean-Benoit LÉVESQUE
Dép. mathématiques et informatique
Université du Québec à Montréal
C. P. 8888, Succ. A, Montréal, Qué.
Canada, H3C 3P8

André LONGTIN
Dép. de Mathématiques
Univ. du Québec à Trois-Rivières
C. P. 500, Trois-Rivières, Qué.
Canada, G9A 5H7

Anne-Marie LORRAIN
9218, Ave. Millen
Montréal, (Québec)
Canada,
H2M 1W7

Diana MARCUS
Dept. of Math., Mesa College
Mesa College Drive
San Diego, CA 92111
U.S.A.

John McKAY
Computer Sc., Concordia University
1455, Boul. de Maisonneuve Ouest
Montréal, Qué.
Canada, H3G 1M8

Guy MELANCON
Dép. mathématiques et informatique
Université du Québec à Montréal
C. P. 8888, Succ. A, Montréal, Qué.
Canada, H3C 3P8

Armel MERCIER
Dép. de Mathématiques, UQAC
555, Boul. Université
Chicoutimi, Qué
Canada, G7H 2B1

Robert MICHAUD
Dép. mathématiques et informatique
Université du Québec à Montréal
C. P. 8888, Succ. A, Montréal, Qué.
Canada, H3C 3P8

Sri G. MOHANTY
Dept. Math. Sci., McMaster Univ.
1280 Main Street West
Hamilton, Ontario
Canada, L8S 4K1

Claire MORAZAIN
Dép. mathématiques et informatique
Université du Québec à Montréal
C. P. 8888, Succ. A, Montréal, Qué.
Canada, H3C 3P8

Tadepalli NARAYANA
Dept. of Mathematics
University of Alberta
Edmonton, Alberta
Canada, T6G 2H1

Oscar NAVA
Dept of Mathematics
Mass. Institute of Technology
Cambridge, MA 02139
U.S.A.

Heinrich NIEDERHAUSEN
Dept. of Mathematics
Florida Atlantic University
Boca Raton, Florida 33431
U.S.A.

Kathy O'HARA
Dept. of Mathematics
Grinnell College
Grinnell, Iowa 50112-0806
U.S.A.

Joseph OLIVEIRA
Dept. of Mathematics
Mass. Institute of Technology
Cambridge, MA 02139
U.S.A.

Peter PAULE
Mathematik, Univ. Bayreuth
Postfach 3008
D 8580 Bayreuth
R.F.A.

Alain PAUTASSO
Computer Sc., Concordia University
1455, Boul. De Maisonneuve Ouest
Montréal, Québec
Canada, H3G 1M8

Francesco PIRAS
Dip. di Mathematica, Un. di Cagliari
Via Ospedale 72
I-09100 Cagliari
Italie

Simon PLOUFFE
Dép. mathématiques et informatique
Université du Québec à Montréal
C. P. 8888, Succ. A, Montréal, Qué.
Canada, H3C 3P8

Robert W. QUACKENBUSH
Dept. of Mathematics and Astronomy
The University of Manitoba
Winnipeg, Manitoba
Canada, R3T 2N2

Don RAWLINGS
Dép. mathématiques et informatique
Université du Québec à Montréal
C. P. 8888, Succ. A, Montréal, Qué.
Canada, H3C 3P8

Eric REGENER
Comp. Sci., Concordia Univ.
1455 De Maisonneuve Ouest
Montréal, Qué.
Canada, H3G 1M8

Jeff REMMEL
Dept. of Mathematics
Univ. Cal. San Diego
La Jolla, CA 92093
U.S.A.

Christophe REUTENAUER
Dép. mathématiques et informatique
Université du Québec à Montréal
C. P. 8888, Succ. A, Montréal, Qué.
Canada, H3C 3P8

Bruce RICHMOND
Dept, of Combin. & Optim.
University of Waterloo
Waterloo, Ont.
Canada, N2L 3G1

Ivan RIVAL
Dept. of Mahematics and Statistics
University of Calgary
Calgary, Alberta
Canada, T2N 1N4

Jean-François ROCHON
Société de Téléinformatique RTC
2050 Mansfield
Montréal (Québec)
Canada, H3A 1Y9

Ivo ROSENBERG
Dép. de Mathématiques et Statistiques
Université de Montréal
C. P. 6128, Succ. A, Montréal, Qué.
Canada, H3C 3J7

Gian-Carlo ROTA
Dept. of Mathematics
Mass. Institute of Technology
Cambridge, MA 02139
U.S.A.

Ernst RUCH
Inst. Quantenchemie Fu Berlin
Holbeinstrasse 48
D-1000, Berlin
R.F.A.

Bruce SAGAN
Department of Mathematics
Univ. of Pennsylvania
Philadelphia, PA 19104
U.S.A.

Richard STANLEY
Dept. of Mathematics
Mass. Institute of Technology
Cambridge, MA 02139
U.S.A.

Dennis STANTON
School of Mathematics
Univ. of Minnesota
Minneapolis, MR 55455
U.S.A.

Volker STREHL
Informatik 1 Univ. Erlangen
Martensstr. 3
D-8520 Erlangen
R.F.A.

Claudette TABIB
1045, rue Michel Moreau
Boucherville (Québec)
Canada
J4B 3Z9

Denis THÉRIEN
Comp. Sci., Mc Gill University
C. P. 6070, Succ. A
Montréal (Québec)
Canada, H3C 3G1

Gabriel THIERRIN
Dept. of Mathematics
University of Western Ontario
London, Ont.
Canada, N6A 5B7

Loys THIMONIER
U.E.R. de Mathematiques
Univ. Amiens
F-80039 Amiens Cedex
France

Pierre TREMBLAY
Dept. of Mathematics
Penn State University
University Park, PA 16802
U.S.A.

Edward VALENTINE
Comp. Sci., Concordia University
1455, Boul. de Maisonneuve Ouest
Montréal, Qué.
Canada, H3G 1M8

Antonietta VENEZIA
Dip. Math., U. di Roma
Piazzale A. Moro
I-00100 Roma
Italie

Gérard X. VIENNOT
UER Math. & Info., Univ. Bordeaux I
351, Cours de la Libération
F-33405 Talence
France

Terry VISENTIN
Dept. of Combinatorics and Optim
University of Waterloo
Waterloo, Ont.
Canada, N2L 3G1

Dennis WHITE
School of Mathematics
University of Minnesota
Minneapolis, MN 55455
U.S.A.

Yeong-Nan YEH
Dép. mathématiques et informatique
Université du Québec à Montréal
C. P. 8888, Succ. A, Montréal, Qué.
Canada, H3C 3P8

Doron ZEILBERGER
Dept. of Mathematical Science
Drexel University
Philadelphia, PA 19104
U.S.A.

Günther ZIEGLER
Dept. of Mathematics
Mass. Institute of Technology
Cambridge, MA 02139
U.S.A.

COLLOQUE DE COMBINATOIRE ÉNUMÉRATIVE, UQAM 1985
DU 28 MAI AU 1er JUIN 1985

CONFÉRENCES (50 min.)

George E. Andrews Pennsylvania State University, State College
1. *SCRATCHPAD and Combinatorics.*
2. *q-Series, Partitions and Physics.*

Richard Askey University of Wisconsin, Madison
Basic Hypergeometric Extensions of the Classical Orthogonal Polynomials.

Dominique Foata Université de Strasbourg
Fonctions symétriques et séries hypergéométriques multivariées.

Adriano M. Garsia University of California, San Diego
Alfred Young revisited.

Ira Gessel Brandeis University, Waltham Mass.
Derangements, Charlier Polynomials and Three-Line Latin Rectangles.

David M. Jackson University of Waterloo
Counting cycles in permutations by group characters.

André Joyal Université du Québec à Montréal
La théorie des espèces de structures.

Adalbert Kerber Universität Bayreuth
Enumeration under Finite Group Action: Symmetry Classes of Mappings.

Gilbert Labelle Université du Québec à Montréal
Méthodes de calcul en théorie des espèces.

Gian-Carlo Rota Massachusetts Institute of Technology, Cambridge
Le pléthysme.

Richard P. Stanley Massachusetts Institute of Technology, Cambridge
Two Poset Polytopes.

Volker Strehl Universität Erlangen-Nürnberg
La combinatoire des configurations de Jacobi.

Gérard X. Viennot Université de Bordeaux I
1. *Empilements I: Lemmes fondamentaux.*
2. *Empilements II: Applications.*

Doron Zeilberger Drexel University, Philadelphia
Towards a Combinatorial Proof of the Jacobian Conjecture?

COMMUNICATIONS (30 min.)

Didier Arquès Institut des Sciences exactes et appl., Mulhouse
Une relation fonctionnelle nouvelle et son application au dénombrement des cartes et hypercartes planaires pointées.

François Bergeron Université du Québec à Montréal
Représentations combinatoires de groupes et algèbres de Lie.

Anders Björner* Massachusetts Institute of Technology, Cambridge
Michelle Wachs University of Miami
Generalized Quotients of Finite Coxeter Groups.

David Bressoud Pensylvania State University, State College
Sur les identités pour les termes constants reliés aux systèmes de racines.

Pierre Damphousse Université de Tours
Classification des cartes cellulaires.

Marie-Pierre Delest Université de Bordeaux I
Enumération de polyominos verticalement convexes.

Serge Dulucq*, Robert Cori,
Gérard X. Viennot Université de Bordeaux I
Chemins dans le plan et permutations de Baxter alternantes.

Omer Egecioglu* University of California, Santa Barbara
Jeff Remmel University of California, San Diego
A Combinatorial Proof of the Giambelli Identity for Schur Functions.

Chris Godsil Simon Fraser University, Vancouver
Generating Latin Rectangles.

D. Gouyou-Beauchamps Université de Bordeaux I
Tableaux de Young et chemins sous-diagonaux.

Werner Hässelbarth Freie Universität Berlin
A Generalisation of the Pólya/de Bruijn Enumeration Theory and its Application to "Chemical Combinatorics".

Germain Kreweras Université Pierre et Marie Curie, Paris
Lois croisées de plusieurs paramètres descriptifs des ponts.

Pierre Leroux* Université du Québec à Montréal
Gérard X. Viennot Université de Bordeaux I
Résolution combinatoire des systèmes d'équations différentielles.

Heinrich Niederhausen Florida Atlantic University, Boca Raton
Polynomial Sequences of Generalized Appell Type with Coefficients of Polynomial Structure.

Jeffrey B. Remmel University of California, San Diego
Q-Rook Theory and Applications.

Christophe Reutenauer Institut de Programmation, Paris
Théorème de Poincaré-Birkhoff-Witt, le logarithme et des représentations du groupe symétrique d'ordre les nombres de Stirling.

Dennis Stanton University of Minnesota, Minneapolis
Applications of q-Hermite Polynomials.

Loys Thimonier* Université d'Amiens
Joffroy Beauquier Université Paris-Sud
Prefix-Free Words of Length n over m Letters: Two-Sided Well-Balanced Parentheses and Palindromes.

Dennis White University of Minnesota, Minneapolis
Hybrid Tableaux.

RÉUNION D'ÉTÉ 1985 DE LA SOCIÉTÉ MATHÉMATIQUE DU CANADA
Université Laval, Québec, du 6 au 8 juin 1985

SESSION SPÉCIALE DE COMBINATOIRE

COMMUNICATIONS (30 min.)

Henry Crapo CRMA, Un. de Montréal et INRIA, Rocquencourt
La topologie géométrique et structurale.

Myriam DeSainte-Catherine* Université de Bordeaux I
Gérard X. Viennot Université de Bordeaux I
Le nombre de tableaux de Young dont les colonnes sont de hauteur paire.

Ira Gessel Brandeis University, Waltham Mass.
Counting Acyclic Digraphs.

Ian Goulden University of Waterloo
Quadratic forms of Schur functions.

Jacques Labelle Université du Québec à Montréal
Décomposition des espèces de structures.

Clement Lam Concordia University
A computer search for projective plane of order 10.

André Longtin Université du Québec à Trois-Rivières
Nombres sécants généralisés: une solution à un problème d'étiquetage d'arbres orientés.

John McKay Concordia University
Computing Galois groups of polynomials over **Q**.

Bruce Sagan Middlebury College, Middlebury Vermont
Shellability of Exponential Structures.

Claudette Tabib CEGEP Édouard-Montpetit, Longueuil Qué.
A propos des inégalités d'Erdös et Moser sur le plus grand sous-tournoi transitif d'un tournoi.

Denis Thérien Université Mc Gill, Montréal
Aspects combinatoires des groupes nilpotents.

Gérard X. Viennot Université de Bordeaux I
Théorie combinatoire des approximants de Padé.

---

* Dans le cas d'un travail conjoint, un astérisque désigne celui des auteurs présentant la communication.

TABLE DES MATIERES

PARTITIONS WITH "N COPIES OF N"

A.K. AGARWAL
Department of Mathematics
The Pennsylvania State University
University Park, PA 16802, USA

Abstract. In this short note we prove a general partition theorem involving partitions with "N copies of N". These partitions arise in the Study of Hard-Hexagon Model and have recently been studied in [1]. To exhibit the importance of our main theorem we present three particular cases which yield elegant partition identities of Rogers-Ramanujan Type. We shall also pose a very significant open problem.

1. The Main Result. We propose to prove the following:

Theorem 1. For $k \geq -3$, let $C_k(n)$ denote the number of partitions with "N copies of N" of n such that each pair of summands m_i, r_j satisfies $|m-r| > i+j+k$. Then

$$\sum_{n=0}^{\infty} C_k(n)q^n = \sum_{n=0}^{\infty} \frac{q^{n\left[1+\frac{(k+3)(n-1)}{2}\right]}}{(q;q)_n\,(q;q^2)_n}, \tag{1.1}$$

here $(a;q)_n$ denote the rising 'q-factorial'.

2. Proof. Let $C_k(m,n)$ denote the number of partitions enumerated by $C_k(n)$ with the added restriction that there be exactly m parts. We shall first prove that

$$\begin{aligned} C_k(m,n) = {} & C_k(m,n-m) + C_k(m-1,n-km-3m+k+2) \\ & + C_k(m,n-2m+1) - C_k(m,n-3m+1). \end{aligned} \tag{2.1}$$

To prove (2.1) we split the partitions enumerated by $C_k(m,n)$ into three classes: (i) those that do not contain k_k as a part, (ii) those that contain 1_1, as a part, and (iii) those that contain $k_k (k > 1)$ as a part. We now transform the partitions in class (i) by deleting 1 from each part ignoring the subscripts. Obviously, this transformation will not disturb the inequalities between the parts and so the transformed partition will be of the type enumerated by $C_k(m,n-m)$. Next we transform the partitions in class

(ii) by deleting the summand 1_1, and then subtracting $k+3$ from all the remaining parts ignoring the subscripts. The transformed partition will be of the type enumerated by $C_k(m-1,n-km-3m+k+2)$. Here we note that k cannot be less than -3. Finally, we transform the partitions in class (iii) by replacing k_k by $(k-1)_{k-1}$ and then subtracting 2 from all the remaining parts. This will produce a partition of $n-1-2(m-1) = n-2m+1$ into m parts. It is important to note here that by this transformation we get only those partitions of $n-2m+1$ into m parts which contain $(k-1)_{k-1}$ as a part. Therefore the actual number of partitions which belong to class (iii) is $C_k(m,n-2m+1) - C_k(m,n-3m+1)$, where $C_k(m,n-3m+1)$ is the number of partitions of $n-2m+1$ into m parts which are free from the parts like k_k. The above transformations clearly establish a bijection between the partitions enumerated by $C_k(m,n)$ and those enumerated by $C_k(m,n-m)$ + $C_k(m-1,n-km-3m+k+2) + C_k(m,n-2m+1) - C_k(m,n-3m+1)$. Thus identity (2.1) is established.]

Let

$$(2.2) \qquad f_k(z,q) = \sum_{n=0}^{\infty} \sum_{m=0}^{\infty} C_k(m,n) z^m q^n .$$

Then (2.1) implies that

$$(2.3) \qquad f_k(z,q) = \sum_{n=0}^{\infty} \sum_{m=0}^{\infty} \Big[C_k(m,n-m)+C_k(m-1,n-km-3m+k+2) + C_k(m,n-2m+1) - C_k(m,n-3m+1)\Big] z^m q^n$$

$$= \sum_{n=0}^{\infty} \sum_{m=0}^{\infty} C_k(m,n-m)(zq)^m q^{n-m} + zq \sum_{n=0}^{\infty} \sum_{m=0}^{\infty} C_k(m-1,n-km-3m+k+2) \cdot (zq^{k+3})^{m-1} q^{n-m(k+3)+k+2}$$

$$+ \frac{1}{q} \sum_{n=0}^{\infty} \sum_{m=0}^{\infty} C_k(m,n-2m+1)(zq^2)^m q^{n-2m+1}$$

$$- \frac{1}{q} \sum_{n=0}^{\infty} \sum_{m=0}^{\infty} C_k(m,n-3m+1)(zq^3)^m q^{n-3m+1}$$

$$= f_k(zq,q) + zqf_k(zq^{k+3},q) + \frac{1}{q} f_k(zq^2,q) - \frac{1}{q} f_k(zq^3,q) .$$

Setting $f_k(z,q) = \sum_{n=0}^{\infty} \lambda_{k,n}(q) z^n$, and then comparing the coefficients of z^n

on each side of (2.3), we see that

$$\lambda_{k,n}(q) = \frac{\lambda_{k,n-1}(q)\ q^{(n-1)(k+3)+1}}{(1-q^n)(1-q^{2n-1})} \tag{2.4}$$

Iterating (2.4) n times and observing that $\lambda_{k,0}(q) = 1$, we find that

$$\lambda_{k,n}(q) = \frac{q^{n\left[1+\frac{(k+3)(n-1)}{2}\right]}}{(q;q)_n\ (q;q^2)_n}\ . \tag{2.5}$$

Therefore

$$f_k(z,q) = \sum_{n=0}^{\infty} \frac{q^{n\left[1+\frac{(k+3)(n-1)}{2}\right]}\ z^n}{(q;q^2)_n(q;q)_n}\ . \tag{2.6}$$

Now

$$\sum_{n=0}^{\infty} C_k(n)q^n = \sum_{n=0}^{\infty}\left\{\sum_{m=0}^{\infty} C_k(m,n)\right\}q^n = f_k(1,q)$$
$$= \sum_{n=0}^{\infty} \frac{q^{n\left[1+\frac{(k+3)(n-1)}{2}\right]}}{(q;q^2)_n(q;q)_n}\ .$$

This completes the proof of the theorem.

3. Particular Cases. If $k=0$, Theorem 1, in view of the identity [2, I(46), p.156]

$$\sum_{n=0}^{\infty} \frac{q^{n(3n-1)/2}}{(q;q)_n(q;q^2)_n} = \frac{1}{(q;q)_\infty} \prod_{n=1}^{\infty} (1-q^{10n})(1-q^{10n-6})(1-q^{10n-4}) \tag{3.1}$$

reduces to

Theorem 3.1. The number of partitions with "N copies of N" of n such that each pair of summands m_i, r_j satisfies $|m-r| > i+j$ equals the number of ordinary partitions of n into parts $\not\equiv 0, \pm 4 \pmod{10}$.

Ex. For $n=6$, we have 8 relevant partitions of each kind, viz., $6_1, 6_2, 6_3, 6_4, 6_5, 6_6, 5_1+1_1, 5_2+1_1$ of the first kind and 51, 3^2, 321, 31^3, 2^3, $2^2 1^2$, 21^4, 1^6 of the second kind.

For $k = -1$, Theorem 1 in view of the identity [2, I(61), p.158]

$$(3.2) \qquad \sum_{n=0}^{\infty} \frac{q^{n^2}}{(q;q)_n (q;q^2)_n} = \frac{1}{(q;q)_\infty} \prod_{n=1}^{\infty} (1-q^{14n})(1-q^{14n-6})(1-q^{14n-8})$$

leads to

Theorem 3.2. The number of partitions with "N copies of N" of n such that each pair of summands m_i, r_j satisfies $|m-r| \geq i+j$ equals the number of ordinary partitions of n into parts $\not\equiv 0, \pm 6 \pmod{14}$.

Example. For $n=6$, we have 10 relevant partitions of each kind, viz., $6_1, 6_2, 6_3, 6_4, 6_5, 6_6, 5_1+1_1, 5_2+1_1, 5_3+1_1, 4_1+2_1$ of the first kind and 51, 42, 41^2, 3^2, 321, 31^3, 2^3, $2^2 1^2$, 21^4, 1^6 of the second kind.

The particular case $k = -2$ of the Theorem 1, in view of the identity [3, Eq. (3.1), p. 219]

$$(3.3) \qquad \sum_{n=0}^{\infty} \frac{q^{\frac{1}{2}(n^2+n)}}{(q;q)_n (q;q^2)_n} = \prod_{n=1}^{\infty} \frac{(1+q^n)(1-q^{7n-2})(1-q^{7n-5})(1-q^{7n})}{(1-q^n)(1+q^{7n-1})(1+q^{7n-6})}$$

corresponds to

Theorem 3.3. The number of partitions with "N copies of N" of n such that each pair of summands m_i, r_j satisfies $|m-r| \geq i+j-1$ equals $\sum_{k=0}^{n} A_{n-k} B_k$, where A_n denote the number of partitions of n into distinct parts $\equiv 3$ or $4 \pmod 7$ and B_n denote the number of ordinary partitions of n into parts $\not\equiv 0, 4, 10 \pmod{14}$.

4. Conclusion. Theorems 3.1, 3.2 and 3.2 are nice combinatorial interpretations of Theorem 1 at $k = 0, -1$ and -2 respectively. theorems 3.1 and 3.2 are the particular cases of the main result of [1]. The most obvious question arising from this work is: Is there a reasonable combinatorial interpretation of Theorem 1 for general value of k?

REFERENCES

1. A.K. Agarwal and G.E. Andrews, Rogers-Ramanujan Identities for Partitions with "N copies of N" (Communicated).

2. L.J. Slater, Further identities of the Rogers-Ramanujan type, Proc. London Math. Soc. 54 (1951-52), pp. 147-167.

3. W.N. Bailey, On the simplification of some identities of the Rogers-Ramanujan type, Proc. London Math. Soc. (3) 1, (1951), pp. 217-221.

RELATIONS FONCTIONNELLES ET DENOMBREMENT DES HYPERCARTES PLANAIRES POINTEES

Didier ARQUES
Institut des Sciences Exactes et Appliquées
4 rue des Frères Lumière, 68093 MULHOUSE-Cédex, France

Abstract

We show here, by using two distinct geometrical decompositions of rooted planar hypermaps, that there exists two functional relations whose unique solution is the generating function enumerating rooted planar hypermaps.
Used together, these two relations allow us to obtain, without any hard formal calculus, a really simple system of parametric equations for the generating series enumerating rooted planar hypermaps by their number of vertices, faces and hyperedges. From this we get the general term of this series.
One of the above cited geometrical decompositions leads us to define a natural notion of the inner hypermap of a rooted planar hypermap. Some enumerations related to this notion are treated.

Introduction

T.R.S. Walsh utilise (cf [7]) les dénombrements sur les cartes eulériennes obtenus à partir de relations de récurrence (cf W.T. Tutte [5]) pour décompter les hypercartes planaires en fonction du nombre de brins et du nombre de faces.
On montre dans cet article, à partir de deux décompositions géométriques différentes, l'existence de deux relations fonctionnelles dont la série génératrice des hypercartes planaires pointées est unique solution. L'une de ces relations est analogue à celle établie par W.T. Tutte pour les cartes planaires pointées (cf [6] et [4]) et est obtenue en contractant une arête. L'autre utilise la décomposition géométrique que nous avons introduite dans [2], et qui consiste à contracter tout un ensemble d'arêtes.

La considération simultanée de ces deux équations permet, en évitant tout calcul formel compliqué, de déterminer très simplement un système d'équations paramétriques pour la série génératrice des hypercartes planaires pointées décomptées en fonction du nombre de sommets, de faces et d'hyperarêtes (cf théorème 3 du III).

La formule de Lagrange permet alors de donner le terme général de cette série génératrice (cf corollaire 1 du théorème 3). Un cas particulier de ce système paramétrique permet de retrouver le dénombrement précité de T.R.S. Walsh. On donne enfin le dénombrement des hypercartes planaires pointées dont tous les sommets appartiennent à la frontière de la face extérieure. Le paragraphe I rappelle les principales définitions utilisées dans la suite.

I. Définitions et notations

Nous rappelons dans ce paragraphe les principales définitions utilisées dans la suite (cf par exemple [3] et [6]).

I.1. Définition

. Une carte planaire est une représentation de la sphère de $\mathbb{R}^3$ comme union d'un nombre fini d'ensembles disjoints appelés cellules. Elles sont de trois types

1 - les sommets qui sont des points.

2 - les arêtes qui sont des arcs simples ouverts de Jordan dont les extrémités (confondues ou non) sont des sommets.

3 - les faces qui sont des domaines simplement connexes dont les frontières sont des réunions de sommets et d'arêtes.

. Deux cellules sont dites incidentes si l'une est dans la frontière de l'autre.

. Le degré d'un sommet est le nombre d'arêtes qui lui sont incidentes. (Une boucle, arête dont les extrémités sont confondues, est comptée pour deux dans le degré de son extrémité).

. Une arête est un isthme si elle est incidente à une seule face.

. Le degré d'une face est le nombre d'arêtes qui lui sont incidentes, les isthmes étant comptés deux fois.

I.2. On appelle brin une arête orientée de la carte planaire et on note B leur ensemble. On associe à tout brin, de façon évidente, son sommet initial, son sommet final, l'arête qui constitue son support, le brin qui lui est opposé.

. On définit la permutation α sur B qui à tout brin associe son brin opposé. α est une involution sans point fixe dont les cycles sont bijectivement associés aux arêtes de la carte.

. On note σ la permutation sur B qui à tout brin b associe le premier brin rencontré en tournant autour du sommet initial de b dans le sens positif choisi sur la sphère. Les cycles de σ sont bijectivement associés aux sommets de la carte.

. On note $\bar{\sigma}$ la permutation $\sigma \circ \alpha$ sur B. Les cycles de $\bar{\sigma}$ sont les circuits orientés constituant les frontières des faces topologiques de la carte. Les cycles de $\bar{\sigma}$ sont donc bijectivement associés aux faces de la carte.
Dans la suite, un sommet (resp. arête, face) sera, suivant le contexte, soit l'objet topologique défini au 1, soit le cycle pour σ (resp. α, $\bar{\sigma}$) qui lui est associé par les définitions ci-dessus.
Un brin est dit incident à une cellule de la carte si il appartient au cycle associé à cette cellule.

. Pour b dans B et τ permutation sur B, on note $\tau^*(b)$ le cycle pour τ engendré par b.
Si A est inclus dans B et si b est dans A, alors $\tau_{|A}(b)$ est le premier brin dans A parmi $\tau(b)$, $\tau^2(b)$, ...

. Une carte planaire est dite pointée si un brin $\tilde{b}$ est choisi. $\tilde{b}$ est appelé le brin pointé de la carte, et son sommet initial $\tilde{s}$ est appelé le sommet pointé de la carte.
On appelle alors face extérieure de la carte, la face $\bar{\sigma}^*(\tilde{b})$ engendrée par le brin pointé $\tilde{b}$. La carte réduite à un sommet est également dite pointée bien qu'elle ne contienne aucun brin.
On appelle circuit, une suite $(b_1,\dots,b_k)$ de brins de la carte tels que l'extrémité finale de b_i soit l'extrémité initiale de b_{i+1} si $1 \leqslant i < k$, de b_1 si $i = k$.

. On représentera dans la suite une carte par une projection stéréographique sur le plan, de façon à envoyer la face extérieure de la carte sur la face infinie de sa représentation dans le plan.

. Deux cartes planaires pointées sont isomorphes s'il existe un homéomorphisme de la sphère, préservant son orientation, appliquant les sommets, arêtes, faces et brin pointé de la première carte

respectivement sur ceux de la seconde.
Une classe d'isomorphie dans l'ensemble des cartes planaires pointées pour la notion d'isomorphie définie ci-dessus sera encore appelée carte planaire pointée dans la suite.

I.3. Hypercarte planaire pointée (cf exemple ci-dessous)

. Une carte planaire est dite deux-coloriable si on peut colorier ses faces avec deux couleurs, toute arête étant incidente à deux faces de couleurs différentes.
La propriété "deux-coloriable" étant compatible avec la relation d'équivalence dont les classes sont les cartes planaires pointées (cf 2) on peut définir une hypercarte planaire pointée comme une carte planaire pointée deux-coloriable contenant au moins un brin. Ce sont ces hypercartes planaires pointées que l'on cherche à dénombrer. Cette définition est équivalente (cf [1] ou [7]) à la définition combinatoire d'une hypercarte (cf [3]).

. Si H est une hypercarte planaire pointée, on note C(H) la carte planaire pointée deux-coloriable qui lui est associée. Les faces de C(H) de la même couleur (resp. de l'autre couleur) que la face extérieure de C(H) sont appelées faces (resp. arêtes) de l'hypercarte et notées h-faces (resp. h-arêtes) dans la suite. Les sommets de C(H) sont les sommets de l'hypercarte H. La face extérieure de C(H) est appelée h-face extérieure de H.
L'ensemble B des brins de C(H) appartenant aux cycles définissant dans C(H) les h-faces, est appelé ensemble des brins de l'hypercarte, notés h-brins dans la suite. On appelle h-degré d'une h-face (resp. d'un sommet) de l'hypercarte H, le nombre de h-brins qui lui sont incidents en tant que cellule de C(H).
Le h-degré d'une h-face coïncide avec son degré comme face de C(H) (évident). Par contre le h-degré d'un sommet de l'hypercarte est moitié de son degré dans C(H).

. <u>Dualité</u> - La dualité (cf [3]) est une bijection dans l'ensemble des hypercartes planaires pointées ; les h-faces (respectivement sommets) d'une hypercarte sont échangés avec les sommets (resp. h-faces) de l'hypercarte duale ; la h-face extérieure est échangée avec le sommet pointé de l'hypercarte duale (les deux ayant même h-degré) ; les deux hypercartes ont mêmes h-arêtes. Cette propriété sera utilisée au théorème 3.

. On définit par convention deux hypercartes planaires pointées associées à la carte planaire pointée réduite à un sommet ; on les note respectivement {p} et {q} suivant que l'unique face de la carte associée est considérée comme h-face ou h-arête.

. Dans l'exemple ci-dessous, l'hypercarte H (dont les h-arêtes sont hachurées) est associée à la carte C(H) dont les brins sont numérotés, le numéro de chaque brin étant placé le long de son support, près de son extrémité initiale. Les deux brins de chaque arête sont numérotés i et -i, celui étiqueté positif étant un h-brin. Le sens positif choisi pour définir les cycles de σ dans C(H) est le sens contraire des aiguilles d'une montre. Le brin pointé est marqué par une flèche.

Dans la figure 1 ci-dessous :

Les sommets S_i, $1 \leqslant i \leqslant 4$ sont dans C(H) identifiés aux cycles de σ donnés par :

$S_1 = (1,-13,13,-14,11,-10,10,-9,12,-12)$, $S_2 = (-1,2,-3,5)$,

$S_3 = (-5,6,-7,7,-8,8,-6,9,-11,14)$, $S_4 = (-2,3,-4,4)$.

Les h-faces f_1 (h-face extérieure), f_i, $2 \leqslant i \leqslant 7$, sont identifiées dans C(H) aux cycles de $\bar{\sigma}$:

$f_1 = (1,2,3,5,6,9,12)$, $f_2 = (14,11)$, $f_3 = (13)$,

$f_4 = (10)$, $f_5 = (4)$, $f_6 = (7)$, $f_7 = (8)$.

Les h-arêtes a_i, $1 \leqslant i \leqslant 5$, sont respectivement dans C(H), les cycles de $\bar{\sigma}$:

$a_1 = (-1,-13,-14,-5)$, $a_2 = (-9,-11,-10)$,

$a_3 = (-12)$, $a_4 = (-3,-4,-2)$, $a_5 = (-6,-7,-8)$.

On obtient alors :

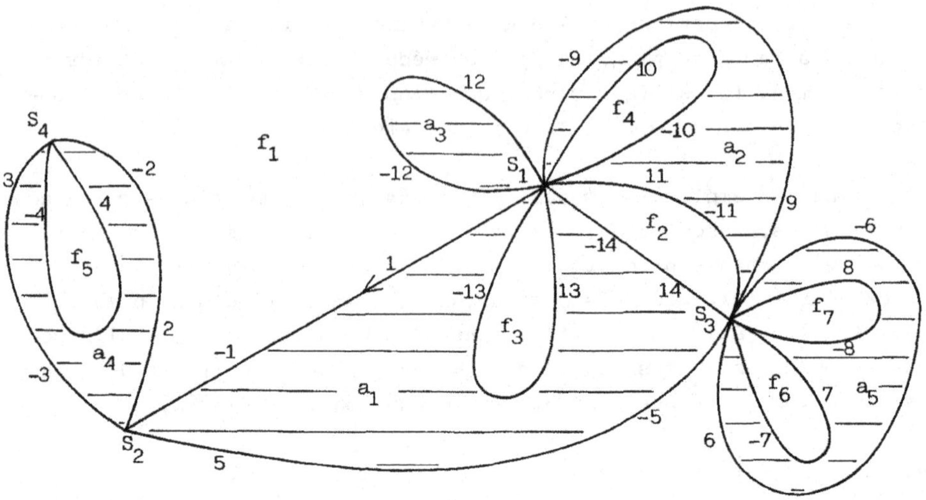

figure 1

II. Relations fonctionnelles pour les hypercartes planaires pointées

II.1. Décompositions d'une hypercarte planaire pointée

. On note $\mathcal{H}$ (resp. $\mathcal{L}$) la famille des hypercartes planaires pointées contenant au moins un brin (resp. et dont l'arête pointée est une boucle). Pour $r \geqslant 1$ (resp. et $\alpha \geqslant 0$, $\beta \geqslant 0$, $\gamma \geqslant 1$) on note $\mathcal{H}_r$ (resp. $\mathcal{H}_{\alpha,\beta,\gamma,r}$) l'ensemble des hypercartes de $\mathcal{H}$ dont le h-degré du sommet pointé est r (resp. qui ont de plus $(\alpha+1)$ sommets, $(\beta+1)$ h-faces et γ h-arêtes).

. On utilise les signes + ou Σ pour noter une réunion disjointe d'ensembles. L'existence d'une bijection entre deux ensembles est indiquée par le signe $\leftrightarrow$.

. On note $\tilde{\mathcal{J}}_1$ l'ensemble des cartes planaires pointées C dont le degré du sommet pointé est égal à un et telles que, en supprimant l'isthme pointé $\tilde{b}$ de C ainsi que son extrémité initiale et en pointant le brin suivant de la face extérieure, $\bar{\sigma}(\tilde{b})$, s'il existe, la sous-carte planaire pointée de C ainsi obtenue soit deux-coloriable et donc appartienne à $\{p\} + \mathcal{H}$. On appelle alors $\mathcal{D}^+$, l'ensemble défini comme la réunion disjointe, pour $k \geqslant 1$, des k-uplets de cartes de $\tilde{\mathcal{J}}_1$, soit

$\mathcal{D}^+ = \sum_{k\geq 1} \tilde{\mathcal{J}}_1^{\,k}$, et on pose $\mathcal{D} = \{p\} + \mathcal{D}^+$

. On note I[n,m] l'ensemble des nombres entiers impairs appartenant à l'intervalle [n,m].

Théorème 1 - On a les bijections suivantes

(a) $\mathcal{H} \leftrightarrow \mathcal{D}^+ \times \left[\{q\} + \sum_{r\geq 1} \mathcal{H}_r \times \mathcal{D}^r\right]$

(b) $\mathcal{L} \leftrightarrow \tilde{\mathcal{J}}_1 \times [\{q\} + \mathcal{H}]$

(c) $\mathcal{H} \leftrightarrow \mathcal{L} + \sum_{r\geq 1} \mathcal{H}_r \times I\,[1,2r-1]$.

Démonstration

1. La bijection (a) consiste à décomposer une hypercarte planaire pointée de $\mathcal{H}$ en son "hypercarte intérieure" dans $\mathcal{H}_r$ (si $r \geq 1$, égale à $\{q\}$ si $r = 0$) et en son "bord", (r+1)-uplet de $\mathcal{D}^+ \times \mathcal{D}^r$ (cf exemple 2 ci-dessous).
Soit donc H dans $\mathcal{H}$ et C(H) la carte planaire pointée deux-coloriable associée, de brin pointé $\tilde{b}$.

α) Un circuit simple élémentaire étant un circuit ne passant pas deux fois par la même arête ou le même sommet, on a le

Lemme - On peut extraire de la suite $\bar{\sigma}^*(\tilde{b})$ des brins du circuit bordant la face extérieure de C(H), une unique sous-suite de brins constituant un circuit simple élémentaire, contenant $\tilde{b}$.

Démonstration - L'arête, support de $\tilde{b}$ est incidente à une h-face et une h-arête de H ; elle n'est donc pas un isthme de C(H). L'existence du circuit simple élémentaire résulte alors du théorème de König. Son unicité est alors due au fait que $\bar{\sigma}^*(\tilde{b})$ est la frontière d'un ouvert connexe. CQFD

On note $P = (b_1 = \tilde{b}, b_2, \ldots, b_k)$ la sous-suite de brins de $\bar{\sigma}^*(\tilde{b})$ définie par le lemme et $(s_1,\ldots,s_k)$ la suite de leurs sommets initiaux. P est un polygone orienté qui partage le plan en deux domaines connexes ouverts qui lui sont intérieur et extérieur. Les brins de la carte planaire C(H) qui sont inclus dans le domaine intérieur (resp. extérieur) à P sont dits brins intérieurs (resp. extérieurs) de la carte C(H).

β) <u>Hypercarte intérieure H_{int} associée à H.</u>

Contractons P et son domaine extérieur en un sommet $\tilde{t}$:
S'il n'est pas vide, l'ensemble des brins intérieurs de C(H) constitue alors une carte planaire connexe deux-coloriable que l'on pointe en distinguant le premier brin qui n'appartient pas à P, de la suite de brins $(\sigma^{-1} \circ \alpha)^*(\tilde{b})$; on note $\tilde{d}$ ce brin et H_{int} l'hypercarte de $\mathcal{H}$, dite intérieure à H, ainsi définie. Le sommet $\tilde{t}$ est le sommet pointé de H_{int}; les brins qui lui sont incidents sont les brins intérieurs de C(H) dont le sommet initial est l'un des sommets s_i, $1 \leqslant i \leqslant k$, du polygone P.
Si 2r, $r \geqslant 1$, est leur nombre (r est alors le h-degré du sommet $\tilde{t}$ dans H_{int}), on les étiquette $d_1,\ldots, d_{2r}$, dans l'ordre où on les rencontre en parcourant P dans le sens contraire des aiguilles d'une montre, en commençant par $d_1 = \tilde{d}$.
Si l'ensemble des brins intérieurs de C(H) est vide, P est alors le bord d'une h-arête ; la contraction de P et de son domaine extérieur en un sommet $\tilde{t}$ donne alors l'hypercarte H_{int} réduite à l'hypercarte {q} (cf convention du I.3.).

γ) <u>Bord de H.</u>

La face extérieure de la carte C(H) étant un ouvert connexe, les brins extérieurs de C(H) constituent k cartes deux-coloriables connexes disjointes, respectivement incidentes aux sommets $s_1,\ldots,s_k$ du polygone P (certaines de ces cartes peuvent éventuellement ne contenir aucune arête et être réduites au sommet s_i correspondant).
Pour i dans {1,...,k} , on définit alors la carte planaire pointée C_i de $\tilde{\mathcal{J}}_1$ en associant au brin b_i de P, que l'on pointe, la carte deux-coloriable précédemment définie, incidente au sommet final s_{i+1} de b_i.
Le (r+1)-uplet $(B_i)_{0\leqslant i\leqslant r}$ dans $\mathcal{D}^+ \times \mathcal{D}^r$, constituant le bord de C est alors obtenu en partageant la suite $\mathcal{C} = (C_1,\ldots,C_k)$ en (r+1) sous-suites (avec la convention qu'une sous-suite vide est l'hypercarte {p}).

. B_0 est la sous-suite des cartes de $\mathcal{C}$ dont les brins pointés sont, sur P, entre le sommet initial s_1 de $b_1 = \tilde{b}$ et le sommet initial de $d_1 = \tilde{d}$ (si $r \geqslant 1$, et s_1 si $H_{int} = \{q\}$).
B_0 est une sous-suite de $\mathcal{C}$ contenant au moins C_1 et donc appartient à $\mathcal{D}^+$.

. Si $r \geqslant 2$, B_i, $1 \leqslant i \leqslant r-1$, est la sous-suite des cartes de $\mathcal{C}$ dont les brins pointés sont entre les sommets initiaux de d_{2i-1} et d_{2i+1}.

Si $r \geq 1$, B_r est la sous-suite des cartes de $\mathcal{C}$ dont les brins pointés sont entre le sommet initial de d_{2r-1} et s_1. Les uplets B_i, $1 \leq i \leq r$, sont dans $\mathcal{D}$ (éventuellement réduits à $\{p\}$ avec la convention ci-dessus).

L'application ainsi définie, qui à une hypercarte de $\mathcal{H}$ associe le uplet $(H_{int}, (B_i)_{0 \leq i \leq r})$ est une bijection (évident) de $\mathcal{H}$ sur

$$\mathcal{D}^+ \times \left[\{q\} + \sum_{r \geq 1} \mathcal{H}_r \times \mathcal{D}^r\right]$$, d'où le résultat annoncé.

Exemple 2 - Si H est l'hypercarte (les h-arêtes sont hachurées) : $k = 5$, $r = 5$.

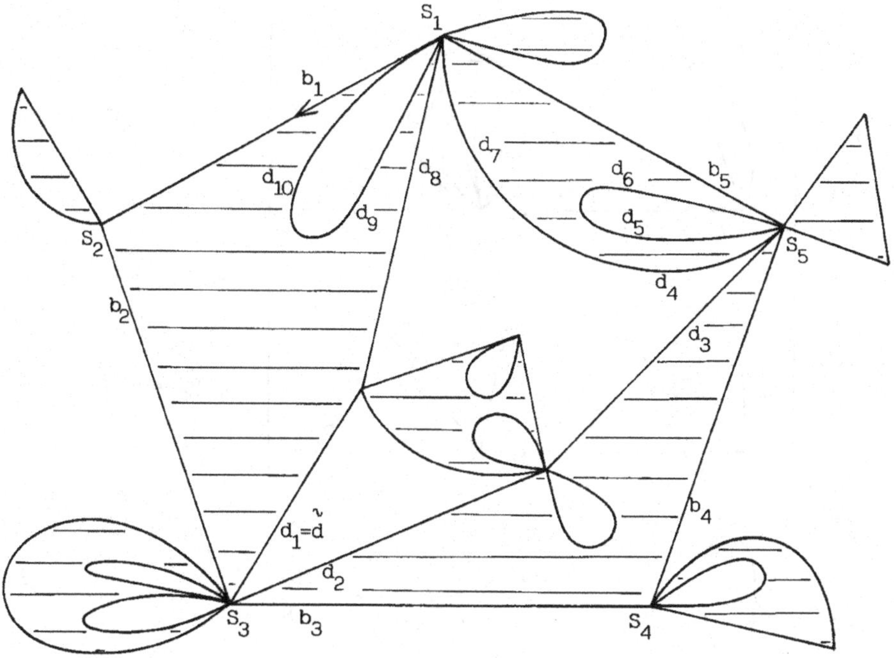

figure 2.1

L'hypercarte planaire pointée H_{int} obtenue en contractant le domaine extérieur au polygone $P = (b_1 = \tilde{b}, b_2, b_3, b_4, b_5)$ en un sommet $\tilde{t}$ et en pointant le brin $d_1 = \tilde{d}$ est

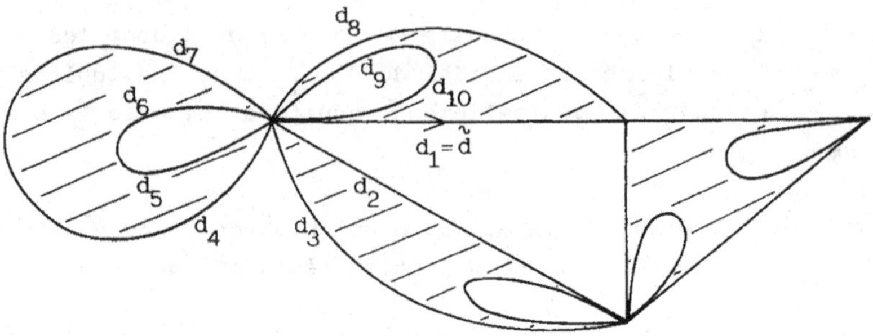

figure 2.2

Son bord $(B_i)_{0 \leqslant i \leqslant r}$ dans $\mathcal{D}^+ \times \mathcal{D}^r$ est constitué par

. B_o = 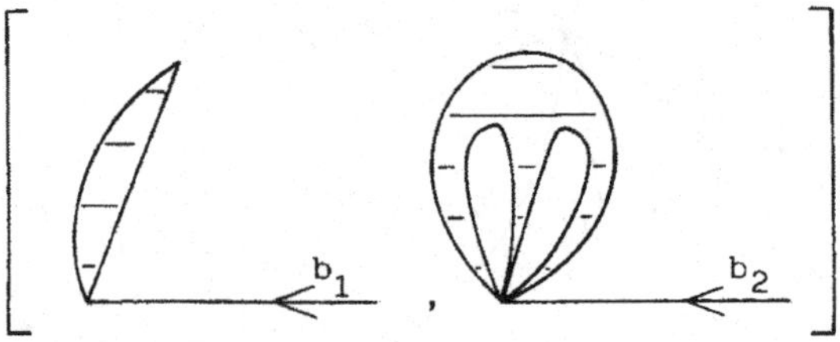

dans $\mathcal{D}^+$

. B_1 = 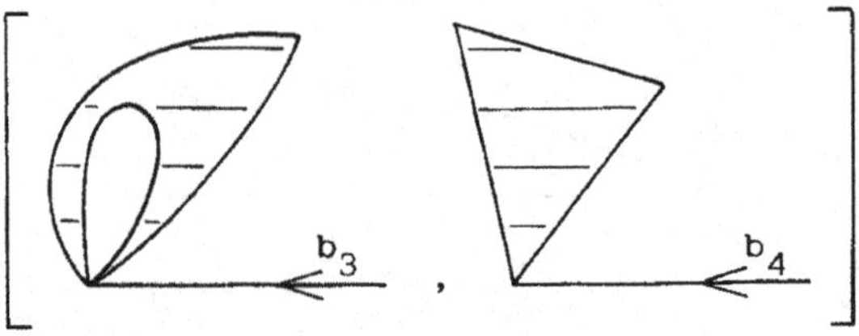

, $B_2 = \{p\}$,

. B_3 = 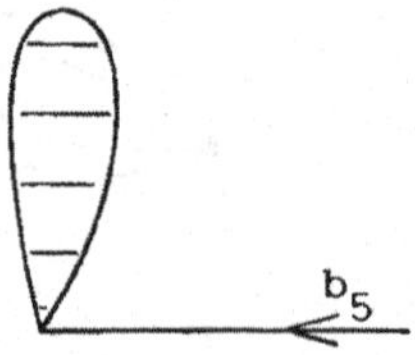

, $B_4 = \{p\}$, $B_5 = \{p\}$.

2. La bijection (b) est un cas particulier de (a) ; le polygone P est constitué d'un seul brin $b_1 = \tilde{b}$, brin pointé de C(H). Le bord de l'hypercarte se réduit à B_o dans $\tilde{\mathcal{J}}_1$ et H_{int} est un élément de $\{q\} + \mathcal{H}$, d'où la bijection (b).

Exemple 3 - Si H est l'hypercarte planaire pointée dans $\mathcal{L}$

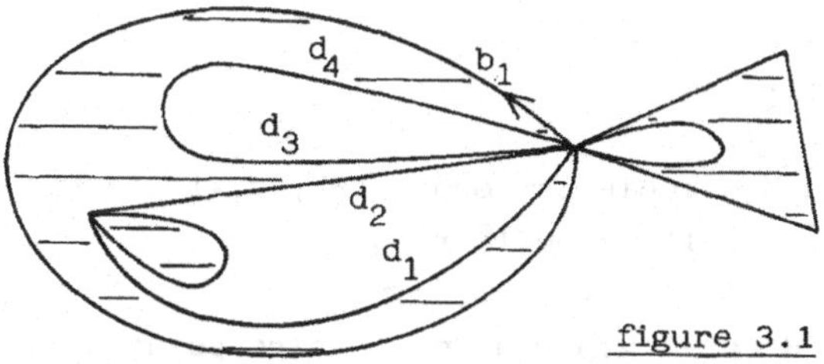

figure 3.1

Le couple (H_{int}, B_o) qui constitue sa décomposition est

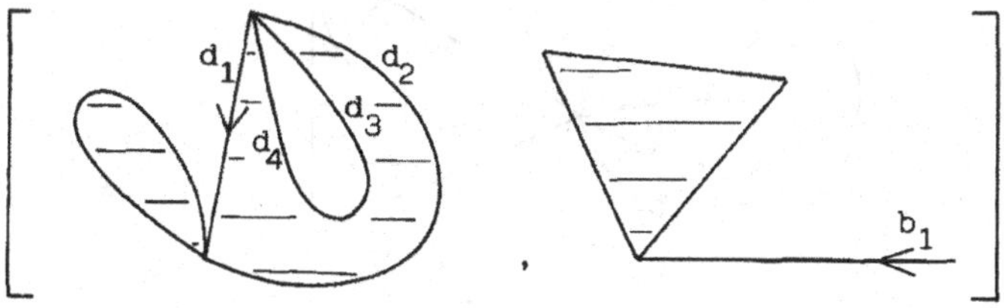

figure 3.2

3. La bijection (c) est (après application de la dualité) celle établie par W.T. Tutte (cf [6], voir également [4]). Nous la décrivons succintement de façon à pouvoir l'utiliser pourl'établissement de l'équation fonctionnelle (3) du théorème 2.

Soit H une hypercarte appartenant à $\mathcal{H}\setminus\mathcal{L}$, de brin pointé $\tilde{b}$. Soit b^* $(= \bar{\sigma}(\tilde{b}))$ le brin suivant $\tilde{b}$ dans la suite des brins bordant la h-face extérieure de H ; il est différent de $\tilde{b}$ car H n'appartient pas à $\mathcal{L}$. Soient d_i et d_f les h-degrés respectivement du sommet initial $\tilde{s}$ de $\tilde{b}$ et de son sommet final s^* (qui est également le sommet initial de b^* et qui est distinct de $\tilde{s}$).
Si l'on contracte le brin pointé $\tilde{b}$ de H et si l'on pointe b^*, on définit une nouvelle hypercarte H^* dont le sommet pointé a pour degré dans $C(H^*)$

$$2r = (2d_f-1) + (2d_i-1)$$

$(2d_f-1)$ représente le nombre de brins dans C(H) incidents au sommet final s^* de $\tilde{b}$ qui lors de la contraction de ce brin sont venus s'ajouter aux $(2d_i-1)$ brins restant incidents au sommet $\tilde{s}$.

Les h-degrés d_i et d_f étant supérieurs ou égaux à un, $(2d_f-1)$ satisfait aux inégalités

$$1 \leqslant 2d_f-1 \leqslant 2r-1.$$

L'application qui à H associe le couple $(H^*, 2d_f-1)$ estune bijection de $\mathcal{H}\text{-}\mathcal{L}$ sur $\sum_{r\geqslant 1} \mathcal{H}_r \times I\,[1,2r-1]$, d'où le résultat.

Exemple 4 - Si H est l'hypercarte planaire pointée

On a $d_i = 3$ et $d_f = 4$

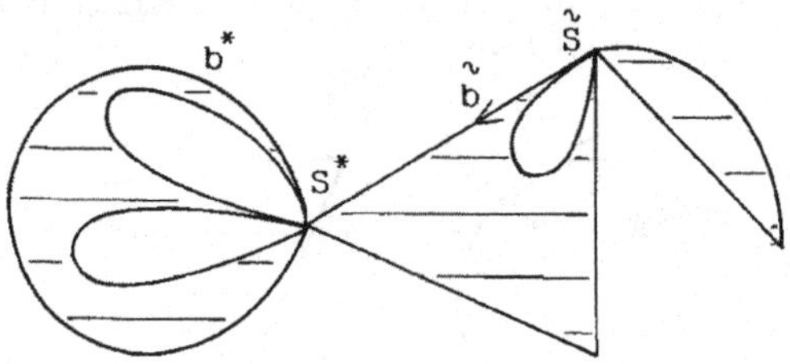

figure 4.1

L'hypercarte H^* qui lui est associée en contractant le brin $\tilde{b}$ est:

figure 4.2:

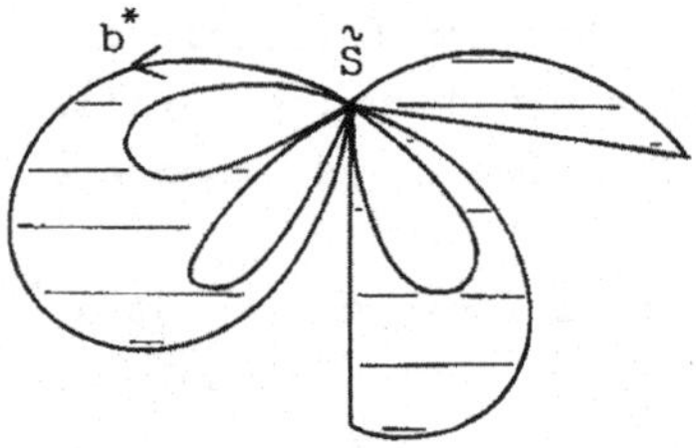

Le degré du sommet pointé de $C(H^*)$ est $2r = 12$. Parmi ces douze brins, les sept $(= 2d_f-1)$ premiers à partir du brin b^* étaient, avant contraction de $\tilde{b}$, incidents dans $C(H)$ à s^*.

II.2. Relations fonctionnelles

A chaque hypercarte planaire pointée de $\mathcal{H}$ on associe un monôme en les variables commutatives u (resp. v), s, f et a.
L'exposant de u (resp. v) donne le nombre de brins (c'est-à-dire le h-degré) de la face extérieure (resp. le h-degré du sommet pointé), l'exposant de s donne le nombre de sommets différents du sommet pointé,

l'exposant de f donne le nombre de h-faces autres que la h-face extérieure, l'exposant de a donne le nombre de h-arêtes.
On note alors, J(u,s,f,a) (resp. K(v,s,f,a)) la série génératrice des hypercartes planaires pointées de $\mathcal{H}$ (c'est-à-dire contenant au moins un brin).
Le théorème classique de dualité dans $\mathcal{H}$(cf [3]), rappelé au I.3., implique les égalités

$$J(u,f,s,a) = K(u,s,f,a) \text{ et } J(1,s,f,a) = J(1,f,s,a)$$

On déduit du théorème 1, le

Théorème 2 - On a les relations fonctionnelles, où l'on note J pour J(u,s,f,a) et K pour K(v,s,f,a)

$$J = \frac{u(1+J)}{1-us(1+J)} \{f\, K(\frac{1}{1-us(1+J)},s,f,a)+a)\} \quad (1)$$

$$K = \frac{v(1+K)}{1-vf(1+K)} \{s\, J(\frac{1}{1-vf(1+K)},s,f,a)+a\} \quad (2)$$

$$K = v(1+K)(fK+a) + vs\,\frac{K-J(1,s,f,a)}{v-1} \quad (3)$$

Démonstration

1. La relation (1) traduit en termes de série génératrice la bijection (a) du théorème 1. La relation (2) se déduit de (1) en substituant v à u, en échangeant s et f et en utilisant les égalités conséquences du théorème de dualité précité.
Démontrons la relation (1) (On reprend les notations utilisées dans la décomposition (a) d'une hypercarte H donnée au théorème 1).

. L'hypercarte {p} est associée au monôme $s^o f^o a^o = 1$
. L'hypercarte {q} est associée au monôme $s^o f^o a^1 = a$
. La série génératrice des cartes de $\tilde{\mathcal{J}}_1$, ensemble auquel appartiennent les cartes planaires pointées C_i, $1 \leq i \leq k$, qui après regroupement permettent de définir les uplets de cartes B_j, $0 \leq j \leq r$ du bord de H est:

$$us(1+J(u,s,f,a)).$$

Le terme (1+J) est associé à l'hypercarte planaire pointée dans $(\{p\}+\mathcal{H})$, incidente à l'extrémité finale du brin pointé de la carte de $\tilde{\mathcal{J}}_1$. Ce brin pointé donne le terme us, car il est incident à la face extérieure de H et ajoute un sommet.

La série génératrice de l'ensemble $\mathcal{D}$ (auquel appartient B_j si $1 \leqslant j \leqslant r$) des p-uplets, p dans $\mathbb{N}$, de cartes de $\tilde{\mathcal{J}}_1$, est donc

$$\sum_{p\geqslant 0} [us(1+J)]^p = \frac{1}{1-us(1+J)} .$$

La série génératrice de l'ensemble $\mathcal{D}^+$ (auquel appartiennent les uplets B_o) des p-uplets, p dans $\mathbb{N}^*$, de cartes de $\tilde{\mathcal{J}}_1$, est alors

$$\frac{us(1+J)}{1-us(1+J)}$$

. La série génératrice des hypercartes H_{int} dans $\mathcal{H}_r$, $r \geqslant 1$, qui sont intérieures aux cartes de $\mathcal{H}$, et donc n'ont pas de brins incidents à la face extérieure (donc pas de terme en u) est :

$$f \sum_{\alpha\geqslant 0,\beta\geqslant 0,\gamma\geqslant 1} \ \sum_{H_{int}\in\mathcal{H}_{\alpha,\beta,\gamma,r}} s^\alpha f^\beta a^\gamma = f \sum_{\alpha\geqslant 0,\beta\geqslant 0,\gamma\geqslant 1} \mathrm{card}(\mathcal{H}_{\alpha,\beta,\gamma,r}) s^\alpha f^\beta a^\gamma$$

où, card $(\mathcal{H}_{\alpha,\beta,\gamma,r})$ est le cardinal de l'ensemble $\mathcal{H}_{\alpha,\beta,\gamma,r}$. Le terme f en facteur est dû au fait que la face extérieure de H_{int}, non décomptée dans l'exposant β du monôme $s^\alpha f^\beta a^\gamma$ associé à H_{int}, devient h-face non extérieure dans l'hypercarte H reconstruite et donc doit être décomptée en plus.

. La série génératrice des cartes de $\mathcal{H}$ est donc (cf (a))

$$J = \frac{1}{s} \cdot \frac{us(1+J)}{1-us(1+J)} \left[a + \sum_{r\geqslant 1} (f. \sum_{\alpha\geqslant 0,\beta\geqslant 0,\gamma\geqslant 0} \mathrm{card}(\mathcal{H}_{\alpha,\beta,\gamma,r}) s^\alpha f^\beta a^\gamma)(\frac{1}{1-us(1+J)})^r\right]$$

Le terme $\frac{1}{s}$ est dû au fait que lorsque l'on concatène les uplets B_j, $0 \leqslant j \leqslant r$, pour reconstituer le bord de H, le produit des séries génératrices associées (déterminées ci-dessus) décompte, par l'exposant de la variable s, tous les sommets du bord de H, y compris le sommet pointé (ce qui est exclu dans la définition de J).

On en déduit $J = \frac{u(1+J)}{1-us(1+J)} \left[a + f\,K(\frac{1}{1-us(1+J)}\,,\,s,f,a)\right]$ CQFD

2. Les deux termes du second membre de (3) correspondent aux deux termes de la décomposition bijective (c) du théorème 1 (On reprend les notations introduites au théorème 1).

Le premier terme de (3) correspond à la décomposition d'une hypercarte de $\mathcal{L}$ (hypercarte dont le brin pointé est une boucle) en son bord B_o dans $\tilde{\mathcal{J}}_1$ et en une hypercarte intérieure H_{int} dans $\{q\} + \mathcal{H}$.
La série génératrice des cartes B_o dans $\tilde{\mathcal{J}}_1$ est $v(1+K)$, la boucle pointée de B_o donnant le terme v et l'hypercarte dans $\{p\} + \mathcal{H}$ incidente à son extrémité, le terme $(1+K)$.
La série génératrice des cartes H_{int} dans $(\{q\} + \mathcal{H})$, qui ont même sommet pointé que l'hypercarte de $\mathcal{L}$ que l'on décompose, est $(a+fK)$: le terme a est associé à l'hypercarte $\{q\}$; pour H_{int} dans $\mathcal{H}$, le terme f en facteur de K est dû au fait que la h-face extérieure de H_{int} devient non extérieure dans l'hypercarte de $\mathcal{L}$ dont H_{int} est l'hypercarte intérieure et qu'il faut alors la décompter.

Le second terme s'obtient en remarquant que pour reconstruire l'hypercarte H de $\mathcal{H}\setminus\mathcal{L}$ à partir d'une hypercarte H^* de $\mathcal{H}_r$, $r \geq 1$, on ajoute un nouveau sommet s^* à H^* (d'où le terme s en facteur ci-dessous), et que pour $1 \leq i \leq r$, les $(2i-1)$ premiers brins (à patir du brin pointé de H^*) incidents au sommet pointé $\tilde{s}$ de H^* sont transférés sur ce nouveau sommet s^*. On crée alors un nouveau brin de $\tilde{s}$ vers s^* qui devient le brin pointé de H. Il reste donc

$$2r-(2i-1)+1 = 2(r-i+1) \text{ brins incidents au sommet pointé } \tilde{s} \text{ de } C(H).$$

La série génératrice des hypercartes de $\mathcal{H}\setminus\mathcal{L}$ est donc

$$s \sum_{\alpha\geq 0,\beta\geq 0,\gamma\geq 1,r\geq 1} \;\sum_{H^*\in\mathcal{H}_{\alpha,\beta,\gamma,r}} s^\alpha f^\beta a^\gamma \sum_{i=1}^{r} v^{r-i+1}$$

soit, en posant $j = r-i$

$$= vs \sum_{\alpha,\beta,\gamma,r} \;\sum_{H^*\in\mathcal{H}_{\alpha,\beta,\gamma,r}} s^\alpha f^\beta a^\gamma \sum_{j=0}^{r-1} v^j$$

$$= vs \sum_{\alpha,\beta,\gamma,r} \sum_{H^*\in\mathcal{H}_{\alpha,\beta,\gamma,r}} s^\alpha f^\beta a^\gamma \frac{v^r-1}{v-1}$$

$$= vs \frac{K(v,s,f,a)-K(1,s,f,a)}{v-1}$$ CQFD

Remarque

Soit A l'anneau des polynômes $\mathbb{Z}[u]$ et A[[s,f,a]] l'algèbre des séries formelles en les variables commutatives s,f,a, à coefficients dans A.

Dans A[[s,f,a]] l'équation (1) en J a un sens (la substitution de u par $\frac{1}{1-us(1+J)}$ dans J(u,s,f,a) y est autorisée) et y admet comme solution la série génératrice des hypercartes planaires pointées, dont on sait que dans tout monôme, le degré de u est majoré par la somme des degrés de s,f et a. Cette série est en fait l'unique solution de (1) dans A[[s,f,a]]. En effet, munissons A[[s,f,a]] de la distance définie pour deux séries S_1 et S_2 par

$$d(S_1,S_2) = 2^{-\nu(S_1S_2)}, \text{ avec :}$$

$\nu(S_1,S_2) = \min\{\alpha+\beta+\gamma/$ les coefficients dans A de S_1 et S_2 pour $s^\alpha f^\beta a^\gamma$ sont distincts$\}$.

A[[s,f,a]] est alors un espace métrique complet. Le second membre de (1) définit un opérateur contractant sur cet espace et y admet un point fixe unique. CQFD.

On a de même, unicité de la solution des équations (2) ou (3).

III. Application aux dénombrements d'hypercartes

III.1. Dénombrement des hypercartes planaires pointées

La considération simultanée des équations (1), (2) et (3) du théorème 2 permet de déterminer très simplement un système d'équations paramétriques dont la solution est la série génératrice sf J(1,s,f,a) des hypercartes pointées contenant au moins un brin, décomptées en fonction du nombre de sommets, de h-faces et de h-arêtes. On donne ensuite le terme général de cette série (corollaire 1 du théorème 3).

Théorème 3 - La série génératrice sf J(1,s,f,a) qui décompte les hypercartes planaires pointées ayant au moins un brin en fonction du nombre de sommets, de h-faces et de h-arêtes est solution du système d'équations paramétriques (où λ,μ,ν sont les paramètres)

$$\Sigma:\begin{cases} s = \lambda(1-\mu-\nu),\ f = \mu(1-\nu-\lambda),\ a = \nu(1-\lambda-\mu), \\ sfJ(1,s,f,a) = \lambda\mu\nu(1-\lambda-\mu-\nu) \end{cases}$$

Démonstration

Dans l'algèbre $\mathbb{Z}[[s,f,a]]$ des séries formelles en les variables commutatives s,f,a, l'équation en u

$$u = 1 + \frac{uf}{1-us(1+J(u,s,f,a))}\left(1 + K\left(\frac{1}{1-us(1+J(u,s,f,a))}, s,f,a\right)\right) \qquad (e)$$

admet une solution unique U = U(s,f,a) (cf. Remarque du théorème 2).

Si l'on pose

$$V = V(s,f,a) = \frac{1}{1-Us(1+J(U,s,f,a))} \qquad (i)$$

l'équation (e) s'écrit

$$U(s,f,a) = \frac{1}{1-Vf(1+K(V,s,f,a))}\ . \qquad (ii)$$

De plus (i) et (ii) entraînent que V(f,s,a) est également solution de (e), et donc est égale (par unicité de cette solution) à U(s,f,a).
En substituant U = U(s,f,a) et V = V(s,f,a) respectivement à u et v et en utilisant les égalités (i) et (ii), les équations (1), (2) et (3) du théorème 2 s'écrivent, en posant :

$$J = J(U,s,f,a) \text{ et } K = K(V,s,f,a).$$

$$J = UV(1+J)(a+fK) \qquad (1)$$

$$K = UV(1+K)(a+sJ) \qquad (2)$$

$$K = V(1+K)(a+fK) + Vs\,\frac{K-J(1,s,f,a)}{V-1} \qquad (3)$$

Or, par (i) et (ii), $1+J = \frac{V-1}{UVs}$ et $1+K = \frac{U-1}{UVf}$

Donc, (1), (2) et (3) donnent en éliminant J et K :

$$s = \frac{V-1}{UV}\,(2-U+UV(f-a))$$

$$f = \frac{U-1}{UV}\,(2-V+UV(s-a))$$

$$U^2V^2fsJ(1,s,f,a) = (U-1)(V-1)(UVa-1) + fUV(V-1) + sUV(U-1) - U^2V^2fs$$

soit :

$$s = \frac{V-1}{UV}\,\left(1 - \frac{U^2Va}{U+V-UV}\right)$$

$$f = \frac{U-1}{UV}\,\left(1 - \frac{UV^2a}{U+V-UV}\right)$$

$$sfJ(1,s,f,a) = \frac{(U-1)(V-1)}{UV}\, a\, \left(1 - \frac{U^2V^2a}{(U+V-UV)^2}\right)$$

Posons $\lambda = \frac{V-1}{V}$, $\mu = \frac{U-1}{U}$, $a = \nu(1-\lambda-\mu)$ dans ce système.
On obtient alors le système d'équations paramétriques annoncé. CQFD

Remarque - La forme paramétrique obtenue pour la série génératrice $sfJ(1,s,f,a)$ montre son invariance par permutation quelconque des variables s,f et a.
On sait déjà (dualité) que $sfJ(1,s,f,a) = fsJ(1,f,s,a)$. D'autre part, l'opération géométrique qui consiste à remplacer le brin pointé d'une hypercarte planaire pointée par son brin opposé est une bijection de $\mathcal{H}$ sur $\mathcal{H}$ qui échange h-faces et h-arêtes. On a donc

$$sfJ(1,s,f,a) = saJ(1,s,a,f). \qquad \text{CQFD}$$

Corollaire 1 - Le nombre d'hypercartes planaires pointées ayant $m_1(\geq 1)$ sommets, $m_2(\geq 1)$ h-faces et $m_3(\geq 1)$ h-arêtes est donné par

$$\sum \prod_{\ell=1}^{3} \binom{m_\ell-1+i_\ell}{m_\ell-1} \binom{i_\ell}{j_\ell} \, . \, C$$

où

$$. \ C = 1 - \sum_{k=1}^{3} \frac{j_k}{m_k-1+i_k} + \frac{j_1j_2j_3}{m_1m_2m_3} - \frac{m_1-1}{m_1-1+i_1} \cdot \frac{j_2}{m_2} \cdot \frac{i_3-j_3}{m_3}$$
$$- \frac{m_2-1}{m_2-1+i_2} \cdot \frac{j_3}{m_3} \cdot \frac{i_1-j_1}{m_1} - \frac{m_3-1}{m_3-1+i_3} \cdot \frac{j_1}{m_1} \cdot \frac{i_2-j_2}{m_2}$$

. la sommation porte sur les $(i_\ell, j_\ell)_{1\leqslant \ell \leqslant 3}$ tels que

$0 \leqslant j_\ell \leqslant i_\ell$, $1 \leqslant \ell \leqslant 3$ et

$m_1-1 = i_2-j_2+j_3$, $m_2-1 = i_3-j_3+j_1$, $m_3-1 = i_1-j_1+j_2$.

Remarque : 1. - Dans le cas où $m_k = 1$, on fait de plus les conventions

$$\frac{j_k}{m_k-1+i_k} = 0 \text{ si } i_k = j_k = 0, \quad \frac{m_k-1}{m_k-1+i_k} = 1 \text{ si } i_k = 0.$$

2. - L'application de la formule de Lagrange à trois variables au système paramétrique Σ donne une formule relativement compliquée dont le corollaire 1 fournit une forme simplifiée. On peut encore très certainement la réduire.

Un cas particulier du résultat obtenu au théorème 3 permet de retrouver le dénombrement présenté par T.R.S. Walsh (cf [7]) :

Corollaire 2 :

1. La série génératrice $S(f,y)$ des hypercartes planaires pointées décomptées en fonction du nombre de h-brins (variable y) et du nombre de h-faces moins une (variable f) satisfait aux équations paramétriques (paramètre p) :

$$(\Sigma_1) : y = \frac{p}{(1+\frac{fp}{1-p})(1+p)^2}, \quad S(f,y) = (1+p) - \frac{p^3}{(1-p)^2} f$$

2. Le nombre d'hypercartes planaires à n h-faces et m h-brins est

$$\frac{2(m-1)!}{(n-1)!(m-n+2)!} \sum_{i=0}^{m-n} \binom{n+i}{i} \binom{2m}{m-n-i} \quad \text{(cf [7]).}$$

Démonstration

1. Par la formule d'Euler pour les hypercartes planaires pointées (cf [3]), on a

$S(f,y) = 1 + J(1,y,fy,y)$ obtenue en substituant fy à f et y à s et a dans $J(1,s,f,a)$.

On déduit alors facilement du système Σ (cf théorème 3), que S(f,y) est solution du système

$$y = \lambda(1-\lambda-\mu)$$

$$fy = \mu(1-2\lambda)$$

$$fy^2 S(f,y) = \lambda\mu\left[(1-\lambda)(1-\mu)-\lambda\right]$$

D'où, en posant $\lambda = \frac{p}{1+p}$ et en éliminant μ le résultat annoncé.

2. L'application de la formule de Lagrange à une variable au système (Σ_1) donne le terme général de S(f,y) sous forme de quatre sommes qui après simplification donnent celle obtenue par T.R.S. Walsh. CQFD.

III.2. En application de la décomposition bijective (a) du théorème 1, on a les résultats suivants

Proposition - 1. Le nombre d'hypercartes planaires pointées ayant une seule face et n h-brins est le $n^{\text{ième}}$ nombre de Catalan : $\frac{1}{n+1}\binom{2n}{n}$, (cf [3]).

2. Le nombre d'hypercartes planaires pointées dont tous les sommets appartiennent à la frontière de la h-face extérieure et qui ont n h-brins, $n \geqslant 1$, est :

$$\frac{1}{n}\sum_{i=0}^{n-1} \binom{n}{i+1}\binom{n}{i} 2^i .$$

Démonstration - 1. Les hypercartes planaires pointées de $\mathcal{H}$ ayant une seule face, ont une hypercarte intérieure (cf théorème 1) réduite à l'hypercarte {q}. Si on note $J_o(u,s,f,a)$ la série génératrice décomptant ces hypercartes, les variables ayant la même signification qu'au II.2., on déduit facilement de la bijection (a) du théorème 1 et de l'équation (1), que J_o est solution de l'équation

$$J_o = \frac{u(1+J_o)}{1-us(1+J_o)} \cdot a \tag{4}$$

Si l'on substitue 1 à u et y à s,f et a dans (4), l'exposant de y donne le nombre de brins de l'hypercarte (par l'équation d'Euler pour les hypercartes planaires) et (4) donne, en posant $j_o = 1 + J_o(1,y,y,y)$:

$$j_o = 1 + y\, j_o^2 \tag{4'}$$

d'où le résultat annoncé.

2. Les hypercartes planaires pointées de $\mathcal{H}$ dont tous les sommets appartiennent à la frontière de la h-face extérieure ont une hypercarte intérieure ayant un seul sommet. Si l'on note $J_1(u,s,f,a)$ leur série génératrice, où les variables u,s,f,a ont la même signification qu'au II.2., on déduit de la bijection (a) du théorème 1, et de l'équation (1), que J_1 est solution de l'équation

$$J_1(u,s,f,a) = \frac{u(1+J_1)}{1-us(1+J_1)}\left(a+f\, J_o\left(\frac{1}{1-us(1+J_1)}, f,s,a\right)\right) \tag{5}$$

Posons $j_1 = J_1(1,y,y,y)$ et $h_o = J_o\left(\frac{1}{1-yj_1}, y,y,y\right)$.

L'exposant de la variable y donne alors le nombre de brins de l'hypercarte et (4) et (5) s'écrivent

$$h_o = \frac{1}{1-y(1+j_1)}\, y(1+h_o)^2 \tag{4}$$

$$j_1 = \frac{y(1+j_1)}{1-y(1+j_1)}\,(1+h_o) \tag{5}$$

D'où, en éliminant h_o, j_1 est solution de

$$j_1 = y(1+j_1)(1+2j_1),$$

ce qui par la formule de Lagrange donne le résultat annoncé. CQFD.

Bibliographie

[1] D. ARQUES, "Les hypercartes planaires sont des arbres très bien étiquetés", Mai 1984. A paraître dans Discrete Mathematics.

[2] D. ARQUES, "Une relation fonctionnelle nouvelle sur les cartes planaires pointées", Septembre 1984. A paraître dans Journal of combinatorial Theory (Series B).

[3] R. CORI, "Un code pour les graphes planaires et ses applications", Astérisque, Société Math. de France 27 (1975).

[4] R. CORI et J. RICHARD, "Enumération des graphes planaires à l'aide des séries formelles en variables non commutatives". Discrete Math. 2 (1972), 115-162.

[5] W.T. TUTTE, "A census of slicings", Can. J. Math. 14 (1962), 708-722.

[6] W.T. TUTTE, "On the enumeration of planar maps", Bull. Amer. Math. Soc. 74 (1968), 64-74.

[7] T.R.S. WALSH, "Hypermaps versus bipartite maps", J. Combinatorial Theory 18 (B) (1975) 155-163.

Didier ARQUES

Institut des Sciences
Exactes et Appliquées

4, rue des Frères Lumière

68093 MULHOUSE Cédex

France.

PREFIX-FREE WORDS OF LENGTH N OVER M LETTERS : TWO-SIDED WELL-BALANCED PARENTHESES AND PALINDROMS

Joffroy BEAUQUIER (*)
Loys THIMONIER (*) (**)

(*) Laboratoire de Recherche en Informatique (LRI) -Université de PARIS-SUD bâtiment 490 - 91405 ORSAY Cédex -France

(**) U.E.R. de Mathematiques -Université d'Amiens - 33 rue St Leu 80039 Amiens Cédex - France

ABSTRACT

Prefix-free words of a given language L can be obtained within the frame of the classical problem of random words. If γ_n is the number of prefix-free words of length n of L, we find here the functional equation verified by the generating function γ of the γ_n's, and solve this problem with exact and asymptotic formulas, for two kinds of languages : the two-sided well-balanced parentheses over the alphabet $A=\{a_1,\ldots a_m, \bar{a}_1,\ldots \bar{a}_m\}$, and the palindroms over $A=\{a_1,\ldots a_m\}$.

I. INTRODUCTION : OBTAINMENT OF PREFIX-FREE WORDS OF A GIVEN LANGUAGE WITHIN THE FRAME OF THE CLASSICAL PROBLEM OF RANDOM WORDS.

Let $A=\{a_1,\ldots a_m\}$ be an alphabet of m letters. At each discrete time t=1,2,3... there is a drawing with replacement of a letter of A, in an equally likely way (each letter has the probability 1/m). So an infinite random word $\omega = a_{i_1}\, a_{i_2} \ldots a_{i_n} \ldots$ is obtained, and let $\omega^{[n]} = a_{i_1} \ldots a_{i_n}$ be the left factor of the n first letters. If $\mathcal{E}$ is an event concerning the factors of ω, a classical problem is to consider the random variable X equal to the first time n at which $\mathcal{E}$ is obtained, in particular to compute the expectation E(X) [Fel 68] [Com 70] (e.g. $\mathcal{E}$ can be the event : "each letter has appeared at the least K times in $\omega^{[n]}$" [ER 61]).

We consider here a language L over A (L is a part of the free monoïd A^* generated by A) and the event :"a first n such that $\omega^{[n]}$ is a member of L is obtained" ; then $\omega^{[n]}$ is a prefix-free word w of L ($w = w_1 w_2$, $w_1 \in L \Rightarrow w_1 = w$, $w_2 = \varepsilon$ the empty word). Problems related to this approach and concerning the probability $Pr(\mathcal{E})$ and the expectation E(X) (when $Pr(\mathcal{E})=1$) were previously studied in the frame of absorbing Markov chains [Fel 68].

In recent works [BT 83] [BT 84], the probability $Pr(\mathcal{E})$ and the conditional expectation $E(X|\mathcal{E})$ were studied for several classes of languages. If γ_n is the number of prefix-free words of length n of L, $Pr(\mathcal{E})$ and $E(X|\mathcal{E})$ can be expressed [BT 84] by means of the generating function γ of the γ_n's ($\gamma(z) = \sum_{n\geq 1} \gamma_n z^n$) : $Pr(\mathcal{E}) = \gamma(1/m)$, $E(X|\mathcal{E}) = \gamma'(1/m)/m\gamma(1/m)$ (this formula is true even if the derivative $\gamma'(1/m)$ is infinite). A non trivial example is the so called Goldstine language [Gol 72] $G=\{a^{n_1}\, b\, a^{n_2}\, b \dots a^{n_k}\, b \mid k\geq 1 / \forall i\, n_i \geq 0 / \exists i,\ i\neq n_i\}$, partially studied in [BT 83]; if Θ is the transcendental theta function ($\Theta(Z)=\sum_{n\geq 0} Z^{\frac{n(n+1)}{2}}$), then we obtain here $\gamma(Z) = 1+\left(\frac{1}{1-Z}-\frac{1}{Z}\right)(\Theta(Z)-1)$ (*) by writing (if P is the language of the prefix-free words of G and $a^* = \bigcup_{n\geq 0}\{a^n\}$) : $P = (a^*\setminus a)b \cup \bigcup_{n\geq 2} ab\, a^2b \dots a^{n-1}\, b(a^*\setminus a^n)b)$;

so $Pr(\mathcal{E}) = \gamma(1/2)=1$, and $E(X|\mathcal{E}) = E(X) = \gamma'(1/2)/2 = 4\,(\Theta(1/2)-1) = 2.56653\dots$ (with an approximation of $\Theta(Z)$ by $\sum_{i=0}^{n-1} Z^{\frac{i(i+1)}{2}}$, with an upper bound $|Z|^{\frac{n(n+1)}{2}}/(1-|Z|)$ for the remainder $\sum_{i\geq n} Z^{\frac{i(i+1)}{2}}$ (here n=8)); the formula (*) allows to obtain asymptotically here :

$\sqrt{2n} - 3/2 + O(1/\sqrt{n}) \leq \gamma_n \leq \sqrt{2n} - 1/2 + O(1/\sqrt{n}) < \sqrt{2n}$, by computing the greatest integer k_n such that $\frac{k_n(k_n+1)}{2} \leq n$ ($k_n = \left\lfloor \frac{-1+\sqrt{1+8n}}{2} \right\rfloor$).

So it appears interesting to compute γ_n, exactly or at the least asymptotically, the more so as the radius of convergence $\rho = 1/\overline{\lim}\, \gamma_n^{1/n}$, verifying $\rho \in [1/m,1]$, occurs in the entropy ($h=\log 1/\rho$ [Kui 70]) associated with the language P of the prefix-free words of L. We will examine the previous problem for two kinds of languages, obtaining functional equations for γ (they arise in enumeration when a recursive procedure is given for "building" the objects that are being counted [Ben 74]). We assume the reader to be familiar with elementary notions of formal languages theory [Har 78].

II. PREFIX-FREE TWO-SIDED WELL-BALANCED PARENTHESES OVER $A=\{a_1,\dots a_m,\bar{a}_1,\dots \bar{a}_m\}$

We consider first m different left parentheses : $(_1, (_2, \dots (_m$, and the associated right parentheses : $)_1,)_2, \dots)_m$; it is convenient to denote these parentheses as the letters of an alphabet $A=\{a_1,\dots a_m, \bar{a}_1,\dots \bar{a}_m\}$. Let L be the language of the two-sided well-balanced parentheses over A, generated by the simple grammar :
$S \to a_i\, S\bar{a}_i\, S + \bar{a}_i\, S\, a_i\, S + \varepsilon$, i=1 to m [Har 78]. The context-free language P of the prefix-free words of L is generated by a slightly more complicated non ambiguous [Har 78] grammar G :

$$S \to \sum_{i=1}^{m} a_i\, S_i\, \bar{a}_i + \sum_{i=1}^{m} \bar{a}_i\, \bar{S}_i\, a_i,\quad S_i \to \varepsilon + \sum_{j=1}^{m} a_j\, S_j\, \bar{a}_j\, S_i + \sum_{j=1, j\neq i}^{m} \bar{a}_j\, \bar{S}_j\, a_j\, S_i \ (i=1 \text{ to } m)$$

$$\bar{S}_i \to \varepsilon + \sum_{j=1}^{m} \bar{a}_j\, \bar{S}_j\, a_j\, \bar{S}_i + \sum_{j=1, j\neq i}^{m} a_j\, S_j\, \bar{a}_j\, \bar{S}_i \quad (i=1 \text{ to } m)\ [\text{Ber } 79] .$$

The generating function of P is algebraic, according to a classical theorem of Chomsky and Schützenberger [CS 63] about the non ambiguous context-free languages ; it is a component of a system of polynomial equations, obtained by the following substitutions in G : rulas become equalities, ε becomes 1, a_i becomes Z, variables become generating functions (the axiom S becomes the generating function $\gamma(Z)$ of P ;

S_i ($\bar{S}_i$) becomes $\beta_i(Z)$ ($\bar{\beta}_i(Z)$) the generating function of the language P_i ($\bar{P}_i$) generated by G with $S_i(\bar{S}_i)$ for the axiom). Because of : on the one hand, the symmetry between the rula beginning by $S_i \rightarrow$, and the corresponding rula beginning by $\bar{S}_i \rightarrow$; on the other hand, the symmetry of the parts of the letters $a_1,...a_m$, $\bar{a}_1,...\bar{a}_m$, the languages P_i and $\bar{P}_i$ have, whatever i, the same number of words of length n, for any n, i.e. $\forall i \in [1,m]$, $\beta_i(Z) = \bar{\beta}_i(Z) = \beta(Z)$. Hence the system :
$\gamma(Z) = 2\,mZ^2\beta(Z)$, $\beta(Z) = 1 + 2(m-1)Z^2[\beta(Z)]^2$, and verifies the polynomial functional equation : $(2m-1)[\gamma(Z)]^2 - 2m\,\gamma(Z) + 4\,m^2 Z^2 = 0.$

$$\gamma(0) = 0\ (\gamma(Z) = \sum_{n\geq 1}\gamma_n Z^n) \Rightarrow \gamma(Z) = \frac{m}{2m-1}\left[1-\sqrt{1-4(2m-1)Z^2}\right] = \sum_{n\geq 1}\gamma_{2n} Z^{2n} ;$$

$\gamma_{2n} = 2m(2m-1)^{n-1} C_n$, with C_n for the nth Catalan number $\binom{2(n-1)}{n-1}/n$, so that asymptotically $\gamma_{2n} \sim m^n 2^{3n-2}/n\sqrt{\pi n}$ (and the radius of convergence ρ is $1/2\sqrt{2m-1}$).

By means of $\gamma(Z)$, we obtain, considering the previous problem of random words :

(1) $Pr(\mathcal{E}) = \gamma(1/2m) = 1/(2m-1)$

(2) $E(X|\mathcal{E}) = \gamma'(1/2m)/2m\gamma(1/2m) = 2 + 1/(m-1)$ if $m>1$, ∞ if $m=1$ ($\gamma'(1/2m) = 2m/(m-1)$).

III. PREFIX-FREE PALINDROMS OVER $A = \{a_1,...a_m\}$

We consider now the language L of the palindroms over m letters $a_1,...a_m$ ($w = a_{i_1} a_{i_2} ... a_{i_l}$ is a palindrom if it is equal to its mirror image $w^T = a_{i_l} ... a_{i_2} a_{i_1}$ and if its length $|w|$ verifies $|w| \geq 2$). The case $m=2$ is particularly simple : the prefix-free words of length n are $a_1 a_2^{n-2} a_1$ and $a_2 a_1^{n-2} a_2$, so $\gamma_n = 2$ and

$$\gamma(Z) = \sum_{n\geq 2}\gamma_n Z^n = \frac{2Z^2}{1-Z},$$ whence $Pr(\mathcal{E}) = \gamma(1/2) = 1$, $E(X|\mathcal{E}) = E(X) = \gamma'(1/2)/2 = 3.$

We consider in the following the case $m>2$: the language P of the prefix-free words of L is no more context-free [Bea 83] and we cannot use the method of II.

We will prove first a recursion formula indicated in [BT 83] :

Theorem 1 : $\gamma_2 = m,\ \gamma_{2n-1} = \gamma_{2n} = m^n - \sum_{i=2}^{n}\gamma_i\, m^{n-i} \quad (n\geq 2)$

Proof : (1) $\gamma_2 = m$ (these palindroms are all the $a_i a_i$'s)

(2) Because of symmetry with respect to the middle letter x ($x \in A$ or $x = \varepsilon$, according to the evenness of n), we obtain :
$card(L \cap A^n) = m^{\lfloor (n+1)/2 \rfloor}$; we assume $n \geq 2$ in the following.

(3) We consider now the complement $\overline{P \cap A^n}$ of $P \cap A^n$ in $L \cap A^n$ and its "building". Let $w \in \overline{P \cap A^n}$: $w \in L$, $w = w_1 w_2$, $w_1 \in P$, $w_2 \neq \varepsilon$; for obtaining w, we can :
- begin with $w_1 \in P \cap A^2$, complete with $\lfloor (n+1)/2 \rfloor - 2$ letters selected in A, and finally complete by symmetry.

- begin again with $w_1 \in P \cap A^3$, etc..., next to $w_1 \in P \cap A^{\lfloor (n+1)/2 \rfloor}$

-possibly go on beyond : $w_1 \in P \cap A^k$, $k > \lfloor (n+1)/2 \rfloor$?

The next combinatorial lemma points out that it is impossible.

(4) <u>Lemma</u> : Let $w \in \overline{P \cap A^n}$: $w \in L$, $w = w_1 w_2$, $w_1 \in P$, $w_2 \neq \varepsilon$; then $|w_1| \leqslant \lfloor (n+1)/2 \rfloor$.

<u>Proof</u> : Let us suppose the contrary ; then : on the one hand, $w \in L \Rightarrow w = fx f^T$ ($x \in A$ or $x = \varepsilon$) ; on the other hand, by hypothesis : $|w_1| \geqslant \lfloor (n+1)/2 \rfloor = |f| + |x|$, whence : $\exists h \neq \varepsilon$, prefix of f^T, and suffix of w_1 , schematized by :

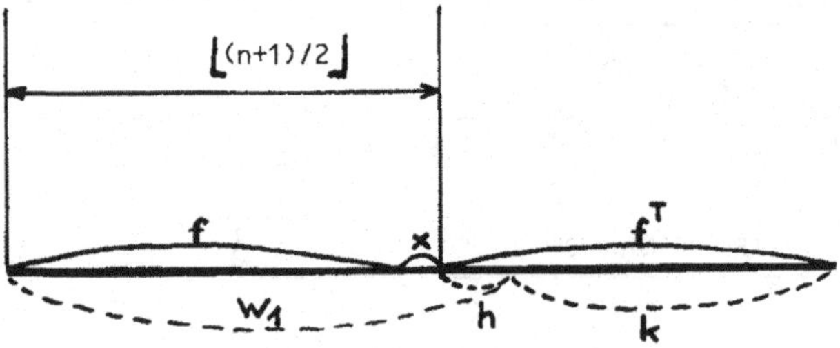

$h \neq f^T$ ($w_2 \neq \varepsilon$), then $f^T = hk$, $k \neq \varepsilon$; $f = (f^T)^T = k^T h^T$, whence $w_1 = fxh = k^T h^T xh$; $w_1 \in L \Rightarrow w_1 = w_1^T = h^T x h k = l k$, $l \in L$, with $k \neq \varepsilon$: this is inconsistent with $w_1 \in P$.

(5) So (3) yields, with (4) :

$$\operatorname{card}(\overline{P \cap A^n}) = \sum_{i=2}^{\lfloor (n+1)/2 \rfloor} \operatorname{card}(P \cap A^i)\; m^{\lfloor (n+1)/2 \rfloor - i}, \text{ whence}$$

with (2): $$\gamma_n = m^{\lfloor (n+1)/2 \rfloor} - \sum_{i=2}^{\lfloor (n+1)/2 \rfloor} \gamma_i\; m^{\lfloor (n+1)/2 \rfloor - i},$$

i.e. $$\gamma_{2n-1} = \gamma_{2n} = m^n - \sum_{i=2}^{n} \gamma_i\; m^{n-i} \;\blacksquare$$

By manipulating formal series, a functional equation for γ can be obtained :

<u>Theorem 2 : $\gamma(Z^2)/1 - mZ^2 = (m^2Z^4/1 - mZ^2) - (Z(\gamma(Z) - mZ^2)/1 + Z)$</u>

<u>Proof</u> : (1) $\gamma(Z) = mZ^2 + \sum_{n \geqslant 2} \gamma_{2n} (Z^{2n-1} + Z^{2n})$ because of $\gamma_{2n-1} = \gamma_{2n}$; it yields : $\gamma(Z) = mZ^2 + (1 + \frac{1}{Z}) (\sum_{n \geqslant 2} \gamma_{2n} Z^{2n})$.

(2) It can be noticed that (if $|Z| < 1/m$) $\gamma(Z)/1 - mZ =$

$$\sum_{n \geqslant 2} \Big(\sum_{i=2}^{n} \gamma_i\; m^{n-i}\Big) Z^n = \sum_{n \geqslant 2} (m^n - \gamma_{2n}) Z^n \quad \text{(theorem 1)},$$

so $\gamma(Z)/1 - mZ = (m^2 Z^2/1 - mZ) - \sum_{n \geqslant 2} \gamma_{2n} Z^n$.

(3) changing Z into Z^2 (so necessarily $|Z| < 1/\sqrt{m}$), the previous equality becomes : $\gamma(Z^2)/1 - mZ^2 = (m^2Z^4/1 - mZ^2) - \sum_{n \geqslant 2} \gamma_{2n} Z^{2n}$, i.e. with (1) :

$$\gamma(Z^2)/1 - mZ^2 = (m^2Z^4/1 - mZ^2) - (Z(\gamma(Z) - mZ^2))/1 + Z) \;\blacksquare$$

This functional equation may be solved by iteration (see also [FRS 84] [FS 85] for analogous equations like $f(Z) = \alpha(Z)f(Z/2) + \beta(Z)$), remarking that $\gamma(Z) = \alpha(Z)\gamma(Z^2) + \beta(Z)$:

Theorem 3 : γ is a transcendental function, having infinitely many singularities $\left(1/m^{1/2^p}\right)$ $p\geq 1$, given by :

$$\gamma(Z) = \sum_{p\geq 0} \left(\beta(Z^{2^p}) \prod_{k=0}^{p-1} \alpha(Z^{2^k})\right), \text{ with } \alpha(Z) =$$

$$-(Z+1)/Z(1-mZ^2), \quad \beta(Z) = mZ^2(1+mZ)/1-mZ^2$$

$$\left(\text{So } \gamma(Z) = \sum_{p\geq 0} m(-1)^p Z^{2^p+1} \frac{1-Z^{2^p}}{1-Z} \frac{1+mZ^{2^p}}{(1-mZ^2)(1-mZ^4)\dots(1-mZ^{2^{p+1}})}\right)$$

Proof :

(1) It is easy to obtain $\alpha(Z) = -(Z+1)/Z(1-mZ^2)$ and $\beta(Z) = mZ^2(1+mZ)/1-mZ^2$, such that $\gamma(Z) = \alpha(Z)\gamma(Z^2) + \beta(Z)$ (*)

(2) By iteration of (*), $\gamma(Z) = \alpha(Z)\gamma(Z^2) + \beta(Z) =$ $\beta(Z) + \alpha(Z)\beta(Z^2) + \alpha(Z)\alpha(Z^2)\gamma(Z^4) = \dots$, so $\gamma(Z) = \sum_{p\geq 0}\left(\beta(Z^{2^p}) \prod_{k=0}^{p-1}\alpha(Z^{2^k})\right)$; this function (if it is converging) is effectively a solution of (*) ($\gamma(Z) = \beta(Z) + \alpha(Z)[\beta(Z^2) + \alpha(Z^2)\beta(Z^4) + \dots] = \beta(Z) + \alpha(Z)\gamma(Z^2)$), so the eventual solution is unique.

(3) We will prove now the convergence of the infinite sum giving $\gamma(Z)$. It is easy to obtain :

$$f_p(Z) = \beta(Z^{2^p}) \prod_{k=0}^{p-1} \alpha(Z^{2^k}) = m(-1)^p Z^{2^p+1} \frac{1-Z^{2^p}}{1-Z} \frac{1+mZ^{2^p}}{(1-mZ^2)(1-mZ^4)\dots(1-mZ^{2^{p+1}})}$$

The convergence of the series $f_p(Z)$ may be reduced to that of the series $\frac{|Z|^{2^p}}{(1-m|Z|^2)(1-m|Z|^4)\dots(1-m|Z|^{2^{p+1}})}$, then to that of the series

$2^p |Z|^{2^p-1}$ $(\exists p_0,\ p>p_0 \Rightarrow m|Z|^{2^p} < 1/2)$ which is converging $(|Z|<1)$ ■

The radius of convergence ρ is $1/\sqrt{m}$, the minimum modulus of the singularities [Hen 77]. The formula of the theorem 3 for γ allows to obtain at any order expansions in $1/m$ for $Pr(\mathcal{E})$ and $E(X|\mathcal{E})$), e.g. $Pr(\mathcal{E}) = 2/m+1/m^2+1/m^3-1/m^4-1/m^5-2/m^6+O(1/m^7)$, $E(X|\mathcal{E}) = 5/2 + 7/14m - 1/8m^2 + O(1/m^3)$.

We will prove now :

Theorem 4 : if $n\geq 2$, $m^n - 2m^{n-1} - 3m^{n-2} < \gamma_{2n} < m^n - 2m^{n-1} + m$ (so $\gamma_{2n} = m^n - 2m^{n-1} + O(m^{n-2})$)

Proof :

(1) $\gamma_{2n} \leqslant m^n$ because of the theorem 1 ; by iteration, using again this theorem, we obtain after a computation :

$$\gamma_{4q} \geqslant m^q \left[1+m - 3m^q + m^{q+1}\right] /(m-1)$$

$$\gamma_{4q+2} \geqslant m^{q+1} \left[2 - 3m^q + m^{q+1}\right] / (m-1)$$; these two formulas give

$$\gamma_{2n} \geqslant m^n - 2m^{n-1} - \frac{2m^{n-1}}{m-1} \geqslant m^n - 2m^{n-1} - 3m^{n-2}.$$

(2) <u>If $n \geqslant 2$, then $\gamma_{2n}/m(m-1)$ is a polynomial of the $(n-2)$th degree in m, of first term m^{n-2}</u>.

Indeed $\gamma_4 = m(m-1)$, then we use recursion arguments and the formula of the theorem1, with $\gamma_2 = m$. The coefficients of this polynomial in m are integers, so its value is an integer, greater than or equals to 1 ($\gamma_{2n} \neq 0$), whence $\gamma_{2n} \geqslant m(m-1)$.

(3) By iteration, with $\gamma_{2n} \geqslant m(m-1)$, using again the theorem 1, we obtain :

$$\gamma_{2n} \leqslant m^n - 2m^{n-1} + m \quad \blacksquare$$

IV. CONCLUSION

A generalization of this problem to any probability distribution $p = \{p_1, \ldots p_m\}$ over $A = \{a_1, \ldots a_m\}$ is studied in [BT 83], [BT 84], [BT 85].

Later results should be obtained with analytic methods relating asymptotic properties and singularities of functions (Darboux's theorem, Mellin transform) [Ben 74] [Sed 83] [Fla 85] or perhaps multivariate generating functions [GJ 83] [FRS 84].

The results in this combinatorial domain have some applications in computer science, e.g. for new paging and decoding probabilistic algorithms [BT 85].

ACKNOWLEDGEMENT

The results of Part III concerning the functional equation were improved after stimulating discussions with P. FLAJOLET and N. SAHEB.

REFERENCES

[Bea 83] J. BEAUQUIER -"Prefix and perfect languages"- Proceedings of the 8th C.A.A.P. (Italy 1983) - Lecture Notes in Computer Science 159, pp 129-140.

[Ben 74] E.A. BENDER "-Asymptotic methods in enumeration"- SIAM Review Vol 16, n°4 October 1974

[Ber 79] J. BERSTEL -"Transductions and context-free languages"- Teubner Verlag (1979)

[BT 83] J. BEAUQUIER, L.THIMONIER -" Formal languages and Bernoulli processes"- To appear in "Algebra, Combinatorics and Logic in Computer science"- Colloquia Mathematica Soc.J. Bolyai (Hungary 1983) vol 42, North Holland.

[BT 84] J. BEAUQUIER, L. THIMONIER -" Computability of probabilistic parameters for some classes of formal languages"- Proceedings of the 11th Symposium Mathematical Foundations of Computer Science (Czechoslovakia 1984) Lecture Notes in Computer Science 176, pp. 194-204

[BT 85] J. BEAUQUIER, L. THIMONIER -"On formal languages, probabilities, paging and decoding algorithms"- Proceedings of the 5th Conference Fundamentals of Computing Theory (German Democratic Republic 1985) - Lecture Notes in Computer Science 199, pp 44-52

[CS 83] N. CHOMSKY, M.P. SCHUTZENBERGER -"The algebraic theory of context-free languages"- Computer programming and formal systems pp. 118-161 North Holland (1963)

[Com 70] L. COMTET -"Analyse Combinatoire"- PUF Paris (1970)

[ER 61] P. ERDOS, A. RENYI -"On a classical problem of probability theory"- Publ. M.I. Hung. Acad. Sci., 6(1961), 215-20

[Fel 68] W.FELLER -"An introduction to probability theory and its applications"- J. Wiley (1968)

[Fla 85] P. FLAJOLET-"Ambiguity and transcendence"- Proceedings of the 12th I.C.A.L.P. (Greece 1985) To appear in Lecture Notes in Computer Science.

[FRS 84] P. FLAJOLET, M. REGNIER, D.SOTTEAU -" Algebraic methods for trie statistics"- To appear in Annuals of Discrete Math. (1984)

[FS 85] P. FLAJOLET, N.SAHEB-" The complexity of generating an exponentially distributed variable"- To appear in Journal of Algorithms (1985)

[GJ 83] I.P.GOULDEN, D.M. JACKSON-"Combinatorial Enumeration"- J. Wiley (1983)

[Gol 72] J. GOLDSTINE-"Substitution and bounded languages"- Journ of Comp. and Syst. Sciences 6 (1972) pp. 9-29

[Har 78] M.A. HARRISON-"Introduction to formal languages theory"- Add. Wesley (1978)

[Hen 77] P.ENRICI-"Applied Computational and Complex analytis"- J.Wiley (vol 1 : 1974 ; vol 2 : 1977)

[Kui 70] W. KUICH-"On the entropy of context-free languages"- Information and control 16, pp 173-200 (1970)

[Sed 83] R. SEDGEWICK -" Mathematical Analysis of Combinatorial Algorithms"- In G. Louchard, G. Latouche ed., Probability theory and Computer Science, Acad. Press (1983)

COMBINATORIAL REPRESENTATIONS OF SOME LIE GROUPS AND LIE ALGEBRAS

F.Bergeron, Dépt. Maths et Info, U.Q.A.M.

1. Introduction

The aim of this paper is to show that interesting combinatorial interpretations can be given to some aspects of Lie group techniques used in the study of symmetries of linear second order partial differential equations (PDE) of mathematical physics. We thus show that group-theoretic methods of special function theory can be reformulated in a combinatorial set-up.

Let **R** be a partial differential operator (of order two) defined on the space of analytic (around the origin) functions F(x,y) in two variables. A linear differential operator of the form:

$$\mathbf{L} = M(x,y)\delta/\delta x + P(x,y)\delta/\delta y + Q(x,y)$$

is said to be a symmetry operator corresponding to **R**, if and only if:

$$[\mathbf{L},\mathbf{R}] = \mathbf{LR} - \mathbf{RL} = F(x,y)\,\mathbf{R}$$

where the function F may depend on **L**. Observe that symmetry operators map solutions T, of the PDE "**R**T=0" , into solutions of the same PDE . Moreover, the set of all such symmetry operators is a (complex) Lie algebra, with the usual Lie bracket. These Lie algebras permit a systematic study of the coordinate systems for which the equation admits separation of variables. For more details on these methods, see W. Miller's book, *Symmetry and separation of variables* . The special functions arising through such separation of variables can then be studied by group-theoretic methods. We will give combinatorial meaning to this approach.

A linear combinatorial operator $\mathfrak{L}$ is a functor from the category of species (in one or more variables, with or without weights) to itself, preserving sums and products by "constants". Constants are species that are empty on all non-empty sets. For a description of the theory of species of structures, see A.Joyal [J1]. For more results involving differential combinatorial operators, see G.Labelle [L1].

The cardinality **Card**($\mathfrak{L}$) of a combinatorial operator is the corresponding operator on power series associated with the species involved. Thus, a combinatorial model for a differential operator **L** is a combinatorial operator $\mathcal{L}$ such that: **Card**($\mathcal{L}$)= **L**. We will give combinatorial representations of Lie algebras of linear differential operators in terms of "algebras" of combinatorial operators. Typical calculations involving Lie algebras and the corresponding Lie groups will have combinatorial interpretations and proofs.

Several authors have worked on combinatorial models of orthogonal families of polynomials, see Bergeron, Foata, Garsia, Gessel, Ismail, J.Labelle, Leroux, Mullin, Remmel, Rota, Stanley, Stanton, Strehl, and Viennot [see bibliography]. All this work points to a more or less uniform combinatorial outlook on models of orthogonal polynomials. This has already been stressed in [B1] , but we believe that a deeper insight into the subject can be gained by the methods first introduced in a special case in [B2], and generalized in the present paper.

2. Algebra of combinatorial operators

Let us simply write $\mathcal{L} = \mathcal{H}$, when two combinatorial operators $\mathcal{L}$ and $\mathcal{H}$ (over the same category of species) are isomorphic as functors. The "sum" $\mathcal{L}+\mathcal{H}$ and the "product" $\mathcal{L}\mathcal{H}$ of $\mathcal{L}$ and $\mathcal{H}$, are defined respectively as pointwise sum and usual composition of the corresponding functors.

We have thus given sense to the concept of algebra of linear combinatorial operators over a category of species. For the moment, let this category be the category of two sorted species. Then we consider the operators $\delta/\delta X$ and $\delta/\delta Y$ which, for given species T and given finite sets A and B, are defined by:

$$(\delta/\delta X\, T)\,[A,B] = T[A+\{*\},B]$$

$$(\delta/\delta Y\, T)\,[A,B] = T[A,B+\{*\}]$$

Consider also, the operator "Q" corresponding to multiplication of two sorted species by a fixed two sorted species Q. This fixed species might be for example one of the species X or Y such that:

$$X[A,B] = \begin{cases} \{A\} & \text{, if B is empty, and A has exacly one element} \\ \emptyset & \text{, otherwise.} \end{cases}$$

$$Y[A,B] = \begin{cases} \{B\} & \text{, if A is empty, and B has exacly one element} \\ \emptyset & \text{, otherwise.} \end{cases}$$

Now, recall that the cardinality of a two sorted species T is:

$$\mathbf{Card}(T) = T(x,y) = \sum t_{n,k}\, x^n y^k / n!\,k!\text{, summed over all n and k in } \mathbb{N}.$$

where $t_{n,k}$ is the number of element of T[A,B] , with A any n element set, and B any k elements set.

Hence, since $\mathbf{Card}(X \bullet T) = x \cdot \mathbf{Card}(T)$, and $\mathbf{Card}(\delta/\delta X\, T) = (\delta/\delta \mathbf{x})\mathbf{Card}(T)$, we conclude that the cardinality of the operators X is x , and that of $\delta/\delta X$ is $\delta/\delta \mathbf{x}$.

Note: In conformity with local traditions, we have adopted here almost the same notation for a combinatorial operator and its cardinality: we use capital letters in the "combinatorial" world and lowercase letters in the "analytical" world.

In the sequel of this paper, an important role will be played by the operators $X \cdot \delta/\delta X$, $Y \cdot \delta/\delta Y$, $X \cdot \delta/\delta Y$ and $Y \cdot \delta/\delta X$, obtained by various compositions of the operators X , Y , $\delta/\delta X$ and $\delta/\delta Y$.

In general, let $\mathcal{L}$ and $\mathcal{H}$ be two linear combinatorial operators. Then, the following identities are easy to verify:

PROPOSITION 1

(1) $\mathbf{Card}(\mathcal{L}+\mathcal{H}) = \mathbf{Card}(\mathcal{L}) + \mathbf{Card}(\mathcal{H})$,

(2) $\mathbf{Card}(\mathcal{L}\mathcal{H}) = \mathbf{Card}(\mathcal{L})\, \mathbf{Card}(\mathcal{H})$,

where the operations on the right-hand side of those identities are respectively point-wise sum, and composition of the linear operators $\mathbf{Card}(\mathcal{L})$ *and* $\mathbf{Card}(\mathcal{H})$.

3. Lie bracket

Throughout this text, a typical structure t of a typical species T will be represented by the following picture:

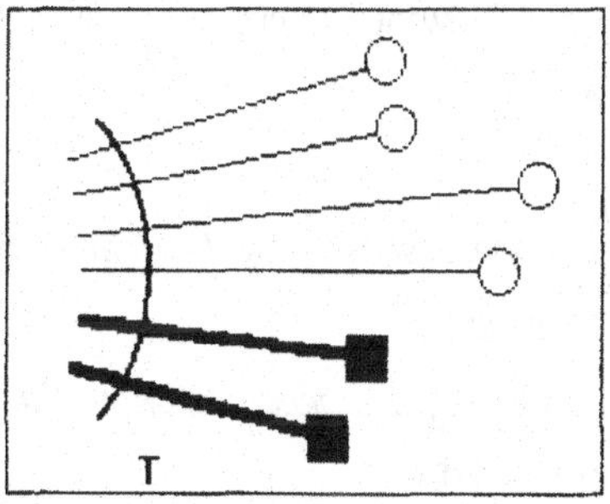

Figure 3.1

Here, the white circles represent labelled points of type "X"; and black squares, labelled points of type "Y". The thick lines are there to help distinguish between points of the two kinds. The labels are not shown for simplicity's sake.

Thus, the effect of a combinatorial operator $\mathfrak{E}$, can be illustrated by showing how typical structures of species $\mathfrak{E}T$, may be constructed out of T-structures (structures of species T). For example, the effect of operator $Y \cdot \delta/\delta X$ is to "replace", in a typical structure, any designated point of type "X" by a point of type "Y":

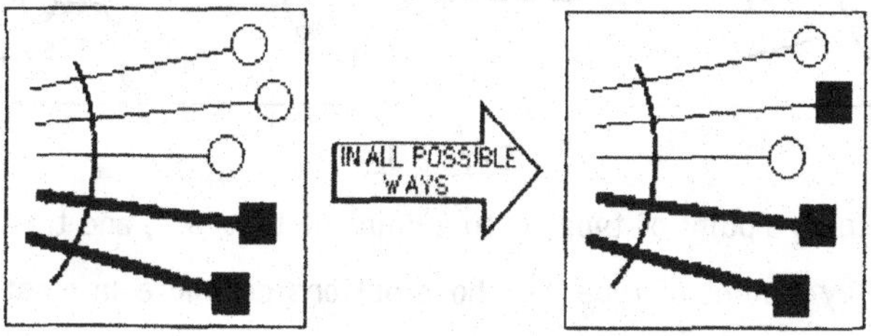

Figure 3.2

Likewise, the effect of $YX \cdot \delta/\delta X$ is to "attach" a point of type "Y" to a selected point of type "X":

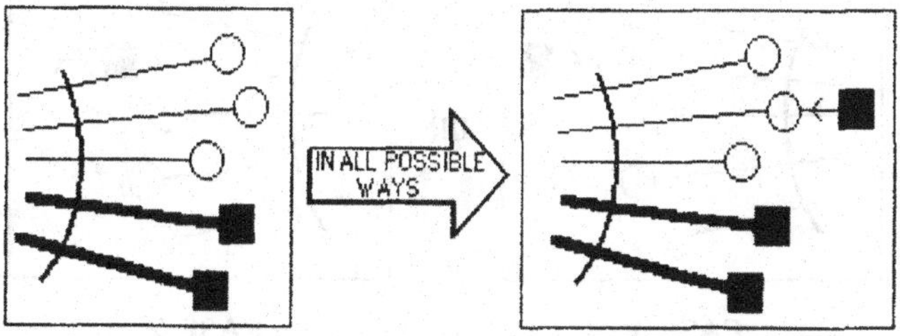

Figure 3.3

We are now ready to show the following, where the brackets [-,-] denote the usual Lie multiplication: $[A,B] = AB - BA$.

PROPOSITION 2

(1) $[\delta/\delta X, Q] = Q'$, *the operator* Q *is multiplication by a fixed species* Q, *and* $Q' = \delta/\delta X\, Q$ *as a species.*

(2) $[Y\cdot\delta/\delta X, X\cdot\delta/\delta X] = Y\cdot\delta/\delta X$, *and hence* $[\delta/\delta X, X\cdot\delta/\delta X] = \delta/\delta X$.

(3) $[X\cdot\delta/\delta X, X^2\cdot\delta/\delta X] = X^2\cdot\delta/\delta X$

(4) $[Y\cdot\delta/\delta X, X^2\cdot\delta/\delta X] = 2\cdot YX\cdot\delta/\delta X$, *and hence* $[\delta/\delta X, X^2\cdot\delta/\delta X] = 2X\cdot\delta/\delta X$

Proof: These identities are all shown easily. As an illustration, let us prove the fourth one. First, set $\mathfrak{I} = Y\cdot\delta/\delta X$ and $\mathfrak{K} = X^2\cdot\delta/\delta X$. Now, we want to compare the operators $\mathfrak{I}\mathfrak{K}$ and $\mathfrak{K}\mathfrak{I}$. They have almost the same effect, with the exclusion of two possibilities, both of the form $YX\cdot\delta/\delta X$. The effect of $\mathfrak{I}\mathfrak{K}$ on a typical T-structure is twofold:

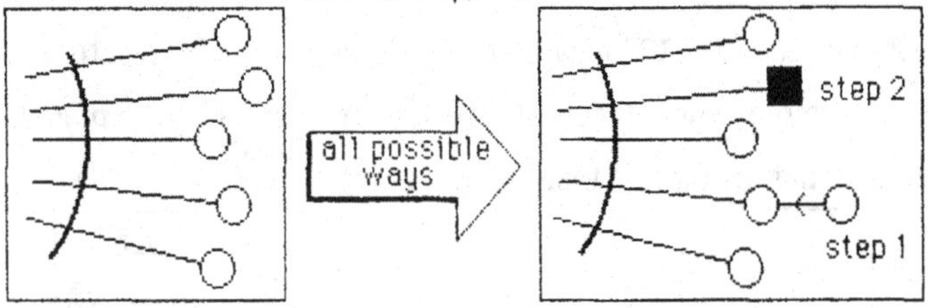

Figure 3.4

We start by attaching a point of type "X" to a point of type "X" , and then, we replace any point of type "X" by a point of type "Y". Those actions commute in most instances, with the exception of the cases when, the replacement is done on one of the two points involved in the first step. In those cases, we obtain structures having the following aspect:

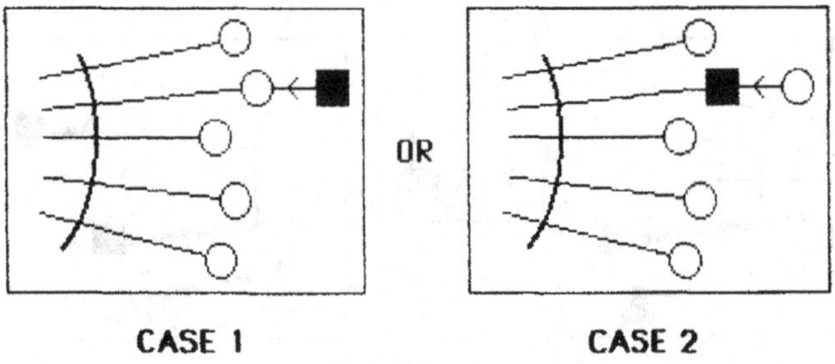

Figure 3.5

Both of which can be tought of as typical structures obtained from T-structures via $YX\cdot\delta/\delta X$. We thus get $[\mathfrak{X},\mathfrak{H}] = 2\cdot YX\cdot\delta/\delta X$, where the "2" expresses the alternative between case 1 and case 2.

4. One parameter groups

Let $\mathfrak{X}$ be a linear combinatorial operator, and write $\mathfrak{exp}[\mathfrak{X}]$ as shorthand for the formal expression:

$$\mathfrak{exp}[\mathfrak{X}] = \sum \mathfrak{X}^n / {}_{n!} \text{ , summed over all n in } \mathbb{N}.$$

Whith $\mathfrak{X}^0$ equal to: $\mathbf{1}$ the identity operator.

Gilbert Labelle in [L1], has given a combinatorial interpretation of the division of $\mathfrak{X}^n$ by n!, for operators of the form $\mathfrak{X} = [Y/(1-Q')]\cdot\delta/\delta X$, with Q a fixed species. This division plays a crucial role in an explicit description of $\mathfrak{exp}[\mathfrak{X}]$. As a special case of Labelle's results, we get a combinatorial version of Taylor's formula, giving the description of $\mathfrak{exp}[Y\cdot\delta/\delta X]$:

$$\mathfrak{exp}[Y\cdot\delta/\delta X]\ T(X) = T(X+Y)$$

Here T(S) is the species obtained by substitution of S into T.

<u>Note:</u> A combinatorial description of $\mathfrak{exp}[Y\cdot\mathfrak{X}]$ is equivalent to the resolution of a combinatorial differential equation. G.Labelle has shown in [L2] that it is not always possible to solve combinatorial differential equations. Thus a combinatorial description of $\mathfrak{exp}[Y\cdot\mathfrak{X}]$ is not always possible in the context of usual species. It is, however, in the context of $\mathbb{L}$-species (see Leroux-Viennot in this volume). For an operator $\mathfrak{X}$ which does not depend on Y, let us observe that when $\mathfrak{exp}[Y\cdot\mathfrak{X}]$ can be given a combinatorial interpretation, then we can give a combinatorial interpretation of the substitution $F[Y\cdot\mathfrak{X}]$ of $Y\cdot\mathfrak{X}$ into any species F. We define $F[Y\cdot\mathfrak{X}]$ by its action on two sorted species T(X,Y) . Let A and B be two finite sets, and define:

$$F[Y\cdot\mathfrak{X}]\ (T)\ [A,B] = F[B] \times \{\mathfrak{exp}[Y\cdot\mathfrak{X}]\ (T)\}[A,B]$$

For the following special cases of operators, we also know how to interpret this exponential:

PROPOSITION 3

(1) $\mathbf{exp}[YX \cdot \delta/\delta X]\, T(X) = T(X \bullet \mathbf{exp}(Y))$

(2) $\mathbf{exp}[X^2 \cdot \delta/\delta X]\, T(X) = T(X/(1-X))$

(3) $\mathbf{exp}[X^2 \cdot \delta/\delta X + YX]\, T(X) = \mathbf{exp}(Y \bullet \mathbf{Cycle}(X)) \bullet T(X/(1-X))$

Where $X/(1-X)$ *is the species of "chains" (total orders) of points of type* X, *and* **Cycle**(X) *is the species of "cycles".*

Proof

(1) is folklore, it is a good exercise for the reader.

(2) is a consequence of the following observations. Note that a structure of species $(X^2 \cdot \delta/\delta X)^n T$ is constructed out of a T-structure "t", by attaching (possibly empty) chains to every points of t, with n equal to the total number of points in these chains; then to all the points in these chains we assign distincts numbers between 1 and n :

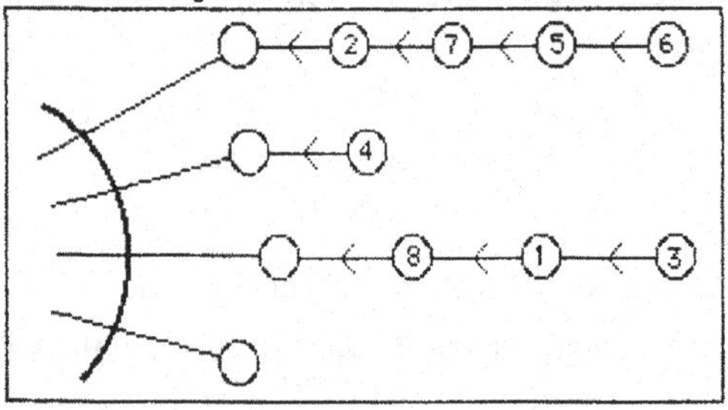

Figure 4.1

This is verified by a recursive argument. In fact, the structures of species $(X^2 \cdot \delta/\delta X)^{n+1} T$ are constructed out those of species $(X^2 \cdot \delta/\delta X)^n T$, by insertion of a new point labeled "n+1" in one the chains as follows: start by selecting any point of the $(X^2 \cdot \delta/\delta X)^n T$-structure. The new point is inserted, immediately "before" this selected point, in the surrounding chain. To divide $(X^2 \cdot \delta/\delta X)^n$ by $n!$, we forget the labels. Finally, if n is arbitrary, we get a description of $\mathbf{exp}[X^2 \cdot \delta/\delta X]$, as an operator identified to the operator that substitutes $X/(1-X)$ to X into species.

(3) is obtained in a similar way, by describing how to successively insert points in a structure of species T , with operator $(X^2 \cdot \delta/\delta X + YX)$. Let us only present geometrically, the result of n (= 16) such insertions:

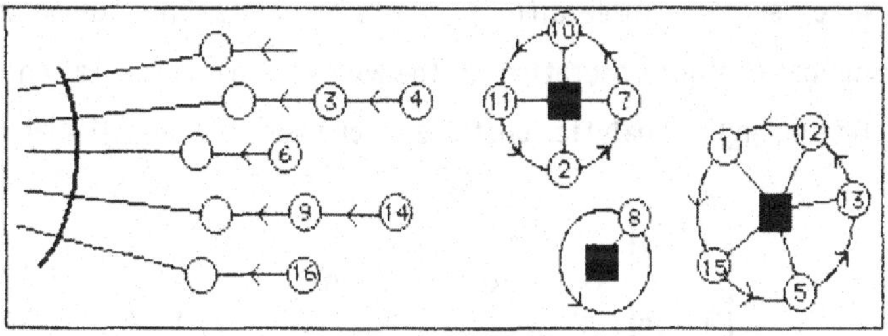

Figure 4.2

The insertion in a cycle is also done just "before" the selected point.

5. LIE ALGEBRAS AND LIE GROUPS

Let $\mathfrak{L}_1, \mathfrak{L}_2, \ldots, \mathfrak{L}_n$ be linear combinatorial operators over the same category of species. Consider the set $K[\mathfrak{L}_1, \mathfrak{L}_2, \ldots, \mathfrak{L}_n]$ of formal linear compositions $a_1\mathfrak{L}_1 + a_2\mathfrak{L}_2 + \ldots + a_n\mathfrak{L}_n$, with coefficients in a ring K with unity. We will work with operators for which it can be shown that all Lie brackets $[\mathfrak{L}_i, \mathfrak{L}_j] = \mathfrak{L}_i\mathfrak{L}_j - \mathfrak{L}_j\mathfrak{L}_i$, for $1 \le i, j \le n$, are linear combinations of $\mathfrak{L}_1, \mathfrak{L}_2, \ldots, \mathfrak{L}_n$. Thus we clearly have a Lie algebra structure on $K[\mathfrak{L}_1, \mathfrak{L}_2, \ldots, \mathfrak{L}_n]$.

EXAMPLE:

Let $\mathfrak{L}_1 = \delta/\delta X$, $\mathfrak{L}_2 = X\delta/\delta X$ and $\mathfrak{L}_3 = X^2\,\delta/\delta X$. And, recall that: $[\mathfrak{L}_1, \mathfrak{L}_2] = \mathfrak{L}_1$, $[\mathfrak{L}_2, \mathfrak{L}_3] = \mathfrak{L}_3$ and $[\mathfrak{L}_1, \mathfrak{L}_3] = 2\mathfrak{L}_2$. The Lie algebra, thus described, is isomorphic to $\mathfrak{sl}(2)$, the algebra of two by two matrices with trace zero. To verify this last assertion, set:

$$L_1 = \begin{bmatrix} 0 & -1 \\ 0 & 0 \end{bmatrix}, \quad L_2 = \begin{bmatrix} 1/2 & 0 \\ 0 & -1/2 \end{bmatrix}, \text{ and } \quad L_3 = \begin{bmatrix} 0 & 0 \\ 1 & 0 \end{bmatrix}$$

which gives a basis of $\mathfrak{sl}(2)$. The isomorphism that sends $\mathfrak{L}_i$ to L_i, $1 \le i \le 3$, preserves Lie brackets.

The elements of the Lie group, corresponding to this algebra, are of the form $\mathit{exp}[a_3\mathfrak{L}_3] \cdot \mathit{exp}[a_1\mathfrak{L}_1] \cdot \mathit{exp}[a_2\mathfrak{L}_2]$ in a neighborhood of the identity of this group. From proposition 3, we easily deduce:

$$\mathit{exp}[W\mathfrak{L}_3] \cdot \mathit{exp}[Z\mathfrak{L}_1] \cdot \mathit{exp}[Y\mathfrak{L}_2]\, T(X) = T(\mathit{exp}(Y) \bullet \{Z + (X/(1-WX))\})$$

with Y,Z, and W representing three different sorts of points. Thus, using proposition 3, we obtain a combinatorial interpretation for the well known representation of SL(2), as a group of operators acting on analytic functions of one variable. Recall that this action A is defined by:

$$A\begin{bmatrix} i & j \\ k & n \end{bmatrix} F(x) = F((ix+k)/(jx+n))$$

6. LAGUERRE POLYNOMIALS

Now that we have described all the basic concepts needed for our discussion, let us study, through examples, how useful results can be obtained by means of combinatorial Lie group representations. Consider the algebra $\mathbb{Z}[\alpha,t][\mathfrak{L}_1,\mathfrak{L}_2,\mathfrak{L}_3,\mathfrak{L}_4]$ with:

$$\mathfrak{L}_1 = (Y/X)\{X\delta/\delta X - t\delta/\delta t\} \qquad \mathfrak{L}_2 = X\delta/\delta X$$

$$\mathfrak{L}_3 = X^2\delta/\delta X + tX\delta/\delta t + tX + \alpha X \qquad \mathfrak{L}_4 = \mathbf{1}$$

In $\mathfrak{L}_1$ we have a formal division by X which becomes a real division by x in the analytical "world". Those operators act on the $\mathbb{Z}[\alpha,t]$-weighted species obtained by substitution of tX for W in any three sorted species F(X,W,Y) , with weight in $\mathbb{Z}[\alpha]$. We write $F_t(X,Y)$ for $F(X,W,Y)_{W=tX}$. A typical $F_t(X,Y)$-structure can be represented as:

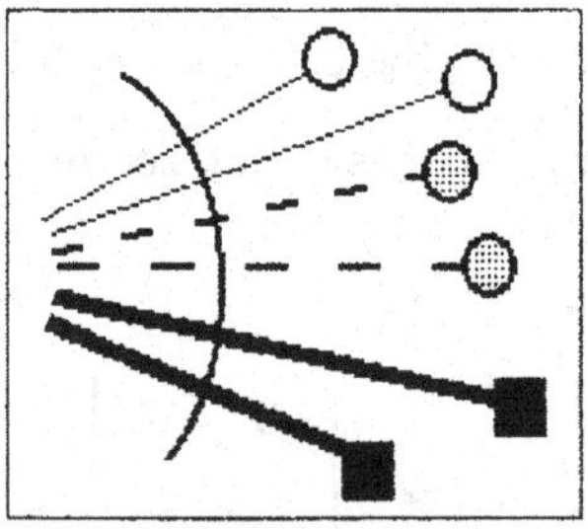

Figure 6.1

The circular points with a dotted pattern, are the points of type X having weight t.

Note that $t\delta/\delta t$, as a combinatorial operator, corresponds to pointing a dotted circular point. Thus we can see that the effect of $\mathfrak{L}_1$, on F_t-structures, is to replace a circular point, without dotted pattern, by a point of type Y. Moreover, the action of $\mathfrak{L}_3$ on typical F_t-structures, is that of adding a new point in one of the four following ways: The first and second one (corresponding to tX and αX), add an isolated point respectively with weight t (dotted circular point) or α . The last two possibilities ($X^2\, \delta/\delta X$ and $tX\delta/\delta t$), both attach a new point to a previously selected one. We have already explained how $X^2\, \delta/\delta X$ acts. The action of $tX\delta/\delta t$ is similar, but with the condition that the point selected for an adjonction is a dotted one. To differentiate between these last two constructions, we represent geometrically $tX\delta/\delta t$ in a distinct manner (see figure 6.2):

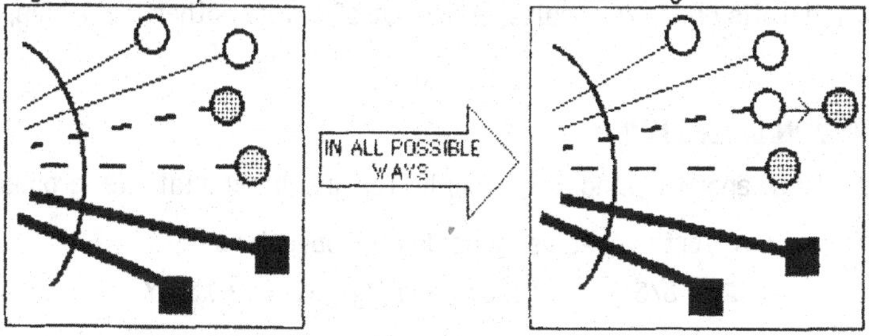

Figure 6.2

The attaching arrow is oriented in the opposite direction to the one selected in figure 3.4 to represent $X^2\delta/\delta X$. Then the dotted pattern (weight) of the selected point is transferred to the new point.

Now, the Lie algebra structure on linear combinations of $\mathcal{D}_1, \mathcal{D}_2, \mathcal{D}_3$ and $\mathcal{D}_4$, is characterized by the identities:

$$[\mathcal{D}_1,\mathcal{D}_4] = [\mathcal{D}_2,\mathcal{D}_4] = [\mathcal{D}_3,\mathcal{D}_4] = 0$$

$$[\mathcal{D}_1,\mathcal{D}_2] = \mathcal{D}_1 \qquad [\mathcal{D}_2,\mathcal{D}_3] = \mathcal{D}_3$$

$$[\mathcal{D}_1,\mathcal{D}_3] = 2\mathcal{D}_2 + \alpha\mathcal{D}_4$$

which can be deduced as in proposition 2.

It is also straightforward to verify that $[(\mathcal{D}_3\mathcal{D}_1+\mathcal{D}_2),\mathcal{D}_3]=[(\mathcal{D}_2{}^2 + \alpha\mathcal{D}_2),\mathcal{D}_3]$. Or in other words, if we write $\mathbb{R}$ for $(\mathcal{D}_3\mathcal{D}_1+\mathcal{D}_2 - \mathcal{D}_2{}^2-\alpha\mathcal{D}_2)$, that $[\mathbb{R},\mathcal{D}_3] = 0$. We conclude that, if T is a species such that $\mathbb{R}T = 0$, then $\mathbb{R}\cdot\exp[\mathcal{D}_3]\,(T) = 0$. Weisner, in *Group - Theoretic origin of certain generating functions* [W1], has shown how generating functions and identities, for families of orthogonal polynomials, can be obtained by studying the action of a Lie group of a differential operator, on the solutions of a PDE "**R**T = 0". Essentially, what we have just outlined, is a proof of:

THEOREM 1

The cardinality of $\mathbf{exp}[\mathfrak{L}_3](1)$ *is a formal power series of the form:*

$$\mathbf{Card}(\mathbf{exp}[\mathfrak{L}_3](1)) = \sum L_n(\alpha;t)\, x^n/n! = \mathbf{exp}\{tx/(1-x) + \alpha \,\mathbf{log}(1/1-x)\}$$

where the $L_n(\alpha;t)$, *are Laguerre polynomials. The* $L_n(\alpha;t)$ *also sastisfies the differential equation:* $nY = tY'' + (\alpha+t)Y'$ (with $Y=L_n(\alpha;t)$ and $Y'=\delta/\delta t\, Y$).

For more details on these results, and proof of several other identities, see [B2].

7. KLEIN-GORDON EQUATION

Let us study another kind of example for combinatorial Lie groups and their manipulation. Over two sorted species, consider the operators:

$$\mathfrak{L}_1 = \delta/\delta X\ ,\ \mathfrak{L}_2 = \delta/\delta Y\ ,\ \text{and}\ \mathfrak{L}_3 = Y\delta/\delta X + X\delta/\delta Y$$

for which, we clearly have:

$$[\mathfrak{L}_1,\mathfrak{L}_2] = 0\ ,\ [\mathfrak{L}_2,\mathfrak{L}_3] = \mathfrak{L}_1\ ,\ \text{and}\ [\mathfrak{L}_1,\mathfrak{L}_3] = \mathfrak{L}_2$$

as well as:

LEMMA 1:

$$[\mathfrak{L}_1{}^2,\mathfrak{L}_3] = [\mathfrak{L}_2{}^2,\mathfrak{L}_3] = 2\mathfrak{L}_1\mathfrak{L}_2$$

Proof: It is easier to study $Z\mathfrak{L}_1$ instead of $\mathfrak{L}_1$, and then set Z equal to 1. The effect of $(Z\mathfrak{L}_1)^2\mathfrak{L}_3$ on typical structures, is first to change the type of a point of type X to Y, or vice-versa. Then, we successively change two different points of type X to type Z. The only possible obstructions to the commutation of $(Z\mathfrak{L}_1)^2$ and $\mathfrak{L}_3$, occur when the points, on which we would want to perform changes, "collide". This can happen only when the first change is a replacement of a point of type Y by a point of type X. And then, when this same point is further replaced by a point of type Z, either in the first or the second application of $\mathfrak{L}_1$.

Thus, $[(Z\mathfrak{L}_2)^2,\mathfrak{L}_3] = 2Z^2\,\mathfrak{L}_1\mathfrak{L}_2$ showing the first identity. By a symmetry argument, we also have: $[\mathfrak{L}_2{}^2,\mathfrak{L}_3] = 2\mathfrak{L}_1\mathfrak{L}_2$.

We conclude that:

THEOREM 2:

The operators $\mathfrak{L}_1$, $\mathfrak{L}_2$, *and* $\mathfrak{L}_3$ *form a basis (as a vector space) of a Lie algebra of symmetry operators for the Klein-Gordon differential operator:*

$$\delta^2/\delta X^2 - \delta^2/\delta Y^2 + \alpha \mathbf{1}$$

Proof:
From lemma 1 we can conclude that $\mathfrak{L}_3$ commutes with the operator $\mathfrak{L}_1{}^2 - \mathfrak{L}_2{}^2$ Since $\mathfrak{L}_3$ also commutes with $\alpha\mathbf{1}$, the theorem follows immediately.

The Lie group corresponding to this algebra, is a representation of Poincaré's group for two-dimensional space-time. That is to say, it is isomorphic to the group of matrices of the form:

$$\begin{bmatrix} \cosh(z) & \sinh(z) & 0 \\ \sinh(z) & \cosh(z) & 0 \\ v & w & 1 \end{bmatrix}$$

More precisely, we show:

PROPOSITION 4

$$\mathbf{exp}[Z\mathfrak{L}_3]\cdot\mathbf{exp}[V\mathfrak{L}_2]\cdot\mathbf{exp}[W\mathfrak{L}_1]\, T(X,Y)$$

$$=$$

$$T(X\bullet\mathbf{cosh}(Z)+Y\bullet\mathbf{sinh}(Z)+V, Y\bullet\mathbf{cosh}(Z)+X\bullet\mathbf{sinh}(Z)+W)$$

where $\mathbf{sinh}(Z)$ *and* $\mathbf{cosh}(Z)$ *are respectively, the species characteristic of odd and even sets, of points of type Z.*

Proof: We only need to describe $\mathbf{exp}[Z\mathfrak{L}_3]$, since everything else has already been done. Now, $\mathfrak{L}_3$ exchanges the sorts X and Y, and the Z in $Z\mathfrak{L}_3$ can be thought of as a counter of exchanges. So, the effect of $\mathbf{exp}[Z\mathfrak{L}_3]$ on typical structures, is to attach to each point of these structures, a set of points of type Z. If this set is even, then the point to which it is attached is of the same type as it was originally, since it has been changed an even number of times. Otherwise, when it is odd, the original point was of the other type. This argument shows that:

$$\mathbf{exp}[Z\mathfrak{L}_3]\, T(X,Y) = T(X\bullet\mathbf{cosh}(Z)+Y\bullet\mathbf{sinh}(Z), Y\bullet\mathbf{cosh}(Z)+X\bullet\mathbf{sinh}(Z))$$

Finally, with similar arguments, one can compute the adjoint action of this representation of Poincaré's group on its Lie algebra. This action permits the determination of inequivalent separable coordinate systems for $\delta^2/\delta X^2 - \delta^2/\delta Y^2 + \alpha\mathbf{1}$. It is characterized by the following identities:

PROPOSITION 5

$$(1)\ \mathfrak{B}_1 \exp[Z\mathfrak{B}_3] = \exp[Z\mathfrak{B}_3]\{\cosh(Z)\cdot\mathfrak{B}_1 + \sinh(Z)\cdot\mathfrak{B}_2\}$$

$$(2)\ \mathfrak{B}_2 \exp[Z\mathfrak{B}_3] = \exp[Z\mathfrak{B}_3]\{\cosh(Z)\cdot\mathfrak{B}_2 + \sinh(Z)\cdot\mathfrak{B}_1\}$$

$$(3)\ \exp[V\mathfrak{B}_1]\,\mathfrak{B}_3 = (V\mathfrak{B}_2 + \mathfrak{B}_3)\exp[W\mathfrak{B}_1]$$

$$(4)\ \exp[W\mathfrak{B}_2]\,\mathfrak{B}_3 = (W\mathfrak{B}_1 + \mathfrak{B}_3)\exp[W\mathfrak{B}_2]$$

Proof: Left to the reader.

8. And more ...

We have only given examples of the kind of results, and computations, that can be worked out in a combinatorial setup for Lie groups and Lie algebras. Other families of orthogonal polynomials can be studied with combinatorial Lie group techniques. Our purpose was only to show why this kind of approach should be given a combinatorial meaning. First, it is easy to implement, and the combinatorial calculations involved are straightforward. But more important, is the possibility of extending known analytical methods to new contexts. Many paths remain to be explored, and will be in the future.

BIBLIOGRAPHY

[A2] R.A.Askey et al. (eds), *Special functions: Group theoretical aspects and applications*, Reidel Publishing Company (1984).

[B1] F.Bergeron, *Modèles combinatoires de polynôme sorthogonaux*, Rapp. Rech. Dép. Math. et Info., N° 2 UQAM (1984).

[B2] F.Bergeron, *Une approche combinatoire de la méthode de Weisner*, dans *Polynômes orthogonaux et applications*, Proceedings Bar-le-Duc (1984), édité par C.Brézinski, A.Draux, A.P.Magnus, P.Maroni, et A.Ronveaux, Lect. Notes in Math No 1171, Springer-Verlag (1985), PP 111-119.

[F1] D.Foata and P.Leroux, *Polynômes de Jacobi, interprétation combinatoire et fonction génératrice*, Proc. Amer. Math. Soc. **87**, (1983), 47-53.

[F2] D.Foata and V.Strehl, *Combinatorics of Laguerre polynomials*, dans *Enumeration and Design*, D.M. Jackson and S.A. Vanstone (ed.), Academic Press (1984), 123-140.

[F3] D.Foata, *Combinatoire des identités sur les polynômes orthogonaux*, Proceedings I.C.M. Warsaw (1983).

[G1] A.M.Garsia and J.Remmel, *A combinatorial interpretation of q-derangement and q-Laguerre numbers*,

Europ. J. Comb. **1** (1980), 47-59.

[I1] E.H.Ismail, D.Stanton and G.Viennot, *The combinatorics of q-Hermite polynomials and the Askey-Wilson integral*, (To appear)

[J1] A.Joyal, *Une théorie combinatoire des séries formelles*, Adv. in Math. **42**, Academic Press (1981), 1-82.

[G1] I.Gessel and R.Stanley, *Stirling polynomials*, J.Comb. Theory, Series A, Vol.24, No.1, 24-33, (1978)

[L1] G.Labelle, *Éclosions combinatoires appliquées à l'inversion multidimensionnelle des séries formelles*, J. of Combinatorial Theory, Vol.39 N^0 1, Series A, Academic Press (1984), 52-82.

[L2] G.Labelle, *On combinatorial differential equations*, J. of Math. Anal. and Appl. , Academic Press, (1985)

[L3] P.Leroux and V.Strehl, *Jacobi polynomials: combinatorics of the basic identities*, Discrete Math, 57 (1985), 167-187.

[L4] P.Leroux et G.Viennot, *Combinatorial resolution of systems of differential equations*, This volume.

[M1] W.Miller, *Symmetry and seperation of variables*, Encyclopedia of Mathematics, Ed. G.-C.Rota, Addison-Wesley (1977).

[R1] G.-C.Rota and R.Mullin, *Finite operator calculus*, Academic Press (1975).

[S1] D.Stanton, *Generalized n-gons and Chebychev polynomials*, J. Comb. Theory, Series A, Vol. 34, No.1, 15-27, (1983)

[S2] D.Stanton, Harmonics on posets, Jour. Comb.Theory, Series A, Vol.40, N°1 (1975).

[V1] G.Viennot, *Une théorie combinatoire des polynômes orthogonaux généraux*, conferences given at UQAM (1983).

[W1] L.Weisner, *Group-theoretic origin of certain generating functions*, Pacific J. Math. **5** (1955), 1033-1039.

Definite Integral Evaluation
by Enumeration,
Partial Results in the Macdonald Conjectures

D. M. Bressoud*
Department of Mathematics
The Pennsylvania State University
University Park, PA 16802

1. Introduction

In this paper we shall present a combinatorial approach to the problem of evaluating definite integrals of the form

$$I(n,S,k(\ )) = \int_{-\pi/2}^{\pi/2} \cdots \int_{-\pi/2}^{\pi/2} (\prod_{\alpha\in S} \sin^{2k(\alpha)}\alpha)\,d\alpha_1\ldots d\alpha_n,$$

as well as their basic or q-analog generalizations, where S is a set of non-zero linear sums of elements of $\{\alpha_1,\ldots,\alpha_n\}$ with non-negative coefficients and k is a mapping from S to the positive integers. An example of such an evaluation is a result conjectured by F. J. Dyson [5] and first proved by Gunson [6] and Wilson [9]:

<u>Proposition 1</u> Let $T = \{\alpha_i+\alpha_{i+1}+\ldots+\alpha_j \mid 1 \le i \le j \le n\}$ and let $k(\ )$ be the constant function $k(\alpha)=k$, then

$$I(n,T,k) = \pi^n 4^{-\binom{n+1}{2}k} \prod_{i=2}^{n+1} \binom{ik}{k}.$$

This proposition was proved combinatorially by D. Zeilberger [10] and the purpose of this paper is to expand his approach.

*Partially supported by N.S.F. grant no. DMS-8404083

The key to nice evaluations of $I(n,S,k(\ ))$ is for S to possess symmetry. It was Ian Macdonald's idea [7] to consider sets S reflecting the symmetries of the root systems associated with Lie algebras.

Definition Let $\langle\ ,\ \rangle$ be the usual inner product on $\mathbb{R}^n$. A subset R of $\mathbb{R}^n$ is a reduced, irreducible, finite root system, or for our purposes simply called a root system, if

i) R spans $\mathbb{R}^n$, $0 \notin R$,

ii) $\alpha \in R$ implies that $n\alpha \in R$ if and only if $n = \pm 1$,

iii) given $\alpha, \beta \in R$, then $\sigma_\alpha(\beta) \in R$ where $\sigma_\alpha(\beta) = \beta - \frac{2\langle\beta,\alpha\rangle}{\langle\alpha,\alpha\rangle}\alpha$ is the reflection of β in the hyperplane thru the origin and perpendicular to α,

iv) given $\alpha, \beta \in R$, then $\frac{2\langle\beta,\alpha\rangle}{\langle\alpha,\alpha\rangle} \in \mathbb{Z}$,

v) R cannot be partitioned into two disjoint, mutually orthogonal, non-empty subsets.

One result from the theory of Lie algebra (see for example Carter [4], Prop. 2.1.2) is the following lemma.

Lemma Let R be a root system in $\mathbb{R}^n$. There exists a subset $\Delta = \{\alpha_1, \ldots, \alpha_n\}$ of R with the property that if $\alpha \in R$ then $\alpha = \Sigma\, c_i\alpha_i$ where the c_i are integers and are either all non-negative or all non-positive.

We call Δ a base for R and let R^+ denote the elements of R which are positive linear sums of elements of Δ, R^- those which are negative linear sums, $R = R^+ \cup R^-$. Before stating Macdonald's conjecture for evaluating $I(n,S,k(\ ))$, we need one more definition.

Definition Given a root system R, let W, the Weyl group, be the group of symmetries of R, that is to say $W = \langle\sigma_\alpha(\ ) \mid \alpha \in R\rangle$. Let $d_1, \ldots, d_n$ be the degrees of the fundamental invariants for W. These are most easily defined as follows. Given $w \in W$, let $N(w) = w(R^+) \cap R^-$, $n(w) = |N(w)|$.

Then $\{d_1,\dots,d_n\}$ is defined by

$$\sum_{w\in W} q^{n(w)} = \prod_{i=1}^{n} \frac{(1-q^{d_i})}{(1-q)} .$$

That such d_i's exist and are distinct integers >1 is a result of the theory of Lie Algebra ([4]; Chapt. 9). Their uniqueness follows from the uniqueness of polynomial factorization.

<u>Conjecture 1 (Macdonald)</u> Let R be a root system in $\mathbb{R}^n$ with base $\Delta = \{\alpha_1,\dots,\alpha_n\}$ and degrees $d_1,\dots,d_n$. Let $k(\ )$ be the constant function $k(\alpha)=k$, $\alpha \in R^+$. Then $I(n,R^+,k) = \pi^n\, 4^{-|R^+|k} \prod_{i=1}^{n} \binom{kd_i}{k}$.

This conjecture has been proved for the root systems A_n (where it becomes Proposition 1), B_n, C_n and D_n, the last three being consequences of A. Selberg's multi-dimensional beta integral evaluation [2,8]. There are strong heuristic arguments for the remaining special root systems G_2, F_4, E_6, E_7 and E_8 but no proofs for arbitrary positive integral k as yet. Macdonald has also conjectured [7] the evaluation of $I(n,R^+,k(\ ))$ where one only requires that if $|\alpha| = |\beta|$ then $k(\alpha) = k(\beta)$. For B_n, C_n and D_n, this conjecture also follows from the Selberg integral evaluation.

2. Restatement of Problem and Basic Analog

We shall restate the evaluation of $I(n,S,k(\ ))$ as a counting problem. By making the substitution

$$\sin\alpha = (e^{i\alpha}-e^{-i\alpha})/2i$$

we see that

$$I(n,S,k)(\ )) = (-4)^{-\Sigma k(\alpha)} \int_{-\pi/2}^{\pi/2} \cdots \int_{-\pi/2}^{\pi/2} \left[\prod_{\alpha\in S} (e^{i\alpha}-e^{-i\alpha})^{2k(\alpha)}\right] d\alpha_1 \dots d\alpha_n .$$

If the inner product is expanded, then all terms yield zero except for the constant term, and thus

$$I(n,S,k(\ )) = \pi^n(-4)^{-\Sigma k(\alpha)}\ ([1]\ \prod_{\alpha\in S}\ (e^{i\alpha}-e^{-i\alpha})^{2k(\alpha)}),$$

where $[x^n]P(x)$ denotes the coefficient of x^n in $P(x)$, $[1]P(x)$ denotes the constant term. Our problem is now to evaluate

$$C(n,S,k(\ )) = [1]\ \prod_{\alpha\in S}\ (e^{\alpha}-e^{-\alpha})^{2k(\alpha)}(-1)^{k(\alpha)}$$

$$= [1]\ \prod_{\alpha\in S}\ (1-e^{\alpha})^{k(\alpha)}(1-e^{-\alpha})^{k(\alpha)}$$

where e^{α} is a purely formal exponential satisfying $e^{\alpha}\cdot e^{\beta} = e^{\alpha+\beta}$. The second form of $C(n,S,k(\ ))$ given above suggests a basic analog

$$C(n,S,k(\ );q) = [1]\ \prod_{\alpha\in S}\ \prod_{i=1}^{k(\alpha)}\ (1-q^i e^{\alpha})(1-q^{i-1}e^{-\alpha}),$$

and Macdonald [7] has made the following conjecture.

Conjecture 2 Let R be a root system in $\mathbb{R}^n$ with base $\Delta = \{\alpha_1,\dots,\alpha_n\}$ and degrees $d_1,\dots,d_n$. Let $k(\ )$ be the constant function $k(\alpha)=k$, $\alpha\in R^+$. Then

$$C(n,R^+,k;q) = \prod_{i=1}^{n} \begin{bmatrix} kd_i \\ k \end{bmatrix}$$

where $\begin{bmatrix} A \\ B \end{bmatrix} = \prod_{j=1}^{B} \frac{(1-q^{A-B+j})}{(1-q^j)}$ is the Gaussian polynomial.

We note that with $q=1$, Conjecture 2 becomes Conjecture 1. Conjecture 2 has only been proved for the root system A_n (see Zeilberger and Bressoud [11]) and small values of k. For $k=1$, it says that

$$C(n,R^+,1;q) = \sum_{w\in W} q^{n(w)}$$

which has been demonstrated combinatorially by Calderbank and Hanlon [3]. Macdonald has also generalized this conjecture to the case where one only requires that if $|\alpha|=|\beta|$ then $k(\alpha)=k(\beta)$.

Before stating what $C(n,S,k(\ );q)$ counts, we need some definitions.

<u>Definition</u> Given a finite sequence of non-negative integers $\omega=c_1c_2\dots c_m$, we define the <u>inversion number</u> of ω, $INV(\omega)$, to be $\Sigma\, \chi(c_i>c_j)$, $1 \le i < j \le m$, where $\chi(A)=1$ if A is true, $\chi(A)=0$ otherwise.

<u>Definition</u> Given $S,k(\ )$, we define a <u>multi-choice set</u> $M=M(S,k(\ ))$ to be an ordered set of finite sequences of zeros and ones, $M=(\omega_\alpha \mid \alpha \in S)$, such that the sequence ω_α has length $2k(\alpha)$ and

$$\sum_{\alpha\in S} \alpha\Pi(\alpha) = 0$$

where $k(\alpha)+\Pi(\alpha)$ is the number of ones in ω_α. Let $\mathfrak{M}=\mathfrak{M}(S,k(\ ))$ be the set of all multi-choice sets for $S,k(\ )$.

<u>Definition</u> Given a multi-choice M, we define its weight, $wt(M)$, to be $(-1)^{\Sigma\Pi(\alpha)}\, q^{\Sigma(INV(\omega_\alpha)+\Pi(\alpha)(\Pi(\alpha)+1)/2)}$, both sums over all $\alpha \in S$.

<u>Proposition 2</u>:

$$C(n,S,k(\ );q) = \sum_{M\in\mathfrak{M}(S,k(\ ))} wt(M).$$

<u>Proof</u>: The q-binomial theorem (see [1], Thm. 3.3) implies that

$$\prod_{i=1}^{k} (1-q^{i-1}e^{-\alpha})(1-q^ie^{\alpha}) = \Sigma(-1)^jq^{j(j+1)/2}\begin{bmatrix}2k\\k+j\end{bmatrix} e^{\alpha j},\ -k \le j \le +k.$$

Thus we have that

$$\prod_{\alpha \in S} \prod_{i=1}^{k(\alpha)} (1-q^{i-1}e^{-\alpha})(1-q^{i}e^{\alpha})$$

$$= \prod_{\alpha \in S} \sum_{j_\alpha} (-1)^{j_\alpha} q^{j_\alpha(j_\alpha+1)/2} \begin{bmatrix} 2k(\alpha) \\ k(\alpha)+j_\alpha \end{bmatrix} e^{\alpha j_\alpha}$$

$$= \sum_{(j_\alpha \mid \alpha \in S)} e^{\Sigma \alpha j_\alpha} \prod_{\alpha \in S} (-1)^{j_\alpha} q^{j_\alpha(j_\alpha+1)/2} \begin{bmatrix} 2k(\alpha) \\ k(\alpha)+j_\alpha \end{bmatrix} .$$

Proposition 2 now follows from the facts that $\begin{bmatrix} 2k \\ k+j \end{bmatrix} = \Sigma q^{INV(\omega)}$ where the sum is over all sequences ω of length $2k$ with $k+j$ ones and $k-j$ zeros (see [1], Thm. 3.6) and that we shall get a contribution to the constant term precisely when $\Sigma \alpha j_\alpha = 0$. □

We now consider what is counted by the right-hand side of Conjecture 2.

<u>Definition</u> Let $\Omega(d,k;\mathfrak{D})$, $\mathfrak{D} = \{d_i \mid 1 \leq i \leq n,\ 1 < d_1 < \ldots < d_n = d\}$, be the set of words ω in d letters such that each letter appears k times and given any letter $e \notin \mathfrak{D}$, the first e comes after the last of each of the letters 1 through $e-1$.

<u>Observation</u> $\prod_{i=1}^{n} \begin{bmatrix} kd_i \\ k \end{bmatrix} = \Sigma\, q^{INV(\omega)}$, the sum being over all $\omega \in \Omega(d_n,k;\mathfrak{D})$.

If for $\omega \in \Omega$, we let the weight of ω, $wt(\omega)$, be $q^{INV(\omega)}$ then Conjecture 2 can be restated as follows.

<u>Conjecture 2'</u> Let R be a root system in $\mathbb{R}^n$ with base $\Delta = \{\alpha_1,\ldots,\alpha_n\}$ and degrees $\mathfrak{D} = \{d_1,\ldots,d_n\}$. Let k be a positive integer. Then there is a one-to-one weight preserving mapping from $\Omega(d_n,k;\mathfrak{D})$ into $\mathfrak{M}(R^+,k)$. Furthermore, if $\mathfrak{G}$ denotes the image of Ω in $\mathfrak{M}$ and $\mathfrak{B}$ denotes the complement of $\mathfrak{G}$ in $\mathfrak{M}$, then

$$\sum_{M \in \mathcal{B}} wt(M) = 0.$$

There is a natural way of constructing a one-to-one weight preserving mapping from Ω into $\mathfrak{M}$ as follows.

We shall specify a partition of R into d_n non-empty and mutually disjoint sets, $R = \dot{\bigcup} S_i$, $1 \le i \le d_n$. First we partition $R^+ = \dot{\bigcup} T_j$, $1 \le j \le n$, such that $|T_j| = d_j - 1$. This is possible since $|R^+|$ equals $d_1 + \ldots + d_n - n$ (see [4], thm. 9.3.4). For each j, $1 \le j \le n$, we order the elements of T_j, say $T_j = \{\beta_1^{(j)}, \beta_2^{(j)}, \ldots, \beta_{d_j - 1}^{(j)}\}$. For $1 \le i \le d_j - 1$, place $-\beta_i^{(j)}$ in S_i. Then place all elements of T_j in S_{d_j}.

<u>Observation</u> The partition $R = \dot{\bigcup} S_i$, $1 \le i \le d_n$, given above satisfies the following conditions:

i) no S_i is empty,

ii) if $\alpha \in R^+$ and $\alpha \in S_j$, then $-\alpha \in S_i$ for some $i < j$,

iii) given $i < j$, if $j \notin \mathcal{D} = \{d_1, \ldots, d_n\}$ then there is no $\alpha \in R^+$ such that $\alpha \in S_j$ and $-\alpha \in S_i$,

iv) given $i < j$, if $j \in \mathcal{D}$, then there is exactly one $\alpha \in R^+$ such that $\alpha \in S_j$, $-\alpha \in S_i$.

Now, given a word ω in Ω, we read it from left to right, one letter at a time. We begin with a multi-choice set in which each sequence of ones and zeros is empty. If the letter read is i, we consider the set S_i. For each $\alpha \in S_i \cap R^+$, we append a 1 to the end of the sequence ω_α. For each $\alpha \in S_i \cap R^-$, we append a 0 to the end of the sequence $\omega_{-\alpha}$. We continue until all letters have been read.

<u>Proposition 3</u> The mapping given above is one-to-one and weight-preserving

from Ω into $\mathfrak{M}$.

Proof Since each letter appears in ω exactly k times and each root $\alpha \in R$ is in exactly one S_i, we end with a multi-choice set M with exactly k ones and k zeros in each sequence ω_α. This is an element of $\mathfrak{M}$. Since $\Pi(\alpha)=0$, we also have that

$$wt(M) = q^{\Sigma INV(\omega_\alpha)}.$$

By the definition of the words in Ω, if $j \notin \mathcal{D}$ and $i < j$ then j must follow i. Thus the only inversions counted in ω involve a $j \in \mathcal{D}$ followed by an i such that $i < j$. Since there is exactly one $\alpha \in R$ such that $\alpha \in S_j$, $-\alpha \in S_i$ and since α is contained in R^+, the inversion j followed by i in ω creates an inversion 1 followed by 0 in ω_α. Conversely, each inversion of a 1 followed by a 0 in ω_α for some α corresponds to an inversion j followed by i, $i < j$, in ω because α is contained in S_j for some $j \in \mathcal{D}$ and $-\alpha$ is in S_i for some $i < j$. Thus we have that

$$INV(\omega) = \Sigma\ INV(\omega_\alpha),$$

and so the mapping is weight-preserving.

The mapping is one-to-one because the original word ω is uniquely reconstructible from its image M. We read the initial entry of each sequence ω_α and construct a choice set C for which $\alpha \in C$ if the first entry of ω_α is 1, $-\alpha \in C$ if the first entry of ω_α is 0, neither is in C if ω_α is empty. We find the smallest i such that $S_i \subseteq C$, i becomes the next letter of the word. We delete the initial entries of all sequences ω_α such that $\pm\alpha \in S_i$ and continue until the multi-choice set is empty. $\square$

3. Killing Bad Guys

The problem of proving conjecture 2 or 2' has thus been reduced to showing that

$$(*) \qquad \sum_{M\in\mathcal{B}} wt(M) = 0$$

where $\mathcal{B}$ (the set of "bad guys") is the complement in $\mathfrak{M}$ of the image of Ω. The equation (*) has been proved by Zeilberger [10] when $R=A_n$, $q=1$ by an involution on $\mathcal{B}$ which matches elements with cancelling weights. It has also been proved by Zeilberger and the author [11] for $R=A_n$, q arbitrary, by a combination of summation and involution arguments.

The equation (*) has at least a heuristic argument for an arbitrary root system when $q=1$. In this case,

$$wt(M) = (-1)^{\Sigma\Pi(\alpha)}$$

where $k+\Pi(\alpha)$ is the number of ones in ω_α. One can choose the partition $R=\dot{\bigcup}S_i$ such that for the choice set C formed by the leading entries of the sequences ω_α, if $C=w(R^+)$ for some $w \in W$, the Weyl group, then $C \supseteq S_i$ for some S_i. Thus if M is an element of $\mathcal{B}$, then as we attempt to encode it as a word in Ω we must eventually reach the point where the choice set C is not of the form $w(R^+)$, $w \in W$. It is a theorem of Calderbank and Hanlon [3] that for any root system R there is an involution on the set of choice sets C which are not of the form $w(R^+)$, $w \in W$, this involution changing the parity of $|C \cap R^+|$ but leaving invariant the sum of the elements of C. This is exactly what we need to change the sign of the weight of M. Unfortunately, the mapping from $\mathcal{B}$ to itself that this defines no longer is self-inverse, and thus does not give us the desired pairing in all cases.

References

1. G. E. Andrews, "The Theory of Partitions", Encyclopedia of Math., vol. 2, Addison-Wesley, Reading, Mass., 1976.

2. R. A. Askey, Some basic hypergeometric extensions of integrals of Andrews and Selberg, SIAM J. Math. Anal., 11(1980), 938-951.

3. R. Calderbank and P. Hanlon, An extension to root systems of a theorem on tournaments, preprint.

4. Roger Carter, "Simple Groups of Lie Type", John Wiley and Sons, London, 1972.

5. F. J. Dyson, Statistical theory of the energy levels of complex systems.I, J. Math. Physics, 3(1962), 140-156.

6. J. Gunson, Proof of a conjecture by Dyson in the statistical theory of energy levels, J. Math. Physics, 3(1962), 752-753.

7. I. G. Macdonald, Some conjectures for root systems, SIAM J. Math. Anal., 13(1982), 988-1007.

8. A. Selberg, Bemerkninger om et Multipelt Integral, Norsk Mat. Tidsskr., 26(1944), 71-78.

9. K. G. Wilson, Proof of a conjecture by Dyson, J. Math. Physics, 3(1962), 1040-1043.

10. D. Zeilberger, A combinatorial proof of Dyson's conjecture, Discrete Math., 41(1982), 317-332.

11. D. Zeilberger and D. Bressoud, A proof of Andrews' q-Dyson conjecture, Discrete Math., 54(1985), 201-224.

ENUMERATION OF CERTAIN YOUNG TABLEAUX WITH BOUNDED HEIGHT

Myriam DESAINTE-CATHERINE and Gérard VIENNOT
Université de Bordeaux I
U.E.R. de Mathématiques et d'Informatique
351, Cours de la Libération
33405 TALENCE - FRANCE -

ABSTRACT - We consider Young tableaux strictly increasing in rows, weakly increasing in columns, and each column having an even number of elements. We show that the number of such tableaux with entries between 1 and n, and having at most 2k rows, is the product $\prod_{1\leq i\leq j\leq n} (i+j+2k)/(i+j)$.
The proof is mainly bijective, using configurations of non-crossing paths. At the end we need the qd-algorithm from Padé approximants theory.

§ 1 - INTRODUCTION.

We consider ***Young tableaux*** with entries from the set {1,...n}, strictly increasing in rows and not decreasing in columns. Note that usually the reverse convention between rows and columns is used.

Gordon [10] proved that the number $a_{n,p}$ of such tableaux having at most p rows is given by the product

$$(1) \qquad a_{n,p} = \prod_{1\leq i\leq j\leq n} \frac{p+i+j-1}{i+j-1} .$$

In fact this product is obtained by setting $q\to 1$ in the product giving the generating function of the corresponding ***plane partitions***, proved by Gordon, and conjectured by Bender and Knuth [1].

We prove here a companion formula for the number $b_{n,k}$ of such Young tableaux having only columns with an even number of elements and bounded by height p=2k.

$$(2) \qquad b_{n,k} = \prod_{1\leq i\leq j\leq n} \frac{2k+i+j}{i+j} .$$

Our motivation comes from the question of Stanley [17] about finding a "natural" proof of (1) . We propose a "90% bijective" proof of (2). We believe that the main interest of this paper is not in the formula (2) but in the techniques used to get it, especially the introduction of the well-known (in numerical analysis) quotient-difference algorithm in order to enumerate configurations of non-intersecting paths. It should be of interest to apply such methods for other "hard" Young tableaux enumeration formulae, written as a ratio of two products, as in (1) or (2).

Our proof is in three steps . The first step uses the Robinson-Schensted correspondence between permutations and pairs of standard Young tableaux. This classical algorithm (for a survey see for example Knuth [14] and more recently Viennot [19]) has been extended by Knuth [13]. An equivalent direct version has been given by Burge [2] for the case of an involution. We will use here a slight modification of Burge's version of Knuth's extension.

The second step is inspired from Viennot's geometric interpretation of the Robinson-Schensted correspondence, with paths and shadows.

Combining steps 1 and 2, we get a bijection between the above Young tableaux and certain configurations of non-crossing paths. In fact these paths are the so-called Dyck paths, enumerated by the Catalan numbers.

We can apply the Gessel-Viennot's methodology [8] [9] about combinatorial interpretation of determinants with non-crossing paths. Here we get a Hankel determinant of Catalan numbers. The third step is the computation of this determinant with the qd-algorithm. Remark that some part of this third step can be again put at the bijective level, as shown by Viennot [21]. In other words, we replace Dyck paths by shorter Dyck paths. In this "compression" process the shorter paths need to be weighted, as in the combinatorial theory of continued fractions and orthogonal polynomials (Flajolet [7], Viennot [20]). The same "compression" is applied to configurations of non-crossing paths and the number of such configurations becomes the weight of a ***single*** configuration. Thus the product (2) is nothing but the product of the weights of all the elementary steps of the paths of this unique configuration.

The computation of these elementary weights, which are rationnal numbers is the only "analytic" step of the proof.

In the final section we give the origin of the formula (2). In fact the above Young tableaux are in bijection with other configurations of non-crossing paths, interpretating certain pfaffians. The analog of Gessel-Viennot's methodology for determinants has been made for pfaffians (see Desainte-Catherine, Viennot [5]). The present paper solves Desainte-Catherine's conjecture [3]. Using bijections closed to other bijections introduced in the combinatorial solution of the Ising model (see for example Fisher [6], Kasteleyn [12]), Desainte-Catherine showed [4] that the product (2) also enumerates certain ***perfect matchings*** of graphs formed with hexagons and pentagons.

§ 2 - PRELIMINARY DEFINITIONS AND NOTATIONS.

We denote by $[0,n]$ the set $\{0,1,\dots,n\}$ and, for short, $[n]=[1,n]$.

A ***partition*** $\lambda = (\lambda_1,\lambda_2,\dots,\lambda_p)$ of n is a sequence $\lambda_1 \geq \dots \geq \lambda_p \geq 0$ such that $n = \lambda_1+\dots+\lambda_p$ and usually visualized by a ***Ferrers diagram*** as in figure 1. The ith row has λ_i cells.

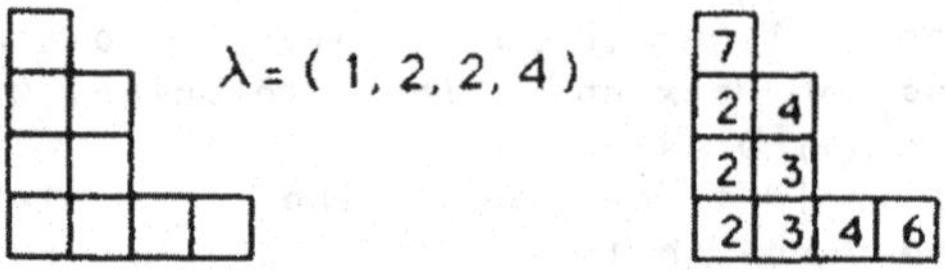

Figure 1. A Ferrers diagram and a Young tableau.

In this paper, we define a ***Young tableau*** of ***shape*** λ as a tableau of integers filling the Ferrers diagram λ such that these integers (or ***entries***) are strictly increasing in each row (from left to right) and weakly increasing in each column (down-up).

Definition 1 – The set of Young tableaux having only columns with an even number of elements, having exactly (resp. at most) 2k rows and with entries in [n] will be denoted by $T_{n,2k}$ (resp. $T_{n,\leq 2k}$).

A ***path*** of $\pi = \mathbb{Z}^2$ is a sequence $\omega = (s_0, s_1, \ldots, s_n)$ of points or vertices of π. We say that ω goes from s_0 (starting point) to s_n (ending point). The ***length*** $|\omega|$ of the path ω is n. An ***elementary step*** is a couple (s_{i-1}, s_i).

Let $\mathbb{K}$ be a commutative ring and v a map (called valuation) v : $\pi^2 \to \mathbb{K}$. The valuation $v(\omega)$ of the path ω is the product of the valuations of the elementary steps

$$v(\omega) = \prod_{i=1}^{n} v(s_{i-1}, s_i).$$

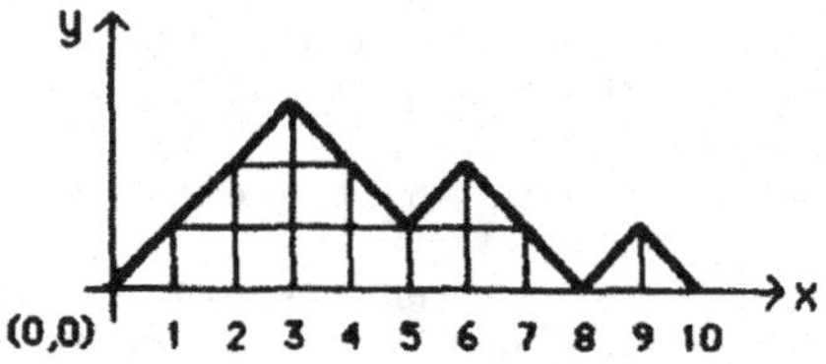

Figure 2. A Dyck path.

A ***Dyck path*** is a path $\omega = (s_0, \ldots, s_{2n})$ of $\mathbb{N} \times \mathbb{N}$ such that $s_0 = (0,0)$, $s_{2n} = (2n,0)$ and having only elementary steps ***North-East*** (i.e. $s_i = s_{i-1} + (1,1)$) or ***South-east*** (i.e. $s_i = s_{i-1} - (1,1)$), see figure 2. The path ω has a ***valley*** in s_i iff (s_{i-1}, s_i) is a South-East step and (s_i, s_{i+1}) is a North-East step.

Definition 2 – A ***fan*** of Dyck paths is a k-uple $\eta = (\omega_1, \ldots, \omega_k)$ of Dyck paths satisfying the two conditions

(i) each path goes from (0,0) to (2n,0),

(ii) for every j, $1 \leq j < k$, ω_j is ***under*** ω_{j+1}, that is for every i, $1 \leq i \leq 2n$, the ordinate (or ***level***) of the ith vertex of ω_j is less or equal than the ordinate of the ith vertex of ω_{j+1}.

We will denote by $Ev_{2n,k}$ the set of such fans (see figure 3).

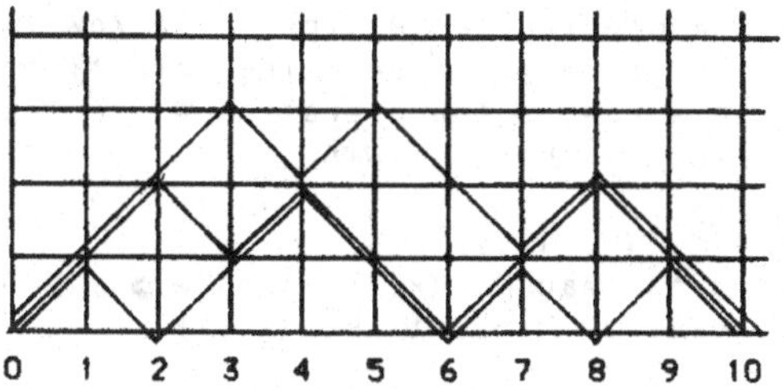

Figure 3. A fan of $Ev_{10,3}$

§ 3 - BIJECTION BETWEEN YOUNG TABLEAUX AND CLOUDS OF POINTS.

Definition 3 - A matrix $M = (a_{ij})_{1\leq i,j\leq n}$ with non negative integers for entries a_{ij}, will be represented as a cloud $\hat{M}$, that is a subset of points of $[n]\times[n]$ where each point (i,j) has the "multiplicity" a_{ij}. The ***number of points*** of the cloud is $\sum_{1\leq i,j\leq n} a_{ij}$. The cloud is ***subdiagonal*** when $a_{ij}=0$ for $i<j$.

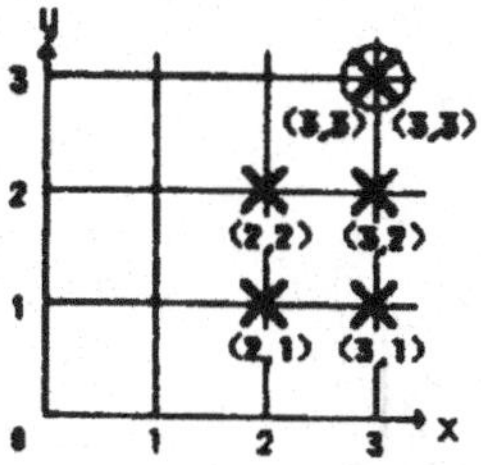

Figure 4. A subdiagonal cloud.

$$A = \begin{bmatrix} 0 & 0 & 0 \\ 1 & 1 & 0 \\ 1 & 1 & 2 \end{bmatrix}.$$

An example is displayed on figure 4. The point (3,3) (denoted ⊛) has multiplicity 2. Remark the reverse notation between the matrix M and the geometric representation.

A ***decreasing subsequence*** extracted from a cloud $\hat{M}$ is a sequence $\sigma = (s_1,\ldots,s_p)$ of points of $\hat{M}$ such that for each i, $0 \leq i \leq p$, s_{i+1} is located at the South-East of s_i (i.e $s_i = (x_i,y_i)$, $s_{i+1} = (x_{i+1}, y_{i+1})$ with $x_{i+1} \geq x_i$ and $y_{i+1} \leq y_i$). The same point may appear several times, but no more than its multiplicity. For example ((2,2),(2,1), (3,1)) or ((3,3),(3,2),(3,1)) or ((3,3),(3,3),(3,2),(3,1)) are three decreasing subsequences extracted from the cloud of figure 4.

The ***depth*** of a cloud is the maximum cardinality of its decreasing subsequences. For example, the cloud of figure 4 has depth 4.

Proposition 4 - There exists a bijection φ between the set $T_{n,2k}$ of Young tableaux (definition 1) and the set of subdiagonal clouds of $[n]\times[n]$ (definition 3) having depth k.

Such bijection comes from Knuth's extension of the Robinson-Schensted correspondence. We use a modification of Burge's presentation in the case of symmetric matrices with zero trace. The bijection φ is described by successive insertions of pairs of integers. First we describe the insertion process of a pair (x,y), $x\geq y$, in the Young tableau P.

Notation - For a non-empty tableau P, the first row is denoted by $\rho_1(P)$, the tableau obtained from P by deleting this first row is denoted by $\delta(P)$. Putting back this row is the operation $*$, so that $P = \rho_1(P) * \delta(P)$. The insertion algorithm is the following.

<u>Procedure</u> Insert (P, (x,y))
  <u>input</u> Young tableau P, (x,y) with $x \geq y \geq 0$
  <u>output</u> tableau Q = Insert (P, (x,y))

<u>begin</u> <u>if</u> P is empty, <u>then</u> $Q = \begin{array}{|c|}\hline x \\ y \\ \hline\end{array}$

<u>elsif</u> y > all the elements of the first row $\rho_1(P)$
<u>then</u> Q is obtained by adding y at the end of $\rho_1(P)$, and adding x at the end of the second row of P (this second row may be empty)
<u>else</u> let y' be the smallest element of $\rho_1(P)$ which is $\geq$y. Let R be the row obtained from $\rho_1(P)$ by replacing y' by y. Then Q = R $*$ Insert (δP, (x,y'))

<u>end</u>

Example 5.

For $P = \begin{array}{lll} 3 & & \\ 2 & & \\ 2 & 3 & \\ 1 & 3 & \\ 1 & 2 & 4 \\ 1 & 2 & 3 \end{array}$, Insert (P, (4,2)) = $\begin{array}{lll} 3 & & \\ 2 & 4 & \\ 2 & 3 & \\ 1 & 2 & \\ 1 & 2 & 4 \\ 1 & 2 & 3 \end{array}$.

Remark that if x is $\geq$ all the elements of P, then Q = Insert (P,(x,y)) is again a Young tableau. The insertion of y in P is nothing but the classical bumping process of the Robinson-Schensted correspondence (with Knuth's extension), but with the modification to get the rows strictly increasing and columns weakly increasing (instead of the reverse usual definition for Young tableaux). This is the only difference from Burge's presentation [2].

Let $\hat{M}$ be a subdiagonal cloud of $[n] \times [n]$. We totally order the points of $\hat{M}$ according to the increasing order of their abscissa x and, for the same abscissa, with decreasing order of their ordinate y. We get a sequence of pairs (x,y), $x \geq y$. For a point (x,y) with multiplicity m, the pair (x,y) is repeated m times. We define the Young tableau $\varphi(\hat{M})$ as the tableau obtained from the empty tableau by successive insertions of these pairs (with respect to their total order).

Example 6 - With the cloud $\hat{M}$ of figure 4, the sequence of pairs is (2,2), (2,1), (3,3), (3,3), (3,2), (3,1). The successive insertions are the following

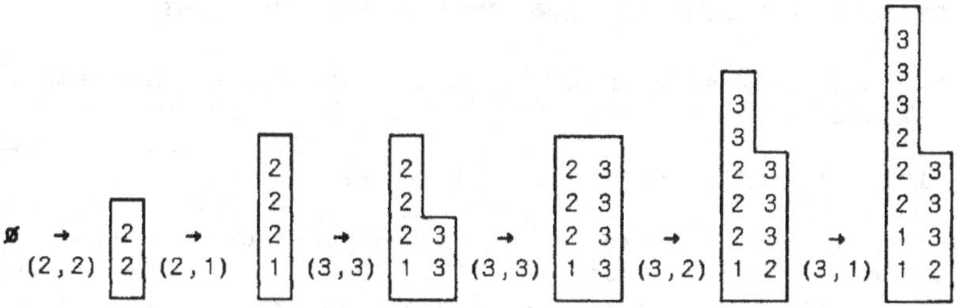

As in Knuth [13] and Burge [2], one can prove that $\varphi(\hat{M})$ is a Young tableau having 2k rows and each column with an even number of cells, iff the cloud has depth k. We get the desired bijection φ □

(pedagogical and historical remark)

Remark 7 – The Robinson-Schensted correspondence is a bijection between permutations σ and pairs (P,Q) of standard Young tableaux (i.e. strictly increasing in rows and columns) having the same shape. If σ is an involution, then P = Q , and the correspondence gives a bijection between involutions and standard Young tableaux. From Schützenberger [16], the number of fixed points of σ is the number of columns having an odd number of elements (odd columns for short).

Knuth's extension [13] is a bijection between clouds $\hat{M}$ of [n] × [n] and pairs (P,Q) of tableaux (weakly increasing in rows and strictly increasing in columns) with same shape. If the matrix M is symmetric, then P = Q and the number of odd columns is the trace of M. Thus one deduce a bijection between symmetric matrices having a zero trace and such tableaux having only even columns (i.e. columns with an even number of elements). This would correspond to "generalized involutions".

In [2], Burge defined an equivalent correspondence for involutions, by successive insertions of fixed points and cycles of length 2. The fixed point insertion increases by one the number of odd columns. This number is invariant under the insertion of a cycle of length 2. This fact explains Schützenberger's property [16].

Knuth's extension can also be visualized as follows. From every cloud, one can associate, with "infinitesimal" moves of the points, another cloud having only one point in each row and column. If the moves are such that the points in each row and column form a "strictly increasing subsequence" of the cloud, then Knuth's extension is obtained by applying the "light and shadows" process of Viennot [18]. The modification used here is to make the moves such that the points of each row and column becomes a "strictly decreasing subsequence" of the cloud (see figure 5). The subdiagonal cloud would be extended by symmetry with corresponding symmetric matrix. The trace is no more zero, but each number of the trace has to be doubled, (i.e. the diagonal elements are not considered as fixed point, but as cycles of length 2).

§ 4 - BIJECTION BETWEEN CLOUDS AND CONFIGURATIONS OF DYCK PATHS.

Proposition 8 - There exists a bijection Ψ between the set of subdiagonal clouds of $[n] \times [n]$ (definition 3) having depth k and fans of k Dyck paths of length 2n+2 (definition 2), such that each path has at least a valley.

The bijection is obtained by applying a sequence of "light and shadow" of the cloud, as in Viennot [18],[19]. The light is now located at the North-West of the cloud. The original process is defined for clouds having at most one point in each row and column of $[n] \times [n]$ Viennot's process can be immediately extended to general clouds, in the same way of Knuth's extension of the Robinson- Schensted correspondence.

More precisely, the ***shadow*** of the point (x,y) (for light coming from North-West) is the set of points (x',y') with $x' \geq x$ and $y' \leq y$. We then consider the ***outstanding line*** formed by the border of the union of the shadows of all the points of the cloud.

This line contains the ***outstanding points*** of the cloud, that is points which are not contained in the shadow of another point. In a second step, we remove from the original cloud these outstanding points. In the case of point with multiplicity $m \geq 2$, it remains a point with multiplicity m-1. Then we light again the remaining cloud and, if not empty, obtain a second outstanding line. Recursively, we thus define a sequence of such lines, until obtaining an empty cloud (see figure 5). By extending these lines with horizontal and vertical lines (in dot lines in figure 5), we get lines visualizing Dyck paths (up to a rotation of 135°). From Viennot [18], the number of such Dyck paths is the depth of the cloud.

Remark that the extension to general clouds of Viennot's construction [18] can be interpretated as transforming the cloud with "infinitesimal" moves of the points such that the points (with multiplicity) in each row and column becomes strictly decreasing subsequences of points.

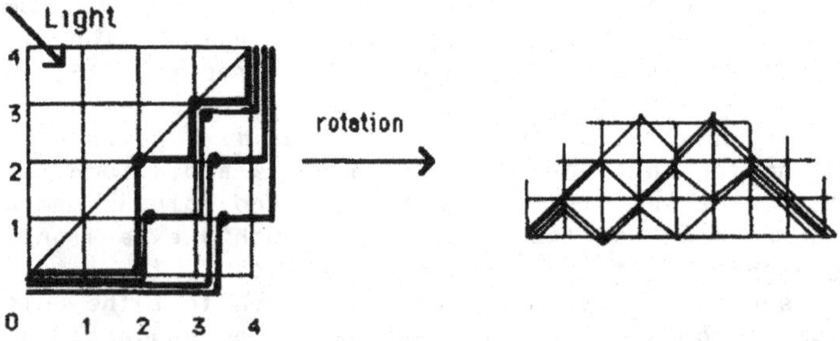

Figure 5. Successive lightings and shadows of the cloud.

Theorem 9 - The set $T_{n,\leq 2k}$ (definition 1) of Young tableaux is in bijection with the set $E_{V_{2n+2,k}}$ of fans (definition 2) of k Dyck paths of length 2n+2.

Let η_{2n+2} be the unique Dyck path of length 2n+2 and not having valleys. From proposition 4 and 8, the map $\Psi\circ\varphi^{-1}$ is a bijection between

Young tableaux of $T_{n,2p}$ (height $2p \leq 2k$) and fans of p Dyck paths of length 2n+2, not reduced to the Dyck path η_{2n+2}. If p< k, it suffices to add k-p times this path η_{2n+2} into the fan to get the desired bijection □

§5 - END OF THE PROOF OF THE FORMULA (2).

A fan of k Dyck paths of length 2n+2 can be transformed bijectively into a configuration $\eta = (\omega_1, \omega_2, \ldots, \omega_k)$ of k Dyck paths, two by two disjoint (no common vertices) and such that for $i, 1 \leq i \leq k$, ω_i goes from (-2i+2,0) to the point (2n+2i-2,0). It suffices to translate to the North successively the paths of the fan 2 steps, 4 steps, etc..., as shown on figure 6, and add at the beginning and the end of each translated path 2,4,6... North-East and South-East steps.

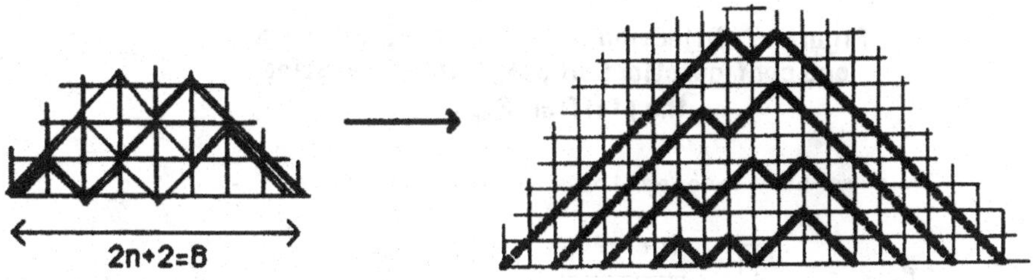

Figure 6. From a fan to non-crossing Dyck paths.

We are thus back to section 6 and 7 of Viennot [21]. Such configurations of non-crossing Dyck paths are interpretated, using Gessel-Viennot [9],[20] methodology, as an Hankel determinant of Catalan numbers. This determinant is computed using the so-called qd-algorithm from Padé approximants theory. Corollary 11 of [21] ends the proof of our formula (2) giving the number of Young tableaux of $T_{n,\leq 2k}$ □
Remark that the number of fans of Dyck paths of $E^{\mathbf{V}}_{2n+2,k}$ can also be given by another determinant due to Kreweras [15].

§6 - RELATIONSHIP WITH PFAFFIANS AND PERFECT MATCHINGS.

For n and k $\geq$ 0, let $A_i = (0, n+2k+1-i)$ for $1 \leq i \leq n+2k$, and $A_i = (i-n-2k, i-n-2k)$ for $n+2k < i \leq 2n + 2k$ (see figure 7). We consider configurations of n+k non-intersecting paths joining two by two these 2(n+k) points. Each path goes from a point A_i to a point A_j with $i < j$ and has elementary steps only North or East (see figure 7). As shown by Desainte-Catherine, Viennot [5], the number of such configurations can be expressed by a certain ***pfaffian*** $Pf_{n,k}$. The term $a_{ij}, i<j<2n+2k$ of this pfaffian is the number of paths going from A_i to A_j that is $a_{ij} = 1$ for $1 \leq i < j \leq n+2k$, $a_{ij} = \binom{n+2k+1-i}{n+2k-j}$ for $1 \leq i \leq n+2k < j \leq 2n+2k$, $a_{ij} = 0$ else.

Such configurations of paths are in bijection with Young tableaux of $T_{n,2k}$. A one-to-one correspondence can easily be found as shown on figure 7.

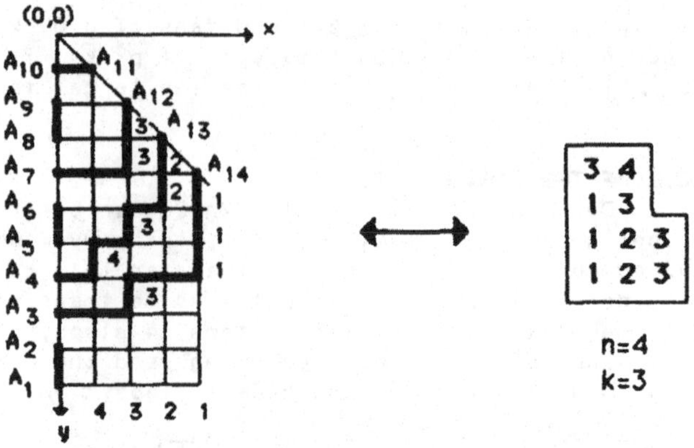

Figure 7. Bijection between Young tableaux and configurations of paths interpretating the pfaffian $Pf_{n,k}$.

Thus we have proved the following formula

$$(3) \qquad Pf_{n,k} = \prod_{1 \leq i \leq j \leq n} \frac{(i+j+2k)}{(i+j)},$$

which was conjectured by Desainte-Catherine [3].

Another interesting corollary is the following. Let $H_{n,k}$ be a graph formed with n-1 rows of hexagons and one row of 2k+n-1 pentagons, as shown on figure 8. A ***perfect matching*** of a graph is a set of two by two disjoint edges covering the graph.

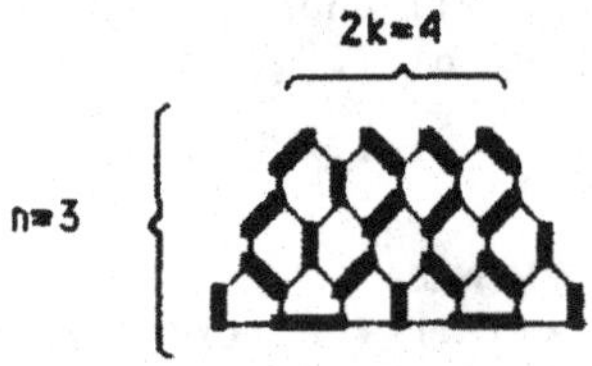

Figure 8. A perfect matching of the graph $H_{3,2}$.

Desainte-Catherine [3], [4] has constructed a bijection between the set of perfect matchings of $H_{n,k}$ and the set of configurations of paths interpretating the pfaffian $Pf_{n,k}$ (figure 7).
This bijection is inspired from physicist's work about the combinatorial solution of the Ising model [6] [12]. Thus, from above, we deduce that the number of perfect matchings of the graph $H_{n,k}$ is again the product $\prod_{1 \leq i \leq j \leq n} (i+j+2k)/(i+j)$. In particular, when k=1, this product becomes the Catalan number C_n.

REFERENCES

[1] E.A. BENDER and D.E. KNUTH , Enumeration of plane partitions, J. Combinatorial Th. (A) 13 (1972) 40-54.

[2] W.H. BURGE , Four correspondences between graphs and generalized Young tableaux, J. Combinatorial Th. (A) 17 (1974) 12-30.

[3] M. DESAINTE-CATHERINE , Couplages et pfaffiens en Combinatoire, Physique et Informatique, thèse 3ème cycle, Univ. of Bordeaux, 1983. See also in "séminaire lotharingien", May 1983, Sainte-Croix-aux-Mines.

[4] M. DESAINTE-CATHERINE , The number of perfect matchings of certain hexagonal graphs, preprint.

[5] M. DESAINTE-CATHERINE and G. VIENNOT, Combinatorial interpretation of Pfaffians, in preparation.

[6] M.E. FISHER , Statistical mechanics of dimers on a plane lattice, Phys. Rev. 124, (1961) 1664-1672.

[7] P. FLAJOLET , Combinatorial aspects of continued fractions, Discrete Maths., 32 (1980) 125-161.

[8] I. GESSEL and G. VIENNOT, Binomial determinants, paths and hook length formulae, Adv. in Maths, 58 (1985) 300-321.

[9] I. GESSEL and G. VIENNOT , Combinatorial interpretation of determinants with weighted paths, in preparation.

[10] B. GORDON , quoted in Stanley [17], p. 265.

[11] I. GOULDEN, D. JACKSON , "Combinatorial enumeration", John Wiley, 1983.

[12] P.W. KASTELEYN ,Graph theory and crystal physics, (1966) in "Graph Theory and Theoretical Physics", ed. F. Harary, p 43-110, Academic Press (1967).

[13] D.E. KNUTH , Permutations , matrices and generalized Young tableaux, Pacific J. Math., 34 (1970) 709-727.

[14] D.E. KNUTH , The art of computer programming, vol.3, sorting and searching §1, Addison-Wesley (1973).

[15] G. KREWERAS, Sur une classe de problèmes de dénombrements liés au treillis de partitions d'entiers,cahiers BURO, 6 (1965) 2-107.

[16] M.P. SCHÜTZENBERGER , Quelques remarques sur une construction de Schensted, Math. Scand. 12 (1963) 117-128.

[17] R. STANLEY , Theory and application of plane partitions, Part 2, Studies in applied Math., 50 (1971) 259-279.

[18] G. VIENNOT , Une forme géométrique de la correspondance de Robinson-Schensted, in "Combinatoire et Représentation du groupe symétrique", ed. D. Foata, Lecture Notes in Maths n°579 , Springer-Verlag, 1976.

[19] G. VIENNOT , Chains and antichains families, Grids and Young Tableaux, Annals of Discrete Maths, 23 (1984) 409-464.

[20] G. VIENNOT , Une théorie combinatoire des polynômes orthogonaux généraux, Lecture Notes,Université du Québec à Montréal 1984,215p.

[21] G. VIENNOT , A combinatorial interpretation of the quotient-difference algorithm, preprint, 1986.

FONCTIONS SYMÉTRIQUES ET SÉRIES HYPERGÉOMÉTRIQUES BASIQUES MULTIVARIÉES, II [1]

PAR

JACQUES DÉSARMÉNIEN et DOMINIQUE FOATA [2]

En l'honneur de Marcel-Paul SCHÜTZENBERGER

RÉSUMÉ. — Dans un article précédent [8], nous avons fait apparaître des fonctions hypergéométriques basiques multivariées dans l'étude de certaines statistiques sur le groupe symétrique. Nous établissons ici les propriétés de symétrie de ces fonctions en faisant appel à l'algèbre des tableaux de Young.

ABSTRACT. — In a previous paper [8] we showed how some multivariate basic hypergeometric functions arose in the study of certain statistics on the symmetric group. Here we establish symmetry properties of those fonctions by means of the Young tableau algebra.

1. Introduction

Pour chaque entier $n \geq 0$, posons :

$$\begin{aligned}
&(a;q)_0 = 1,\\
&(a;q)_n = (1-a)(1-aq)\ldots(1-aq^{n-1}) \quad (n \geq 1),\\
&(a;q)_\infty = \lim_n (a;q)_n = \prod_{n\geq 0}(1-aq^n).
\end{aligned}$$

D'autre part, pour chaque paire d'entiers positifs r, s, adoptons la notation :

$$\begin{aligned}
(u;q_1,q_2)_{r,s} &= 1, \qquad \text{si } r \text{ ou } s \text{ est nul};\\
&= \prod_{1\leq i\leq r}\ \prod_{1\leq j\leq s}(1-uq_1^{i-1}q_2^{j-1}), \qquad \text{si } r,s\geq 1;
\end{aligned}$$

et

$$\begin{aligned}
(u;q_1,q_2)_{\infty,\infty} &= \lim_{r,s}(u;q_1,q_2)_{r,s}\\
&= \prod_{i\geq 1}\prod_{j\geq 1}(1-uq_1^{i-1}q_2^{j-1}).
\end{aligned}$$

---

[1] Ce texte a été composé par le laboratoire de typographie informatique de l'Université Louis-Pasteur à Strasbourg, au moyen du préprocesseur *STRATEC*. Le fichier obtenu a été ensuite traité par le logiciel *TEX/SM 90*.

[2] Département de Mathématique, Université Louis-Pasteur de Strasbourg, 7, rue René-Descartes, F-67084 Strasbourg Cedex.

Dans [8] nous avons établi que l'identité :

$$(1.1)\qquad \sum_n C_n \frac{u^n}{(t_1;q_1)_{n+1}\,(t_2;q_2)_{n+1}} = \sum_{r,s} {t_1}^r {t_2}^s \frac{(-zu;q_1,q_2)_{r+1,s+1}}{(u;q_1,q_2)_{r+1,s+1}},$$

définissait une suite de polynômes

$$C_n = \sum C(n;m,r,s,i,j) z^m t_1^r t_2^s q_1^i q_2^j$$

à cinq variables z, t_1, t_2, q_1, q_2, où les coefficients $C(n;m,r,s,i,j)$ étaient des entiers positifs, de somme égale à $n!\,2^n$.

Cette identité contenait comme cas particuliers, d'une part, les formules classiques sur les q-séries (par exemple, la formule q-binomiale [1, p. 17, 2, p. 66]), d'autre part, les identités sur les distributions multivariées de statistiques sur le groupe symétrique.

Soient $(a_1, a_2, \ldots, a_k)$ une suite d'entiers positifs de somme n et $W = W(a_1, a_2, \ldots, a_k)$ l'ensemble de tous les réarrangements du mot $1^{a_1} 2^{a_2} \cdots k^{a_k}$. Si $w = x_1 x_2 \cdots x_n$ est un tel réarrangement, sa *ligne de route*, notée $\mathrm{Ligne}\, w$, est définie comme l'ensemble des entiers i tels que $1 \le i \le n-1$ et $x_i > x_{i+1}$, tandis que le *nombre de descentes* $\mathrm{Des}\, w$ et l'*indice majeur* $\mathrm{Maj}\, w$ sont définis par :

$$\mathrm{Des}\, w = |\mathrm{Ligne}\, w| \qquad \mathrm{Maj}\, w = \sum\{i : i \in \mathrm{Ligne}\, w\}.$$

On doit à MACMAHON d'avoir introduit la notion d'indice majeur ("major index"), d'avoir également calculé sa fonction génératrice sur tout ensemble W de réarrangements, enfin d'avoir montré qu'elle était la même que celle du nombre des inversions (*cf.* [22, 23 § 104, 24, 25]).

Lorsque tous les a_i sont égaux à 1 (et donc k à n), chaque réarrangement w dans W est une permutation de $12\cdots n$. On peut alors définir les quantités :

$$\mathrm{Ides}\, w = \mathrm{Des}\, w^{-1} \qquad \mathrm{Imaj}\, w = \mathrm{Maj}\, w^{-1},$$

où w^{-1} est l'inverse de w dans le groupe (symétrique) W. Les distributions univariées ou multivariées des statistiques Des, Ides, Maj, Imaj sur V ont fait l'objet de nombreuses études et ont été calculées avec succès. Par exemple, le polynôme générateur de Des sur W n'est autre que le *polynôme eulérien* (*cf.* [10]), les q-nombres eulériens donnent la distribution du couple (Des, Maj) (*cf.* [3, 4, 5]). La fonction génératrice du quadruplet (Des, Ides, Maj, Imaj), toujours sur le groupe symétrique W, fut calculée par GARSIA–GESSEL [13] et RAWLINGS [26], tandis que le groupe de symétrie de la distribution de ce quadruplet fut obtenu dans le contexte des tableaux de Young (*cf.* [11]). D'autres résultats sur ces statistiques sont dues à CARLITZ [6], CHEEMA–MOTZKIN [7], GESSEL [14], RAWLINGS [27], ROSELLE [29], STANLEY [33]. Voir également [9].

Dans [8], nous avons étendu la définition des statistiques Des, Ides, Maj, Imaj à l'ensemble des permutations colorées, ensemble de cardinal $n!\,2^n$, et fait apparaître les polynômes C_n de la formule (1.1) comme des fonctions fait apparaître les polynômes C_n de la formule (1.1) comme des fonctions génératrices d'un 5-vecteur sur cet ensemble. Nous avons ainsi pu démontrer que, par spécialisation, on obtenait toutes les formules sur le groupe symétrique faisant intervenir les quatre statistiques ci-dessus.

Rappelons également que (1.1) se particularise en la formule :

$$\sum_n C_n(z,q_1,q_2)\frac{u^n}{(q_1,q_1)_n\,(q_2,q_2)_n}=\frac{(-zu;q_1,q_2)_{\infty,\infty}}{(u;q_1,q_2)_{\infty,\infty}},$$

(les $C_n(z,q_1,q_2)$ étant des polynômes), formule considérée comme un analogue de la formule q-binomiale au cas de deux bases q_1, q_2.

Dans notre article, nous n'avions cependant pas respecté le principe de RIORDAN, qui veut que toute définition nouvelle de polynômes ou de suites de nombres soit nécessairement accompagnée de la table des premières valeurs (permettant ainsi au lecteur de vérifier aisément les relations de récurrence dans les cas initiaux). Nous nous proposons ici de réparer cette offense et de calculer les premières valeurs de C_n pour $n=1,2,3,4,5,6$. On trouvera celles-ci dans l'annexe 2, qui contient les tables des coefficients $C(n;m,r,s,i,j)$.

Comme le lecteur peut le constater ces tables présentent plusieurs symétries, suivant la diagonale principale, à l'intérieur de chaque bloc correspondant à une valeur fixée de i,j, entre blocs,... Le but principal de cet article sera de prouver ces symétries, de façon plus essentielle, de dégager le groupe de symétrie sous-jacent. Le résultat prouvé s'exprime analytiquement sous la forme suivante :

THÉORÈME 1.1. — *Pour tout 5-vecteur $v=(m,r,s,i,j)$, on a les relations :*

(1.2) $C(n;m,r,s,i,j)=C(n;m,s,r,j,i)$,

(1.3) $C(n;m,r,s,i,j)=C(n;m,r,s,i,ns-j)$,

(1.4) $C(n;m,r,s,i,j)=C(n;m,n-1-r,n-1-s,\binom{n}{2}-i,\binom{n}{2}-j)$,

(1.5) $C(n;m,r,s,i,j)=C(n;n-m,r,n-1-s,i,\binom{n}{2}-j)$.

Le calcul effectif des polynômes C_n repose sur la manipulation des (t,q)-*tableaux* $F_{\nu/\theta}(t,q)$. Ceux-ci sont introduits dans la prochaine section comme polynômes générateurs de tableaux gauches d'une forme donnée ν/θ. L'algèbre des fonctions de Schur permet d'exprimer les polynômes C_n en fonction des (t,q)-tableaux (formule (3.5)). De façon équivalente, C_n s'exprime comme fonction génératrice de paires de tableaux gauches par une certaine statistique V (formule (3.8)). La section 4 contient des indications pour le calcul effectif des (t,q)-tableaux. Dans la section 5, nous donnons la construction de trois involutions sur les tableaux de forme $\lambda\otimes\mu$, permettant dans la section suivante de dégager le groupe de symétrie d'ordre 32 de la distribution V.

2. Les (t,q)-tableaux

Désignons par *partition* toute suite finie décroissante $\nu = (\nu_1, \nu_2, \ldots, \nu_p)$ d'entiers supérieurs ou égaux à 1. Si la somme $\nu_1 + \nu_2 + \cdots + \nu_p$ de ces entiers est égale à n, on dit que ν est une *partition de n* et on pose $|\nu| = n$. Le *diagramme de Ferrers* associé à ν est l'ensemble des couples (i,j) du plan euclidien satisfaisant à $1 \leq i \leq \nu_j$, $1 \leq j \leq p$. Il est commode de l'identifier à la partition elle-même.

Soient $\nu = (\nu_1, \nu_2, \ldots, \nu_p)$ et $\theta = (\theta_1, \theta_2, \ldots, \theta_r)$ deux diagrammes de Ferrers. Si $\nu \supset \theta$, la différence ensembliste $\nu - \theta$, qu'on note le plus souvent ν/θ, est appelée *diagramme gauche.* On s'intéressera plus particulièrement aux diagrammes gauches ν/θ de la forme suivante : on part de deux diagrammes de Ferrers *quelconques* $\lambda = (\lambda_1, \lambda_2, \ldots, \lambda_p)$ et $\mu = (\mu_1, \mu_2, \ldots, \mu_r)$ et l'on considère l'ensemble, noté $\lambda \otimes \mu$, de tous les translatés $(\lambda_1 + i, j)$ $(1 \leq i \leq \mu_j\ ;\ 1 \leq j \leq r)$ et $(i, r+j)$ $(1 \leq i \leq \lambda_j\ ;\ 1 \leq j \leq p)$.

Par exemple, avec $\lambda = (2,1)$ et $\mu = (3,1)$, on obtient pour $\lambda \otimes \mu$ le diagramme gauche matérialisé par les croix :

```
×
× ×
    ×
    × × ×
```

Soit I un sous-ensemble de cardinal n et ν/θ un diagramme gauche contenant n points. Supposons que l'on écrive les n entiers de I sur les n points du diagramme ν/θ de façon à obtenir une croissance dans chaque ligne (de gauche à droite) et chaque colonne (de bas en haut). La configuration obtenue est appelée *tableau standard, de contenu I et de forme* ν/θ. Lorsque $I = [n]$, on remplace "de contenu I" par "d'ordre n". Dans la suite, on utilisera essentiellement les *tableaux standard d'ordre n, de forme* ν/θ (et plus particulièrement ceux de forme $\lambda \otimes \mu$ pour λ et μ quelconques tels que $|\lambda| + |\mu| = n$) et les *tableaux standard, de contenu I et de forme* λ. On dira aussi qu'un tableau est *droit* (resp. *gauche*), si sa forme est un diagramme de Ferrers (resp. diagramme gauche).

Par exemple,

$$P_1 = \begin{matrix} 6\ 8 \\ 4\ 5\ 9 \end{matrix}\ ;\quad Q_1 = \begin{matrix} 2\ 8 \\ 1\ 6\ 7 \end{matrix}\ ;\quad P_2 = \begin{matrix} 3 \\ 1\ 2\ 7 \end{matrix}\ ;\quad Q_2 = \begin{matrix} 4 \\ 3\ 5\ 9 \end{matrix}\ ;$$

sont des tableaux *standard*, de forme $\lambda = (3,2)$, pour les deux premiers et $\mu = (3,1)$ pour les deux derniers. Ils ont des contenus différents les uns des autres.

Les deux tableaux

$$R_1 = \begin{matrix} 6\ 8 & \\ 4\ 5\ 9 & \\ & 3 \\ & 1\ 2\ 7 \end{matrix} \qquad R_2 = \begin{matrix} 7 & \\ 6\ 8 & \\ 1\ 2 & \\ & 4 \\ & 3\ 5\ 9 \end{matrix}$$

sont standard, d'*ordre* 9 ; le premier est de forme $\lambda\otimes\mu$, le second $\lambda'\otimes\mu$ (notant ici λ' le diagramme *transposé* déduit de λ). Ils sont formés au moyen des précédents tableaux de façon claire. La notation

$$R_1 = P_1 \otimes P_2 \qquad \text{et} \qquad R_2 = Q'_1 \otimes Q_2$$

est alors évidente.

La *ligne inverse de route* (*cf.* [11]) d'un tableau standard R, d'ordre n, de forme ν/θ, est l'ensemble des entiers k tels que $1 \le k \le n-1$ et tels que $(k+1)$ soit écrit *plus haut* que k dans R. Cette ligne inverse de route est notée Iligne R. On pose également :

$$\text{(2.1)} \qquad \operatorname{Ides} R = |\operatorname{Iligne} R| \qquad \text{et} \qquad \operatorname{Imaj} R = \sum\{i : i \in \operatorname{Iligne} R\}.$$

Dans l'exemple précédent, on a Iligne $R_1 = \{2,3,5,7\}$, de sorte que Ides $R_1 = 4$ et Imaj $R_1 = 2+3+5+7 = 17$.

La statistique Imaj introduite ici sur les tableaux gauches reprend l'information contenue dans la *charge* et la *cocharge* des tableaux, deux notions introduites par LASCOUX–SCHÜTZENBERGER [17, 18]. Voir également MACDONALD [21, p. 129].

Pour chaque diagramme gauche ν/θ de cardinal n, on introduit le polynôme générateur du couple (Ides, Imaj) sur l'ensemble de tous les tableaux standard R, d'ordre n, de forme ν/θ, à savoir le polynôme :

$$\text{(2.2)} \qquad F_{\nu/\theta}(t,q) = \sum t^{\operatorname{Ides} R} q^{\operatorname{Imaj} R} \qquad (R \text{ standard, de forme } \nu/\theta).$$

Ce polynôme sera désigné par (t,q)-*tableau.*

Dans le présent article, on considèrera essentiellement les (t,q)-tableaux $F_{\lambda\otimes\mu}(t,q)$ correspondant aux diagrammes $\lambda\otimes\mu$.

3. Fonctions de Schur

Nous rappelons brièvement ici comment les polynômes C_n (*cf.* (1.1) et (1.2)) s'expriment en fonction des (t,q)-tableaux et comment l'algèbre des fonctions de Schur fournit tous les éléments de calcul nécessaires.

On note $S_{\nu/\theta}(x)$ la *fonction de Schur gauche* associée au diagramme gauche ν/θ, avec x comme ensemble de variables (*cf.* [21, p. 42]). Lorsque $\nu/\theta = \lambda\otimes\mu$ (avec λ et μ diagrammes de Ferrers, comme indiqué dans la section précédente), il résulte de la définition même des fonctions de Schur (*cf.* [21, p. 42]) que l'on a :

$$\text{(3.1)} \qquad S_{\lambda\otimes\mu}(x) = S_\lambda(x) S_\mu(x) = S_{\mu\otimes\lambda}(x).$$

Par ailleurs, les produits de fonctions de Schur s'expriment comme combinaisons linéaires d'autres fonctions de Schur :

$$\text{(3.2)} \qquad S_\lambda(x) S_\mu(x) = \sum_\nu g_{\lambda\mu\nu} S_\nu(x),$$

où les coefficients $g_{\lambda\mu\nu}$ sont des entiers positifs, qu'on peut évaluer par l'algorithme de Littlewood-Richardson.

Partons alors des deux formules de Cauchy (*cf.* [21, p. 33 et 35]) :

$$\sum_{\lambda} S_\lambda(x)S_\lambda(y) = \prod_{i,j}(1 - x_i y_j)^{-1},$$
$$\sum_{\lambda} S_\lambda(x)S_{\lambda'}(y) = \prod_{i,j}(1 + x_i y_j),$$

et multiplions les membre à membre compte tenu de (3.1) tout en introduisant une variable d'homogénéité u. Nous obtenons :

$$\sum_n u^n \sum_{\lambda,\mu} z^{|\lambda|} S_{\lambda\otimes\mu}(x)S_{\lambda'\otimes\mu}(y) = \prod_{i,j} \frac{(1 + zux_iy_j)}{(1 - ux_iy_j)},$$

où, pour chaque $n \geq 0$ fixé, la seconde sommation est sur les paires de partitions (λ, μ) telles que $|\lambda| + |\mu| = n$. En prenant pour x (resp. y) un ensemble fini $\{x_1, \ldots, x_{r+1}\}$ (resp. $\{y_1, \ldots, y_{s+1}\}$) de variables et en faisant les substitutions $x_i \leftarrow q_1^{i-1}$, $y_j \leftarrow q_2^{j-1}$, on en déduit la formule :

$$\sum_n u^n \sum_{\lambda,\mu} z^{|\lambda|} S_{\lambda\otimes\mu}(1, \ldots, q_1^r) S_{\lambda'\otimes\mu}(1, \ldots, q_2^s) = \frac{(-zu; q_1, q_2)_{r+1,s+1}}{(u; q_1, q_2)_{r+1,s+1}}.$$

Multipliant par $t_1^r t_2^s$ et sommant par rapport à r et s, on obtient, comme second membre, le second membre de l'identité (1.1). Le premier membre, lui, s'écrit :

$$\sum_n u^n \sum_{\lambda,\mu} z^{|\lambda|} \sum_r t_1^r S_{\lambda\otimes\mu}(1, \ldots, q_1^r) \sum_s t_2^s S_{\lambda'\otimes\mu}(1, \ldots, q_2^s).$$

Comparant avec (1.1), on voit donc que C_n est égal à l'expression :

$$(3.3) \qquad C_n = \sum_{\lambda,\mu} z^{|\lambda|} (t_1; q_1)_{n+1} (t_2; q_2)_{n+1} \sum_{r,s} t_1^r t_2^s S_{\lambda\otimes\mu}(1, \ldots, q_1^r) S_{\lambda'\otimes\mu}(1, \ldots, q_2^s).$$

Le lemme suivant, énoncé et démontré dans [8, théorème 4.1], permet non seulement de prouver que C_n est un polynôme, mais fournit aussi une interprétation combinatoire pour C_n, compte tenu des propriétés bien connues des fonctions de Schur. Remmel [28] a utilisé récemment ce lemme pour exploiter combinatoirement plusieurs formules classiques sur les fonctions de Schur. Notre collègue Richard Stanley, dans une correspondance privée, nous a fait savoir que ce lemme pouvait se déduire de la proposition 8.3, p. 24 de sa thèse [32], pourvu que l'on sache faire le rapprochement souhaité entre (P, ω)-partitions et (t, q)-tableaux.

LEMME. — *Soit ν/θ un diagramme gauche de n éléments, alors le (t,q)-tableau $F_{\nu/\theta}(t,q)$, tel qu'il est défini en (2.2) est donné par :*

$$F_{\nu/\theta}(t,q) = (t;q)_{n+1} \sum_r t^r S_{\nu/\theta}(1,q,q^2,\ldots,q^r). \tag{3.4}$$

Comparant (3.3) et (3.4), on en déduit que C_n est un *polynôme* et qu'il peut être exprimé au moyen de la formule :

$$C_n = \sum_{\lambda,\mu} z^{|\lambda|} F_{\lambda\otimes\mu}(t_1,q_1) F_{\lambda'\otimes\mu}(t_2,q_2), \tag{3.5}$$

où la somme est étendue sur l'ensemble des couples de diagrammes de Ferrers tels que $|\lambda| + |\mu| = n$.

Donnons enfin une autre expression pour C_n, qui prend en charge la définition (2.2) des (t,q)-tableaux. Soient R_1, R_2 deux tableaux standard, d'ordre n, de forme $\lambda_1 \otimes \mu_1$ et $\lambda_2 \otimes \mu_2$, respectivement. On dit qu'ils sont *jumelables* si $\lambda_1 = \lambda'_2$ et $\mu_1 = \mu_2$. La paire $\lambda_1\otimes\mu_1, \lambda_2\otimes\mu_2$ ($= \lambda_1\otimes\mu_1, \lambda'_1\otimes\mu_1$) est appelée *forme* de R_1R_2. Le *V-vecteur* de la paire R_1R_2 est, par définition, le vecteur :

$$V(R_1R_2) = (|\lambda_1|, \mathrm{Ides}\,R_1, \mathrm{Ides}\,R_2, \mathrm{Imaj}\,R_1, \mathrm{Imaj}\,R_2). \tag{3.6}$$

On note aussi $v(R_1R_2)$ le *monôme* :

$$v(R_1R_2) = z^{|\lambda|} t_1^{\mathrm{Ides}\,R_1} t_2^{\mathrm{Ides}\,R_2} q_1^{\mathrm{Imaj}\,R_1} q_2^{\mathrm{Imaj}\,R_2}. \tag{3.7}$$

Enfin, on désigne par $\mathcal{T}'_n$ l'ensemble des couples jumelables de tableaux standard d'ordre n. Il résulte de (3.5) et de (2.1) que C_n est le *polynôme générateur des paires de tableaux standard d'ordre n, jumelables, par le vecteur V*, ou encore que l'on a :

$$C_n = \sum_{R_1R_2} v(R_1R_2) \qquad (R_1R_2 \in \mathcal{T}'_n). \tag{3.8}$$

4. Le calcul des polynômes

Nous montrons ici comment on peut simplement calculer les polynômes $F_\lambda(t,q)$, puis en déduire l'expression de $F_{\lambda\mu}(t,q)$, enfin déterminer C_n au moyen de la formule (3.5).

Chaque (t,q)-tableau $F_\lambda(t,q)$ sera représenté par son diagramme de Ferrers sous-jacent. Par exemple :

$F_{4,2}(t,q) =$ 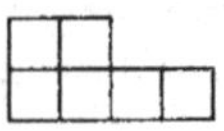 .

De la même manière, le symbole 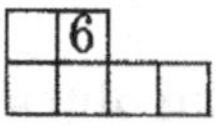 désigne la fonction génératrice, par Ides et Imaj (*cf.* (2.1)), des tableaux standard de forme $\lambda = 4, 2$, ayant l'entier 6 dans le coin supérieur droit.

En se reportant à la définition même de la ligne inverse d'un tableau et de Ides et Imaj, on a :

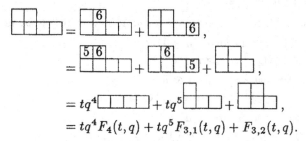

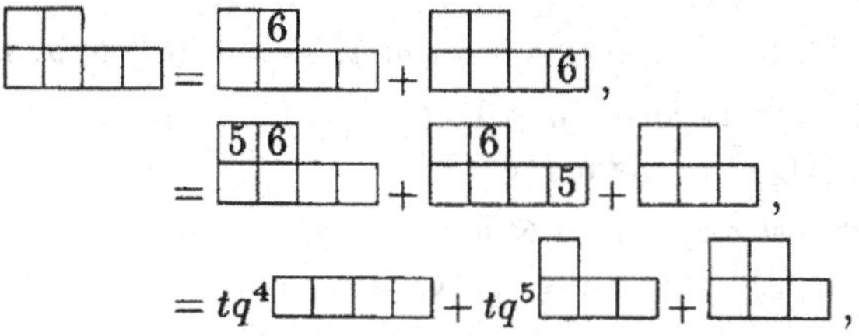

$$= tq^4 F_4(t,q) + tq^5 F_{3,1}(t,q) + F_{3,2}(t,q).$$

Si on connait déjà l'expression des polynômes $F_\lambda(t,q)$ pour les λ tels que $|\lambda| \le 5$, on obtient donc celle de $F_{4,2}(t,q)$. Dans l'annexe 1, c'est ainsi que les $F_\lambda(t,q)$ ont été calculés jusqu'à l'ordre 6.

Maintenant, les identités (3.1) et (3.2) sur les fonctions de Schur entraînent les formules :

$$F_{\lambda\otimes\mu}(t,q) = F_{\mu\otimes\lambda}(t,q) \tag{4.1}$$

et

$$F_{\lambda\otimes\mu}(t,q) = \sum_\nu g_{\lambda\mu\nu} F_\nu(t,q). \tag{4.2}$$

La première de ces formules permet de ne calculer $F_{\lambda\otimes\mu}(t,q)$ que pour les couples λ, μ tels que $\lambda \le \mu$ (pour un ordre total donné sur les partitions). La seconde dit qu'une bonne table des coefficients $g_{\lambda\mu\nu}$ de Littlewood-Richardson, telle qu'elle est donnée dans JAMES-KERBER [15] ou WYBOURNE [34], suffit pour déduire l'expression de $F_{\lambda\otimes\mu}(t,q)$ de celle des $F_\lambda(t,q)$.

Muni de la table des $F_{\lambda\otimes\mu}(t,q)$, on peut utiliser directement la formule (3.5) pour calculer les polynômes C_n. On peut aussi faire usage de la formule :

$$C_n = \sum_{\lambda,\mu} z^{|\lambda|} \sum_{\nu_1,\nu_2} g_{\lambda\mu\nu_1} g_{\lambda'\mu\nu_2} F_{\nu_1}(t_1,q_1) F_{\nu_2}(t_2,q_2). \tag{4.3}$$

5. Les involutions

Soit $\mathcal{T}_n$ l'ensemble de tous les tableaux standard d'ordre n dont la forme est un produit $\lambda \otimes \mu$ (éventuellement réduite à un diagramme de Ferrers). Nous nous proposons de montrer qu'on peut construire trois *involutions* J, S et T de $\mathcal{T}_n$ ayant les propriétés suivantes :

Si $P \otimes Q$ est un tableau standard d'ordre n, de forme $\lambda \otimes \mu$, alors

(5.1) $(P \otimes Q)^S$ *est de forme* $\mu \otimes \lambda$ *et*

(5.1′) $\operatorname{Iligne}(P \otimes Q)^S = \operatorname{Iligne} P \otimes Q$;

(5.2) $(P \otimes Q)^J$ *est de forme* $\lambda \otimes \mu$ *et*

(5.2′) $\operatorname{Iligne}(P \otimes Q)^J = n - \operatorname{Iligne} P \otimes Q$;

(5.3) $(P \otimes Q)^T$ *est de forme* $\mu' \otimes \lambda'$ *et*

(5.3′) $\operatorname{Iligne}(P \otimes Q)^T = [n-1] \setminus \operatorname{Iligne} P \otimes Q$.

De plus,

(5.4) *S, J et T commutent deux à deux.*

L'involution T est simplement la *transposition* des tableaux, de sorte que (5.3) et (5.3′) sont des propriétés immédiates.

La construction des deux autres involutions repose sur les propriétés du *jeu de taquin* et sur les propriétés de l'opération de *vidage-remplissage* J des tableaux droits (*cf.* [30, 31, 16, p. 48–73]). Rappelons qu'à tout tableau gauche (par exemple un tableau $P \otimes Q$ de forme $\lambda \otimes \mu$), on peut faire correspondre un tableau droit de même contenu et ayant la même ligne inverse de route. Ce tableau droit, que l'on notera $\operatorname{Taq}(P \otimes Q)$ (si l'on part du tableau gauche $P \otimes Q$), s'obtient à partir de $P \otimes Q$ en appliquant un nombre suffisant de fois les mouvements de base du jeu de taquin (*cf.* [31, 20]). On a donc :

(5.5) $$\operatorname{Iligne} P \otimes Q = \operatorname{Iligne} \operatorname{Taq}(P \otimes Q).$$

Considérons, par exemple, les deux tableaux :

$$P \otimes Q = \begin{array}{cccccc} 6 & 8 & & & & \\ 4 & 5 & 9 & & & \\ & & & 3 & & \\ & & & 1 & 2 & 7 \end{array} \qquad \text{et} \qquad R = \begin{array}{ccc} 6 & & \\ 4 & 8 & \\ 3 & 5 & 9 \\ 1 & 2 & 7 \end{array}$$

On vérifie qu'ils ont la même ligne inverse de route $\{2,3,5,7\}$ et que l'on a $R = \operatorname{Taq}(P \otimes Q)$.

Rappelons aussi que le vidage-remplissage des tableaux droits est une involution $R \mapsto R^J$, conservant le contenu et la forme et satisfaisant à

(5.6) $$\operatorname{Iligne} R^J = n - \operatorname{Iligne} R,$$

si R est d'ordre n (*cf.* [11]).

Par exemple, le vidé-rempli R^J du tableau R ci-dessus est donné par :

$$R^J = \begin{matrix} 8 \\ 7\,9 \\ 3\,5\,6 \\ 1\,2\,4 \end{matrix}$$

et l'on a Iligne $R^J = \{2,4,6,7\} = 9 - \{2,3,5,7\} = n -$ Iligne R.

Soient maintenant λ, μ, ν trois partitions de l, m, n, respectivement, telles que $l+m=n$ et $\lambda, \mu \subset \nu$ et soit R_0 un tableau standard, d'ordre n, de forme ν. On note $W_1 = W(\lambda, \mu, R_0)$ l'ensemble des tableaux standard $P \otimes Q$, d'ordre n, de forme $\lambda \otimes \mu$, tels que $\mathrm{Taq}(P \otimes Q) = R_0$. Le résultat remarquable dû à SCHÜTZENBERGER (*cf.* [31, p. 95]) est que le cardinal de $W(\lambda, \mu, R_0)$ ne dépend que de la *paire non ordonnée* $\{\lambda, \mu\}$ et de la *forme* ν de R_0 ; il est, de plus, égal au coefficient $g(\lambda, \mu, \nu)$ de Littlewood-Richardson. On a, enfin,

$$g(\lambda, \mu, \nu) = g(\lambda', \mu', \nu'). \tag{5.7}$$

Comme le vidage-remplissage conserve la forme des tableaux droits, on conclut immédiatement que les huit ensembles W_1, $W_s = W(\mu, \lambda, R_0)$, $W_j = W(\lambda, \mu, {R_0}^J)$, $W_{sj} = W(\mu, \lambda, {R_0}^J)$, $W_t = W(\mu', \lambda', {R_0}^T)$, $W_{st} = W(\lambda', \mu', {R_0}^T)$, $W_{jt} = W(\mu', \lambda', {R_0}^{JT})$, $W_{sjt} = W(\lambda', \mu', {R_0}^{JT})$ ont tous même cardinal, égal à $g(\lambda, \mu, \nu)$.

Supposons que le triplet (λ, μ, ν) soit tel que $\nu \leq \nu'$ (par exemple, par rapport à l'ordre lexicographique inverse des partitions), le tableau droit R_0 étant toujours supposé de forme ν. Au quadruplet (W_1, W_s, W_j, W_{js}) faisons correspondre quatre bijections *arbitraires*, comme indiqué dans le diagramme :

$$\begin{array}{ccc} W(\lambda, \mu, R_O) & \xrightarrow{S} & W(\mu, \lambda, R_0) \\ \downarrow J & & \downarrow J \\ W(\lambda, \mu, {R_0}^J) & \xrightarrow{S} & W(\mu, \lambda, {R_0}^{JS}) \end{array}$$

Si $P \otimes Q$ est dans W_1, notons $(P \otimes Q)^S$, $P \otimes Q)^J$, $(P \otimes Q)^{JS}$ les images de $P \otimes Q$ par les applications de ce diagramme, soit :

$$\begin{array}{ccc} P \otimes Q & \longrightarrow & (P \otimes Q)^S \\ \downarrow & & \downarrow \\ (P \otimes Q)^J & \longrightarrow & (P \otimes Q)^{JS} \end{array}$$

Ce diagramme étant construit, formons ensuite le diagramme obtenu en remplaçant les quatre tableaux par leurs transposés :

$$\begin{array}{ccc} (P \otimes Q)^T & \longrightarrow & (P \otimes Q)^{ST} \\ \downarrow & & \downarrow \\ (P \otimes Q)^{JT} & \longrightarrow & (P \otimes Q)^{JST} \end{array}$$

Comme on sait (*cf.* [11, 31]) que, si $\mathrm{Taq}(P \otimes Q) = R_0$, alors $\mathrm{Taq}((P \otimes Q)^T) = R_0^T$, il est clair que ces quatre tableaux transposés $(P \otimes Q)^T$, $(P \otimes Q)^{ST}$, $(P \otimes Q)^{JT}$ et $(P \otimes Q)^{JST}$ appartiennent respectivement à W_t, W_{st}, W_{jt}, W_{jst}. Ceci prouve les propriétés (5.1) et (5.2). La propriété (5.4) résulte de la construction même de ces involutions.

Enfin, les propriétés (5.1′) et (5.2′) sont automatiquement vérifiées, puisque l'on a pour tout $P \otimes Q \in W(\lambda, \mu, R_0)$, la relation $\mathrm{Taq}(P \otimes Q) = \mathrm{Taq}\, R_0$, d'où $\mathrm{Iligne}\, P \otimes Q = \mathrm{Iligne}\, R_0$. Par définition même des ensembles W, on en tire :

$$\mathrm{Iligne}(P \otimes Q)^S = \mathrm{Iligne}\, R_0 = \mathrm{Iligne}\, P \otimes Q\ ;$$
$$\mathrm{Iligne}(P \otimes Q)^J = \mathrm{Iligne}\, {R_0}^J = n - \mathrm{Iligne}\, R_0 = n - \mathrm{Iligne}\, P \otimes Q.$$

Remarque. — On peut obtenir une construction explicite des involutions S, J et T en utilisant les deux lemmes 3.7 et 4.5 de SCHÜTZENBERGER [31]. Cependant les bijections entre les ensembles W dépendent de deux tableaux standard P_0 et Q_0, de forme λ et μ, choisis arbitrairement.

6. Les symétries

Nous disposons de tous les éléments pour démontrer le théorème 1.1. Pour visualiser les quatre propriétés à établir, il est bon de se reporter aux tables de l'annexe 2, où sont reproduites les tables des valeurs $C(n; v)$ pour tous les vecteurs v et les valeurs n de 1 à 6.

La relation (1.2) dit que chaque table est symétrique par rapport à sa diagonale principale.

La relation (1.3) exprime le fait que, dans chaque table, pour tout couple de valeurs (r, s), le bloc correspondant à (r, s) a un axe de symétrie vertical. Il a donc aussi un axe de symétrie horizontal en conjuguant les propriétés (1.2) et (1.3).

La relation (1.4) dit que le centre de la table est un centre de symétrie.

Enfin, (1.5) affirme que la table correspondant à la première valeur $n - m$ du vecteur V se déduit de la table correspondant à m par une symétrie par rapport à son axe vertical. De ce fait, on peut représenter en une seule fois les tables m et les tables $n - m$ en disposant les variables de façon adéquate.

Notons $\mathcal{T}_n'$ l'ensemble de toutes les paires R_1R_2 jumelables (*cf.* section 3). Pour démontrer (1.2), on considère l'involution $\mathtt{i}$ de $\mathcal{T}_n'$ définie par :

$$\mathtt{i}(R_1R_2) = R_2R_1. \tag{6.2′}$$

Cette involution envoie bien chaque paire R_1R_2, de forme $\lambda \otimes \mu, \lambda' \otimes \mu$, de V-vecteur (m, r, s, i, j) sur R_2R_1 de forme $\lambda' \otimes \mu, \lambda \otimes \mu$ et de V-vecteur (m, s, r, j, i).

Pour obtenir (1.3), on considère la bijection de $\mathcal{T}_n'$ sur lui-même définie par :

$$\mathtt{j}(R_1R_2) = R_1{R_2}^J. \tag{6.3′}$$

D'après (3.5) et (5.5), on définit bien là une bijection, qui envoie chaque R_1R_2 de forme $\lambda\otimes\mu, \lambda'\otimes\mu$, de V-vecteur (m,r,s,i,j) sur $R_1{R_2}^J$, de forme $\lambda\otimes\mu, \lambda'\otimes\mu$ et de V-vecteur $(m,r,s,i,ns-j)$.

Avec $\mathbf{t}$ définie par :

$$(6.4') \qquad \mathbf{t}(R_1R_2) = {R_1}^{ST}{R_2}^{ST}.$$

on tient une involution de $\mathcal{T}'_n$, qui d'après (5.3) envoie R_1R_2 sur ${R_1}^{ST}{R_2}^{ST}$, de forme $\lambda'\otimes\mu', \lambda\otimes\mu'$ et de V-vecteur

$$(m, n-1-r, n-1-s, \tbinom{n}{2}-i, \tbinom{n}{2}-j).$$

La dernière involution, notée $\mathbf{s}$, est définie par :

$$(6.5') \qquad \mathbf{s}(R_1R_2) = {R_1}^S{R_2}^T.$$

Elle envoie la paire R_1R_2, de forme $\lambda\otimes\mu, \lambda'\otimes\mu$ et de V-vecteur (m,r,s,i,j) sur une paire ${R_1}^S{R_2}^T$, de forme $\mu\otimes\lambda, \mu'\otimes\lambda$, de V-vecteur

$$(n-m, r, n-1-s, i, \tbinom{n}{2}-j).$$

Ceci achève la démonstration du théorème 1.1.

Soit G le groupe engendré par les involutions $\mathbf{i}$, $\mathbf{j}$, $\mathbf{t}$ et $\mathbf{s}$. On vérifie immédiatement les relations :

$$\mathbf{i}^2 = \mathbf{j}^2 = \mathbf{t}^2 = \mathbf{s}^2 = (\mathbf{ij})^4 = (\mathbf{is})^4 = 1,$$
$$\mathbf{it} = \mathbf{ti}, \quad \mathbf{jt} = \mathbf{tj}, \quad \mathbf{st} = \mathbf{ts}, \quad \mathbf{js} = \mathbf{sj}.$$

Par conséquent, le groupe G contient le groupe diédral $D_4(\mathbf{i},\mathbf{j})$, d'ordre 8, engendré par $\{\mathbf{i},\mathbf{j}\}$, ainsi que le produit de ce groupe par le groupe $\{1,\mathbf{t}\}$ d'ordre 2.

Soit R_1R_2 un élément de $\mathcal{T}'_n$. Les éléments de l'orbite de R_1R_2, par rapport au groupe G, sont de la forme ${R_k}^\alpha {R_l}^\beta$, avec $\{k,l\} = \{1,2\}$ et α, β des monômes de degré au plus égal à 1 en chacune des variables S, J, T. Les éléments ${R_k}^\alpha {R_l}^\beta$ de l'orbite ne contenant ni S, ni T dans les exposants α et β sont au nombre de huit. Ce sont en fait les huit éléments de l'orbite de R_1R_2 par rapport au sous-groupe $D_4(\mathbf{i},\mathbf{j})$:

$$R_1R_2,\ R_2R_1,\ R_2{R_1}^J,\ {R_1}^JR_2,\ {R_1}^J{R_2}^J,\ {R_2}^J{R_1}^J,\ {R_2}^JR_1,\ R_1{R_2}^J.$$

Les éléments ${R_k}^\alpha {R_l}^\beta$ tels que α et β sont divisibles par ST sont au nombre de huit :

$$ {R_1}^{ST}{R_2}^{ST},\ {R_2}^{ST}{R_1}^{ST},\ {R_2}^{ST}{R_1}^{JST},\ {R_1}^{JST}{R_2}^{ST},$$
$${R_1}^{JST}{R_2}^{JST},\ {R_2}^{JST}{R_1}^{JST},\ {R_2}^{JST}{R_1}^{ST},\ {R_1}^{ST}{R_2}^{JST}.$$

Les seize éléments écrits constituent l'orbite de R_1R_2 par rapport au produit $D_4(\mathbf{i},\mathbf{j}) \times \{1,\mathbf{t}\}$.

À cause de la définition de $\mathbf{s}$, les seules autres paires ${R_k}^\alpha {R_l}^\beta$ possibles doivent satisfaire l'une des deux conditions :

(i) S divise α, T ne divise pas α, T divise β, S ne divise pas β,
(ii) T divise α, S ne divise pas α, S divise β, T ne divise pas β.

Ces paires forment l'orbite de $R_1{}^S R_2{}^T$ par rapport au groupe $D_4(\mathrm{i},\mathrm{j}) \times \{1, \mathrm{t}\}$ et sont au nombre de seize :

$$\begin{gathered}R_1{}^S R_2{}^T,\ R_2{}^T R_1{}^S,\ R_2{}^T R_1{}^{JS},\ R_1{}^{JS} R_2{}^T,\\ R_1{}^{JS} R_2{}^{JT},\ R_2{}^{JT} R_1{}^{JS},\ R_2{}^{JT} R_1{}^S,\ R_1{}^S R_2{}^{JT},\\ R_1{}^T R_2{}^S,\ R_2{}^S R_1{}^T,\ R_2{}^S R_1{}^{JT},\ R_1{}^{JT} R_2{}^S,\\ R_1{}^{JT} R_2{}^{JS},\ R_2{}^{JS} R_1{}^{JT},\ R_2{}^{JS} R_1{}^T,\ R_1{}^T R_2{}^{JS}.\end{gathered}$$

Il n'y a pas d'autres paires possibles. Le groupe G est donc d'ordre 32.

BIBLIOGRAPHIE

[1] ANDREWS (George E.). — *The Theory of Partitions.* — Reading, Mass., Addison-Wesley, 1976 (*Encyclopedia of Math. and Its Appl.*, **2**).

[2] BAILEY (W.N.). — *Generalized Hypergeometric Series.* — Cambridge University Press, 1935.

[3] CARLITZ (Leonard). — q-Bernoulli and Eulerian numbers, *Trans. Amer. Math. Soc.*, t. **76**, 1954, p. 332–350.

[4] CARLITZ (Leonard). — Eulerian numbers and polynomials, *Math. Magazine*, t. **33**, 1959, p. 247–260.

[5] CARLITZ (Leonard). — A combinatorial property of q-Eulerian numbers, *Amer. Math. Monthly*, t. **82**, 1975, p. 51–54.

[6] CARLITZ (Leonard). — The Expansion of certain Products, *Proc. Amer. Math. Soc.*, t. **7**, 1956, p. 558–564.

[7] CHEEMA (M.S.) and MOTZKIN (T.S.). — Multipartitions and Multipermutations, *Combinatorics* [Los Angeles. 1968], p. 39–70. — Providence, Amer. Math. Soc., 1971 (*Proc. Symposia in Pure Math.*, **19**).

[8] DÉSARMÉNIEN (Jacques) et FOATA (Dominique). — Fonctions symétriques et séries hypergéométriques basiques multivariées, *Bull. Soc. Math. France*, t. **113**, 1985, p. 3–22.

[9] FOATA (Dominique). — Distributions eulériennes et mahonniennes sur le groupe des permutations, *Higher Combinatorics* [M. Aigner, ed., Berlin. 1976], p. 27–49. Amsterdam, D. Reidel, 1977 (Proc. NATO Adv. Study Inst.).

[10] FOATA (Dominique) et SCHÜTZENBERGER (Marcel-Paul). — *Théorie géométrique des polynômes eulériens.* — Berlin, Springer-Verlag, 1970 (*Lecture Notes in Math.*, **138**).

[11] FOATA (Dominique) et SCHÜTZENBERGER (Marcel-Paul). — Major Index and Inversion of Permutations, *Math. Nachr.*, t. **83**, 1978, p. 143–159.

[12] FOULKES (Herbert). — Enumeration of Permutations with Prescribed Up-down and Inversion Sequences, *Discrete Math.*, t. **15**, 1976, p. 235–252.

[13] GARSIA (Adriano M.) and GESSEL (Ira). — Permutation Statistics and Partitions, *Advances in Math.*, t. **31**, 1979, p. 288–305.

[14] GESSEL (Ira). — Generating functins and enumeration of sequences, Ph.D. thesis, department of mathematics, M.I.T., Cambridge, Mass., 111 p., 1977.

[15] JAMES (Gordon) and KERBER (Adalbert). — *The Representation Theory of the Symmetric Group.* — Reading, Mass., Addison-Wesley, 1981 (*Encyclopedia of Math. and Its Appl.*, **16**).

[16] KNUTH (Donald E.). — *The Art of Computer Programming*, vol. 3, Sorting and Searching. — Don Mills, Ontario, Addison-Wesley, 1972.

[17] LASCOUX (Alain) et SCHÜTZENBERGER (Marcel-Paul). — A new statistics on words, *Combinatorial Mathematics, Optimal Designs and their Applications* [J. Srivastava, ed., Fort Collins, Colorado. 1978], p. 251–255. — Amsterdam, North-Holland, 1980 (*Annals of Discrete Math.*, **6**).

[18] LASCOUX (Alain) et SCHÜTZENBERGER (Marcel-Paul). — Sur une conjecture de H.O. Foulkes, *C.R. Acad. Sc. Paris*, t. **286A**, 1978, p. 385–387.

[19] LASCOUX (Alain) et SCHÜTZENBERGER (Marcel-Paul). — Formulaire raisonné des fonctions symétriques, L.I.T.P., U.E.R. Math., Univ. Paris VII, 138 p., 1984.

[20] LASCOUX (Alain) et SCHÜTZENBERGER (Marcel-Paul). — Le monoïde plaxique, *Non-commutative Structures in Algebra and geometric Combinatorics* [A. de Luca, ed., Napoli. 1978], p. 129–156. — Roma, Consiglio Nazionale delle Ricerche, 1981 (*Quaderni de "La Ricerca Scientifica"*, **109**).

[21] MACDONALD (Ian G.). — *Symmetric Functions and Hall Polynomials*. — Oxford, Clarendon Press, 1979.

[22] MACMAHON (Percy Alexander). — The indices of permutations and the derivation therefrom of functions of a single variable associated with the permutations of any assemblage of objects, *Amer. J. Math.*, t. **35**, 1913, p. 314–321.

[23] MACMAHON (Percy Alexander). — *Combinatory Analysis*, vol. 1. — Cambridge, Cambridge Univ. Press, 1915 (Réimprimé par Chelsea, New York, 1955).

[24] MACMAHON (Percy Alexander). — Two applications of general theorems in combinatory analysis, *Proc. London Math. Soc.*, t. **15**, 1916, p. 314–321.

[25] MACMAHON (Percy Alexander). — *Collected Papers*, vol. 1 [George E. ANDREWS, ed.]. — Cambridge, Mass., The M.I.T. Press, 1978.

[26] RAWLINGS (Don). — Generalized Worpitzky Identities with Applications to Permutation Enumeration, *Europ. J. Comb.*, t. **2**, 1981, p. 67–78.

[27] RAWLINGS (Don). — The Combinatorics of certain Products, *Proc. Amer. Math. Soc.*, t. **83**, 1983, p. 560–562.

[28] REMMEL (Jeff). — Symmetric functions and q-series, preprint, Univ. Calif. San Diego, 1984.

[29] ROSELLE (David P.). — Coefficients associated with the Expansion of certain Products, *Proc. Amer. Math. Soc.*, t. **45**, 1974, p. 144–150.

[30] SCHÜTZENBERGER (Marcel-Paul). — Quelques remarques sur une construction de Schensted, *Math. Scand.*, t. **12**, 1963, p. 117–128.

[31] SCHÜTZENBERGER (Marcel-Paul). — La correspondance de Robinson, *Combinatoire et représentation du groupe symétrique* [Actes Table Ronde C.N.R.S., Strasbourg. 1976], p. 59–113. — Berlin, Springer-Verlag, 1977 (*Lecture Notes in Math.*, **579**).

[32] STANLEY (Richard P.). — *Ordered Structures and Partitions*. — Providence, R.I., Amer. Math. Soc., 1972 (*Memoirs Amer. Math. Soc.*, **119**).

[33] STANLEY (Richard P.). — Binomial posets, Möbius inversion, and permutation enumeration, *J. Combinatorial Theory Ser. A*, t. **20**, 1976, p. 336–356.

[34] WYBOURNE (Brian G.). — *Symmetry principles and atomic spectroscopy*. — New York, Wiley 1970.

Annexe 1

TABLE DES (t, q)-TABLEAUX

Lorsque λ est un diagramme de Ferrers, le calcul du (t, q)-tableau $F_\lambda(t, q)$ a été expliqué dans la section 4. Pour un diagramme gauche $\lambda \otimes \mu$, on part des tables des (t, q)-tableaux $F_\nu(t, q)$ et on utilise la relation (3.4), ainsi qu'une table des coefficients de Littlewood-Richardson (*cf.* [15, 34]).

Dans les tables suivantes, les rubriques des colonnes sont les exposants r de t, et i de q du polynôme $F_{\lambda\mu}(t, q)$. La rubrique de ligne est le diagramme $\lambda \otimes \mu$.

Par exemple, dans la table $n = 5$, sur la ligne $2 \otimes 21$, on trouve les coefficients :

| |1 2 2 1|1 3 4 3 1| 1 1 | |

Se reportant au haut de la table, on obtient donc pour le polynôme $F_{2\otimes 21}(t, q)$ la valeur :

$$t(q + 2q^2 + 2q^3 + q^4) + t^2(q^3 + 3q^4 + 4q^5 + 3q^6 + q^7) + t^3(q^7 + q^8).$$

$n = 1$

$r \to$	0
$i \to$	0
1	1

↑ $\lambda \otimes \mu$

$n = 2$

$r \to$	0	1
$i \to$	0	1
2	1	
11		1
$1 \otimes 1$	1	1

↑ $\lambda \otimes \mu$

$n = 3$

$r \to$	0	1		2
$i \to$	0	1	2	3
3	1			
21		1	1	
111				1
$1 \otimes 2$	1	1	1	
$1 \otimes 11$		1	1	1
$2 \otimes 1$	1	1	1	
$11 \otimes 1$		1	1	1

↑ $\lambda \otimes \mu$

$n = 4$

$r \to$	0	1			2			3
$i \to$	0	1	2	3	3	4	5	6
4	1							
31		1	1	1				
22			1			1		
211					1	1	1	
1111								1
$1 \otimes 3$	1	1	1	1				
$1 \otimes 21$		1	2	1	1	2	1	
$1 \otimes 111$					1	1	1	1
$2 \otimes 2$	1	1	2	1		1		
$2 \otimes 11$		1	1	1	1	1	1	
$11 \otimes 2$		1	1	1	1	1	1	
$11 \otimes 11$			1		1	2	1	1
$3 \otimes 1$	1	1	1	1				
$21 \otimes 1$		1	2	1	1	2	1	
$111 \otimes 1$					1	1	1	1

↑ $\lambda \otimes \mu$

$n = 5$

$r \rightarrow$	0	1				2					3				4
$i \rightarrow$	0	1	2	3	4	3	4	5	6	7	6	7	8	9	10
5	1														
41		1	1	1	1										
32			1	1			1	1	1						
311						1	1	2	1	1					
2211							1	1	1			1	1		
2111											1	1	1	1	
11111															1
$1 \otimes 4$	1	1	1	1	1										
$1 \otimes 31$		1	2	2	1	1	2	3	2	1					
$1 \otimes 22$			1	1			2	2	2			1	1		
$1 \otimes 211$						1	2	3	2	1	1	2	2	1	
$1 \otimes 1111$											1	1	1	1	1
$2 \otimes 3$	1	1	2	2	1		1	1	1						
$2 \otimes 21$		1	2	2	1	1	3	4	3	1		1	1		
$2 \otimes 111$						1	1	2	1	1	1	1	1	1	
$11 \otimes 3$		1	1	1	1	1	1	2	1	1					
$11 \otimes 21$			1	1		1	3	4	3	1	1	2	2	1	
$11 \otimes 111$							1	1	1		1	2	2	1	1
$3 \otimes 2$	1	1	2	2	1		1	1	1						
$21 \otimes 2$		1	2	2	1	1	3	4	3	1		1	1		
$111 \otimes 2$						1	1	2	1	1	1	1	1	1	
$3 \otimes 11$		1	1	1	1	1	1	2	1	1					
$21 \otimes 11$			1	1		1	3	4	3	1	1	2	2	1	
$111 \otimes 11$							1	1	1		1	2	2	1	1
$4 \otimes 1$	1	1	1	1	1										
$31 \otimes 1$		1	2	2	1	1	2	3	2	1					
$22 \otimes 1$			1	1			2	2	2			1	1		
$211 \otimes 1$						1	2	3	2	1	1	2	2	1	
$1111 \otimes 1$											1	1	1	1	1

$\uparrow$
$\lambda \otimes \mu$

$n = 6$

$r \rightarrow$	0	1					2							3							4					5
$i \rightarrow$	0	1	2	3	4	5	3	4	5	6	7	8	9	6	7	8	9	10	11	12	10	11	12	13	14	15
6	1																									
51		1	1	1	1	1																				
42			1	1	1			1	1	2	1	1														
411							1	1	2	2	2	1	1													
33				1					1	1	1						1									
321								1	2	2	2	1			1	2	2	2	1							
222										1						1	1	1					1			
3111														1	1	2	2	2	1	1						
2211															1	1	2	1	1			1	1	1		
21111																					1	1	1	1	1	
111111																										1
$1 \otimes 5$	1	1	1	1	1	1																				
$1 \otimes 41$		1	2	2	2	1	1	2	3	4	3	2	1													
$1 \otimes 32$			1	2	1			2	4	5	4	2			1	2	3	2	1							
$1 \otimes 311$							1	2	4	4	4	2	1	1	2	4	4	4	2	1						
$1 \otimes 221$								1	2	3	2	1			2	4	5	4	2			1	2	1		
$1 \otimes 2111$														1	1	2	2	2	1	1	1	2	2	2	1	
$1 \otimes 11111$																					1	1	1	1	1	1
$2 \otimes 4$	1	1	2	2	2	1		1	1	2	1	1														
$2 \otimes 31$		1	2	3	2	1	1	3	6	7	6	3	1		1	2	3	2	1							
$2 \otimes 22$			1	1	1			2	3	5	3	2			1	3	3	3	1				1			
$2 \otimes 211$							1	2	4	4	4	2	1	1	3	5	6	5	3	1		1	1	1		
$2 \otimes 1111$														1	1	2	2	2	1	1	1	1	1	1	1	
$11 \otimes 4$		1	1	1	1	1	1	1	2	2	2	1	1													
$11 \otimes 31$			1	1	1		1	3	5	6	5	3	1	1	2	4	4	4	2	1						
$11 \otimes 22$				1				1	3	3	3	1			2	3	5	3	2			1	1	1		
$11 \otimes 211$								1	2	3	2	1		1	3	6	7	6	3	1	1	2	3	2	1	
$11 \otimes 1111$															1	1	2	1	1		1	2	2	2	1	1
$3 \otimes 3$	1	1	2	3	2	1		1	2	3	2	1					1									
$3 \otimes 21$		1	2	2	2	1	1	3	5	6	5	3	1		1	2	2	2	1							
$3 \otimes 111$							1	1	2	2	2	1	1	1	1	2	2	2	1							
$21 \otimes 3$		1	2	2	2	1	1	3	5	6	5	3	1		1	2	2	2	1							
$21 \otimes 21$			1	2	1		1	4	8	10	8	4	1	1	4	8	10	8	4	1		1	2	1		
$21 \otimes 111$								1	2	2	2	1		1	3	5	6	5	3	1	1	2	2	2	1	
$111 \otimes 3$							1	1	2	2	2	1	1	1	1	2	2	2	1							
$111 \otimes 21$								1	2	2	2	1		1	3	5	6	5	3	1	1	2	2	2	1	
$111 \otimes 111$										1					1	2	3	2	1		1	2	3	2	1	1

$\uparrow$

$\lambda \otimes \mu$

Annexe 2

TABLES DES NOMBRES $C(n; m, r, s, i, j)$

Les polynômes C_n définis par la relation (1.1) sont des polynômes à cinq variables :

$$C_n = C_n(z, t_1, t_2, q_2, q_2) = \sum C(n; m, r, s, i, j) z^m t_1^r t_2^s q_1^i q_2^j,$$

où $m = 1, 2, \ldots, n$; $r, s = 0, 1, \ldots, n-1$ et $i, j = 0, 1, \ldots, \binom{n}{2}$. Les tables qui suivent donnent les valeurs des coefficients $C(n; m, r, s, i, j)$ pour $n = 1, 2, 3, 4, 5$ et toutes les valeurs de m, ainsi que pour $n = 6$ et $m = 0, 1, 5, 6$. À chaque couple (n, m) tel que $0 \leq m \leq n/2$ et $1 \leq n \leq 6$ est associée une table des valeurs de $C(n; m, r, s, i, j)$. Les tables correspondant à $n = 6$ et $m = 2, 3, 4$ font apparaître des coefficients à trois chiffres et deviennent ainsi trop volumineuses pour la reproduction. De plus, la capacité mémoire du logiciel TEX est dépassée! On notera que r et i sont des indices de *ligne* et s et j des indices de *colonne*.

Se reportant, par exemple, à la table $n = 5, m = 1$, on constatera qu'à l'intersection de la *ligne* $r = 2$, $i = 5$ et de la *colonne* $s = 1$, $j = 3$, on trouve le nombre 8. Donc $C(5; 1, 2, 1, 5, 3) = 8$.

Lorsque $n/2 < m \leq n$, les coefficients $C(n; m, r, s, i, j)$ sont lus dans la table associée au couple $(n, n-m)$. Les indices de ligne r et i restent les mêmes, mais les indices de colonne s et j doivent être lus de droite à gauche, en prenant la numérotation des deux dernières lignes de la table.

Enfin, l'avant-dernière ligne du tableau (n, m) contient la somme des coefficients de chaque colonne, divisée par le coefficient binomial $\binom{n}{m}$:

$$C(n; m, \cdot, s, \cdot, j) / \binom{n}{m} = (1/\binom{n}{m}) \sum_{r,s} C(n; m, r, s, i, j).$$

Comme on peut le vérifier facilement (par exemple en utilisant l'interprétation de C_n en termes de permutations colorées décrite dans notre article précédent [8]), ces sommes divisées ne sont autres que les coefficients des *nombres q-Eulériens* $A_{n,s}(q)$ définis par CARLITZ [3, p. 336]) :

$$A_{n,s}(q) = \sum_j \left(C(n; m, \cdot, s, \cdot, j) / \binom{n}{m}\right) q^j.$$

La dernière ligne du tableau (n, m) contient la suite familière des nombres Eulériens :

$$A_{n,s} = A_{n,s}(1) = \sum_j C(n; m, \cdot, s, \cdot, j) / \binom{n}{m}.$$

$n = 1, m = 0$

$s \to$		0	r	i
$j \to$		0	↓	↓
0	0	1	0	0
↑	↑	0	$\leftarrow j$	
r	i	0	$\leftarrow s$	
$\Sigma \to$		1		
$\Sigma \to$		1		

$n = 1, m = 1$

$n = 2, m = 0$

$s \to$		0	1	r	i
$j \to$		0	1	↓	↓
0	0	1		0	0
1	1		1	1	1
↑	↑	1	0	$\leftarrow j$	
r	i	1	0	$\leftarrow s$	
$\Sigma \to$		1	1		
$\Sigma \to$		1	1		

$n = 2, m = 2$

$n = 2, m = 1$

$s \to$		0	1	r	i
$j \to$		0	1	↓	↓
0	0	1	1	0	0
1	1	1	1	1	1
↑	↑	1	0	$\leftarrow j$	
r	i	1	0	$\leftarrow s$	
$\Sigma \to$		1	1		
$\Sigma \to$		1	1		

$n = 2, m = 1$

$n = 3, m = 0$

$s \to$		0	1		2	r	i
$j \to$		0	1	2	3	↓	↓
0	0	1				0	0
1	1		1	1		1	1
	2		1	1		2	
2	3				1	3	2
↑	↑	3	2	1	0	$\leftarrow j$	
r	i	2	2		0	$\leftarrow s$	
$\Sigma \to$		1	1	1	1		
$\Sigma \to$		1	2		1		

$n = 3, m = 3$

$n = 3, m = 1$

$s \to$		0	1		2	r	i
$j \to$		0	1	2	3	↓	↓
0	0	1	1	1		0	0
1	1	1	2	2	1	1	1
	2	1	2	2	1	2	
2	3		1	1	1	3	2
↑	↑	3	2	1	0	$\leftarrow j$	
r	i	2	2		0	$\leftarrow s$	
$\Sigma \to$		1	1	1	1		
$\Sigma \to$		1	2		1		

$n = 3, m = 2$

$n = 4, m = 0$

$s \to$		0	1			2			3	r	i
$j \to$		0	1	2	3	3	4	5	6	↓	↓
0	0	1								0	0
	1		1	1	1					1	
1	2		1	2	1		1			2	1
	3		1	1	1					3	
	3					1	1	1		3	
2	4			1		1	2	1		4	2
	5					1	1	1		5	
3	6								1	6	3
↑	↑	6	5	4	3	3	2	1	0	$\leftarrow j$	
r	i	3	2			1			0	$\leftarrow s$	
$\Sigma \to$		1	3	5	3	3	5	3	1		
$\Sigma \to$		1	11			11			1		

$n = 4, m = 4$

$n = 4, m = 1$

$s \to$		0	1			2			3	r	i
$j \to$		0	1	2	3	3	4	5	6	↓	↓
0	0	1	1	1	1					0	0
	1	1	2	3	2	1	2	1		1	
1	2	1	3	5	3	2	4	2		2	1
	3	1	2	3	2	1	2	1		3	
	3		1	2	1	2	3	2	1	3	
2	4		2	4	2	3	5	3	1	4	2
	5		1	2	1	2	3	2	1	5	
3	6					1	1	1	1	6	3
↑	↑	6	5	4	3	3	2	1	0	$\leftarrow j$	
r	i	3	2			1			0	$\leftarrow s$	
$\Sigma \to$		1	3	5	3	3	5	3	1		
$\Sigma \to$		1	11			11			1		

$n = 4, m = 3$

$n = 4,\ m = 2$

	s →	0	1			2			3	r	i
	j →	0	1	2	3	3	4	5	6	↓	↓
0	0		1	1	1	1	1	1		0	0
	1	1	2	4	2	2	4	2	1	1	
1	2	1	4	6	4	4	6	4	1	2	1
	3	1	2	4	2	2	4	2	1	3	
	3	1	2	4	2	2	4	2	1	3	
2	4	1	4	6	4	4	6	4	1	4	2
	5	1	2	4	2	2	4	2	1	5	
3	6		1	1	1	1	1	1		6	3
↑	↑	6	5	4	3	3	2	1	0	← j	
r	i	3	2			1			0	← s	
Σ →		1	3	5	3	3	5	3	1		
Σ →		1	11			11			1		

$n = 4,\ m = 2$

$n = 5,\ m = 0$

	s →	0	1				2					3				4	r	i
	j →	0	1	2	3	4	3	4	5	6	7	6	7	8	9	10	↓	↓
0	0	1															0	0
	1		1	1	1	1											1	
1	2		1	2	2	1		1	1	1							2	1
	3		1	2	2	1		1	1	1							3	
	4		1	1	1	1											4	
	3						1	1	2	1	1						3	
	4			1	1		1	3	4	3	1		1	1			4	
2	5			1	1		2	4	6	4	2		1	1			5	2
	6			1	1		1	3	4	3	1		1	1			6	
	7						1	1	2	1	1						7	
	6											1	1	1	1		6	
3	7							1	1	1		1	2	2	1		7	3
	8							1	1	1		1	2	2	1		8	
	9											1	1	1	1		9	
4	10															1	10	4
↑	↑	10	9	8	7	6	7	6	5	4	3	4	3	2	1	0	← j	
r	i	4	3				2					1				0	← s	
Σ →		1	4	9	9	4	6	16	22	16	6	4	9	9	4	1		
Σ →		1	26				66					26				1		

$n = 5,\ m = 5$

$n = 5$, $m = 1$

	$s \rightarrow$	0	1				2					3				4	r	i
	$j \rightarrow$	0	1	2	3	4	3	4	5	6	7	6	7	8	9	10	↓	↓
0	0	1	1	1	1	1											0	0
1	1	1	2	3	3	2	1	2	3	2	1						1	1
	2	1	3	6	6	3	2	6	8	6	2		1	1			2	
	3	1	3	6	6	3	2	6	8	6	2		1	1			3	
	4	1	2	3	3	2	1	2	3	2	1						4	
2	3		1	2	2	1	2	4	6	4	2	1	2	2	1		3	2
	4		2	6	6	2	4	12	16	12	4	2	6	6	2		4	
	5		3	8	8	3	6	16	22	16	6	3	8	8	3		5	
	6		2	6	6	2	4	12	16	12	4	2	6	6	2		6	
	7		1	2	2	1	2	4	6	4	2	1	2	2	1		7	
3	6						1	2	3	2	1	2	3	3	2	1	6	3
	7			1	1		2	6	8	6	2	3	6	6	3	1	7	
	8			1	1		2	6	8	6	2	3	6	6	3	1	8	
	9						1	2	3	2	1	2	3	3	2	1	9	
4	10											1	1	1	1	1	10	4
↑	↑	10	9	8	7	6	7	6	5	4	3	4	3	2	1	0	$\leftarrow j$	
r	i	4	3				2					1				0	$\leftarrow s$	
$\Sigma \rightarrow$		1	4	9	9	4	6	16	22	16	6	4	9	9	4	1		
$\Sigma \rightarrow$		1	26				66					26				1		

$n = 5$, $m = 4$

$n = 5$, $m = 2$

	$s \rightarrow$	0	1				2					3				4	r	i
	$j \rightarrow$	0	1	2	3	4	3	4	5	6	7	6	7	8	9	10	↓	↓
0	0		1	1	1	1	1	1	2	1	1						0	0
1	1	1	2	4	4	2	2	5	7	5	2	1	2	2	1		1	1
	2	1	4	8	8	4	5	12	17	12	5	2	5	5	2		2	
	3	1	4	8	8	4	5	12	17	12	5	2	5	5	2		3	
	4	1	2	4	4	2	2	5	7	5	2	1	2	2	1		4	
2	3	1	2	5	5	2	2	8	10	8	2	2	5	5	2	1	3	2
	4	1	5	12	12	5	8	22	30	22	8	5	12	12	5	1	4	
	5	2	7	17	17	7	10	30	40	30	10	7	17	17	7	2	5	
	6	1	5	12	12	5	8	22	30	22	8	5	12	12	5	1	6	
	7	1	2	5	5	2	2	8	10	8	2	2	5	5	2	1	7	
3	6		1	2	2	1	2	5	7	5	2	2	4	4	2	1	6	3
	7		2	5	5	2	5	12	17	12	5	4	8	8	4	1	7	
	8		2	5	5	2	5	12	17	12	5	4	8	8	4	1	8	
	9		1	2	2	1	2	5	7	5	2	2	4	4	2	1	9	
4	10						1	1	2	1	1	1	1	1	1		10	4
↑	↑	10	9	8	7	6	7	6	5	4	3	4	3	2	1	0	$\leftarrow j$	
r	i	4	3				2					1				0	$\leftarrow s$	
$\Sigma \rightarrow$		1	4	9	9	4	6	16	22	16	6	4	9	9	4	1		
$\Sigma \rightarrow$		1	26				66					26				1		

$n = 5$, $m = 3$

$n = 6,\ m = 0$

s →		0	1					2							3							4					5	r ↓	i ↓
j →		0	1	2	3	4	5	3	4	5	6	7	8	9	6	7	8	9	10	11	12	10	11	12	13	14	15		
0	0	1																										0	0
1	1		1	1	1	1	1																					1	1
	2		1	2	2	2	1		1	1	2	1	1															2	
	3		1	2	3	2	1		1	2	3	2	1					1										3	
	4		1	2	2	2	1		1	1	2	1	1															4	
	5		1	1	1	1	1																					5	
2	3							1	1	2	2	2	1	1														3	2
	4			1	1	1		1	3	5	6	5	3	1		1	2	2	2	1								4	
	5			1	2	1		2	5	10	11	10	5	2		2	4	5	4	2								5	
	6			2	3	2		2	6	11	14	11	6	2		2	5	6	5	2				1				6	
	7			1	2	1		2	5	10	11	10	5	2		2	4	5	4	2								7	
	8			1	1	1		1	3	5	6	5	3	1		1	2	2	2	1								8	
	9							1	1	2	2	2	1	1														9	
3	6														1	1	2	2	2	1	1							6	3
	7								1	2	2	2	1		1	3	5	6	5	3	1		1	1	1			7	
	8								2	4	5	4	2		2	5	10	11	10	5	2		1	2	1			8	
	9				1				2	5	6	5	2		2	6	11	14	11	6	2		2	3	2			9	
	10								2	4	5	4	2		2	5	10	11	10	5	2		1	2	1			10	
	11								1	2	2	2	1		1	3	5	6	5	3	1		1	1	1			11	
	12														1	1	2	2	2	1	1							12	
4	10																					1	1	1	1	1		10	4
	11															1	1	2	1	1		1	2	2	2	1		11	
	12										1					1	2	3	2	1		1	2	3	2	1		12	
	13															1	1	2	1	1		1	2	2	2	1		13	
	14																					1	1	1	1	1		14	
5	15																										1	15	5
↑	↑	15	14	13	12	11	10	12	11	10	9	8	7	6	9	8	7	6	5	4	3	5	4	3	2	1	0	← j	
r	i	5	4					3							2							1					0	← s	
Σ →		1	5	14	19	14	5	10	35	66	80	66	35	10	10	35	66	80	66	35	10	5	14	19	14	5	1		
Σ →		1	57					302							302							57					1		

$n = 6,\ m = 6$

$n = 6,\ m = 1$

$s \to$		0	1					2							3							4					5	r	i
$j \to$		0	1	2	3	4	5	3	4	5	6	7	8	9	6	7	8	9	10	11	12	10	11	12	13	14	15	↓	↓
0	0	1	1	1	1	1	1																					0	0
	1	1	2	3	3	3	2	1	2	3	4	3	2	1														1	
	2	1	3	6	7	6	3	2	6	10	13	10	6	2		1	2	3	2	1								2	
1	3	1	3	7	9	7	3	2	8	14	18	14	8	2		2	4	6	4	2								3	1
	4	1	3	6	7	6	3	2	6	10	13	10	6	2		1	2	3	2	1								4	
	5	1	2	3	3	3	2	1	2	3	4	3	2	1														5	
	3		1	2	2	2	1	2	4	7	8	7	4	2	1	2	4	4	4	2	1							3	
	4		2	6	8	6	2	4	13	24	29	24	13	4	2	8	16	19	16	8	2		1	2	1			4	
	5		3	10	14	10	3	7	24	45	54	45	24	7	4	16	32	38	32	16	4		2	4	2			5	
2	6		4	13	18	13	4	8	29	54	66	54	29	8	4	19	38	46	38	19	4		3	6	3			6	2
	7		3	10	14	10	3	7	24	45	54	45	24	7	4	16	32	38	32	16	4		2	4	2			7	
	8		2	6	8	6	2	4	13	24	29	24	13	4	2	8	16	19	16	8	2		1	2	1			8	
	9		1	2	2	2	1	2	4	7	8	7	4	2	1	2	4	4	4	2	1							9	
	6							1	2	4	4	4	2	1	2	4	7	8	7	4	2	1	2	2	2	1		6	
	7			1	2	1		2	8	16	19	16	8	2	4	13	24	29	24	13	4	2	6	8	6	2		7	
	8			2	4	2		4	16	32	38	32	16	4	7	24	45	54	45	24	7	3	10	14	10	3		8	
3	9			3	6	3		4	19	38	46	38	19	4	8	29	54	66	54	29	8	4	13	18	13	4		9	3
	10			2	4	2		4	16	32	38	32	16	4	7	24	45	54	45	24	7	3	10	14	10	3		10	
	11			1	2	1		2	8	16	19	16	8	2	4	13	24	29	24	13	4	2	6	8	6	2		11	
	12							1	2	4	4	4	2	1	2	4	7	8	7	4	2	1	2	2	2	1		12	
	10														1	2	3	4	3	2	1	2	3	3	3	2	1	10	
	11								1	2	3	2	1		2	6	10	13	10	6	2	3	6	7	6	3	1	11	
4	12								2	4	6	4	2		2	8	14	18	14	8	2	3	7	9	7	3	1	12	4
	13								1	2	3	2	1		2	6	10	13	10	6	2	3	6	7	6	3	1	13	
	14														1	2	3	4	3	2	1	2	3	3	3	2	1	14	
5	15																					1	1	1	1	1	1	15	5
↑	↑	15	14	13	12	11	10	12	11	10	9	8	7	6	9	8	7	6	5	4	3	5	4	3	2	1	0	$\leftarrow j$	
r	i	5	4					3							2							1					0	$\leftarrow s$	

$n = 6,\ m = 5$

RAISING OPERATORS AND YOUNG'S RULE

A. M. Garsia (*).

ABSTRACT: In a few mysterious lines of QSA VI Alfred Young introduced the notion of *Raising operator*. Very sketchily he goes through some rather remarkable manipulations to derive what is now sometimes referred to as Young's rule. In previous joint work [3] we have made rigorous a portion of Young's argument by interpreting these operators as acting on Ferrers' diagrams. Other authors, somewhat later have presented similar interpretations (see [8] and [10]). In the present work we bring some evidence to suggest that in Young's interpretation, raising operators acted on Tableaux rather than shapes. With this view, we can finally put together a rigorous version of the remaining unexplained portion of Young's treatment. This effort has also led us to a remarkably elementary and very combinatorial proof of Pieri's rule.

Introduction

One should only read the few sentences in the introduction to G. James book [7] to appreciate some of our contemporaries attitude towards the work of A. Young. Indeed, Young's style of writing is quite forbidding. His definitions are vague, his proofs are sketchy and sometimes only carried out in a few simple cases. One gets the impression that he may have discovered the result by experimentation, convinced himself of its general validity but only succeded in proving it in the cases presented. Matters are made even worse by the fact he often expects (without explicit mention) the reader to be familiar with definitions and arguments given in his previous writings on the subject.

The best attitude to have in regards to Young's work is to view it not as a place to learn the subject (a much less painful introduction to QSA can be found in Rutheford [9]) but rather regard it as a collection of hints and a source of combinatorial inspiration. With this view, one discovers that Young could do with the algebra of the symmetric group pretty much what he wanted or for that matter whatever anyone else would ever want to do.

This work is concerned with only a few pages of Young's writings. (QSA IV p. 259-261 and QSA VI p. 196-201). Starting from the cryptic sentences that we find there, we have been led to remarkably simple proofs of two basic results in the representation theory of the symmetric group. Namely: *Young's rule* and *Pieri's rule*.

Perhaps we should recall that Young's rule is concerned with the decomposition of the character of the permutation representation induced by the action of S_n on the left cosets of a Young subgroup (a product of smaller symmetric groups). Pieri's rule is traditionally stated as a combinatorial recipe for obtaining the Schur function expansion of the product of a Schur function by a homogeneous symmetric function. In Young's context it simply gives the decomposition of the representation of S_{h+k} obtained by inducing from $S_k \times S_h$

1980 Mathematics Classification. 05A15, 05A19, 20C30, 20C35, 68-04, 68C05.
(*) Work supported by NSF grant at the Univ. of Cal. San Diego and by ONR grant at the MIT.

the outer tensor product of an irreducible representation of S_k by the trivial representation of S_h .

The contents are divided into three sections. In the first we review some notation and give Young's *natural units* in a form which is convenient for our developments. In the second section give our derivation of Pieri's rule and derive some remarkable identities concerning Young's idempotents. In the final section we we present our new interpretation of Young's *proof* of Young's rule and complete Young's argument by means of the identities obtained in the second section.

Whether or not our interpretation of those pages of QSA is in agreement with what Young had in mind, what matters is that it is in reading those pages that we have been led to the present developments. We therefore hope that our experience will stimulate further readings of his remarkable works.

We wish to acknowledge here our gratitude towards Jeff Remmel and Luc Favreau who patiently participated in endless discussions as to the possible interpretations of Young's cryptic statements. Their specific mathematical contributions to this paper will be quoted later and in context.

1. The natural units.

Let G be a finite group and let $N(G)$, $k(G)$ denote respectively the order and the number of conjugacy classes of G . We shall also denote by $A(G)$ the corresponding group algebra.

Let Λ denote an index set of cardinality $k(G)$ and

$$e^{\lambda}_{i,j} \quad \text{for } \lambda \in \Lambda \qquad \text{and for} \quad i\,,j = 1,2,\ldots,m_\lambda$$

be elements of $A(G)$ satisfying the following conditions

1) each $e^{\lambda}_{i,j}$ is different from zero.

2) $e^{\lambda}_{i,j}\, e^{\mu}_{r,s} = 0$ if $\lambda \neq \mu$ or $j \neq r$.

1.1

3) $e^{\lambda}_{i,j}\, e^{\lambda}_{j,k} = e^{\lambda}_{i,k}$ for all i,j,k and λ .

4) $\sum_{\lambda\in\Lambda} m_\lambda^2 = N(G)$

Then it is quite elementary to show that every element $f \in A(G)$ can be expanded in the form

$$f = \sum_{\lambda\in\Lambda} h_\lambda \sum_{i,j=1}^{m_\lambda} f\, e^{\lambda}_{j,i} \big|_\epsilon \; e^{\lambda}_{i,j} \,, \tag{1.2}$$

where $|$ denotes the operation of taking a coefficient, ϵ denotes the identity of G and h_λ can be computed from the identities:

$$\frac{1}{h_\lambda} = e^{\lambda}_{i,i} \big|_\epsilon \qquad i=1,2,..,m_\lambda \;. \tag{1.3}$$

It is also quite easy to show that we must have

$$e^{\lambda}_{i,j} \big|_\epsilon = 0 \qquad \text{for all } \lambda\in\Lambda \text{ and all } i\neq j \;. \tag{1.4}$$

Finally, we can easily show that the coefficients $a^{\lambda}_{i,j}(\sigma)$ in the expansion

$$\sigma = \sum_{\lambda} \sum_{i,j} a^{\lambda}_{i,j}(\sigma)\, e^{\lambda}_{i,j} \qquad (\sigma\in G) \tag{1.5}$$

give a complete set of irreducible representations of G . Moreover, the fundamental characters of G are given by the expressions

$$\chi^{\lambda} = h_\lambda \sum_{i=1}^{n_\lambda} e^{\lambda}_{i,i} \;. \tag{1.6}$$

Such units as the e_{ij}^{λ} are in principle all that we need to have to study the representations of G. Of course for an arbitrarily given G these units are hard to come by. But remarkably, for the symmetric group S_n, Young was able to write down explicitly two different sets of them. They are referred to now as the *natural* and the *semi orthogonal units*. The latter are so named since they give rise to a set of unitary irreducible representations.

Young succeded in establishing quite a number of basic properties of the representations of S_n simply by working with the natural units. In particular his approach to the proof of Young's rule is based on some remarkable properties of these units.

Nevertheless, his constructions are unduly intricate in this case and we shall have to give here a simpler and more convenient definition. To this end we need to introduce some notation.

We shall follow as closely as possible Young's notation, with some exceptions. For typographic reasons it is more convenient to represent Ferrers' diagrams as left justified rows of dots, of lengths weakly increasing from top to bottom. Rows will be numbered from bottom to top. As customary we shall identify partitions with Ferrers' diagrams. The notation $\lambda \vdash n$ as usual indicates that λ is a partition of n, or equivalently that the corresponding Ferrers' diagram has n dots. A tableau of shape λ is obtained by replacing the dots of the Ferrrers' diagram corresponding to λ by arbitrary integers. We shall follow the French notation and call a tableau *standard*, when the dots are replaced (if $\lambda \vdash n$) by the successive integers $1,2,...,n$ so that they result in increasing order from left to right on the rows and from bottom to top on the columns. Column strict tableaux are analogously defined. That is they are tableaux with entries weakly increasing in rows from left to right and strictly increasing in columns from bottom to top. With these exceptions, whatever concept we shall use here undefined, its definition may be found in [4], [5], or [6].

In the picture below we depict the Ferrers' diagram corresponding to the partition $\lambda = (1,3,3,4)$, a standard tableau T of shape λ and a column strict tableau S of the same shape.

$$\begin{array}{llll} \bullet & & & \\ \bullet & \bullet & \bullet & \\ \bullet & \bullet & \bullet & \\ \bullet & \bullet & \bullet & \bullet \end{array} \qquad T = \begin{array}{llll} 6 & & & \\ 4 & 8 & 10 & \\ 2 & 5 & 9 & \\ 1 & 3 & 7 & 11 \end{array} \qquad S = \begin{array}{llll} 4 & & & \\ 3 & 3 & 4 & \\ 2 & 2 & 3 & \\ 1 & 1 & 1 & 2 \end{array}$$

The multiset of entries of a tableau is usually referred to as the *content of the tableau*. For instance, the content of the tableau S given above is $1^3\, 2^3\, 3^3\, 4^2$.

For a given set A of integers we shall denote by $[A]$ the formal sum of all permutations of A. In other words, if S_A denotes the symmetric group of A then

$$[A] = \sum_{\sigma \in S_A} \sigma . \qquad 1.7$$

It is convenient to write these permutations as products of cycles. For instance, we write

$$[248] = \epsilon + (2,4) + (2,8) + (4,8) + (2,4,8) + (2,8,4) .$$

We also set

$$[A]' = \sum_{\sigma \in S_A} \text{sign}(\sigma)\ \sigma . \qquad 1.8$$

In particular

$$[248]' = \epsilon - (2,4) - (2,8) - (4,8) + (2,4,8) + (2,8,4) .$$

This given, if the rows of a tableau T are $R_1, R_2, \cdots, R_h$ and its columns are $C_1, C_2, \cdots, C_k$ then Young sets

$$P(T) = [R_1]\,[R_2] \cdots [R_h] \quad , \quad N(T) = [C_1]'\,[C_2]' \cdots [C_k]' .$$

Thus for the tableau T given above we have

$$P(T) = [1\ 3\ 7\ 11]\ [2\ 5\ 9]\ [4\ 8\ 10] \quad \text{and} \quad N(T) = [1\ 2\ 4\ 6]'\ [3\ 5\ 8]'\ [7\ 9\ 10]' .$$

For an *injective* (all entries distinct) tableau T of shape λ, we shall set here

$$\gamma(T) = \frac{P(T)\,N(T)}{h_\lambda} \qquad 1.9$$

where h_λ is the *product of the hooks* of the Ferrers diagram of λ. The constant h_λ was first computed by Young in QSA II (pp. 364-367), it is precisely the integer that makes $\gamma(T)$ into an idempotent. This latter fact is not trivial (see [4] for a proof). If $\lambda \vdash n$ and h_λ is defined as the ratio

$$h_\lambda = \frac{n!}{n_\lambda}$$

where n_λ denotes the number of standard tableaux of shape λ, then the idempotency of $\gamma(T)$ can be shown with a reasonable amount of work (see [4]).

For two standard tableaux T_1 and T_2 (not necessarily of the same shape or size) we shall let $T_1 \cap T_2$ denote their common subtableau. That is, if the integers $1,2,..,i$ occupy the same positions in T_1 and T_2 but $i+1$ does not, then $T_1 \cap T_2$ is the subtableau of T_1 (or T_2) containing $1,2,..,i$. We shall also call $i+1$ the *first letter of disagreement* between T_1 and T_2.

This given, we shall say that T_1 precedes T_2 if the first letter of disagreement between T_1 and T_2 is higher in T_2 than it is in T_1. This total order of standard tableaux is introduced in QSA IV (p. 258). We shall thus refer to it as the *Young order*. Let then

$$T_1^\lambda,\ T_2^\lambda,\ \cdots,\ T_{n_\lambda}^\lambda,$$

denote the standard tableaux of shape λ arranged in the Young order. It is also convenient to denote by σ_{ji}^λ the permutation that sends the tableau T_i^λ into the tableau T_j^λ. For, simplicity we shall sometimes omit the superscript λ whenever dealing only with tableaux of the same shape. Of course we intend a permutation σ to act on a tableau T by simply replacing the entry i with its image σ_i. With these conventions, we can easily derive that we have

$$P(\sigma T) = \sigma\, P(T)\, \sigma^{-1}\ ,\quad N(\sigma T) = \sigma\, N(T)\, \sigma^{-1}\ ,\quad \gamma(\sigma T) = \sigma\, \gamma(T)\, \sigma^{-1}\ . \qquad 1.11$$

Now, a very elementary argument, (see for instance [9]) shows that we have

$$N(T_j)\, P(T_i) = 0\ ,\quad \text{for all } j > i\ , \qquad 1.12$$

and therefore we must also have

$$\gamma(T_j)\, \gamma(T_i) = 0\ ,\quad \text{for all } j > i\ , \qquad 1.13$$

In general, this may not hold for $j < i$. Nevertheless, for some shapes (such as *hooks* for instance) 1.12 does hold for all $i \neq j$. In this case it is easy to show, using 1.11, that the units

$$e_{ij}^\lambda = P(T_i^\lambda)\, \sigma_{ij}^\lambda\, N(T_j) / h_\lambda$$

do satisfy the conditions 2) and 3) of 1.1.

To compensate for the fact that 1.12 does not always hold as desired, Young sets

$$e_{ij}^\lambda = P(T_i^\lambda)\, \sigma_{ij}^\lambda\, N(T_j)\, M_j^\lambda / h_\lambda\ , \qquad 1.14$$

where M_j^λ are elements of the group algebra put together precisely to make the relations 2) and 3) valid without exception.

The only thing that remains to be verified is condition 1.1 4). However, the identity

$$n! = \sum_{\lambda \vdash n} n_\lambda^2$$

when n_λ gives the number of standard tableaux of shape λ is well known and easy to prove. Indeed, a very elementary and quite straightforward proof is given by Young in QSA II.

Young's definition of the M_j^λ is rather messy and difficult to use. Yet in several occasions Young uses these factors in a very surprising manner, indicating that he knew a lot more about them than he ever explicitely stated. Now, it develops that if we set

$$M_j^\lambda = (1-\gamma(T_{j+1}^\lambda))\,(1-\gamma(T_{j+2}^\lambda))\,(1-\gamma(T_{n_\lambda}^\lambda)) \ , \qquad 1.15$$

then conditions 1), 2) and 3) of 1.1 are easily shown (using 1.11 and 1.13) to hold without exception.

This will be our construction of the natural units. We shall see that this definition makes the natural units a very convenient tool for studying the representations of the symmetric groups.

In Young's terminology, $P(T)$ is the *row* group of T and $N(T)$ is the *column* group. We shall indicate, as it has become customary, by α and β generic permutations of the row and column groups respectively. Now a very basic fact (see [9] for a proof) is that for any two tableaux T_1 and T_2 of the same shape the inequality

$$N(T_1)\,P(T_2) \neq 0$$

implies the existence of permutations α_1, β_1 and α_2, β_2 respectively of the row and column groups of T_1 and T_2 giving

$$T_1 = \beta_1\,\alpha_2\,T_2 = \alpha_1\,\beta_1\,T_2 = \alpha_2\,\beta_2\,T_2 \ ,$$

from which we derive that

$$\begin{aligned} P(T_1)\,N(T_1)\,P(T_2)\,N(T_2) &= \mathrm{sign}(\beta_1)\,P(T_1)\,N(T_1)\,\beta_1\,\alpha_2\,P(T_2)\,N(T_2) = \\ &= P(T_1)\,N(T_1)\,P(T_1)\,N(T_1)\,\beta_1\,\alpha_2 \ . \end{aligned}$$

Thus using the idempotency of every $\gamma(T)$ we get

$$\gamma(T_1)\,\gamma(T_2) = \mathrm{sign}(\beta_1)\,\gamma(T_1) \ .$$

Now omitting the superscript λ let us set as Young does

$$E_{ij} = P(T_i)\,\sigma_{ij}\,N(T_j) \qquad 1.19$$

To show that our units are the same as those given by Young we need only show the basic

Lemma 1.1

The following identities hold for the natural units corresponding to shape λ *:*

$$E_{rs} = h_\lambda \sum_{j=s}^{n_\lambda} b_{js}\,e_{rj} \ . \qquad 1.20$$

where b_{js} *is different from zero only if the permutation* σ_{js} *is expressible in the form*

$$\sigma_{js} = \beta_{js}\,\alpha_{js} \qquad 1.21$$

with the last two factors elements of the column and row groups of T_j *and in the latter case we have*

$$b_{js} = \mathrm{sign}(\beta_{js}) \qquad 1.22$$

Proof

To simplify our notation let us set

$$P_i = P(T_i^\lambda) \ , \quad N_i = N(T_i^\lambda) \ , \quad \gamma_i = \gamma(T_i^\lambda) \ , \quad e_i = e_{i,i}^\lambda \ .$$

Then from 1.15 and 1.2 with $f = E_{rs}$ we get (by repetitive uses of 1.11)

$$\begin{aligned} e_i\,E_{rs}\,e_j &= h_\lambda\,E_{rs}\,e_{ji}\,|_\epsilon\,e_{ij} = \\ &= P_r\,\sigma_{rs}\,N_s\,P_j\,\sigma_{ji}\,N_i\,(1-\gamma_{i+1})\cdots(1-\gamma_{n_\lambda})\,|_\epsilon\,e_{ij} = \\ &= h_\lambda^2\,\gamma_r\,\sigma_{rs}\,\sigma_{ji}\,\gamma_i\,(1-\gamma_{i+1})\cdots(1-\gamma_{n_\lambda})\,|_\epsilon\,e_{ij} = \end{aligned}$$

$$= h_\lambda^2 \sigma_{ji} \gamma_i (1-\gamma_{i+1}) \cdots (1-\gamma_{n_\lambda}) \gamma_r \sigma_{rs} |_\epsilon e_{ij} .$$

Now using 1.12 it is easy to derive that

$$\gamma_i(1-\gamma_{i+1}) \cdots (1-\gamma_{n_\lambda}) \gamma_r = \begin{cases} \gamma_r & \text{if } i = r , \\ 0 & \text{otherwise} \end{cases}$$

This gives

$$\begin{aligned} e_r E_{rs} e_j &= h_\lambda^2 \sigma_{jr} \gamma_r \sigma_{rs} |_\epsilon e_{rj} = h_\lambda^2 \gamma_j \sigma_{jr} \sigma_{rs} |_\epsilon e_{rj} = \\ &= h_\lambda^2 \gamma_j \sigma_{js} |_\epsilon e_{rj} = h_\lambda P_j N_j \sigma_{js} |_\epsilon e_{rj} = \\ &= h_\lambda b_{js} e_{rj} . \end{aligned}$$

Now clearly either σ_{js} is of the form 1.21 and b_{js} is given by 1.22 or this term will necessarily be equal to zero. Thus the lemma follows from 1.2 .

Comparing our formula 1.20 with the expressions obtained for the natural units in Young's QSA IV (p. 258) or Rutheford ([9] P. 51) we can easily derive that our units are the same as Young's.

Remark 1.1

Of course we see from the definition of the b_{js} that $b_{ss} = 1$. Thus the matrix

$$B = \| b_{js} \| \qquad (b_{js} = 0 \quad \text{for} \quad j < s)$$

has determinant one, and therefore is invertible over the integers.

Introducing the matrices

$$e = \| e_{ij} \| \quad , E = \| E_{ij} \| \quad , L = \| L_{ij} \| = B^{-1} , \tag{1.23}$$

we can write 1.22 in the form

$$E = h_\lambda e B \tag{1.24}$$

or equivalently

$$e = \frac{1}{h_\lambda} E L \tag{1.25}$$

and in component form:

$$e_{ij} = \frac{1}{h_\lambda} \sum_{k=j}^{n_\lambda} E_{ik} L_{kj} . \tag{1.26}$$

Remark 1.2

The presence of denominators in some of these formulas has led several workers in this area to the mistaken belief that Young's developments concerning the natural representation are only valid in characteristic zero. Actually, it develops that both Specht work on the so called *Specht modules* and Rota's *Straightening Formula*, which are generally regarded as characteristic free developments can be related to the natural units. In particular, the derivation of Specht Modules that is given in [7], in each of its significant steps, can be identified (see [1]) with Young's derivation of the natural. Thus, when James' polytabloids are taken as basis the resulting representation matrices are of course identical with those originally given by Young. Familiarity with these two derivations, reveals that Young's is not only more elementary but also simpler. It should be mentioned that the natural representation can also be used to put together an efficient algorithm for carrying out Rota's straightening.

The present development should help making all this a bit less surprising. For, we can at the least deduce from formula 1.25 that there are no denominators in the final expression giving the natural representation. In fact, formulas 1.5 and 1.25 combined give

$$a_{ij}^{\lambda}(\sigma) = \sum_{k=i}^{n_\lambda} E_{jk} L_{ki} \Big|_{\sigma^{-1}} . \tag{1.27}$$

We shall not dwell any further into this here since we come back to it with greater detail in [1].

Remark 1.3

There is an interesting partial order of standard tableaux which underlies the Young order, it is worthwhile studying it on its own merits. Given two standard tableaux T_1 , T_2 let us set

$$T_1 \rightarrow T_2 \tag{1.28}$$

if and only if

$$N(T_1)\, P(T_2) \neq 0. \tag{1.29}$$

Note that if T_2 comes before T_1 in the Young order then there is a pair of letters that is vertical in T_1 and horizontal in T_2 and the product in 1.29 is then equal to zero. Thus the Young order is a linear extension of the relation $T_1 \rightarrow T_2$. Note further that in a chain

$$T_1 \rightarrow T_2 \rightarrow T_3 \rightarrow \cdots \rightarrow T_k \rightarrow \cdots \tag{1.30}$$

the common subtableaux

$$T_1 \cap T_2 \,,\; T_1 \cap T_3 \,,\; \cdots \,,\; T_1 \cap T_k$$

form a weakly decreasing sequence. Indeed we see, for instance, that if $T_1 \cap T_2 \supseteq T_2 \cap T_3$ then the first letter of disagreement between T_3 and T_2 is the same as that between T_3 and T_1 and is higher in T_3 that in T_2 and a fortiori than in T_1 . This forces $T_1 \cap T_2 \supseteq T_1 \cap T_3$. On the other hand, if $T_1 \cap T_2 \subset T_2 \cap T_3$ (strictly) then T_3 as far as T_1 is concerned behaves exactly like T_2 and $T_1 \cap T_3 = T_1 \cap T_2$.

Thus the only way that a chain as in 1.30 can loop back on itself (say $T_k = T_1$) is that all of the intermediate tableaux contain T_1 as a subtableau. This means that for tableaux of the same size we can define a partial order by taking the transitive closure of " $\rightarrow$ ".

Remark 1.4

A useful property of tableaux idempotents is that they do reflect in some weak sense tableau containment. Indeed, if T_1 is a subtableau of T_2 then using the fact that the row and column groups of T_1 are subgroups of those of T_2 we can (by means of two coset decompositions) write

$$P(T_2)\, N(T_2) = P(T_1)\, R\, C\, N(T_1)$$

with R and C sums of coset representatives respectively lying in the row and column groups of T_2

2. Pieri's rule.

Before we can proceed we need to introduce some notation. Let us go back for a moment to the general case of a finite group G . For a given element f of the group algebra $A(G)$ set

$$\Gamma_G f = \sum_{\sigma \in G} \sigma f \sigma^{-1} \tag{2.1}$$

It is immediate that the operator Γ_G is linear and maps $A(G)$ into its center. Moreover, we see that for any two elements $f\,,\, g \in A(G)$ we have

$$\Gamma_G f g = \Gamma_G g f .$$

Using this simple fact we can easily verify that for any set of units e_{ij}^{λ} satisfying 1.1 we have:

$$\Gamma_G e_{ij}^{\lambda} = 0 \qquad \text{(for } i \neq j \text{)} \tag{2.2}$$

$$\Gamma_G e_{i,i}^{\lambda} = \chi^{\lambda} \qquad \text{(for } i = 1,..,m_\lambda \text{)} \tag{2.3}$$

Going back to the symmetric groups, it will be convenient to use Γ_n rather than Γ_G for the case when $G = S_n$. Now, it is easy to see that 2.3 in the case of the natural units reduces to the statement that

$$\Gamma_n \gamma(T) = \chi^\lambda$$

holds for any tableau T of shape $\lambda \vdash n$. Indeed, it is clear that the left hand side of this identity is independent of the particular T that is chosen. But in the case that $T = T^\lambda_{n_\lambda}$ we have

$$\gamma(T^\lambda_{n_\lambda}) = e_{n_\lambda . n_\lambda}$$

Here and in the rest of this section, for $n = a + b$ we shall let T_A denote a standard tableau of shape λ in the integers 1,2,..,a and B denote the set of consecutive integers $a+1, a+2, \ldots, n$. We shall say that a standard tableau T, in the integers 1,2,..,n is *B-pieri over* T_A, if and only if T can be obtained by adding the elements of B to T_A so that no two of them fall in the same column. Similarly a shape μ is said to be *b-pieri over* λ if the Ferrers diagram of μ can be obtained by adding to the diagram of λ b dots no two on the same column.

For a given $\lambda \vdash a$, let A^λ denote the irreducible representation of the symmetric group [1..a] corresponding to λ. Furthermore, let the symbol $A^\lambda\, a \uparrow n$ denote the representation of the symmetric group [1..n] obtained by inducing from [1..a]×[a+1..n] the tensor product of A^λ by the trivial representation of [B].

This given, the following result is basic in the representation theory of the symmetric groups.

PIERI'S RULE

The irreducible constituents of $A^\lambda\, a \uparrow n$ *are the representations* A^μ *corresponding to the partitions* μ *that are b-pieri over* λ *and each occurs with multiplicity one.*

In Young's work this result appears as an identity involving characters of symmetric groups. It is expressed there in terms of the operator Γ_n. Namely, Young states (see QSA IV p. 260) that for any $\lambda \vdash a$ and any tableau T_A of shape λ as indicated above we have

$$\Gamma_n \gamma(T_A) \frac{[B]}{b!} = \sum_\mu {}^b \chi^\mu . \qquad 2.4$$

where the b is to indicate that the summation is to be carried out only over the partitions μ that are b-pieri over λ.

It is not difficult to see that this identity implies Pieri's rule. Indeed, the character version of the rule is simply the identity

$$\Gamma_n \frac{\chi^\lambda}{a!} \frac{[B]}{b!} = \sum_\mu {}^b \chi^\mu . \qquad 2.5$$

However, since

$$\chi^\lambda = \Gamma_a \gamma(T_A) ,$$

and

$$\Gamma_n \frac{(\Gamma_a \gamma(T_A))}{a!} \frac{[B]}{b!} = \Gamma_n \gamma(T_A) \frac{[B]}{b!} ,$$

we see that 2.4 and 2.5 are equivalent identities.

We have found Young's proof of 2.4 difficult to decipher. Nevertheless using his sentences as a collection of hints, and taking 1.14 and 1.15 as the definition of the natural units we have put together the argument given below.

We start with formula 1.2 with

$$f = \frac{\gamma(T_A)}{h_\lambda} \frac{[G]}{b!}$$

In view of 2.2 and 2.3 the identity in 1.2 yields that

$$\Gamma_n f = \sum_\mu h_\mu \sum_{r=1}^{n_\mu} f e^\mu_{rr} |_\epsilon \chi^\mu . \qquad 2.6$$

We are thus reduced to calculating the expressions

$$h_\mu f e^\mu_{rr} |_\epsilon = \frac{P(T_A) N(T_A)}{h_\lambda} \frac{[B]}{b!} P(T^\mu_r) N(T^\mu_r) M^\mu_r |_\epsilon . \qquad 2.7$$

If we expand the factor M^μ_r here according to its definition 1.15 we are led to summands of the form

$$P(T_A) N(T_A) [B] P(T^\mu_r) N(T^\mu_r) P(T^\mu_{i_1}) N(T^\mu_{i_1}) P(T^\mu_{i_2}) N(T^\mu_{i_2}) \cdots P(T^\mu_{i_k}) N(T^\mu_{i_k}) |_\epsilon .$$

with

$$r < i_1 < i_2 < \cdots < i_k \leqslant n_\mu .$$

Now, this may be rewritten as

$$N(T_A)P(T^\mu_r) N(T^\mu_r) P(T^\mu_{i_1}) N(T^\mu_{i_1}) P(T^\mu_{i_2}) N(T^\mu_{i_2}) \cdots P(T^\mu_{i_k}) N(T^\mu_{i_k})P(T_A) [B] |_\epsilon , \qquad 2.8$$

and the non vanishing of such a term yields us the relations

$$T_A \to T^\mu_r \to T^\mu_{i_1} \to T^\mu_{i_2} \to \cdots \to T^\mu_{i_k} \to T_A .$$

In view of Remark 1.3 we must conclude that T_A must then be a subtableau of all the successive ones. Moreover, owing to the presence of the factor [B] in 2.8 we derive that no two elements of B can be in the same column of $T^\mu_{i_k}$. In other words $T^\mu_{i_k}$ itself must be B-pieri over T_A, and thus also that μ must be b-pieri over λ. Since the tableaux

$$T^\mu_r , T^\mu_{i_1} , T^\mu_{i_2} , \cdots , T^\mu_{i_k} ,$$

have all the same shape, they must all necessarily be B-pieri over T_A. However, a moments reflection reveals that this last condition forces a term such as in in 2.8 to vanish identically. The reason for this is that if two different standard tableaux T_1 and T_2 of same shape are both B-pieri over the same tableau T_A then some element of B must be in a higher row in T_1 than it is in T_2 and for some other element of B the opposite must hold. This will give pairs of elements in the same column of T_1 that are in the same row of T_2 and vice versa. Now this fact implies that

$$P(T_1) N(T_2) = 0 = P(T_2) N(T_1)$$

We must then conclude that the only surviving term in the expansion of M^μ_r is the trivial identity term. In other words we can omit M^μ_r altogether in the identity 2.7

We are thus reduced to calculating terms such as

$$\frac{P(T_A)N(T_A)}{h_\lambda} P(T^\mu_r)N(T^\mu_r) \frac{[B]}{b!} |_\epsilon$$

With T^μ_r B-pieri over T_A. However using Remark 1.4 we can reduce this to

$$P(T^\mu_r)N(T^\mu_r) \frac{[B]}{b!} |_\epsilon$$

Finally we can get rid of the last factor [B] by the following argument (typical in Young's work). Assume that B splits in to the subsets $B_1 , B_2 , \cdots , B_k$ in successive rows of T^μ_r. Letting

$$[B] = \sum_i \tau_i [B_1][B_2] \cdots [B_k]$$

be the left coset decomposition of [B] we can rewrite our term in the form

$$\sum_i \tau_i [B_1][B_2] \cdots [B_k] P(T_f^\mu)N(T_f^\mu) |_\epsilon .$$

Now the "B" factors get absorbed by $P(T_f^\mu)$ yielding the numerical factor

$$b_1!\, b_2! \cdots b_k! ,$$

b_i indicating the cardinality of B_i .

We are finally reduced to evaluating terms of the form

$$\tau_i P(T_f^\mu)N(T_f^\mu) |_\epsilon = N(T_f^\mu) P(\tau_i T_f^\mu)\tau_i |_\epsilon .$$

Now since each τ_i (with the exception of the identity) moves around the elements of B in T_f^μ bringing some of them down and some of them up, there will necessarily be pairs of entries in of the columns of T_f^μ that are in the same row of $\tau_i T_f^\mu$. But this forces all these terms to be zero, leaving us with

$$\frac{b_1!\, b_2! \cdots b_k!}{b!} P(T_f^\mu)N(T_f^\mu) |_\epsilon = \frac{b_1!\, b_2! \cdots b_k!}{b!} .$$

Note that the multinomial coefficient

$$\frac{b!}{b_1!\, b_2! \cdots b_k!}$$

gives precisely the number of standard tableaux of shape μ that are B-pieri over T_A . Denoting this coeffiecent by b_μ we can recapitulate these findings as follows:

$$\gamma(T_A) \frac{[B]}{b!} |_{e_{ff}^\mu} = \begin{cases} \frac{1}{b_\mu} & \text{if } T_f^\mu \text{ is B-pieri over } T_A , \\ 0 & \text{otherwise} \end{cases} \qquad 2.9$$

combining this with formula 2.6 we see that we must have 2.4 as asserted.

3. Young's rule.

Let

$$\mathbf{a} = \{ a_1 , a_2 , \cdots , a_k \}$$

be a composition of n and let $A_1 = \{1,2,..,a_1\}$ and

$$A_i = \{a_1+a_2+ \cdots +a_{i-1} , \cdots , a_1+a_2+ \cdots +a_i \} \qquad \text{(for } i=2,..,k \text{)}$$

be the corresponding decomposition of the interval [1,n] into successive disjoint intervals.

For a partition λ and a composition **a** it is customary to denote by $K_{\lambda,a}$ the number of column strict tableaux of shape λ and content $1^{a_1} 2^{a_2} \cdots k^{a_k}$

The character version of Young's rule can be stated as follows

$$\Gamma_n \frac{[A_1][A_2] \cdots [A_k]}{a_1!\, a_2! \cdots a_k!} = \sum_\lambda \chi^\lambda K_{\lambda,a} . \qquad 3.1$$

It is well known and easy to show that the left hand side gives the character of the permutation representation induced by the action of S_n on the left cosets of the Young subgroup $[A_1][A_2] \cdots [A_k]$. Thus this identity simply says that the multiplicity of A^λ in this representation is given precisely by the number $K_{\lambda,a}$.

Young states 3.1 in a very curious manner. Under the assumption that $a_1 \geq a_2 \geq \cdots \geq a_k$ he writes

$$\Gamma_n \frac{[A_1][A_2] \cdots [A_k]}{a_1!\, a_2! \cdots a_k!} = \prod_{i<j} \frac{1}{1-S_{ij}} \chi^{(a_1,\ldots,a_k)} . \qquad 3.2$$

Where he says that $S_{i,j}$ is the operation of " *moving one letter from the* j^{th} *row to the* i^{th} ".

He then states that this relation *may be inverted* to

$$\chi^{(a_1,\ldots,a_k)} = \prod_{i<j} (1 - S_{i,j})\, \Gamma_n \frac{[A_1][A_2]\cdots[A_k]}{a_1!\, a_2! \cdots a_k!} . \qquad 3.3$$

This development is extremely puzzling, for it is clear from Young's use of words that he intends $S_{i,j}$ to act on *tableaux*, yet it appears (see [3]) that the only way to make simultaneous sense out of 3.2 and 3.3 is to interpret $S_{i,j}$ as an operation on *shapes*. That is, we replace the word *letter* in the quoted sentence above by the word *dot*.

However, since we have

$$\chi^a = h_a \sum_{r=1}^{n_a} e^a_{r,r} ,$$

and each unit $e^a_{r,r}$ corresponds to a uniquely determined tableau, it is quite possible that Young really meant $S_{i,j}$ to act on tableaux. This possibility should be further reinforced by the proof of Young's Rule that we are about to present.

Our argument stems from the few cryptic sentences at the end of page 196 of QSA VI. The starting point is again formula 1.2 with the $e^\lambda_{i,j}$ the natural units. For the case

$$f = \frac{[A_1][A_2]}{a_1!\, a_2!}$$

let us write

$$f \frac{\chi^\lambda}{h_\lambda} = \sum_{r=1}^{n_\lambda} A^\lambda_{r,r}(f)\, e^\lambda_{r,r} + \text{NDT}$$

where "NDT" stands here for *non diagonal terms*, that is terms in $e^\lambda_{i,j}$ with $i \neq j$. The nature of these terms is of no consequence here since (in view of 2.3) they are sent to zero by Γ_n.

Now Young states that when $\lambda = (a_1+b, a_2-b)$ with $b \geq 0$ the coefficient of $e^\lambda_{r,r}$ is zero unless T^λ_r has the letters $1,2,..,a_1$ all in the first row and in this case

$$A^\lambda_{r,r}(f) = \frac{1}{\binom{a_2}{b}}$$

From this assertion, 3.1 (in the case of two part partitions) can be easily derived since the binomial coefficient $\binom{a_2}{b}$ gives precisely the number of standard tableaux $T^\lambda_{r,r}$ that have $1,2,..,a_1$ in the first row and the coefficient $K_{\lambda,a}$ is precisely equal to one in this case. Young then goes on to say that the general case can be proved in the "*same way*". No justification whatsoever is given for the assertion even in the case of two part partitions.

Actually it is not too difficult to verify the validity of Young's assertion in the case of two part partitions. The real puzzle starts when we try to interpret the cryptic *same way* and state the basic identity in the general case.

To formulate the most natural and tempting interpretation we need some notation. Given a column strict tableau S of shape λ and content $1^{a_1} 2^{a_2} \cdots k^{a_k}$ let us say that a standard tableau T *fits* S if and only if T is obtained by replacing in S the 1's by the elements of A_1, the 2's by the elements of A_2, $\cdots$, the k's by the elements of A_k. Finally, let $n(S)$ denote the number these tableaux.

This given, we may conjecture that

$$\frac{[A_1][A_2]\cdots[A_k]}{a_1!\, a_2! \cdots a_k!} = \sum_\lambda \sum_{r=1}^{n_\lambda} A^\lambda_{r,r}(\mathbf{a})\, e^\lambda_{r,r} + \text{NDT} . \qquad 3.4$$

where $A^{\lambda}_{r,r}(\mathbf{a})$ is equal to zero unless T^{λ}_{r} fits some column strict tableau S of content $1^{a_1}2^{a_2}\cdots k^{a_k}$ and in this case.

$$A^{\lambda}_{r,r}(\mathbf{a}) = \frac{1}{n(S)} . \tag{3.5}$$

In a joint effort with J. Remmel we succeeded in proving this conjecture for the case of three part partitions, however the general case presents insurmountable difficulties. Nevertheless the effort was not totally fruitless since it inspired the argument we used in this paper for the proof of Pieri's Rule.

We should note though that, for the sole purpose of proving Young's Rule, it is sufficient to prove 3.4 with $A^{\lambda}_{r,r}(\mathbf{a})$ given by 3.5 as indicated above and "NDT" replaced by *terms annihilated by* Γ_n. Now it develops that we can indeed prove this version of the conjecture.

More precisely, given two elements f, g of the group algebra of S_n let us write $f \equiv_n g$ if and only if $\Gamma_n(f-g) = 0$. This given, our basic identity may be stated as follows.

Lemma 3.1

For any composition $\mathbf{a} = (a_1, a_2, \cdots, a_k)$ *we have*

$$\frac{[A_1][A_2]\cdots[A_k]}{a_1!\,a_2!\cdots a_k!} \equiv_n \sum_{S} \frac{1}{n(S)} \sum_{T \text{ fits } S} \gamma(T) , \tag{3.6}$$

where the first sum is to be carried out over all column strict tableaux of content $1^{a_1}2^{a_2}\cdots k^{a_k}$.

Proof

We proceed by induction on k. The assertion is trivial for $k=1$. For $k-1$ we can write it in the form

$$\frac{[A_1]\cdots[A_{k-1}]}{a_1!\cdots a_{k-1}!} = \sum_{S}{}^{(k-1)} \sum_{T_A \text{ fits } S} \frac{\gamma(T_A)}{n(S)} + E , \tag{3.7}$$

where the superscript $(k-1)$ is to indicate that now the first sum runs over all column strict tableaux of content $1^{a_1}\cdots(k-1)^{a_{k-1}}$. Furthermore, assuming that

$$a_1 + \cdots + a_{k-1} = a ,$$

the error term E satisfies

$$\Gamma_a E = 0 .$$

Aiming to prove the identity for k let us set

$$B = \{a+1, a+2, \cdots, a+b\} \qquad (\text{with } a+b=n) .$$

Multiplying 3.7 by $\frac{[B]}{b!}$ we get

$$\frac{[A_1]\cdots[A_{k-1}]}{a_1!\cdots a_{k-1}!}\frac{[B]}{b!} = \sum_{S}{}^{(k-1)} \sum_{T_A \text{ fits } S} \frac{\gamma(T_A)}{n(S)}\frac{[B]}{b!} + E\frac{[B]}{b!} \tag{3.8}$$

Note now that the summand is precisely of the form studied in our proof of Pieri's rule. Indeed, equation 2.9 can be written as

$$\gamma(T_A)\frac{[B]}{b!} = \sum_{\mu} \sum_{T^{\mu}_{r} \text{ B-pieri over } T_A} \frac{e^{\mu}_{rr}}{b_{\mu}} + \text{NDT} . \tag{3.9}$$

Moreover, from 1.20 we derive that

$$e^{\mu}_{rr} = \gamma(T_r^{\mu}) + \text{NDT} .$$

and thus 3.9 yields

$$\gamma(T_A)\,\frac{[B]}{b!} = \sum_{\mu}\ \sum_{T_r^{\mu}\ B\text{-pieri over } T_A} \frac{\gamma(T_r^{\mu})}{b_{\mu}} + \text{NDT}\,. \qquad 3.10$$

Now this equation gives us precisely what is needed to complete the induction argument. For, substituting it in 3.8 and observing that

$$\Gamma_n(E\,[B]) = \Gamma_n\Big(\frac{\Gamma_a(E)}{a!}\,[B]\Big) = 0\,,$$

we finally deduce the identity

$$\frac{[A_1]\cdots[A_{k-1}]}{a_1!\cdots a_{k-1}!}\,\frac{[B]}{b!} \equiv_n \sum_{S}{}^{(k-1)} \sum_{T_A \text{ fits } S}\ \sum_{\mu}\ \sum_{T_r^{\mu}\ B\text{-pieri over } T_A} \frac{\gamma(T_r^{\mu})}{n(S)\, b_{\mu}}\,.$$

which is easily seen to be merely a more complicated way of writing our identity 3.6 .

Remark 3.1

Starting from our identity 3.6 it is possible to justify Young's expression 3.2 with the raising operators acting on tableau idempotents. Just as in [3] we do not have associativity and we shall have to define the action of a compound raising operator

$$Q = \prod_{i<j} S_{i,j}^{\alpha_{ij}} \qquad 3.11$$

rather than the action of the individual $S_{i,j}$'s . To this end, for a given column strict tableau S let Q(S) be the compound operator obtained by letting α_{ij} in 3.11 be equal to the number of j's that are in the ith row of S . For instance if (in English notation)

$$S = \begin{array}{l} 1\ 1\ 1\ 1\ 2\ 3\ 4 \\ 2\ 3\ 3\ 4 \\ 3\ 4\ 4 \\ 4 \end{array}\ , \qquad 3.12$$

then

$$Q(S) = S_{1,2}\, S_{1,3}\, S_{2,3}^2\, S_{1,4}\, S_{2,4}\, S_{3,4}^2\,.$$

We may say for instance that Q(S) *raises 2 letters from the fourth row to the third.* We may also write the equation

$$Q(S)\ \begin{array}{l} 1\ 1\ 1\ 1 \\ 2\ 2 \\ 3\ 3\ 3\ 3 \\ 4\ 4\ 4\ 4\ 4 \end{array} = \begin{array}{l} 1\ 1\ 1\ 1\ 2\ 3\ 4 \\ 2\ 3\ 3\ 4 \\ 3\ 4\ 4 \\ 4 \end{array}\ .$$

Or even more generally we may write equations such as

$$S_{1,2}\, S_{2,3}^2\, S_{3,4}\ \begin{array}{llll} 1 & 2 & 3 & 4 \\ 5 & 6 & & \\ 8 & 9 & 10 & \\ 12 & 13 & & \end{array} = \begin{array}{lllll} 1 & 2 & 3 & 4 & 6 \\ 5 & 9 & 10 & & \\ 8 & 13 & & & \\ 12 & & & & \end{array}$$

In other words if Q is given by 3.11 then Q T is obtained by raising in succession for each j and for i=1 to j−1 the $\alpha_{i,j}$ last letters of the jth row of T up to the end of the ith row.

Now, the basic rule in defining the action of a compound raising operator Q on an idempotent $\gamma(T)$ is to set the result equal to zero if Q T is not a tableau or if two letters which are in the same row in T end up in the same column of Q T . If neither of this two events occurs then we set

$$Q\gamma(T) = \gamma(QT)\,.$$

We can easily see then that if T is a tableau of shape $(a_1, a_2, \ldots, a_k)$ then $Q\gamma(T) = 0$ unless $Q = Q(S)$ for some column strict tableau of content $1^{a_1} 2^{a_2} \cdots k^{a_k}$, and then of course Q T is precisely the tableau obtained by replacing the j 's of S from left to right by the elements of the jth row of T.

Note now that since for any tableau T of shape $\lambda \vdash n$ we have

$$\Gamma_n \gamma(T) = \chi^\lambda$$

equation 3.6 can be rewritten in the form

$$\frac{[A_1][A_2]\cdots[A_k]}{a_1!\,a_2!\cdots a_k!} \equiv_n \sum_S \gamma(T(S)) \ ,$$

where T(S) is the tableau obtained by replacing for each j in succession the j's of S from left to right by the elements of A_j in their natural order. However, due to our conventions concerning raising operators we can rewrite this last equation as

$$\frac{[A_1][A_2]\cdots[A_k]}{a_1!\,a_2!\cdots a_k!} \equiv \prod_{i<j} \frac{1}{1-S_{i,j}} \gamma(A) \ , \qquad 3.13$$

where (assuming $a_1 \geqslant a_2 \geqslant \cdots \geqslant a_k$) the tableau A is simply that whose j^{th} row consists of the elements of A_j in their natural order.

Now it is easy to see that since a raising operator acts on the position of a letter and a permutation acts on the value these two actions commute. Thus, for any Q and any permutation σ, we do have that

$$Q\sigma T = \sigma Q T$$

and a fortiori we do have as well that

$$\sigma Q \gamma(T) \sigma^{-1} = Q \gamma(\sigma T) \ .$$

We thus deduce from 3.13

$$\sigma \frac{[A_1][A_2]\cdots[A_k]}{a_1!\,a_2!\cdots a_k!} \sigma^{-1} \equiv_n \prod_{i<j} \frac{1}{1-S_{i,j}} \gamma(\sigma A) \ , \qquad 3.14$$

If we now sum these identities for all σ in S_n we derive that

$$\Gamma_n \frac{[A_1][A_2]\cdots[A_k]}{a_1!\,a_2!\cdots a_k!} \equiv_n \prod_{i<j} \frac{1}{1-S_{i,j}} \chi^{(a_1,a_2,\ldots,a_k)} \ ,$$

But now the congruence becomes equality since both sides are central elements of the group algebra of S_n. Thus 3.2 holds true as Young asserted even with this interpretation of the raising operators.

Remark 3.2

We have no idea whatsoever how Young could justify *inverting* 3.2 to get 3.3. This is done in [3] with a completely different definition of raising operators. The closest we can come in tying 3.3 to 3.2 via the present definition of raising operators is by working not with idempotents but rather with words in the free monoid. However, since this topic would lead us far out of the present context, we have to refer the reader to [2] for further details.

We are thus left with two open problems. Namely, whether or not 3.4 (with 3.5) holds true as generally as Young seem to indirectly assert and the derivation of 3.3 from 3.2 with the present definition of Raising operators.

Bibliography

[1] A. M. Garsia and L. Favreau, Characteristic free aspects of Young's natural representation, (to appear).

[2] A. M. Garsia and T. McLarnan, A non commutative version of the determinantal formula for Schur functions, (to appear).

[4] A. M. Garsia and J. Remmel, On the Raising operators of Alfred Young, Proceedings of Symposia in Pure Math., V. 34, (1979), pp. 181-198.

[3] A. M. Garsia and J. Remmel, Symmetric functions and Raising operators, J. Linear and Multilinear Algebra, 10 (1981), 15-43.

[5] A. M. Garsia and J. Remmel, Shuffles of permutations and the Kronecker product. (To appear in the Asian J. of Combinatorics)

[6] A. M. Garsia and J. Remmel, Algorithms for Plethysm, Contemporary Mathematics, V 34 (1941) "Combinatorics and Algebra", edited by C. Greene.

[7] G. James, The representation Theory of the Symmetric groups, Springer Lecture notes in Math. #682, N. Y. 1978.

[8] I. G. MacDonald, Symmetric functions and Hall Polynomials, Oxford University Press, New York (1979).

[9] Rutherford, Substitutional Analysis, Edinburgh 1948.

[10] G. Thomas, A note on Young's Raising operator, C. R. Math. Rep. Acad. Sci. Canada 2 (1980), #1 , pp. 35-36.

[11] A. Young, The collected Papers of Alfred Young 1873-1940, University of Toronto Press, Math. Expositions # 21.

Counting Three-Line Latin Rectangles

Ira M. Gessel*
Department of Mathematics
Brandeis University
Waltham, MA 02254

A $k \times n$ Latin rectangle is a $k \times n$ array of numbers such that (i) each row is a permutation of $[n] = \{1, 2, \ldots, n\}$ and (ii) each column contains distinct entries. If the first row is $12 \cdots n$, the Latin rectangle is said to be *reduced*. Since the number $k \times n$ Latin rectangles is clearly $n!$ times the number of reduced $k \times n$ Latin rectangles, we shall henceforth consider only reduced Latin rectangles. It is known [7, exercise 4.5.10, p. 288; solution, p. 507] that the number of (reduced) $3 \times n$ Latin rectangles is the coefficient of $x^n/n!$ in

$$e^{2x} \sum_{n=0}^{\infty} n! \frac{x^n}{(1+x)^{3n+3}}. \tag{1}$$

For other work on the enumeration of $3 \times n$ Latin rectangles, see [1],[2],[8]-[12], and [13, pp. 204-210].

A $3 \times n$ Latin rectangle may be identified with the pair (π, σ) of permutations of $1, \ldots, n$ which are its second and third rows. A pair (π, σ) of permutations corresponds to a $3 \times n$ Latin rectangle if and only if for each i in $[n]$, $\pi(i) \neq i$, $\sigma(i) \neq i$, and $\pi(i) \neq \sigma(i)$; in other words, π, σ, and $\pi\sigma^{-1}$ are derangements.

We generalize (1) to count π and σ by their numbers of cycles, and we obtain the following result:

Theorem 1. *The number of pairs (π, σ) of permutations of $[n]$ such that π, σ, and $\pi\sigma^{-1}$ are derangements, π has j cycles, and σ has k cycles, is the coefficient of $\alpha^j \beta^k x^n/n!$ in*

$$e^{2\alpha\beta x} \sum_{n=0}^{\infty} \frac{(\alpha)_n (\beta)_n}{n!} \frac{x^n}{(1+\alpha x)^{n+\beta}(1+\beta x)^{n+\alpha}(1+x)^{n+\alpha\beta}}, \tag{2}$$

where $(\alpha)_n = \alpha(\alpha+1)\ldots(\alpha+n-1)$.

Our approach is similar to that taken by Foata and others [3]-[6] in their combinatorial study of orthogonal polynomials. We work with digraphs corresponding to permutations. We may identify the permutation π of $[n]$ with the digraph on $[n]$ having an edge from i to $\pi(i)$ for each i.

* partially supported by NSF grant DMS-8504134

We define a *Latin configuration* to be a digraph on $[n]$ with edges in three colors, yellow, blue, and green, with the following properties:

1) The yellow edges form a derangement π on $[n]$.
2) The blue edges form a derangement σ on $[n]$.
3) If there is a green edge from i to j, then there must be both a yellow and a blue edge from i to j.

The *weight* of a Latin configuration is defined to be $\alpha^{|\pi|}\beta^{|\sigma|}\gamma^{g}$, where $|\pi|$ is the number of cycles of π, $|\sigma|$ is the number of cycles of σ, and g is the number of green edges. Let $L_n(\alpha,\beta,\gamma)$ be the sum of the weights of all Latin configurations on $[n]$ and let $K_n(\alpha,\beta,\gamma)$ be the sum of $\alpha^{|\pi|}\beta^{|\sigma|}\gamma^{|I(\pi,\sigma)|}$ over all pairs (π,σ) of derangements of $[n]$, where $I(\pi,\sigma)$ is the set of values of i for which $\pi(i)=\sigma(i)$.

Lemma 1.

$$L_n(\alpha,\beta,\gamma)=K_n(\alpha,\beta,\gamma+1).$$

Proof. Let $D(n)$ be the set of derangements of $[n]$. Then

$$K_n(\alpha,\beta,\gamma+1)=\sum_{\pi,\sigma\in D(n)}\alpha^{|\pi|}\beta^{|\sigma|}(\gamma+1)^{|I(\pi,\sigma)|}=\sum_{\substack{\pi,\sigma\in D(n)\\ G\subseteq I(\pi,\sigma)}}\alpha^{|\pi|}\beta^{|\sigma|}\gamma^{|G|}.$$

But a pair (π,σ) of derangements together with a subset G of $I(\pi,\sigma)$ corresponds to a Latin configuration in which the green edges are those from i to $\pi(i)=\sigma(i)$ for $i\in G$.

It follows that $K_n(\alpha,\beta,\gamma)=L_n(\alpha,\beta,\gamma-1)$. We now determine the generating function

$$L(x)=\sum_{n=0}^{\infty}L_n(\alpha,\beta,\gamma)\frac{x^n}{n!}$$

for Latin configurations.

First we may split the vertices of a Latin configuration into two classes: those in green cycles and all others. A green cycle is coextensive with a blue cycle and a yellow cycle, and therefore can have no edges connecting it with any other vertices.

The set of green cycles of a Latin configuration constitutes a derangement, since the only restriction on them is that there be no fixed points. Since the generating function for derangements is $e^{-x}/(1-x)$, the generating function for green cycles, together with their associated blue and yellow cycles, is $\left(e^{-\gamma x}/(1-\gamma x)\right)^{\alpha\beta}$ because every edge is weighted γ and every cycle is weighted $\alpha\beta$. Thus $L(x)=\left(e^{-\gamma x}/(1-\gamma x)\right)^{\alpha\beta}R(x)$, where $R(x)$ is the generating function for *green-acyclic* Latin configurations, that is, Latin configurations with no green cycles.

We shall count green-acyclic Latin configurations by constructing them in two steps, with each step translating into a generating function operation. We first insert the green edges, obtaining a set of green paths and a set of isolated vertices. We then "mark" certain vertices with indeterminates. Finally, based on the marks, we insert the yellow and blue edges.

First we make two observations which are the basis for the derivation of our generating function.

1) The green edges constitute a set of disjoint paths, each of which has at least two vertices.
2) If the green edges are contracted (together with their associated yellow and blue edges), what remains is a digraph consisting of a yellow permutation and a blue permutation. Any yellow or blue loop in the contracted graph must be attached to a vertex that was contracted.

Now we do the construction. We start with a set of green paths, each of at least two vertices, and a set of isolated vertices. We mark each isolated vertex with A and B. We mark the head of each green path with either A or α and with B or β.

The generating function for the configurations we have constructed so far is

$$\exp\left(\frac{(\mathrm{A}+\alpha)(\mathrm{B}+\beta)\gamma x^2}{1-\gamma x}+\mathrm{AB}x\right), \tag{3}$$

where each green edge is weighted γ.

Next, we put in the yellow and blue edges as follows:

1. Construct a yellow derangement through all the vertices marked A and a blue derangement through all the vertices marked B.
2. For each vertex v which is the head of a green path, do the following: Let u be the tail of the path. If v is marked A, there is a yellow edge (w, v). Replace this edge with the yellow edge (w, u). If v is marked α, add the yellow edge (v, u).
3. Repeat step 2 with A, α, and "yellow" replaced by B, β, and "blue."
4. Add a yellow edge and blue edge parallel to every green edge.

It is clear that only Latin configurations are obtained, and each is obtained exactly once.

We now describe the operation on the generating function (3) which corresponds to the insertion of yellow and blue edges. Let $D_n(s) = \sum_\pi s^{|\pi|}$, where the sum is over all derangements π of $[n]$. Then the Latin configurations coming from a term $\mathrm{A}^i\mathrm{B}^j\alpha^k\beta^l\gamma^m x^n/n!$ in (3) will have total weight $D_i(\alpha)D_j(\beta)\alpha^k\beta^l\gamma^m x^n/n!$. This is because each vertex marked α yields a yellow cycle, each vertex marked β yields a blue cycle, and the contribution from the vertices marked A and B is $D_i(\alpha)D_j(\beta)$.

It is convenient to adopt the "symbolic" or "umbral" convention

$$\mathrm{A}^i = D_i(\alpha)$$

$$\mathrm{B}^j = D_j(\beta),$$

by which we mean that henceforth after an expression is expanded, any occurrence of A^i is to be replaced by $D_i(\alpha)$ and any occurrence of B^j is to be replaced by $D_j(\beta)$. As shown by Rota [14], this means formally that we apply to all our formulas a linear operator Φ defined by

$$\Phi\left(\mathrm{A}^i\mathrm{B}^j\alpha^k\beta^l\gamma^m\frac{x^n}{n!}\right) = D_i(\alpha)D_j(\beta)\alpha^k\beta^l\gamma^m\frac{x^n}{n!}.$$

Thus we need only evaluate (3) with this interpretation. First we study some properties of these "umbral variables." We note that

$$\sum_{i=0}^{\infty} D_i(\alpha)\frac{x^i}{i!} = \left(\frac{e^{-x}}{1-x}\right)^{\alpha},$$

and thus

$$e^{\mathrm{A}x} = \left(\frac{e^{-x}}{1-x}\right)^{\alpha}.$$

Now we introduce umbral variables a and b defined by $\mathrm{a} = \alpha + \mathrm{A}$ and $\mathrm{b} = \beta + \mathrm{B}$, that is,

$$\mathrm{a}^n = (\alpha + \mathrm{A})^n$$

and

$$\mathrm{b}^n = (\beta + \mathrm{B})^n$$

for all n. Then we have

$$\sum_{n=0}^{\infty} \mathrm{a}^n\frac{x^n}{n!} = e^{\mathrm{a}x} = e^{\alpha x}e^{\mathrm{A}x} = (1-x)^{-\alpha} = \sum_{n=0}^{\infty}(\alpha)_n\frac{x^n}{n!}.$$

Thus $\mathrm{a}^n = (\alpha)_n$, and similarly, $\mathrm{b}^n = (\beta)_n$.

Lemma 2. *For all* k,

$$e^{\mathrm{a}x}\mathrm{a}^k = (1-x)^{-\alpha}\left(\frac{\mathrm{a}}{1-x}\right)^k.$$

Proof. We have

$$\begin{aligned} e^{\mathrm{a}x}\mathrm{a}^k &= \sum_{n=0}^{\infty}\mathrm{a}^{n+k}\frac{x^n}{n!} = \sum_{n=0}^{\infty}(\alpha)_{n+k}\frac{x^n}{n!} \\ &= (\alpha)_k\sum_{n=0}^{\infty}(\alpha+k)_n\frac{x^n}{n!} = \frac{(\alpha)_k}{(1-x)^{\alpha+k}} = (1-x)^{-\alpha}\left(\frac{\mathrm{a}}{1-x}\right)^k. \end{aligned}$$

It follows by linearity that for any series f for which $f(\mathrm{a})$ makes sense,

$$e^{\mathrm{a}x}f(\mathrm{a}) = (1-x)^{-\alpha}f\left(\frac{\mathrm{a}}{1-x}\right).$$

Similarly, we have

$$e^{\mathrm{b}x}f(\mathrm{b}) = (1-x)^{-\beta}f\left(\frac{\mathrm{b}}{1-x}\right).$$

Putting these together, we have

Lemma 3.

$$e^{au+bv}f(a,b)=(1-u)^{-\alpha}(1-v)^{-\beta}f\left(\frac{a}{1-u},\frac{b}{1-v}\right).$$

Now we write (3) as

$$\exp\left(\frac{ab\gamma x^2}{1-\gamma x}+abx-a\beta x-\alpha bx+\alpha\beta x\right)=e^{\alpha\beta x}e^{-a\beta x-\alpha bx}\exp\left(\frac{abx}{1-\gamma x}\right).$$

It then follows from Lemma 3 that this is

$$e^{\alpha\beta x}(1+\beta x)^{-\alpha}(1+\alpha x)^{-\beta}\exp\left(\frac{abx}{(1+\beta x)(1+\alpha x)(1-\gamma x)}\right)$$
$$=e^{\alpha\beta x}\sum_{n=0}^{\infty}\frac{(\alpha)_n(\beta)_n}{n!}\frac{x^n}{(1+\alpha x)^{n+\beta}(1+\beta x)^{n+\alpha}(1-\gamma x)^n},$$

so multiplying by the generating function for green cycles, we get the generating function for Latin configurations,

$$\sum_{n=0}^{\infty}L_n(\alpha,\beta,\gamma)\frac{x^n}{n!}=e^{\alpha\beta(1-\gamma)x}\sum_{n=0}^{\infty}\frac{(\alpha)_n(\beta)_n}{n!}\frac{x^n}{(1+\alpha x)^{n+\beta}(1+\beta x)^{n+\alpha}(1-\gamma x)^{n+\alpha\beta}}.$$

Then changing γ to $\gamma-1$, we have

$$\sum_{n=0}^{\infty}K_n(\alpha,\beta,\gamma)\frac{x^n}{n!}=e^{\alpha\beta(2-\gamma)x}\sum_{n=0}^{\infty}\frac{(\alpha)_n(\beta)_n}{n!}\frac{x^n}{(1+\alpha x)^{n+\beta}(1+\beta x)^{n+\alpha}(1-(\gamma-1)x)^{n+\alpha\beta}}$$

as the generating function for pairs (π,σ) of derangements by cycles of π, cycles of σ, and equal values of π and σ. Setting $\gamma=0$ yields Theorem 1.

Note in particular that setting $\gamma=1$ yields

$$\sum_{n=0}^{\infty}D_n(\alpha)D_n(\beta)\frac{x^n}{n!}=e^{\alpha\beta x}\sum_{n=0}^{\infty}\frac{(\alpha)_n(\beta)_n}{n!}\frac{x^n}{(1+\alpha x)^{n+\beta}(1+\beta x)^{n+\alpha}},\tag{4}$$

and setting $\alpha=\beta=1$ in (4) yields

$$\sum_{n=0}^{\infty}D_n^2\frac{x^n}{n!}=e^x\sum_{n=0}^{\infty}n!\frac{x^n}{(1+x)^{2n+2}},$$

where $D_n=D_n(1)$ is the derangement number.

The methods used in this paper can also be used to count pairs (π,σ) of permutations by the number of cycles of π and σ and the number of fixed points of π, σ, and $\pi\sigma^{-1}$. If we omit from this generating function the number of fixed points of $\pi\sigma^{-1}$ then we obtain a formula equivalent to the bilinear generating function for Charlier polynomials (of which (4) is a specialization).

The first six values of the polynomials $K_n(\alpha,\beta,\gamma)$ are as follows:

$$K_0 = 1$$

$$K_1 = 0$$

$$K_2 = \alpha\beta\gamma^2$$

$$K_3 = 2\alpha\beta(1+\gamma^3)$$

$$K_4 = 3\alpha\beta(2+2\alpha+2\beta+2\alpha\beta+8\gamma+4\alpha\gamma^2+4\beta\gamma^2+2\gamma^4+\alpha\beta\gamma^4)$$

$$K_5 = 4\alpha\beta(48+30\alpha+30\beta+30\alpha\beta+30\gamma+60\alpha\gamma+60\beta\gamma+30\alpha\beta\gamma + 60\gamma^2+35\alpha\beta\gamma^2+30\alpha\gamma^3+30\beta\gamma^3+6\gamma^5+5\alpha\beta\gamma^5)$$

References

1. K. B. Athreya, C. R. Pranesachar, and N. M. Singhi, On the number of Latin rectangles and chromatic polynomials of $L(K_{r,s})$, *Europ. J. Combinatorics* **1** (1980), 9-17.
2. K. P. Bogart and J. Q. Longyear, Counting 3 by n Latin rectangles, *Proc. Amer. Math. Soc.* **54** (1976), 463-467.
3. D. Foata, A combinatorial proof of the Mehler formula, *J. Combinatorial Theory, Ser. A* **24** (1978), 367-376.
4. D. Foata and J. Labelle, Modèles combinatoires pour les polynômes de Meixner, *Europ. J. Combinatorics* **4** (1983), 305-311.
5. D. Foata and P. Leroux, Combinatoire et fonction génératrice, *Proc. Amer. Math. Soc.* **87** (1983), 47-53.
6. D. Foata and V. Strehl, Combinatorics of Laguerre polynomials, in *Enumeration and Design*, ed. D. M. Jackson and S. A. Vanstone, Academic Press, 1984, pp. 123-140.
7. I. P. Goulden and D. M. Jackson, *Combinatorial Enumeration*, Wiley, 1983.
8. S. M. Jacob, The enumeration of the Latin rectangle of depth three by means of a formula of reduction, with other theorems relating to non-clashing substitutions and Latin squares, *Proc. London Math. Soc.* **31** (1930), 329-354.
9. S. M. Kerewala, The enumeration of the Latin rectangle of depth three by means of difference equations, *Bull. Calcutta Math. Soc.* **33** (1941), 119-127.
10. C. R. Pranesachar, Enumeration of Latin rectangles via SDR's, in *Combinatorics and Graph Theory*, ed. S. B. Rao, *Lecture Notes in Mathematics*, vol. 885, Springer-Verlag, 1981, pp. 380-390.
11. J. Riordan, Three-line Latin rectangles, *Amer. Math. Monthly* **51** (1944), 450-452.
12. J. Riordan, Three-line Latin rectangles—II, *Amer. Math. Monthly* **53** (1946), 18-20.
13. J. Riordan, *An Introduction to Combinatorial Analysis*, Wiley, 1958.
14. G.-C. Rota, The number of partitions of a set, *Amer. Math. Monthly* **71** (1964), 498-504.

CHEMINS SOUS-DIAGONAUX ET TABLEAUX DE YOUNG

Dominique GOUYOU-BEAUCHAMPS
Université de Bordeaux I
U.E.R. de Mathématiques et d'Informatique
33405 TALENCE, France

Abstract :

We consider path in the lattice of positive integer coordinate where the possible "moves" are of four kinds : (1) increasing the x coordinate by 1, (2) decreasing the x coordinate by 1, (3) increasing the y coordinate by 1, (4) decreasing the y coordinate by 1. The number of such paths of length ℓ, from (0,0) to any point whose y-coordinate is 0, lying below or touching the main diagonal, is $C_n C_{n+1}$ for $\ell=2n$ and $C_{n+1} C_{n+1}$ for $\ell=2n+1$ where C_n is the Catalan number. We give a bijective proof of this result. As corollary we give exact formulas for the number of standard Young tableaux having n cells and a most k rows in the cases $k=4$ and $k=5$.

I - Introduction.

Les nombres de Catalan $C_n = \frac{(2n)!}{n!(n+1)!}$ interviennent dans de très nombreuses formules d'énumération. Citons, entre autres, le nombre de façons de partager un polygone en triangles (Euler [8], Segner [21]), le nombre de manières d'effectuer un produit de n facteurs (Rodrigues [21], Catalan [3]), le problème du scrutin (Bertrand [2]), les arbres dessinés (Harary, Prins, Tutte [15]), les chemins minimaux sous-diagonaux (André [1]). Plus récemment des formules d'énumération ont fait intervenir le produit de deux nombres de Catalan consécutifs $C_n C_{n+1}$ et le carré d'un nombre de Catalan $C_n C_n$. On rencontre ces nombres dans l'énumération de certaines familles de cartes planaires (Mullin [17], Tutte [18]) et dans l'énumération des permutations de Baxter alternantes (Dulucq [6], Cori, Dulucq, Viennot [5]).

Le principal résultat de cet article est d'établir que $C_n C_n$ et $C_n C_{n+1}$ énumèrent aussi une classe de chemins sous-diagonaux.

On appelle chemin une ligne polygonale dans le quart de plan des coordonnées entières constituée par une suite de points

$A_0, A_1, \ldots, A_n$ tels que A_0 soit l'origine et que les coordonnées (p_i, q_i) de A_i et (p_{i+1}, q_{i+1}) de A_{i+1} vérifient l'une des quatres conditions suivantes $(i=0,1,\ldots,n-1)$:

(1) $p_{i+1}=p_i+1$ et $q_{i+1}=q_i$ (pas Est) .
(2) $p_{i+1}=p_i$ et $q_{i+1}=q_i+1$ (pas Nord).
(3) $p_{i+1}=p_i-1$ et $q_{i+1}=q_i$ (pas Ouest).
(4) $p_{i+1}=p_i$ et $q_{i+1}=q_i-1$ (pas Sud).

La longueur d'un tel chemin est n. Il peut être considéré comme non minimal car ce n'est pas forcément le chemin le plus court qui relie l'origine au point A_n. Il est dit sous-diagonal si pour $i=1,2,\ldots,n$ p_i est supérieur ou égal à q_i . Ces chemins sous-diagonaux ont été introduits par G. Viennot qui a conjecturé le résultat suivant qui est démontré ici :

<u>Théorème 1</u> - Le nombre de chemins sous-diagonaux de longueur $2n$ dont l'extrémité terminale est sur l'axe des x est égal à C_nC_{n+1}. Le nombre de tels chemins de longueur $2n+1$ est égal à $C_{n+1}C_{n+1}$.

La preuve donnée ici est purement combinatoire. Elle utilise la notion classique de chemin minimal ou chemin de Dyck. Ce sont les chemins précédents qui n'utilisent pas les pas Ouest et Sud. Ainsi, dans le troisième paragraphe on construit une bijection entre les chemins sous-diagonaux et des paires de chemins minimaux sous-diagonaux et des paires de chemins minimaux sous-diagonaux reliant l'origine à un même point et ne se coupant pas.

Le quatrième paragraphe démontre l'égalité du nombre de paires de chemins minimaux sous-diagonaux ne se coupant pas et du nombre de paires de chemins minimaux sous-diagonaux dont les longueurs sont les mêmes ou diffèrent de 2. Ces deux bijections prouvent le théorème 1 car le nombre de chemins minimaux sous-diagonaux reliant l'origine au point (n,n) est égal à C_n.

Un corollaire de ce résultat est la démonstration combinatoire des deux identités suivantes :

$$(1) \quad C_nC_{n+1} = \sum_{k=0}^{n} \frac{(2n)!(2n+2)!(2k+3)!}{(n-k)!(n-k+1)!(2k)!(n+k+2)!(n+k+3)!}$$

$$(2) \quad C_nC_n = \sum_{k=0}^{n-1} \frac{(2n-1)!(2n+1)!(2k+4)!}{(n-k-1)!(n-k)!(2k+1)!(n+k+2)!(n+k+3)!}$$

Ces identités sont à rapprocher ce celle-ci (cf [5]) bien connue : $C_nC_{n+1} = \sum_{k=0}^{n} \binom{2n}{2k} C_{n-k}C_k$. Mais cette dernière peut se démontrer très facilement par le calcul en utilisant la convolution de Vandermonde alors que la démonstration par le calcul des deux premières n'est pas immédiate.

Un deuxième corollaire du théorème 1 est l'énumération des tableaux de Young standard ayant n cases et au plus k lignes pour $k=4$ et $k=5$. En effet, il existe une bijetion ([13] ,[14]) entre les tableaux de Young standard ayant n cases et au plus 4 lignes et les chemins sous-diagonaux de longueur n dont l'extrémité terminale est sur l'axe des x.

II - Définitions

On utilisera des ensembles finis appelés alphabets dont les éléments seront appelés des lettres. Les principaux alphabets utilisés seront $Z=\{x,\bar{x},y,\bar{y}\}$ et $A=\{a,\bar{a}\}$. Un mot est une suite finie de lettres que l'on notera $f=a_1a_2\ldots a_n$. Le mot vide (la suite vide) sera notée ε . L'ensemble X^* de tous les mots sur l'alphabet X (monoide libre engendré par X) est muni classiquement de l'opération binaire de concaténation qui met bout-à-bout 2 mots.

La longueur d'un mot, notée $|f|$, est le nombre de lettres de f. Pour une lettre x, $|f|_x$ note le nombre de lettres x contenues dans f. Un mot f' est un facteur gauche d'un mot f si il existe un mot f'' tel que $f=f'f''$.

On définit les deux morphismes δ_x et δ_y de Z^* dans $\mathbb{N}$ par:

$$\delta_x(x)=1,\ \delta_x(\bar{x})=-1,\ \delta_x(y)=\delta_x(\bar{y})=0 \quad \text{et}$$

$$\delta_y(x)=\delta_y(\bar{x})=0\ ,\ \delta_y(y)=1\ ,\ \delta_y(\bar{y})=-1.$$

De même on définit le morphisme δ de A^* dans $\mathbb{N}$ par :

$$\delta(a)=1\ ,\ \delta(\bar{a})=-1.$$

On appelle langage de Dyck (parfois appelé langage restreint de Dyck) le langage suivant :

$$\{f\in A^*\mid \delta(f)=0 \text{ et } \forall f' \text{ facteur gauche de } f \text{ alors } \delta(f')\geq 0\}.$$

On le note D. Il est bien connu que $|D\cap A^{2n}|=C_n$. L'ensemble des facteurs gauches de D de longueur l et d'image p par δ est noté $F_{l,p}$ (l et p sont de même parité). Le nombre

$|F_{l,p}|=(p+1)\dfrac{l\,!}{[\frac{l-p}{2}]\,!\ [\frac{l+p}{2}+1]}$ a souvent été calculé (cf Comtet [4], André [1], Kreweras [16], Narayana [18], Gouyou-Beauchamps [12]).

Les mots de $F_{l,p}$ codent les chemins minimaux sous-diagonaux. Il suffit de coder les pas Est par a et les pas Nord par $\bar{a}$. Les chemins de

Dyck, cas particulier des chemins de Motzkin définis par Viennot (cf. [26] [9] et [10]), sont des chemins minimaux sous-diagonaux à qui on fait subir une rotation. Les pas Est deviennent alors des pas Nord-Est joignant un point de coordonnées (i,j) au point de coordonnées $(i+1, j+1)$ et les pas Nord des pas Sud-Est joignant (i,j) au point $(i+1, j-1)$. Le chemin au lieu de devoir rester au-dessous de la diagonale principale doit ne pas passer au-desous de l'axe des x. La figure 1 montre le chemin minimal sous-diagonal et le chemin de Dyck correspondant codés par $aa\bar{a}aaa\bar{a}\bar{a}\bar{a}\bar{a}aaaa\bar{a}a$.

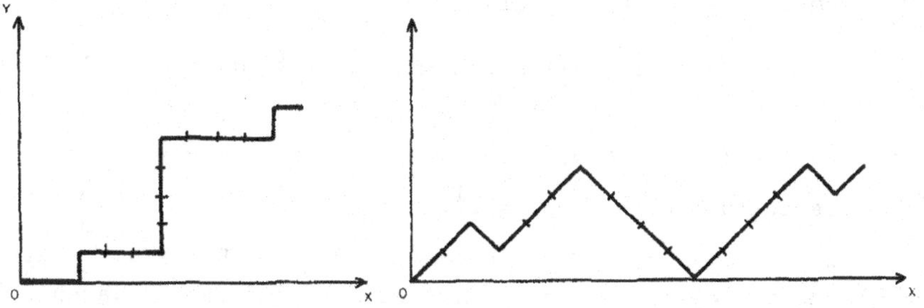

FIGURE 1

III - Chemins de Dyck ne se coupant pas.

Soit V le langage composé des mots f de Z^* vérifiant les deux propriétés suivantes :

i) $\delta_x(f) = 0$.

ii) $\forall f'$, facteur gauche de $f, \delta_y(f') \geq \delta_x(f') \geq 0$.

Il est facile de constater que les mots du langage V codent les chemins sous-diagonaux dont l'extrémité terminale est sur l'axe des x. Les pas Nord, Sud, Est et Ouest sont codés respectivement $x, \bar{x}$, y et $\bar{y}$. La contrainte $\delta_y(f') \geq \delta_x(f')$ force le chemin à être sous-diagonal la contrainte $\delta_x(f') \geq 0$ empêche le chemin de passer au-dessous de l'axe des x, et la contrainte $\delta_x(f) = 0$ oblige le chemin à finir sur l'axe des x.

La paire (g,h) de $F_{1,p} \times F_{1,p}$ est formée de mots qui ne se coupent pas si pour tout g' (resp. h') facteur gauche de g (resp. h) tel que $|g'| = |h'|$ on a $\delta(h') \geq \delta(g')$. On note $V_{1,p}$ l'ensemble des paires de mots de $F_{1,p} \times F_{1,p}$ qui ne se coupent pas. On peut remarquer que si (g,h) est formée de mots qui ne se coupent pas, alors $|f| = |g|$ et $\delta(f) = \delta(g)$. La figure 2 donne un exemple d'une paire de mots qui ne se coupent pas et d'une paire de mots qui se coupent.

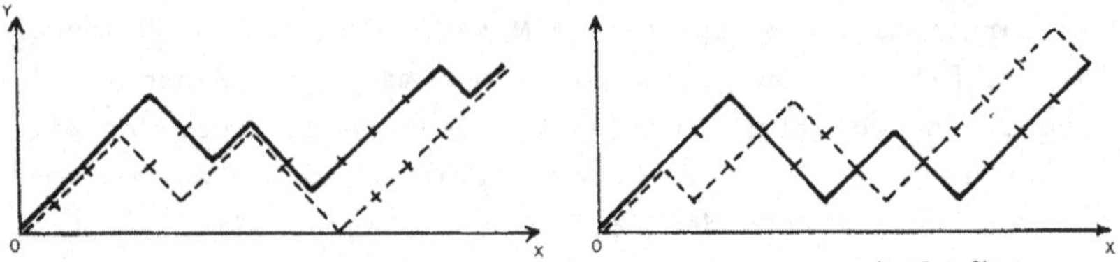

Mots qui ne se coupent pas ($l=15, p=5$)
($g=aa\bar a\bar a a\bar a\bar a\bar a aaaaaaa, h=aaaaaaa\bar a\bar a a\bar a\bar a aa\bar a$)

Mots qui se coupent ($l=15, p=5$)
($g=aa\bar a aa\bar a\bar a\bar a aaaaaa\bar a, h=aaaaaaa\bar a\bar a\bar a a\bar a\bar a aa$)

FIGURE 2

<u>Propriété 2</u> - L'ensemble des mots de V de longueur l est en bijection avec l'ensemble des paires de mots (g,h) de $\bigcup_{p=0}^{l} V_{l,p}$ qui ne coupent pas.

Démonstration - On définit le morphisme E de Z^* dans $A^* \times A^*$ de la façon suivante :

$$E(y) = (a,a) \ ; \ E(\bar y) = (\bar a, \bar a) \ ; \ E(x) = (\bar a, a) \ ; \ E(\bar x) = (a, \bar a).$$

E est clairement une bijection de Z^l dans $A^l \times A^l$.

Soit f un mot de V de longueur l. Posons $E(f) = (g,h)$. Par construction $|f| = |g| = |h|$. Soient f',g' et h' les facteurs gauches de f,g et h de longueur $m (0 \le m \le l)$. on constate facilement que $\delta(h') = \delta_x(f') + \delta_y(f')$ et que $\delta(g') = \delta_y(f') - \delta_x(f')$. La propriété ii des mots de V implique alors que $\delta(h') \ge \delta(g') \ge 0$ et la propriété i que $\delta(h) = \delta(g)$. Ainsi E(f) est une paire de mots qui ne se coupent pas.

E est donc bien la bijection annoncée.

<u>Remarque</u> -

La paire (g,h) de $F_{l,p} \times F_{l+2,p+2}$ est formée de mots qui ne se touchent pas si, pour tout g' (resp. h') facteur gauche de g (resp. h) tel que $|g'| = |h'| - 2$ on a $\delta(h') \quad \delta(g')$. Notons que l et p doivent être de même parité, que les deux premières lettres de h doivent être des a et que $|h| = |g| + 2$ et $\delta(h) = \delta(g) + 2$.

La figure 3 donne un exemple d'une paire de mots qui ne se touchent pas et d'une paire de mots qui se touchent

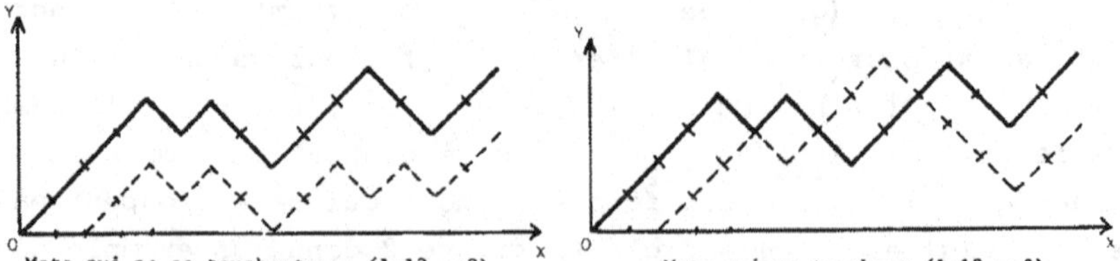

Mots qui ne se touchent pas ($l=13, p=3$)
($g=aa\bar a\bar a\bar a aa\bar a\bar a aaaa, h=aaaaa\bar a\bar a\bar a aa\bar a\bar a aaa$)

Mots qui se touchent ($l=13, p=3$)
($g=aaa\bar a aaa\bar a\bar a\bar a\bar a aa, h=aaaaa\bar a a\bar a\bar a a\bar a\bar a aaa$)

FIGURE 3

Soit $T_{l,p}$ l'ensemble des paires de mots de $F_{l,p} \times F_{l+2,p+2}$ qui ne se touchent pas. On remarque que $|T_{l,p}| = |V_{l,p}|$ puisque (g,h) appartient à $V_{l,p}$ si et seulement si (g,aah) appartient à $T_{l,p}$.

Propriété 3 - $|V_{l,p}| = |T_{l,p}| = |F_{l,p}||F_{l+2,p+2}| - |F_{l+2,p}||F_{l,p+2}|$.

Démonstration -

On utilise une technique désormais classique de Gessel et Viennot [11] (cf. aussi [20],[22], [27], [7] et [19] qui consiste, pour évaluer la différence du nombre d'éléments de 2 classes de n-uplets de chemins, à mettre en bijection les n-uplets de chemins qui se coupent de chacune des deux classes et ainsi à ne garder que les n-uplets de chemins qui ne se coupent pas.

Soit (g,h) un élément de $F_{l,p} \times F_{l+2,p+2}$. Notons $g_3, g_4, \ldots, g_{l+2}$ les lettres de A composant g et $h_1, h_2, \ldots, h_{l+2}$ celles de h. Deux cas peuvent se produire :

- Soit (g,h) est une paire de mots ne se touchant pas.
- Soit (g,h) est une paire de mots qui se touchent, c'est-à-dire qu'il existe un entier i $(3 \le i \le l+2)$ tel que $\delta(h_1 h_2 \ldots h_i) \le \delta(g_3 g_4 \ldots g_i)$

Supposons que (g,h) soit une paire de mots qui se touchent. Soit j le plus petit indice $(2 \le j \le l+2)$ tel que $\delta(h_1 h_2 \ldots h_j) = \delta(g_3 g_4 \ldots g_j)$. Si j=2 on prend $g_3 g_4 \ldots g_j = \varepsilon$. L'indice j existe car si deux mots se touchent les chemins de Dyck correspondants passent au moins une fois par le même point puisque d'une part $|\delta(h_1 h_2 \ldots h_i) - \delta(g_3 g_4 \ldots g_i)|$ est toujours un nombre pair et d'autre part $|\delta(h_1 h_2 \ldots h_i) - \delta(h_1 h_2 \ldots h_{i-1})|$ est toujours égal à 1 (ainsi que $|\delta(g_3 g_4 \ldots g_i) - \delta(g_3 g_4 \ldots g_{i-1})|$).

Construisons les deux mots $g'' = h_1 h_2 \ldots h_j g_{j+1} g_{j+2} \ldots g_{l+2}$ et $h'' = g_3 g_4 \ldots g_j h_{j+1} h_{j+2} \ldots h_{l+2}$. Il est clair que (g",h") appartient à $F_{l+2,p} \times F_{l,p+2}$.

Par le même procédé, on peut faire correspondre à un élément (g",h") de $F_{l+2,p} \times F_{l,p+2}$ un élément (g,h) de l'ensemble des mots qui se touchent. En effet les chemins de Dyck correspondants à g" et h" se croisent et donc se touchent obligatoirement au moins une fois . Il suffit d'inverser les deux chemins à partir du premier point où ils se rencontrent. La correspondance est bijective.

Le nombre d'éléments de $T_{l,p}$ et donc de $V_{l,p}$ est bien égal à $|F_{l,p}||F_{l+2,p+2}| - |F_{l+2,p}||F_{l,p+2}|$.

<u>Corollaire</u> :

a) $|V\cap Z^{2n}| = \sum_{k=0}^{n} |V_{2n,2k}| = \sum_{k=0}^{n} \dfrac{(2n)!(2n+2)!(2k+3)!}{(n-k)!(n-k+1)!(2k)!(n+k+2)!(n+k+3)!}$

b) $|V \cap Z^{2n-1}| = \sum_{k=0}^{n-1} |V_{2n-1,2k+1}| =$

$$= \sum_{k=0}^{n-1} \frac{(2n-1)!(2n+1)!(2k+4)!}{(n-k-1)!(n-k)!(2k+1)!(n+k+2)!(n+k+3)!} .$$

<u>Preuve</u> : Comme on l'a vu dans l'introduction

$|F_{l,p}| = (p+1) \dfrac{l\,!}{[\frac{l-p}{2}]!\,[\frac{l+p}{2}+1]!}$ si l et p sont de même parité et

$|F_{l,p}| = 0$ sinon.

Donc $|V_{l,p}| = \dfrac{l!(l+2)!(p+3)!}{[\frac{l-p}{2}]!\,[\frac{l-p}{2}+1]!\,p!\,[\frac{l+p}{2}+2]!\,[\frac{l+p}{2}+3]!}$ et ainsi, en reportant les valeurs de $|V_{l,p}|$ pour $l=2n$ et $2n-1$ on démontre le corollaire. La figure 4 donne le tableau des premières valeurs de $V_{l,p}$.

l p	0	1	2	3	4	5	6	7	8	9	10
0	1										
1	0	1									
2	1	0	1								
3	0	3	0	1							
4	3	0	6	0	1						
5	0	14	0	10	0	1					
6	14	0	40	0	15	0	1				
7	0	84	0	90	0	21	0	1			
8	84	0	300	0	175	0	28	0	1		
9	0	594	0	825	0	308	0	36	0	1	
10	594	0	2475	0	1925	0	504	0	45	0	1

FIGURE 4

IV- Le nombre de chemins sous-diagonaux se terminant sur l'axe des x.

On appelle M_{2l} l'ensemble des mots $f=f_1f_2\dots f_{2l}$ de $F_{2l,0}$ tels que $f_{l-1}=f_l=a$. De même on appelle D_{2l} l'ensemble des mots $f=f_1f_2\dots f_{2l}$ de $F_{2l,0}$ tels que $f_{l-1}=f_l=\bar{a}$.

On note $\bar{f}$ l'image de $f \in A^*$ par le morphisme qui change les a en $\bar{a}$ et les $\bar{a}$ en a.

$\tilde{\bar{f}}$ est l'image miroir du mot $\bar{f}$, c'est-à-dire que si $f=f_1f_2\dots f_n$ alors $\tilde{\bar{f}}=\bar{f}_n\dots\bar{f}_2\bar{f}_1$.

Propriété 4 - Le nombre de mots de V de longueur l est égal à $|M_{2l+4}|-|D_{2l+4}|$.

Démonstration : Soit (g,h) un élément de $F_{l,p}\times F_{l+2,p+2}$. Le mot $gaa\tilde{h}$ appartient à $F_{2l+4,0}$ et donc à M_{2l+4}. Inversement, pour tout mot $f=f_1f_2\dots f_{2l+4}$ de M_{2l+4}, il existe un entier p tel que $f_1f_2\dots f_l$ appartienne à $F_{l,p}$ et $f_{2l+4}\bar{f}_{2l+3}\dots\bar{f}_{l+3}$ à $F_{l+2,p+2}$. Donc

$$|M_{2l+4}| = \sum_{p=0}^{l} |F_{l,p}||F_{l+2,p+2}| .$$

De même si (g,h) est un élément de $F_{l,p+2}\times F_{l+2,p}$ le mot $g\bar{a}\bar{a}\tilde{h}$ appartient à $F_{2l+4,0}$ et donc à D_{2l+4}. Inversement pour tout mot f de D_{2l+4} il existe un entier p tel que $f_1f_2\dots f_l$ appartienne à $F_{l,p+2}$ et $\bar{f}_{2l+4}$ $\bar{f}_{2l+3}\dots\bar{f}_{l+3}$ à $F_{l+2,p}$. Ainsi $|D_{2l+4}|=\sum_{p=0}^{l}|F_{l,p+2}||F_{l+2,p}|$.

Comme $M_{2l+4}\cap D_{2l+4}=\phi$, on a bien l'égalité, d'après la propriété 3 :

$$|V \cap Z^l|=\sum_{p=0}^{l}(|F_{l,p}||F_{l+2,p+2}|-|F_{l,p+2}||F_{l+2,p}|)=|M_{2l+4}|-|D_{2l+4}| .$$

Propriété 5 : a) $|M_{4n+4}|-|D_{4n+4}|= C_nC_{n+1}$.

b) $|M_{4n+2}|-|D_{4n+2}|= C_nC_n$.

Démonstration :

a) $l=2n$

Soit f un élément de F_{2l+4}. Le mot f peut s'écrire $f=f_1f_2f_3$ avec $|f_1|= |f_3|=l$ et $|f_2|=4$. Soit 2p l'image de f_1 par δ. Si de plus f appartient à M_{2l+4}, il est dans l'une des quatre classes suivantes définies par la valeur de f_2 :

α) $f_2 = aa\bar{a}\bar{a}$, $\delta(f_1f_2)=2p$ avec $0 \le p \le n$.

β) $f_2 == aa\bar{a}a$, $\delta(f_1f_2)=2p+2$ avec $0 \le p \le n-1$.

γ) $f_2 = aaa\bar{a}$, $\delta(f_1f_2)=2p+2$ avec $0 \le p \le n-1$.

δ) $f_2 = aaaa$, $\delta(f_1f_2)=2p+4$ avec $0 \le p \le n-2$.

Si f est un élément de D_{2l+4}, suivant la valeur de f_2, il appartient à l'une des quatre classes suivantes :

α') $f_2 = \bar{a}\bar{a}aa$, $\delta(f_1)=2p$ avec $1 \le p \le n$.

β') $f_2 = \bar{a}\bar{a}a\bar{a}$, $\delta(f_1)=2p+2$ avec $0 \le p \le n-1$.

γ') $f_2 = \bar{a}\bar{a}\bar{a}a$, $\delta(f_1)=2p+2$ avec $1 \le p \le n-1$

δ') $f_2 = \bar{a}\bar{a}\bar{a}\bar{a}$, $\delta(f_1)=2p+4$ avec $0 \le p \le n-2$.

Soit P l'application de Z^{2l+4} dans Z^{2l+4} définie par :

$$P(f_1f_2f_3) = \tilde{\bar{f}}_3\bar{f}_2\tilde{\bar{f}}_1 \text{ avec } |f_1|=|f_3|=1 \text{ et } |f_2|= 4 .$$

Si on décompose les classes α et γ de M_{2l+4} en les 3 classes suivantes:

α_1) $f_2=aa\bar{a}\bar{a}$, $\delta(f_1f_2)=2p$ avec $1 \le p \le n$,

γ_1) $f_2=aaa\bar{a}$, $\delta(f_1f_2)=2p+2$ avec $1 \le p \le n-1$,

ε) soit $f_2=aa\bar{a}\bar{a}$ et $\delta(f_1f_2)=0$, soit $f_2=aaa\bar{a}$ et $\delta(f_1f_2)=2$,

il est clair que P est une bijection entre la classe α_1 (resp. β, γ_1 et δ) de M_{2l+4} et la classe α' (resp. β', γ' et δ') de D_{2l+4} . Donc le nombre d'éléments de la classe ε est égal à $|M_{4n+4}|-|D_{4n+4}|$ (cf. figure 5).

Or un élément de la classe ε est égal soit a $f_1aa\bar{a}\bar{a}f_3$ avec $\delta(f_1)=\delta(f_3)=0$ soit $f_1aaa\bar{a}f_3$ avec $\delta(f_1)=\delta(aaa\bar{a}f_3)=0$.

Notons soit $f_4=a\bar{a}f_3$ dans le premier cas, soit $f_4=aaf_3$ dans le second cas. Le couple (f_1,f_4) appartient alors a $F_{2n,0} \times F_{2n+2,0}$. Inversement à tout couple de $F_{2n,0} \times F_{2n+2,0}$. On peut associer un élément de la classe ε de M_{2l+4} .

Ainsi $C_nC_{n+1}=|M_{4n+4}|-|D_{4n+4}|$.

b) $l=2n-1$.

La démonstration se fait de la même manière.

Soit f un élément de F_{2l+4}. Le mot f peut s'écrire $f=f_1f_2f_3$ avec $|f_1|=|f_3|=l$ et $|f_2|=4$. Soit $2p-1$ l'image de f_1 par δ .

Si de plus f appartient à M_{2l+4}, il est dans l'une des quatre classes suivantes définies par la valeur de f_2 :

α) $f_2=aa\bar{a}\bar{a}$, $\delta(f_1f_2)=2p-1$ avec $1\leq p\leq n$.

β) $f_2=aa\bar{a}a$, $\delta(f_1f_2)=2p+1$ avec $1\leq p\leq n-1$.

γ) $f_2=aaa\bar{a}$, $\delta(f_1f_2)=2p+1$ avec $1\leq p\leq n-1$.

δ) $f_2=aaaa$, $\delta(f_1f_2)=2p+3$ avec $1\leq p\leq n-2$.

Par contre si f appartient à D_{2l+4}, il est dans l'une des quatre classes suivantes définies selon les valeurs de f_2 :

α') $f_2=\bar{a}\bar{a}aa$, $\delta(f_1)=2p-1$ avec $2\leq p\leq n$.

β') $f_2=\bar{a}\bar{a}a\bar{a}$, $\delta(f_1)=2p+1$ avec $1\leq p\leq n-1$.

γ') $f_2=\bar{a}\bar{a}\bar{a}a$, $\delta(f_1)=2p+1$ avec $1\leq p\leq n-1$.

δ') $f_2=\bar{a}\bar{a}\bar{a}\bar{a}$, $\delta(f_1)=2p+3$ avec $1\leq p\leq n-2$.

Si on décompose la classe α en les deux classes suivantes :

α_1) $f_2=aa\bar{a}\bar{a}$, $\delta(f_1f_2)=2p-1$ avec $2\leq p\leq n$,

ε) $f_2=aa\bar{a}\bar{a}$, $\delta(f_1f_2)=1$,

il est clair que l'application P décrite plus haut est aussi une bijection de la classe α_1 (resp. β,γ et δ) de M_{2l+4} dans la classe α' (resp. β',γ' et δ') de D_{2l+4}. Et donc le nombre d'éléments de la classe ε est égal à $|M_{4n+2}|-|D_{4n+2}|$ (cf. figure 6).

Mais un élément f de la classe ε s'écrit $f_1aa\bar{a}\bar{a}f_3$ avec la condition $\delta(f_1)=\delta(\bar{\tilde{f}}_3)=1$. La classe ε est donc en bijection avec l'ensemble $F_{2n-1}\times F_{2n-1,1}$. Or $F_{2n-1.1}=C_n=\dfrac{(2n)\ !}{n!\,(n+1)\ !}$.

Ainsi $C_nC_n = |M_{4n+2}| - |D_{4n+2}|$.

Ceci termine la preuve de la propriété 5.

Le théorème 1 est une conséquence directe des propriétés 4 et 5.

Les premières valeurs des coefficients de la série énumératrice sont : 1, 1, 2, 4, 10, 25, 70, 196, 588, 1764, 5544, 17424, 56628, 184041.

Les identités (1) et (2) résultent des propriétées 3 et 5.

Remarque 1 :

On voit facilement que le langage V n'est pas algébrique. Les séries :

$$h(z) = \sum_{n=0}^{\infty} C_nC_n z^n = 4/\pi \int_0^{\pi/2} (1-\sqrt{1-16z\sin^2\varphi})(8z\sin^2\varphi)^{-1}\cos^2\varphi \, d\varphi$$

$$\text{et} \quad g(z) = \sum_{n=0}^{\infty} C_nC_{n+1} z^n = 2/\pi \int_0^{\pi/2} (1-\sqrt{1-16z\sin^2\varphi})\, z^{-1}\cos^2\varphi \, d\varphi$$

ne sont pas algébriques et donc la série génératrice des mots de V ne l'est pas non plus. Par contre elles sont "differentiably finite" au sens de Stanley [24] puisque leurs coefficients vérifient les relations :

$$(n+2)^2\, C_{n+1}C_{n+1} - 4(2n+1)^2 C_nC_n = 0$$

$$\text{et} \quad (n+3)(n+2)C_{n+1}C_{n+2} - 4(2n+3)(2n+1)C_nC_{n+1} = 0.$$

Remarque 2 :

On appelle R-chemins les chemins sous-diagonaux qui sont colorés de **la façon** suivante : les pas Ouest (resp. Nord, resp. Sud) dont le point de départ a des coordonnées de parité différente sont colorés avec quatre (resp. deux, resp. deux) couleurs. Par les mêmes techniques que précédemment (cf. [13]), on peut démontrer le résultat suivant : le nombre de R-chemins de longueur 2n aboutissant à l'origine est égal à $(R_nR_{n+2} - R_{n+1}^2)/2$ où R_n est le nombre de Schröder

$$\sum_{i\geq 0} \binom{2n-i}{i} C_{n-i} .$$

Ce résultat peut être rapproché du fait que le nombre de chemins sous-diagonaux de longueur 2n retournant à l'origine est égal à $C_n C_{n+2} - C_{n+1}^2 = \frac{3!(2n)!(2n+2)!}{n!(n+1)!(n+2)!(n+3)!}$.

Les premiers nombres de R-chemins de longueur 2n aboutissant à l'origine sont $(0 \leq n \leq 6)$: 1, 4, 28, 284, 3652, 55108, 932476.

Remarque 3 :

Il existe une bijection ([13],[14]) entre les tableaux de Young standard ayant n cases, p colonnes de hauteur impaire et au plus 4 lignes et les chemins sous-diagonaux de longueur n aboutissant au point de coordonnées (p,0). On a alors les résultats suivants : Le nombre de tableaux de Young standard ayant l cases et au plus 4 lignes est égal à $C_n C_n$ pour $l=2n-1$ et $C_n C_{n+1}$ pour $l=2n$.
Le nombre de tableaux de Young standard ayant n cases et au plus 5 lignes est égal à $\sum_{i=0}^{n/2} \frac{3!n!(2i+2)!}{(n-2i)!i!(i+1)!(i+2)!(i+3)!}$

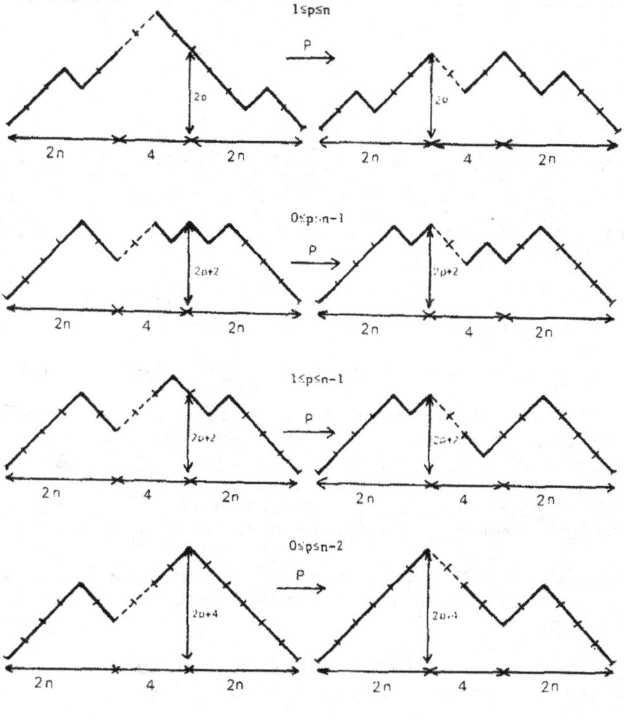

FIGURE 5

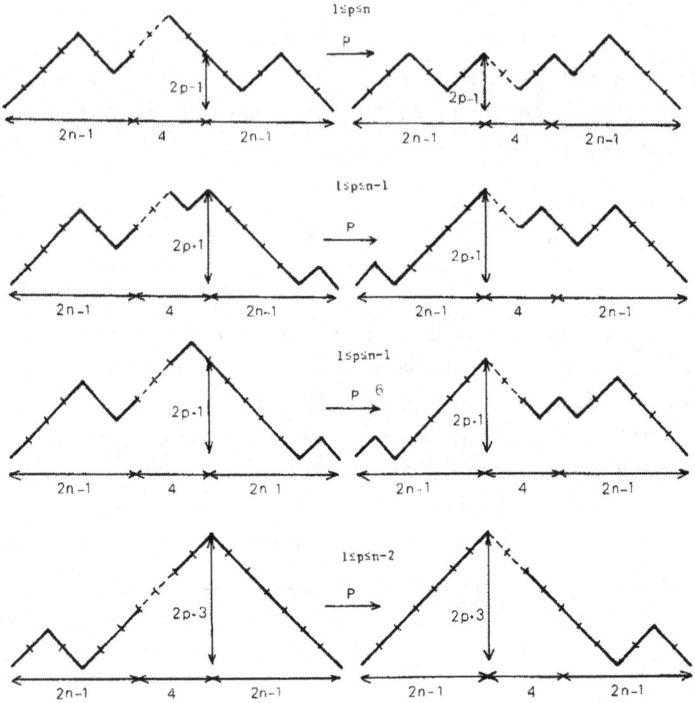

FIGURE 6

Références :

[1] ANDRE, Solution... . C.R. Acad. Sci. Paris, 105(1887), 436-437.

[2] BERTRAND, Solution d'un problème. C.R. Acad. Sci. Paris, 105 (1887), 369.

[3] E. CATALAN, Notes sur une équation aux différences finies. J. Math. Pures Appl., 3(1838), 508-516.

[4] L. COMTER, Analyse combinatoire, P.U.F., Paris, 1970.

[5] R. CORI, S. DULUCQ, G. VIENNOT , Shuffle of parenthesis systems and Baxter permutations. Rapport CNRS LA 226, Bordeaux (1984), à paraitre dans J.C.T. (A).

[6] S. DULUCQ, Equations avec opérateurs : un outil combinatoire. Thèse 3ème cycle, Univ. Bordeaux 1 (1981).

[7] S. DULUCQ, G. VIENNOT, Un théorème généralisant l'invertion des matrices et le "théorème Maitre" de Mac-Mahon. Manuscrit (1982).

[8] L. EULER, Novi Comentarii Academiae Scientarium Imperialis Petropolitanae, 7(1758-1759), 13-14.

[9] P. FLAJOLET, Combinatorial aspects of continued fractions. Discrete math., 32 (1980), 125-161.

[10] J. FRANCON, G. VIENNOT , Permutations selon leurs pics, creux, doubles montées et doubles descentes, nombres d'Euler et nombres de Genocchi. Discrete Math., 28 (1979),21-35.

[11] I.M. GESSEL, G. VIENNOT, Binomial determinants, Paths and Hook lengths formulae, Rapport Bordeaux I n°8502, à paraitre dans Adv. in Maths.

[12] D. GOUYOU-BEAUCHAMPS, Deux propriétés combinatoires du langage de Lukasiewicz, R.A.I.R.O., 3 (1975), 13-24.

[13] D. GOUYOU-BEAUCHAMPS, Codages par des mots et des chemins : problème combinatoires et algorithmiques, Thèse d'état Université de Bordeaux I, 1985.

[14] D. GOUYOU-BEAUCHAMPS, Standard Young tableaux of height 4 and 5, soumis à Europ. J. Combinatorics.

[15] F. HARARY, G. PRINS, W.T. TUTTE, The number of plane trees. Indag. Math., 26 (1964), 319-329.

[16] KREWERAS, Dénombrement des chemins minimaux à sauts imposés. C.R. Acad. Sci. Paris, 263 (1966), 1-3.

[17] R.C. MULLIN, The enumeration of hamiltonian polygons in triangular maps. Pacific J. Math., 16(1966), 139-145.

[18] T.V. NARAYANA, A partial order and its applications to probality theory. Shankhya, 21 (1959) 91-98.

[19] J.B. REMMEL, Bijective proofs of formulae for the number of standard Young tableaux. Linear and Multilinear Algebra 11 (1982), 45-100.

[20] J.B. REMMEL, R. WHITNEY, A bijective proof of the hook formula for the number of column strict tableaux with bounded entries. Europ. J. Combinatorics, 4 (1983), 45-63.

[21] O. RODRIGUES, Sur le nombre de manières d'effectuer un produit de n facteurs. J. Math. Pures et Appl., 3(1838), 549.

[22] M. de SAINTE-CATHERINE, Couplage et Pfaffien en combinatoire et et physique. Thèse 3ème cycle, Université de Bordeaux I 1983.

[23] J.A. SEGNER, Enumeralis modurum quibus figurae planae rectilinae perdiagonales dividantur in triangula. Novi Comentarii Academiae Scientatium Imperialis Petrpolitanae, 7 (1758-1759), 203-209.

[24] R.P. STANLEY, Differentiably finite power series. Europ. J. Comb., 1 (1980), 175-188.

[25] W.T. TUTTE, A census of hamiltonian polygons. Canad. J. Maths. 14 (1962), 402-417.

[26] G. VIENNOT, Interpretations combinatoires des nombres d'Euler et de Genocchi. Séminaire de théorie des nombres de l'Université de Bordeaux I, exposé 11 (1981-1982).

[27] G. VIENNOT, Théorie combinatoire des polynômes orthogonaux généraux. Notes de séminaires de l'université du Québec à Montréal, (1983).

Foncteurs analytiques et espèces de structures

par

André Joyal
Université du Québec à Montréal

Introduction

Le but de ce travail est d'exposer la théorie des espèces de structures en relation avec la théorie des foncteurs analytiques. Ceci permet de donner à la théorie des opérations sur les espèces de structures une base conceptuelle satisfaisante. Nous développerons ensuite la théorie des espèces tensorielles en gardant en vue des applications aux algèbres de Lie libres.

Chapitre 1. Foncteurs analytiques

1.0. La construction du monoïde librement engendré par un ensemble A est succinctement résumée par la série géométrique

$$L(A) = \sum_{n\geq 0} A^n$$

De même, la série exponentielle

$$\exp(A) = \sum_{n\geq 0} A^n / \mathfrak{S}_n$$

nous donne la construction du monoïde commutatif libre sur A (le quotient $A^n/\mathfrak{S}_n$ est pris relativement à l'action naturelle du groupe symétrique $\mathfrak{S}_n$ sur A^n).

Une troisième construction est celle de l'ensemble C(A) des mots circulaires sur un alphabet A:

$$C(A) = \sum_{n\geq 0} A^n / C_n$$

où C_n est le sous-groupe de $\mathfrak{S}_n$ engendré par la permutation $i \mapsto i + 1 \pmod{n}$.

Avec l'appui financier du programme FCAR, Québec, EQ 1608, et du CRSNG, Canada.

Ces trois constructions sont en fait des foncteurs de la catégorie des ensembles vers elle-même,

$$L\,,\ \exp\,,\ C\ :\ \mathrm{Ens} \longrightarrow \mathrm{Ens}$$

Plus généralement, soit

$$\mathfrak{S}_n \times F[n] \longrightarrow F[n] \qquad (n \geq 0)$$

une suite de représentations ensemblistes des groupes symétriques.

Définition 1. Un foncteur F: Ens $\longrightarrow$ Ens est *analytique* s'il possède un développement en série de Taylor

$$F(A) = \sum_{n \geq 0} A^n \underset{\mathfrak{S}_n}{\times} F[n]$$

le n-ième terme de cette série étant le quotient du produit cartésien $A^n \times F[n]$ par la relation d'équivalence identifiant $(f, \sigma t)$ avec $(f\sigma, t)$ pour tout $f \in A^n, \sigma \in \mathfrak{S}_n, t \in F[n]$.

Remarque. Nous avons fait agir $\mathfrak{S}_n$ à droite sur A^n en posant $f\sigma = (f_1, \ldots, f_n)\sigma = (f_{\sigma(1)}, \ldots, f_{\sigma(n)})$. On peut aussi faire agir $\mathfrak{S}_n$ à gauche en posant $\sigma.f = f\,\sigma^{-1}$. On a alors une bijection.

$$A^n \underset{\mathfrak{S}_n}{\times} F[n] \cong F[n] \times A^n \,/_{\mathfrak{S}_n}$$

Nous dirons que $(F[n] \mid n \geq 0)$ est la suite de *coefficients* de Taylor du foncteur F. Cette terminologie présume de l'unicité du développement en série de Taylor d'un foncteur analytique. Nous allons énoncer un résultat à cet effet. Remarquons d'abord qu'une suite $\theta = (\theta_n \mid n \geq 0)$ d'applications $\mathfrak{S}_n$–équivariantes

$$\theta_n : F[n] \longrightarrow G[n] \qquad (n \geq 0)$$

induit une transformation naturelle

$$\underline{\theta} : F \longrightarrow G$$

entre les foncteurs analytiques correspondants. Nous dirons que la suite θ est un *morphisme* de la suite $(F[n] \mid n \geq 0)$ vers la suite $(G[n] \mid n \geq 0)$.

Théorème 1. L'application $\theta \rightarrow \underline{\theta}$ est une bijection entre l'ensemble des isomorphismes de $(F[n] \mid n \geq 0)$ vers $(G[n] \mid n \geq 0)$ et l'ensemble des isomorphismes naturels de F vers G.

Autrement dit, la série de Taylor d'un foncteur analytique est unique à isomorphisme canonique près. Ce résultat se généralise comme suit. Un carré commutatif

$$\begin{array}{ccc} A & \xrightarrow{f} & B \\ {\scriptstyle g}\downarrow & & \downarrow{\scriptstyle v} \\ C & \xrightarrow[u]{} & D \end{array}$$

est *régulier* si pour tout $(b, c) \in B \times C$ tel que $v(b) = u(c)$ il existe $a \in A$ tel que $(f(a), g(a)) = (b, c)$.

Définition 2. Soient F, G des foncteurs Ens $\longrightarrow$ Ens. Une transformation naturelle α: F $\longrightarrow$ G est *régulière* si pour tout f: A $\longrightarrow$ B le carré

$$\begin{array}{ccc} F(A) & \xrightarrow{F(f)} & F(B) \\ {\scriptstyle \alpha_A}\downarrow & & \downarrow{\scriptstyle \alpha_B} \\ G(A) & \xrightarrow[G(f)]{} & G(B) \end{array}$$

est régulier.

Proposition 1. L'application $\theta \mapsto \underline{\theta}$ est une bijection entre l'ensemble des morphismes de $(F[n] \mid n \geq 0)$ vers $(G[n] \mid n \geq 0)$ et l'ensemble des transformations naturelles régulières de F vers G.

La démonstration est contenue dans l'appendice. Le théorème 1 est une conséquence de cette proposition puisque tout isomorphisme naturel F $\longrightarrow$ G est régulier.

Il est possible de caractériser les foncteurs analytiques par des conditions simples: bicontinuité et régularité (appendice).

La proposition entraîne que pour tout foncteur analytique F il existe une et une seule transformation naturelle régulière (le *support*):

$$\text{supp: } F \longrightarrow \exp$$

En effet, comme $\exp[n] = 1$ pour tout $n \geq 0$, il y a un et un seul morphisme de suites

$$\theta: (F[n] \mid n \geq 0) \longrightarrow (\exp[n] \mid n \geq 0)$$

et par suite une et une seule transformation naturelle régulière $\underline{\theta}$ = supp.

1.1. Les foncteurs analytiques de plusieurs variables sont définis au moyen des séries de Taylor en plusieurs variables. Soit $d \in \mathbb{N}$. Pour tout $n = (n_1, .. , n_d) \in \mathbb{N}^d$ et pour tout $A = (A_1, .. , A_d) \in \text{Ens}^d$, posons

$$A^n = A_1^{n_1} \times \ldots \times A_d^{n_d} \qquad\qquad \mathfrak{S}_n = \mathfrak{S}_{n_1} \times \ldots \times \mathfrak{S}_{n_d}$$

Soit $(F[n] \mid n \in \mathbb{N}^d)$ une famille de $\mathfrak{S}_n$–ensembles

$$\mathfrak{S}_n \times F[n] \longrightarrow F[n] \qquad\qquad (n \in \mathbb{N}^d)$$

Un foncteur F: $\text{Ens}^d \longrightarrow$ Ens est *analytique* s'il possède un développement en série de Taylor

$$F(A_1, \ldots, A_d) = \sum_{n \in \mathbb{N}^d} F[n] \times A^n / \mathfrak{S}_n$$

On démontre que la suite de coefficients $(F[n] \mid n \in \mathbb{N}^d)$ est uniquement déterminée (à isomorphisme canonique près) par le foncteur F.

Notation. Nous utiliserons, pour désigner les foncteurs, les conventions usuelles de l'Analyse pour les fonctions d'une ou de plusieurs variables. Nous écrirons souvent F(X) plutôt que F pour désigner un foncteur d'une variable X. Nous écrirons de même F(X, Y) pour désigner un foncteur de deux variables. Avec ces notations, X pourra dénoter aussi bien le foncteur identité $Id: Ens \longrightarrow Ens$ que le foncteur projection $p_1: Ens^2 \longrightarrow Ens$.

1.2. Soit Γ un groupe fini agisssant sur un ensemble fini I. Pour tout ensemble A, le groupe Γ agit naturellement sur $A^I = \{f: I \longrightarrow A\}$:

$$(\sigma.f)(i) = f(\sigma^{-1}i) \qquad (i \in I, \sigma \in \Gamma)$$

La *puissance symétrique* de A par le Γ-ensemble I est le quotient $A^I/_\Gamma$. Vérifions que l'opération de puissance symétrique est un foncteur analytique. En effet, soit $n = \text{Card } I$ et soit $\Delta(I, n)$ l'ensemble des bijections de I vers $\{1, 2, ..., n\}$. La composition $(v,f) \rightarrow f \circ v$ définit une application

$$\Delta(I, n) \times A^n \longrightarrow A^I$$

Les groupes Γ et $\mathfrak{S}_n$ opèrent sur $\Delta(I, n)$ et ces actions commutent entre elles. On obtient, par passage au quotient, une application

$$\Delta(I, n)/_\Gamma \times A^n \longrightarrow A^I/_\Gamma$$

et, finalement, une bijection

$$\Delta(I, n)/_\Gamma \times A^n/_{\mathfrak{S}_n} \xrightarrow{\sim} A^I/_\Gamma$$

Les puissances symétriques se généralisent au cas de plusieurs variables: soit Γ un groupe fini agissant sur des ensembles finis $I_1, ..., I_d$. La puissance symétrique d'un multi-ensemble $(A_1, ..., A_d)$ par $(I_1, ..., I_d)$ est le quotient

$$A_1^{I_1} \times ... \times A_d^{I_d}/_\Gamma$$

On vérifie qu'il y a un isomorphisme

$$A_1^{I_1} \times ... \times A_d^{I_d}/_\Gamma \longrightarrow \Delta(I, n)/_\Gamma \times A^n/_{\mathfrak{S}_n}$$

où $n = (n_1, ..., n_d) = (\text{card } I_1, ..., \text{card } I_d)$ et $\Delta(I, n) = \Delta(I_1, n_1) \times ... \times \Delta(I_d, n_d)$.

1.3. Nous allons maintenant décrire des opérations sur les foncteurs. Nous démontrerons plus tard que ces opérations préservent l'analycité.

(i) la *somme*

$$(F + G)(A) = F(A) + G(A)$$

et, plus généralement, la combinaison linéaire

$$(\sum_{i \in I} C_i \times F_i)(A) = \sum_{i \in I} C_i \times F_i(A)$$

où $(C_i \mid i \in I)$ est une famille d'ensembles et $(F_i \mid i \in I)$ est une famille de foncteurs;

(ii) le *produit* (fini)

$$(F{\cdot}G)(A) = F(A) \times G(A)$$

et, plus généralement,

$$(\prod_{i\in I} F_i)(A) = \prod_{i\in I} F_i(A)$$

et aussi la puissance

$$F^I(A) = F(A)^I$$

(iii) la *composition* (ou la *substitution*)

$$(F{\circ}G)(A) = F(G(A))$$

et, plus généralement,

$$F(G_1, .. ,G_r)(A) = F(G_1(A), .. ,G_r(A))$$

où $F: Ens^r \longrightarrow Ens$ et $G_i: Ens^d \longrightarrow Ens$ $(1 \le i \le r)$;

(iv) le *quotient* par l'action d'un groupe

$$(F/_\Gamma)(A) = F(A)/_\Gamma$$

où Γ agit sur F par transformation naturelle.

Théorème 2. Les foncteurs analytiques sont clos sous les opérations suivantes:

i) les combinaisons linéaires finies ou infinies
ii) les produits finis
iii) la composition
iv) le quotient par l'action d'un groupe.

Pour démontrer ce théorème, on peut utiliser les propriétés caractérisant l'analycité (appendice) ou encore procéder par calcul direct des coefficients des séries de Taylor. Cependant, il appert que ces coefficients ont une forme compliquée quand on choisit de les décrire, ainsi que l'avons fait jusqu'à présent, comme des représentations ensemblistes des groupes symétriques. Nous obtenons des calculs plus transparents en prenant un autre modèle: les espèces de structures. La démonstration du théorème est par suite reportée en 2.1.

Chapitre 2. Espèces de structures

2.0. La théorie des espèces de structures est essentiellement une théorie des coefficients des foncteurs analytiques. Elle entretient une relation étroite avec la théorie des séries formelles et avec la combinatoire. Grâce à elle, il devient possible de développer une vision combinatoire des opérations définies sur les foncteurs analytiques.

Définition 1. Une *espèce de structures* F est un foncteur

$$F[\ ]: \mathbf{B} \longrightarrow \mathrm{Ens}$$

où **B** est la catégorie des ensembles finis et bijections. Un élément $\alpha \in F[S]$ est une *structure d'espèce* F sur S; nous dirons aussi que c'est une F-*structure* sur S. Si $\beta = F[u](\alpha)$ où $u: S \xrightarrow{\sim} T$, nous dirons que β est obtenue en *transportant* α le long de la bijection u et que u est un *isomorphisme* de α vers β.

Les F-structures forment une catégorie (un groupoïde): la catégorie *el* F des *éléments* de F; un objet de *el* F est un couple (α, S) où $\alpha \in F[S]$; un morphisme $u: (\alpha, S) \longrightarrow (\beta, T)$ est un isomorphisme de α vers β. Les composantes connexes de ce groupoïde sont les *types* de structures: deux structures ont le même type si et seulement si elles sont isomorphes. Nous désignerons l'ensemble des types de l'espèce F par

$$\tilde{\sum_S} F[S]$$

Pour chaque entier $n \in \mathbb{N}$, on définit, par transport de structures, une action du groupe symétrique $\mathfrak{S}_n$ sur $F[n] = F[\{1,2, \ldots,n\}]$.

Proposition 1. On a une bijection canonique

$$\tilde{\sum_S} F[S] = \sum_{n \geq 0} F[n] /_{\mathfrak{S}_n}$$

Preuve. Toute structure $\alpha \in F[S]$ est isomorphe à une structure $\alpha' \in F[n]$ où $n = \mathrm{Card}\ S$. De plus, α' et α'' sont isomorphes à α si et seulement si elles sont isomorphes entre-elles, c'est-à-dire si et seulement si il existe $\sigma \in \mathfrak{S}_n$ tel que $\sigma.\alpha' = \alpha''$. C.Q.F.D.

Une espèce de structure de plusieurs variables est un foncteur

$$F[\ ]: \mathbf{B}^d \longrightarrow \mathrm{Ens}$$

où $d \in \mathbb{N}$ est le nombre de variables. On définit sans difficulté les concepts d'*isomorphisme* et de *type* de F-structures. Nous dénoterons

$$\tilde{\sum_{S_1, \ldots, S_d}} F[S_1, \ldots, S_d]$$

l'ensemble des types de l'espèce F. Il est aussi possible de définir des sommations partielles; ainsi

$$\tilde{\sum_T} F[S, T]$$

désignera, pour chaque $S \in \mathbf{B}$, l'ensemble des types de l'espèce $T \mapsto F[S, T]$.

Un exemple d'espèce de deux variables est l'espèce Δ. Pour tout $(S, T) \in \mathbf{B}^2$ on a

$$\Delta[S, T] = \{f: S \longrightarrow T \mid f \text{ est bijective}\}$$

Nous allons maintenant dresser une liste d'identités gouvernant l'utilisation de $\tilde{\sum}$ et de Δ. Ces identités seront utiles dans les calculs à venir.

a) associativité

$$\tilde{\sum_S}\ \tilde{\sum_T} F[S, T] = \tilde{\sum_{S,T}} F[S, T]$$

b) commutativité

$$\tilde{\sum_{S,T}} F[S, T] = \tilde{\sum_{T,S}} F[S, T]$$

c) distributivité

$$\tilde{\sum_T} G[S] \times F[S, T] = G[S] \times \tilde{\sum_T} F[S, T]$$

d) remplacement

$$F[S] \times \Delta[S, T] = F[T] \times \Delta[S, T]$$

e)

$$\tilde{\sum_T} \Delta[S, T] = 1$$

Le symbole d'égalité dans ces identités signifie qu'il y a un isomorphisme naturel entre les deux membres. Les espèces figurant dans ces formules peuvent comporter d'autres variables que celles décrites explicitement. Cependant, pour la distributivité, il est nécessaire de supposer que G[S] est indépendante de la variable T. Ajoutons à cette liste quelques relations:

f)

$$\tilde{\sum_{S,T}} F[S, T] = \tilde{\sum_R} \sum_{S+T=R} F[S, T]$$

où la sommation ordinaire est indexée par les couples (S, T) tels que $S \cap T = \varnothing$ et $S \cup T = R$.

g)

$$\tilde{\sum_S} \sum_{i \in I} A_i \times \Delta[T_i, S] = \sum_{i \in I} A_i$$

h)

$$\Delta[U+V, R] = \sum_{S+T=R} \Delta[U, S] \times \Delta[V, T]$$

Voyons qu'une espèces de structures F est entièrement déterminée (à isomorphisme canonique près) par la suite $(F[n] \mid n \geq 0)$.

Proposition 2. On a un développement en série

$$F[S] = \sum_{n \geq 0} F[n] \times \Delta[S, n] /_{\mathfrak{S}_n}$$

Preuve. On a successivement

$$\begin{aligned} F[S] &= F[S] \times \tilde{\sum_T} \Delta[S, T] \\ &= \tilde{\sum_T} F[S] \times \Delta[S, T] \\ &= \tilde{\sum_T} F[T] \times \Delta[S, T] \\ &= \sum_{n \geq 0} F[n] \times \Delta[S, n] /_{\mathfrak{S}_n} \end{aligned}$$

où nous avons utilisé e), c), d) et la proposition 1. On peut aussi faire une démonstration directe que nous laissons au lecteur.

Le développement en série de la proposition 2 nous permet de voir que toute suite $(F[n] \mid n \geq 0)$ de représentations des groupes symétriques définit une espèce

$$S \mapsto \sum_{n \geq 0} F[n] \times \Delta[S, n] /_{\mathfrak{S}_n}$$

Nous dirons que c'est l'*extension* de la suite $(F[n] \mid n \geq 0)$.

Pour tout entier $n \geq 0$, soit $Ens^{\mathfrak{S}_n}$ la catégorie des $\mathfrak{S}_n$-ensembles. Posons

$$Ens^{\mathfrak{S}_*} = \prod_{n \geq 0} Ens^{\mathfrak{S}_n}$$

D'autre part, soit $Ens^{\mathbf{B}}$ la catégorie des espèces (les morphismes étant les transformations naturelles). on a le foncteur *restriction* $r : F \mapsto (F[n] \mid n \geq 0)$

$$r : Ens^{\mathbf{B}} \longrightarrow Ens^{\mathfrak{S}_*}$$

et le foncteur *extension*

$$e : Ens^{\mathfrak{S}_*} \longrightarrow Ens^{\mathbf{B}}$$

Proposition 3. Les foncteurs r et e sont des équivalences de catégories (quasi-)inverses l'une de l'autre.

Nous pouvons maintenant donner une description combinatoire du foncteur analytique associé à une espèce F. Pour tout ensemble A, considérons l'espèce

$$S \mapsto F[S] \times A^S$$

Nous dirons qu'un élément $(\alpha, p) \in F[S] \times A^S$ est une *F-structure étiquetée* par A; la fonction $p: S \longrightarrow A$ est un *étiquetage* de α par A; un élément de A est une *étiquette* . Posons

$$F(A) = \tilde{\sum_S} F[S] \times A^S$$

Un élément de F(A) est un *type* de F-structures étiquetées par A. Pour toute fonction $f: A \longrightarrow B$, nous noterons $F(f): F(A) \longrightarrow F(B)$ l'application obtenue en composant les étiquetages par f. Nous avons un foncteur

$$F(\ ): Ens \longrightarrow Ens$$

Proposition 4. Le foncteur F( ) est analytique et l'on a

$$F(A) = \sum_{n \geq 0} F[n] \times A^n /_{\mathfrak{S}_n}$$

Preuve. C'est une conséquence de la proposition 1.

Les résultats démontrés jusqu'ici nous permettent maintenant d'affirmer qu'il y a équivalence entre les trois concepts suivants:

A) Une suite de représentation ensemblistes des groupes symétriques

$$\mathfrak{S}_n \times F[n] \longrightarrow F[n] \qquad (n \geq 0)$$

B) Un foncteur analytique

$$F(\ ): \text{Ens} \longrightarrow \text{Ens}$$

C) Une espèce de structures

$$F[\ ]: \mathbf{B} \longrightarrow \text{Ens}$$

Notation. Nous utiliserons indifféremment F ou F(X) pour désigner l'un ou l'autre de ces concepts. L'usage de parenthèses nous permettra de résoudre, au besoin, l'ambiguité. Autrement dit,

$$F[S],\ F[n],\ F(X)[S],\ F(X)[n]$$

sont des *coefficients* de F alors que F(A) est le résultat de l'*évaluation* du foncteur analytique F(X) en $A \in \text{Ens}$.

2.1. Dans ce paragraphe, nous allons donner une liste de formules pour le calcul de la somme, du produit et de la substitution d'espèces de structures. Chacune des formules est une recette combinatoire permettant de construire les structures de l'espèce désirée.

(i) la *somme*

$$(F+G)[S] = F[S] + G[S]$$

et la combinaison linéaire

$$(\sum_{i\in I} C_i \times F_i)[S] = \sum_{i\in I} C_i \times F_i[S]$$

(ii) le *produit fini*

$$(F\cdot G)[S] = \sum_{U+V=S} F[U] \times G[V]$$

$$(\prod_{i\in I} F_i)[S] = \sum_{f\in I^S} \prod_{i\in I} F_i[f^{-1}(i)]$$

$$F^I[S] = \sum_{f\in I^S} \prod_{i\in I} F[f^{-1}(i)]$$

(iii) la *composition* $F\circ G$ (si $G[\varnothing] = \varnothing$)

$$(F\circ G)[S] = \sum_{R\in Eq[S]} F[S/R] \times \prod_{C\in S/R} G[C]$$

où Eq[S] est l'ensemble des relations d'équivalence sur S.

(iv) la *composition* $F\circ G$ (cas général). Considérons l'espèce en deux variables

$$W[S, T] = \sum_{f:S\to T} F[T] \times \prod_{i\in T} G[f^{-1}(i)]$$

On a

$$(F\circ G)[S] = \tilde{\sum_{T}} W[S, T]$$

(v) la *composition* $F(G_1, \dots, G_r)$ (si $G_i[\varnothing] = \varnothing$, $1\leq i\leq r$)

$$F(G_1, \dots, G_r)[S] = \sum_{(R, t)\in Eq^r[S]} F[S/R] \times \prod_{C\in |S|_{/R}} G_{t(c)}[\underline{C}]$$

où $S = (S_1, \dots, S_d)$, $|S| = S_1 + \dots + S_d$ et où

$$Eq^r[S] = \{(R, t) \mid R\in Eq[|S|],\ t: |S|_{/R} \longrightarrow \{1, 2, \dots, r\}\}$$

Nous avons posé

$$S/R = (t^{-1}(1), \dots, t^{-1}(r)) \in \mathbf{B}^r \quad \text{et} \quad \underline{C} = (C\cap S_1, \dots, C\cap S_d) \in \mathbf{B}^d$$

(vi) l'**évaluation** (partielle). Si $F = F(X, Y)$ et $A\in$ Ens, alors

$$\begin{aligned} F(A, Y)[T] &= \tilde{\sum_{S}} F[S, T] \times A^S \\ &= \sum_{n\geq 0} F[n, T] \times A^n / \mathfrak{S}_n \end{aligned}$$

(vii) le **quotient** par un groupe

$$(F/_\Gamma)[S] = F[S]/_\Gamma$$

Preuve. Les formules i), vi) et vii) sont évidentes.

ii)
$$\begin{aligned} F(A) \times G(A) &= (\tilde{\sum_{S}} F[S] \times A^S) \times (\tilde{\sum_{T}} G[T] \times A^T) \\ &= \tilde{\sum_{S,T}} F[S] \times G[T] \times A^{S+T} \\ &= \tilde{\sum_{R}} \sum_{S+T=R} F[S] \times G[T] \times A^{S+T} \\ &= \tilde{\sum_{R}} (\sum_{S+T=R} F[S] \times G[T])A^R \end{aligned}$$

iv)
$$\begin{aligned} F(G(A)) &= \tilde{\sum_{T}} F[T] \times G(A)^T \\ &= \tilde{\sum_{T}} F[T] \times G^T(A) \\ &= \tilde{\sum_{T}} F[T] \times \tilde{\sum_{S}} G^T[S] \times A^S \\ &= \tilde{\sum_{T}} \tilde{\sum_{S}} F[T] \times G^T[S] \times A^S \end{aligned}$$

$$= \tilde{\sum_S} \tilde{\sum_T} F[T] \times G^T[S] \times A^S$$

$$= \tilde{\sum_S} (\tilde{\sum_T} F[T] \times G^T[S]) \times A^S$$

$$= \tilde{\sum_S} (\tilde{\sum_T} W[S, T]) \times A^S$$

puisque l'on a $W[S, T] = F[T] \times G^T[S]$.

(iii) L'hypothèse que $G[\varnothing] = \varnothing$ entraîne que

$$\prod_{i \in T} G[f^{-1}(i)] = \varnothing$$

pour toute fonction $f: S \longrightarrow T$ non surjective. On a donc

$$W[S, T] = \sum_{f:S \twoheadrightarrow T} F[T] \times \prod_{i \in T} G[f^{-1}(i)]$$

Comme tout élément $(\alpha, \beta, f) \in W[S, T]$ est isomorphe à un et un seul élément $(\alpha', \beta', p) \in W[S, S/R]$ où $p: S \rightarrow S/R$ est la projection canonique, on a

$$\tilde{\sum_T} W[S, T] = \sum_{R \in Eq[S]} F[S/R] \times \prod_{C \in S/R} G[C]$$

(v) est laissé au lecteur diligent.

Remarque. Toute substitution se ramène à une substitution sans terme constant suivie d'une évaluation partielle. Par exemple pour calculer $F(G_1, \ldots, G_r)$, calculons d'abord l'espèce

$$H = H(X, Y) = F(Y_1 G_1(X), \ldots, Y_r G_r(X))$$

où $X = (X_1, \ldots, X_d)$, $Y = (Y_1, \ldots, Y_r)$. Soient $S = (S_1, \ldots, S_d) \in \mathbf{B}^d$ et $T = (T_1, \ldots, T_r) \in \mathbf{B}^r$. Un élément de $H[S, T]$ est un triplet (f, α, β) où

i) $f: S_1 + \ldots + S_d \longrightarrow T_1 + \ldots + T_r$

ii) α est une fonction qui associe à chaque élément $x \in T_i$ une structure d'espèce G_i sur la fibre $f^{-1}(x)$

iii) $\beta \in F[T_1, \ldots, T_r]$

Nous dirons que (f, α, β) est une *fibration* $(G_1, \ldots, G_r)$*-enrichie* sur les fibres et F-*enrichie* sur la base. Pour obtenir $F(G_1, \ldots, G_r)$, il suffit de substituer $Y_1 = 1, \ldots, Y_r = 1$ dans $H(X, Y)$. Autrement dit,

$$F(G_1, \ldots, G_r)[S] = \tilde{\sum_T} H[S, T]$$

2.2. Dans cette partie, nous allons examiner les opérations de produit *cartésien* et de produit *scalaire* d'espèces et aussi les opérations de base du calcul *différentiel*.

Le produit *cartésien* $F \Diamond G$ de deux espèces $F, G: \mathbf{B}^d \longrightarrow$ Ens:

$$(F \Diamond G)[S] = F[S] \times G[S] \qquad\qquad (S \in \mathbf{B}^d)$$

Le produit *scalaire* $<F, G>$:

$$<F, G> = \tilde{\sum_S} F[S] \times G[S]$$

Si la sommation est restreinte à certaines variables, on obtient le produit scalaire *partiel* :

$$<F(X, Y), G(X, Y)>_X = \tilde{\sum_S} F[S, T] \times G[S, T]$$

Remarque.

$$<F(X, Y), G(X, Y)>_X = (F \Diamond G)(1, Y)$$

$$<F(X, Y), G(X, Y)>_Y = (F \Diamond G)(X, 1)$$

$$<F(X, Y), G(X, Y)> = (F \Diamond G)(1, 1)$$

Proposition 5 Soit $F(X)$ une espèce de structure. On a

$$F(XY) = F(X) \Diamond \exp(XY)$$
$$= F(Y) \Diamond \exp(XY)$$

Preuve. En effet, un calcul simple nous montre que

$$\exp(XY)[S, T] = \Delta[S, T]$$

et

$$F(XY)[S, T] = F[S] \times \Delta[S, T]$$
$$= F[T] \times \Delta[S, T]$$

Proposition 6. [Y.N. Yeh] Soit $F(X)$ une espèce de structures et soit A un ensemble. On a

$$F(AX) = F(X) \Diamond \exp(AX)$$

Preuve. Il suffit de constater que

$$F(AX)[S] = F[S] \times A^S$$

et

$$\exp(AX)[S] = A^S$$

C.Q.F.D.

Pour le calcul différentiel, on pose

$$(d/dx)^T F(X)[S] = F^{(T)}(X)[S] = F[S+T]$$

On vérifie les identités suivantes:

a) la formule de Leibniz

$$(FG)^{(T)} = \sum_{A+B=T} F^{(A)} G^{(B)}$$

b) la formule de Taylor

$$F(X+Y) = \sum_{n\geq 0} F^{(n)}(X) \times Y^n / \mathfrak{S}_n$$

c) la "règle de dérivation en chaîne"

$$F(G(X))' = F'(G(X))G'(X)$$

Pour toute espèce H(X), on peut définir un opérateur différentiel H(d/dx).

$$H(d/dx)F(X)[S] = \tilde{\sum_{T}} H[T] \times F[T+S]$$

L'opérateur H(d/dx) est adjoint à l'opérateur de multiplication par H(X):

$$\langle H(d/dx)F, G\rangle = \langle F, HG\rangle$$

en effet,

$$\begin{aligned} \langle H(d/dx)F, G\rangle &= \tilde{\sum_{T}} \tilde{\sum_{S}} H[T] \times F[T+S] \times G[S] \\ &= \tilde{\sum_{R}} \sum_{S+T=R} H[T] \times G[S] \times F[T+S] \\ &= \tilde{\sum_{R}} (\sum_{S+T=R} H[T] \times G[S]) \times F[R] \\ &= \tilde{\sum_{R}} (HG)[R] \times F[R] \\ &= \langle HG, F\rangle \end{aligned}$$

Chapitre 3 Espèces virtuelles

3.0. Le but de ce chapitre est d'introduire le concept d'espèce *virtuelle* pour pallier à l'absence de l'opération de soustraction en théorie des espèces. La définition des opérations de somme et de produit d'espèces virtuelles ne pose pas de difficultés car il suffit d'imiter les règles de l'algèbre. Par contre, le prolongement de l'opération de substitution est plus délicat et nous introduisons pour cela une Règle des signes qui permet de résoudre le problème de façon satisfaisante.

Définition 1. Une espèce de structures F: $\mathbf{B}^d \longrightarrow$ Ens est *finitaire* si

$$\text{Card } F[n] < \infty, \quad \text{pour tout } n \in \mathbb{N}^d$$

Si de plus

$$\text{Card } F[n] = 0$$

sauf pour un nombre fini de $n \in \mathbb{N}^d$, nous dirons que F est *polynômiale.*

Définition 2. Une espèce *virtuelle* V est une différence formelle V = F-G entre des espèces finitaires. La relation d'égalité F-G = M-N entre espèces virtuelles signifie que F+N = M+G.

La transitivité de la relation d'égalité entre espèces virtuelles est conséquence de la propriété de simplification des espèces finitaires:

$$F+M = G+M \Rightarrow F = G$$

Pour démontrer cette propriété, nous utiliserons le principe suivant (équivalent au principe d'involution de Garsia-Milne [G.1]). Soient A, B, C des ensembles finis et soit

$$\varphi: A+C \xrightarrow{\sim} B+C$$

une bijection. Alors on peut construire une bijection *résiduelle*

$$\psi: A \xrightarrow{\sim} B$$

ne dépendant que de φ. En effet, pour tout $x \in A$ la suite $x, \varphi(x), \varphi^2(x), \ldots$ doit se terminer en un élément $y = \psi(x) \in B$. De même, pour tout $y \in B$, la suite $y, \varphi^{-1}(y), \varphi^{-2}(y), \ldots$ doit se terminer en un élément $x = \psi^{-1}(y) \in A$.

Supposons maintenant qu'un groupe Γ agisse sur A, B, C et que φ soit Γ-équivariante. On voit immédiatement que ψ est Γ-équivariante. De même, si on suppose que φ est une transformation naturelle de foncteurs (à valeurs dans les ensembles finis), on voit que ψ est aussi une transformation naturelle.

La somme et le produit d'espèces virtuelles sont définis comme suit:

$$(F-G)+(M-N) = (F+M)-(G+N)$$

$$(F-G)\cdot(M-N) = (FM+GN)-(GM+FN)$$

Ces opérations donnent à l'ensemble des espèces virtuelles en les variables $X_1, \ldots, X_d$ une structure d'anneau commutatif que nous désignerons par $\mathbf{B}[[X_1, \ldots, X_d]]$. Les espèces virtuelles polynômiales constituent un sous-anneau que nous désignerons par $\mathbf{B}[X_1, \ldots, X_d]$.

Pour tout groupe fini Γ, soit $B(\Gamma)$ l'*anneau de Burnside* du groupe Γ. Rappelons sa construction. Les éléments de $B(\Gamma)$ sont des différences formelles entre Γ-ensembles finis avec la même relation d'égalité que dans la définition 2. Nous dirons que les éléments de $B(\Gamma)$ sont des *représentations ensembliste virtuelles* du groupe Γ. La somme et le produit sont définis comme ci-haut, sauf qu'il faut plutôt utiliser le produit cartésien de Γ-ensembles.

La proposition 3 du chapitre 2 entraîne que l'on a un isomorphisme de groupes additifs

$$\mathbf{B}[X_1, \ldots, X_d] \simeq \bigoplus_{n \in \mathbb{N}^d} B(\mathfrak{S}_n)$$

En outre, puisque

$$B(\mathfrak{S}_n)\cdot B(\mathfrak{S}_m) \subseteq B(\mathfrak{S}_{n+m})$$

pour le produit des espèces, on voit que $\mathbf{B}[X_1, \ldots, X_d]$ est un anneau gradué. Le complété de

$\mathbf{B}[X_1, \ldots, X_d]$ s'identifie avec $\mathbf{B}[[X_1, \ldots, X_d]]$:

$$\mathbf{B}[[X_1, \ldots, X_d]] \simeq \prod_{n \in \mathbb{N}^d} \mathbf{B}(\mathfrak{S}_n)$$

Tout élément $F \in \mathbf{B}[[X_1, .., X_d]]$ s'écrit sous la forme d'une série formelle

$$F = \sum_{n \in \mathbb{N}^d} F_n$$

où F_n est la partie homogène de degré $n \in \mathbb{N}^d$. La partie homogène de degré 0 est le terme constant $F_0 = F(0) = F(0, \ldots, 0)$.

Lemme (Classique). Si $F(0) = 1$, alors F est inversible dans l'anneau $\mathbf{B}[[X_1, \ldots, X_n]]$.

Preuve. Posons $F = 1+T$. La série géométrique

$$(1+T)^{-1} = \sum_{n \geq 0} (-1)^n T^n$$

converge puisque T est sans terme constant.

En vertu de ce lemme, l'espèce exponentielle $\exp(X) \in \mathbf{B}[[X]]$ est inversible puisque $\exp(0) = 1$. Nous allons calculer l'espèce virtuelle $\exp(X)^{-1}$. Posons, pour tout $S \in \mathbf{B}$,

$$\varepsilon_0[S] = \{f: S \twoheadrightarrow (2k) \mid k \in \mathbb{N}\}$$

$$\varepsilon_1[S] = \{f: S \twoheadrightarrow (2k+1) \mid k \in \mathbb{N}\}$$

où $(n) = \{1, 2, \ldots, n\}$ pour tout $n \in \mathbb{N}$.

Proposition 1. On a

$$\exp(X)^{-1} = \varepsilon_0(X) - \varepsilon_1(X)$$

Preuve. Posons $\exp(X) = 1 + T(X)$. On trouve que pour tout $n \in \mathbb{N}$,

$$T^n[S] = \{f: S \twoheadrightarrow (n)\}$$

ce qui donne le résultat.

Il faut souligner ici que l'égalité

$$\exp(X)^{-1} = \sum (-1)^n X^n / \mathfrak{S}_n$$

est fausse. Nous sommes maintenant en mesure de définir exp(-X):

Définition 3.

$$\exp(-X) = \exp(X)^{-1}$$

Cette définition suggère la notation suivante:

$$\varepsilon^S = \varepsilon_0[S] - \varepsilon_1[S] \qquad (S \in \mathbf{B})$$

Il est peut-être bon de souligner ici que ε^S n'est pas la différence entre deux ensembles mais la différence entre deux foncteurs. Ainsi, ε^n est la représentation ensembliste virtuelle obtenue en effectuant la différence entre $\varepsilon_0[n]$ et $\varepsilon_1[n]$. Nous allons maintenant énoncer la règle des signes. Soit F(X) une espèce virtuelle.

Règle des signes [J.1]

$$F(-X)[n] = \varepsilon^n F[n]$$

Dans cette formule $\varepsilon^n F[n]$ désigne le produit de ε^n et de F[n] dans l'anneau de Burnside $B(\mathfrak{S}_n)$. Dans le cas de plusieurs variables, nous aurons

$$F(-X, Y)[n,m] = \varepsilon^n\, F[n,m]$$

où ε^n désigne l'élément de $B(\mathfrak{S}_n \times \mathfrak{S}_m)$ obtenu en composant la représentation ε^n de $\mathfrak{S}_n$ avec la projection $\mathfrak{S}_n \times \mathfrak{S}_m \longrightarrow \mathfrak{S}_n$. Nous aurons de même

$$F(X, -Y)[n, m] = \varepsilon^m\, F[n, m]$$

$$F(-X, -Y)[n, m] = \varepsilon^n\, \varepsilon^m\, F[n, m]$$

$$= \varepsilon^{n+m}\, F[n, m]$$

où ε^{n+m} est l'élément de $B(\mathfrak{S}_n \times \mathfrak{S}_m)$ obtenue par restriction de l'élément ε^{n+m} de $B(\mathfrak{S}_{n+m})$ au sous-groupe $\mathfrak{S}_n \times \mathfrak{S}_m \subset \mathfrak{S}_{n+m}$.

La Règle des signes permet de résoudre le problème de la substitution. Par exemple, supposons que l'on veuille substituer M–N dans F(X). On calcule d'abord l'espèce virtuelle F(X–Y):

$$F(X-Y)[n, m] = \varepsilon^m\, F[n+m]$$

ce qui donne

$$F(X-Y) = F_0(X, Y) - F_1(X, Y)$$

et on pose ensuite

$$F(M-N) = F_0(M, N) - F_1(M, N)$$

On montre que cette opération est bien définie, c'est à dire que

$$F_0(M, N) - F_1(M, N) = F_0(P, Q) - F_1(P, Q)$$

si $M-N = P-Q$.

Pour éliminer les problèmes de convergence, nous allons limiter la substitution F(G) aux deux cas suivants:

a) F est polynômiale b) le terme constant de G est nul

Théorème 1 [J.2, Y.1] La substitution des espèces virtuelles, telle que décrite avec la Règle des signes, est une opération bien définie. De plus, elle donne lieu à une loi de composition associative.

Remarque. Il est intéressant de reformuler la Règle des signes en utilisant le produit cartésien des espèces virtuelles. On pose

$$(F - G) \Diamond (M - N) = (F\Diamond M + G\Diamond M) - (F\Diamond N + G\Diamond M)$$

La règle des signes devient

$$F(-X) = F(X) \Diamond \exp(-X)$$

$$F(-X, Y) = F(X, Y) \Diamond \exp(-X)$$

3.1. Soit $B(\Gamma)$ l'anneau de Burnside d'un groupe fini Γ. Nous allons voir que chaque élément $F \in \mathbf{B}[X_1, \dots, X_d]$ définit une opération

$$F: B(\Gamma)^d \longrightarrow B(\Gamma)$$

Remarquons d'abord que si F est une espèce polynômiale ordinaire (nous dirons aussi que F est *positive*) et $A_1, \dots, A_d$ sont des Γ-ensembles finis alors $F(A_1, \dots, A_d)$ est un Γ-ensemble fini. Cette opération s'étend aux espèces virtuelles:

$$(F - G)(A_1, \dots, A_d) = F(A_1, \dots, A_d) - G(A_1, \dots, A_d)$$

Nous allons prolonger ces opérations aux Γ-ensembles virtuels. Soit $(A_1 - B_1, .., A_d - B_d) \in B(\Gamma)^d$. La Règle des signes nous permet de calculer l'espèce virtuelle

$$F(X_1, \dots, X_d ; Y_1, \dots, Y_d) = F(X_1 - Y_1, \dots, X_d - Y_d)$$

On pose ensuite

$$F(A_1 - B_1, \dots, A_d - B_d) = F(A_1, \dots, A_d ; B_1, .., B_d)$$

Pour s'assurer que cette opération est bien définie, il faut s'assurer que

$$F(A_1, \dots, A_d ; B_1, \dots, B_d) = F(U_1, \dots, U_d ; V_1, \dots, V_d)$$

dès que $A_i - B_i = U_i - V_i \qquad (1 \le i \le d)$

Théorème 2 [J.2]. Les éléments $F \in \mathbf{B}[X_1, \dots, X_d]$ donnent lieu à des opérations bien définies

$$F: B(\Gamma)^d \longrightarrow B(\Gamma)$$

De plus, on a l'associativité

$$F(G_1, \dots, G_d)(b) = F(G_1(b), \dots, G_d(b))$$

où $\quad G_i \in \mathbf{B}[X_1, \dots, X_r] \quad (1 \le i \le d) \quad$ et $\quad b = (b_1, \dots, b_r) \in B(\Gamma)^r$

3.2. Le théorème des fonctions implicites se transpose aisément en théorie des espèces. Soient $X = (X_1, \dots, X_d)$, $Y = (Y_1, \dots, Y_r)$ et $F(X, Y) = (F_1(X, Y), \dots, F_r(X, Y))$ des espèces virtuelles. Nous dirons que les espèces virtuelles $N(X) = (N_1, \dots, N_r)$ sont *solutions* du système d'équations

$$F(X, Y) = 0$$

si l'on a $N(0) = 0$ et

$$F(X, N(X)) = 0$$

Théorème 3. Si la matrice jacobienne $(\partial F/\partial Y)(0, 0)$ est inversible, alors le système d'équations

$$F(X, Y) = 0$$

possède une et une seule solution $Y = N(X)$.

La démonstration est laissée en exercice. Le théorème montre en particulier que si $F(X)$ est une espèce telle que $F(0) = 0$ et $F'(0) = 1$ alors $F(X)$ possède un inverse $G(X)$ pour l'opération de substitution:

$$F(G(X)) = G(F(X)) = X$$

On peut donner une description combinatoire des coefficients de $G(X)$ et, plus généralement, des coefficients de $H(G(X))$ où H est une espèce quelconque. Soit S un ensemble fini. L'ensemble $Eq[S]$ des relations d'équivalence sur S est ordonné. Notons $\hat{0}$ l'élément minimum de $Eq[S]$ et $\hat{1}$ l'élément maximum. Soit $R = (\hat{0} = R_0 < R_1 < R_2 < \ldots < R_n)$ une chaîne strictement croissante d'éléments de $Eq[S]$. On a une suite de projections canoniques

$$S = S/_{R_0} \xrightarrow{p} S/_{R_1} \xrightarrow{p} \ldots \xrightarrow{p} S/_{R_n}$$

Posons

$$f(R) = \{p^{-1}(x) \mid x \in S/_{R_k},\ 1 \leq k \leq n\}$$

Nous dirons que $f(R)$ est l'ensemble des *fibres* de R. Posons

$$S/_R = S/_{R_n}$$

Nous dirons que $S/_R$ est la *base* de R. Soit

$$Q_n[S] = \{R = (\hat{0} = R_0 < R_1 < \ldots < R_n) \mid R_i \in Eq[S],\ 1 \leq i \leq n\}$$

$$W_n[S] = \sum_{R \in Q_n[S]} H[S/_R] \times \prod_{C \in f[R]} F[C]$$

Proposition 2. On a

$$H(G(X)) = \sum_{n=0}^{\infty} (-1)^n W_n(X)$$

Preuve. Nous allons appliquer aux espèces de structures une méthode d'inversion des séries formelles due à G. Labelle [L.1]. Considérons l'opérateur additif

$$\delta\colon \mathbf{B}[[X]] \longrightarrow \mathbf{B}[[X]]$$

qui transforme H(X) dans H(F(X)) − H(X). On a par définition

$$\delta(H)[S] = \sum_{\substack{R\in Eq[S]\\ R\neq\hat{0}}} H[S_{/R}] \times \prod_{A\in S_{/R}} F[A]$$

$$= W_1[S]$$

Un raisonnement par récurrence nous donne

$$\delta^n(H)[S] = W_n[S]$$

Ce qui entraîne que $\delta^n(H)[S] = \varnothing$ dès que n est assez grand et, par suite, que la série

$$\sum_{n\geq 0} (-1)^n\, \delta^n(H) = \sum_{n\geq 0} (-1)^n\, W_n(X)$$

converge vers un élément $W(X) \in \mathbf{B}[[X]]$. On a

$$W(F) = (1+\delta)(W)$$

$$= (1+\delta) \sum_{n\geq 0} (-1)^n\, \delta^n(H)$$

$$= H$$

C.Q.F.D.

Considérons maintenant le cas où H(X) = X. Pour tout ensemble fini S et pour tout $n\geq 0$ posons

$$P_n[S] = \{(\hat{0} = R_0 < R_1 < .. < R_n = \hat{1}) \mid R_i\in Eq[S],\ 1\leq i\leq n\}$$

$$V_n[S] = \sum_{R\in P_n[S]} \prod_{C\in f(R)} F[C]$$

On a alors

$$G(X) = \sum_{n\geq 0} (-1)^n\, V_n(X)$$

Dans le cas où F(X) = exp(X) − 1, la proposition 2 nous donne

Corollaire.

$$\log(1+X) = \sum_{n=0}^{\infty} (-1)^n\, P_n(X)$$

Chapitre 4. Espèces tensorielles

4.0. Dans cette partie, nous développerons certains aspects de la théorie des foncteurs analytiques définis sur la catégorie des espaces vectoriels. Cette théorie est en quelque sorte le prolongement de la théorie des foncteurs polynômiaux formalisée par I.G. Macdonald [M.1]. Du point de vue algébrique, c'est la théorie des λ-anneaux, des fonctions symétriques et des fonctions de Schur [K.1]. Les espèces tensorielles sont l'analogue linéaire des espèces de structures: ce sont les coefficients des foncteurs analytiques à valeurs dans la catégorie des espaces vectoriels. Certaines espèces tensorielles sont équipées d'une structure d'algèbre, en particulier d'algèbre de Lie: ce sont les algèbres de Lie *tordues* de M.G. Barratt [B.1]. Nous utiliserons la théorie des espèces tensorielles pour démontrer que la construction des algèbres de Lie libres s'obtient en évaluant un foncteur analytique Lie (X). Les coefficients de Lie (X) sont des représentations du groupe symétrique isomorphes aux représentations de Reutenauer [R.1]. Nous exprimerons ces coefficients au moyen de l'homologie du treillis des partitions. Ces calculs permettent de jeter une lumière nouvelle sur certains travaux de R. Stanley [S.1] et de P. Hanlon [H.1].

4.1. Soit k un corps de caractéristique 0. Nous désignerons par Vect ou Vect_k la catégorie des k-espaces vectoriels.

Définition 1. Une *espèce tensorielle* F est un foncteur

$$F[\ ]: \mathbf{B} \longrightarrow \text{Vect}$$

A toute espèce tensorielle F est associée une suite (F[n] | n≥0) de représentations linéaires des groupes symétriques

$$\mathfrak{S}_n \times F[n] \longrightarrow F[n] \qquad (n\geq 0)$$

Définition 2. Un foncteur F: Vect ⟶ Vect est *analytique* s'il possède un développement en série de Taylor

$$F(V) = \bigoplus_{n\geq 0} F[n] \otimes V^{\otimes n} / \mathfrak{S}_n$$

On démontre que la suite des coefficients (F[n] | n≥0) est déterminée (à isomorphisme canonique près) par le foncteur F et que de plus elle détermine (à isomorphisme canonique près) l'espèce tensorielle F. Autrement dit, les concepts suivants sont équivalents:

i) une espèce tensorielle F = F[ ]

ii) une suite (F[n] | n≥0) de représentations linéaires des groupes symétriques

iii) un foncteur analytique F = F( ): Vect ⟶ Vect

Nous adopterons pour les espèces tensorielles, les mêmes conventions d'écriture que celles que nous avons adoptées au Chapitre 2 pour les espèces de structures.

Remarque. Nous dirons qu'un élément $t \in F(V)$ est un *tenseur d'espèce* F sur V. Cette terminologie est conforme à l'usage du mot tenseur en physique, en géométrie et en algèbre.

Exemples.

$$L(V) = \bigoplus_{n \geq 0} V^{\otimes n} = \text{algèbre tensorielle } T(V)$$

$$\exp(V) = \bigoplus_{n \geq 0} V^{\otimes n} / \mathfrak{S}_n = \text{algèbre symétrique } S(V)$$

$$\Lambda(V) = \bigoplus_{n \geq 0} \Lambda[n] \otimes V^{\otimes n} / \mathfrak{S}_n = \bigoplus_{n \geq 0} \Lambda^n(V) = \text{algèbre extérieure}$$

Dans ce dernier exemple, le coefficient $\Lambda[n]$ est la représentation *alternée* de $\mathfrak{S}_n$:

$$\Lambda[n] = k\, e_n \qquad \sigma. e_n = (\operatorname{sgn} \sigma)\, e_n$$

La catégorie $\mathrm{Vect}^{\mathbf{B}}$ des espèces tensorielles est une catégorie abélienne semi-simple. Elle est munie de *sommes directes quelconques*

$$(\bigoplus_{i \in I} F_i)[S] = \bigoplus_{i \in I} F_i[S]$$

et d'un *produit tensoriel*

$$(F \otimes G)[S] = \bigoplus_{A+B=S} F[A] \otimes G[B]$$

associatif, symétrique et unitaire [M.1].

Les foncteurs analytiques sont clos sous la composition, ce qui se traduit par l'existence d'une loi de composition sur les espèces tensorielles. Dans le cas où $G[\varnothing] = 0$, les coefficients du composé $F \circ G$ sont donnés par la formule

$$(F \circ G)[S] = \bigoplus_{R \in Eq[S]} F[S/_R] \otimes \bigotimes_{C \in S/_R} G[C]$$

Pour tout ensemble A, nous noterons k(A) ou kA l'espace vectoriel librement engendré par A. De même, pour toute espèce de structure F nous noterons k(F) ou kF l'espèce tensorielle obtenue en composant les foncteurs

$$\begin{array}{ccc} \mathbf{B} & \xrightarrow{F} & \mathrm{Ens} \\ & {\scriptstyle kF} \searrow & \downarrow {\scriptstyle k(\,)} \\ & & \mathrm{Vect} \end{array}$$

S'il n'y a pas d'ambiguïté, nous désignerons par F l'espèce tensorielle kF. On vérifie que le foncteur

$$k(\,): \mathrm{Ens}^{\mathbf{B}} \longrightarrow \mathrm{Vect}^{\mathbf{B}}$$

préserve les opérations de somme, de produit et de composition.

Soit V un espace vectoriel. L'évaluation en V définit un foncteur $F \mapsto F(V)$

$$e(V)\colon \mathrm{Vect}^{\mathbf{B}} \longrightarrow \mathrm{Vect}$$

Nous allons voir qu'il possède un adjoint à droite [K.1]

$$\{V,\ \}\colon \mathrm{Vect} \longrightarrow \mathrm{Vect}^{\mathbf{B}}$$

Pour tout $W \in \mathrm{Vect}$ et pour tout $S \in \mathbf{B}$, posons

$$\{V, W\}[S] = \mathrm{Hom}(V^{\otimes S}, W)$$

On a

$$\begin{aligned} \mathrm{Hom}(F(V), W) &= \prod_{n\geq 0} \mathrm{Hom}(F[n] \underset{\mathfrak{S}_n}{\otimes} V^{\otimes n}, W) \\ &= \prod_{n\geq 0} \mathrm{Hom}_{\mathfrak{S}_n}(F[n], \mathrm{Hom}(V^{\otimes n}, W)) \\ &= \mathrm{Hom}(F, \{V, W\}) \end{aligned}$$

Ce qui montre que $e(V)$ et $\{V,\ \}$ sont adjoints l'un de l'autre. De plus, $e(V)$ préserve le produit tensoriel:

$$(F\otimes G)(V) = F(V) \otimes G(V)$$

$$k1\,(V) = k$$

et par suite on a des transformations naturelles

$$\{V, W_1\} \otimes \{V, W_2\} \longrightarrow \{V, W_1 \otimes W_2\}$$

$$k \longrightarrow \{V, k\}$$

4.2. Dans cette partie, nous allons étudier les espèces tensorielles munies d'une structure algébrique.

Définition 3. [B.1] Une algèbre *tordue* est une espèce tensorielle F munie d'une structure d'algèbre:

$$F\otimes F \longrightarrow F$$

Il y a plusieurs variétés d'algèbres tordues: associative, commutative, algèbre de Lie, etc.

Exemples. Pour toute espèce G, l'espèce

$$T(G) = \bigoplus_{n\geq 0} G^{\otimes n}$$

est une algèbre tordue associative. De même,

$$\exp(G) = S(G) = \bigoplus_{n\geq 0} G^{\otimes n} / \mathfrak{S}_n$$

est une algèbre tordue associative et commutative. Voici une description plus détaillée du concept d'algèbre de Lie tordue: le crochet

$$F\otimes F \xrightarrow{[\,,\,]} F$$

doit satisfaire aux identités

i) $[\,,\,] + [\,,\,]\tau = 0$

ii) $[\,,[\,,\,]] + [\,,[\,,\,]]\sigma + [\,,[\,,\,]]\sigma^2 = 0$

où τ: $F\otimes F \xrightarrow{\sim} F\otimes F$ est la symétrie du produit tensoriel et où σ: $F\otimes F\otimes F \longrightarrow F\otimes F\otimes F$ est la permutation cyclique des facteurs. Les parenthèses $[\,,[\,,\,]]$ désignent le morphisme composé

$$F\otimes F\otimes F \xrightarrow{F\otimes[\,,\,]} F\otimes F \xrightarrow{[\,,\,]} F .$$

Pour tout espace vectoriel V, soit Lie(V) l'algèbre de Lie libre sur V.

Théorème 1. Le foncteur Lie: Vect $\longrightarrow$ Vect est analytique:

$$\mathrm{Lie}(V) = \bigoplus_{n\geq 0} \mathrm{Lie}[n] \otimes V^{\otimes n} / \mathfrak{S}_n$$

Remarque. Ce résultat et sa démonstration sont valables pour une variété quelconque d'algèbres et pas seulement pour les algèbres de Lie.

Preuve. Nous avons vu, à la fin du paragraphe précédent, que pour tout $V \in$ Vect, le foncteur évaluation $e(V)$ possède un adjoint à droite $\{V, \}$. Remarquons d'abord que si W est une algèbre de Lie, alors $\{V, W\}$ est une algèbre de Lie tordue: en effet, le composé

$$\begin{array}{ccc} \{V, W\} \otimes \{V, W\} & \longrightarrow & \{V, W\otimes W\} \\ & \searrow & \downarrow \{V, [\,,\,]\} \\ & & \{V, W\} \end{array}$$

est une structure d'algèbre de Lie sur $\{V, W\}$. D'autre part, soit $F = F(X)$ l'algèbre de Lie tordue libre sur l'espèce tensorielle $X = kX$. Le foncteur évaluation en $V \in$ Vect préservant le produit tensoriel, on constate que $F(V)$ est munie d'une structure d'algèbre de Lie. Pour toute algèbre de Lie W on a alors des bijections naturelles entre les morphismes suivants

$$F(V) \longrightarrow W \qquad (1)$$

$$F \longrightarrow \{V, W\} \qquad (2)$$

$$X \longrightarrow \{V, W\} \qquad (3)$$

$$V = X(V) \longrightarrow W \qquad (4)$$

$$\mathrm{Lie}(V) \longrightarrow W \qquad (5)$$

où les morphismes (1), (2) et (5) sont des morphismes d'algèbres. Ceci montre que l'on a

$$\mathrm{Lie}(V) = F(V)$$

C.Q.F.D.

Remarque. La suite (Lie[n] | n≥0) de représentations des groupes symétriques est isomorphe à celle obtenue par C. Reutenauer [R.1].

Théorème 2. (Poincaré-Birkhoff-Witt). Soit U(F) l'algèbre enveloppante d'une algèbre de Lie tordue F. Le morphisme canonique $F \longrightarrow U(F)$ est injectif et l'on a de plus un isomorphisme canonique

$$U(F) \simeq \exp(F)$$

Preuve. On vérifie d'abord, par un argument de foncteur adjoint, que pour tout $V \in$ Vect on a

$$U(F)(V) = U(F(V))$$

Remarquons [B.2] ensuite que pour toute algèbre de Lie L on a un isomorphisme

$$\eta: U(L) \simeq S(L)$$

naturel entre les foncteurs U et S. Ceci entraîne que l'on a des isomorphismes naturels

$$U(F(V)) \simeq S(F(V))$$

et finalement que

$$U(F) \simeq \exp(F)$$

C.Q.F.D.

Proposition 1.

$$\exp(\mathrm{Lie}\,(X)) = 1 / 1-X$$

Preuve. Appliquant le théorème précédent, on a successivement

$$\begin{aligned} \exp(\mathrm{Lie}\,(X)) &= U(\mathrm{Lie}\,(X)) \\ &= T(X) \\ &= 1 / 1-X \end{aligned}$$

4.3. Dans cette partie, nous allons discuter de la théorie des espèces tensorielles *virtuelles*. Celles-ci sont obtenues en prenant des différences formelles entre espèces tensorielles finitaires. La transitivité de la relation d'égalité

$$F-G = M-N \Leftrightarrow F \oplus N \simeq M \oplus G$$

est conséquence du fait que $\mathrm{Vect}^{\mathbf{B}}$ est une catégorie semi-simple. Les espèces tensorielles virtuelles en d-variables $X_1, \dots, X_d$ forment un anneau que nous désignerons par

$$R[[X_1, \dots, X_d]]$$

Les espèces polynômiales forment un sous-anneau

$$R[X_1, \dots, X_d] \subset R[[X_1, \dots, X_d]]$$

On a un isomorphisme de groupes additifs

$$R[X_1, \dots, X_d] \simeq \bigoplus_{n \in \mathbb{N}^d} R(\mathfrak{S}_n)$$

où $R(\mathfrak{S}_n)$ est le groupe des représentations virtuelles de $\mathfrak{S}_n$. Ceci donne à $R[X_1, ..., X_d]$ une structure d'anneau gradué puisque

$$R(\mathfrak{S}_n) \cdot R(\mathfrak{S}_m) \subseteq R(\mathfrak{S}_{n+m})$$

pour la multiplication des espèces. L'anneau $R[[X_1, ... ,X_n]]$ s'identifie au complété de l'anneau gradué $R[X_1, ... ,X_n]$:

$$R[[X_1, ... ,X_d]] = \prod_{n \in \mathbb{N}^d} R(\mathfrak{S}_n)$$

L'opération de linéarisation se prolonge aux espèces virtuelles et on obtient ainsi un homomorphisme d'anneaux

$$k(\ ): B[[X_1, ... ,X_n]] \longrightarrow R[[X_1, ... ,X_n]]$$

S'il n'y a pas d'ambiguité, nous désignerons parfois l'espèce tensorielle kF par F.

La Règle des signes prend une forme simplifiée en théorie des espèces tensorielles. Posons

$$\Lambda(X) = \sum_{n\geq 0} \Lambda[n] \otimes X^n /_{\mathfrak{S}_n} = \sum_{n\geq 0} \Lambda^n(X)$$

Proposition 2.

$$\exp(X)^{-1} = \sum_{n\geq 0} (-1)^n \Lambda^n(X)$$

Preuve. Pour tout $S \in \mathbf{B}$ et pour tout $n \geq 0$, posons

$$L_n[S] = \{f: S \twoheadrightarrow (n)\}$$

La proposition 1 du chapitre 3 nous donne

$$\exp(X)^{-1}[S] = \sum_{n\geq 0} (-1)^n L_n[S]$$

Nous allons décrire un complexe de chaînes

$$kL_0[S] \overset{0}{\longleftarrow} kL_1[S] \overset{\partial}{\longleftarrow} kL_2[S] \overset{\partial}{\longleftarrow} ...$$

dont l'homologie est donnée par la suite

$$\Lambda^0[S]\ ,\ \Lambda^1[S]\ ,\ \Lambda^2[S]\ , ...$$

Une identité classique sur l'homologie des complexes nous donnera:

$$\sum_{n\geq 0} (-1)^n L_n[S] = \sum_{n\geq 0} (-1)^n \Lambda^n[S]$$

Si Card $S \leq 1$, alors $\partial = 0$ et $\Lambda^n[S] = kL_n[S]$. Si Card $S > 1$, posons, pour tout $n \geq 0$,

$$C_n[S] = \{(\varnothing < A_0 < ... < A_n < S) \mid A_i \subset S,\ 1 \leq i \leq n\}$$

Il y a une bijection naturelle évidente

$$C_n[S] \simeq L_{n+2}[S]$$

Soit $\Delta°S$ le complexe simplicial des parties propres non vides de S. Géométriquement, $\Delta°S$ est le bord du simplexe dont les sommets sont les éléments de S; la réalisation géométrique de $\Delta°S$ est une sphère de dimension égale à Card S−2. Par définition, $C_n[S]$ est l'ensemble des simplexes de dimension n de la première subdivision barycentrique de $\Delta°S$. L'homologie réduite de $\Delta°S$ peut donc se calculer au moyen du complexe de chaîne

$$k \leftarrow kC_0[S] \leftarrow kC_1[S] \leftarrow \dots$$

Comme on a (résultat classique de topologie):

$$\tilde{H}_n(\Delta°S) = \Lambda^{n+2}[S] \qquad (n\geq 0)$$

la démonstration est terminée.

Utilisant les notations du chapitre 3, cette proposition montre que

$$\varepsilon^n = (-1)^n \Lambda[n]$$

dans l'anneau $R(\mathfrak{S}_n)$.

Pour tout ensemble S, soit $\Pi[S]$ l'ensemble ordonné des relations d'équivalences *propres* sur S:

$$\Pi[S] = \{R \in Eq[S] \mid \hat{0} < R < \hat{1}\}$$

Dans ce qui suit, $\tilde{H}_*(\Pi[S])$ désignera l'homologie réduite du complexe simplicial des chaînes non vides de $\Pi[S]$.

Théorème 3.

$$\begin{aligned} \log(1+X)[n] &= k.1 && \text{si } n = 1 \\ &= -k.1 && \text{si } n = 2 \\ &= (-1)^{n+1}\tilde{H}_{n-3}(\Pi[n]) && \text{si } n > 2 \end{aligned}$$

Preuve. D'après le corollaire de la proposition 2 du chap. 3,

$$\log(1+X) = \sum_{n\geq 0} (-1)^n P_n(X)$$

où

$$P_n[S] = \{(\hat{0}=R_0 < \dots < R_n=\hat{1}) \mid R_i \in Eq[S],\ 1\leq i\leq n\}$$

Pour tout $n\geq 0$, soit $C_n(\Pi[S])$ l'ensemble des chaînes de longueur n de l'ensemble ordonné $\Pi[S]$. On a $C_n(\Pi[S]) = P_{n+2}[S]$. Considérons le complexe augmenté

$$kP_1[S] \leftarrow kC_0(\Pi[S]) \leftarrow kC_1(\Pi[S]) \leftarrow \dots$$

dont l'homologie est (si Card S>2)

$$0\ ,\ \tilde{H}_0(\Pi[S]),\ \tilde{H}_1(\Pi[S]), \dots$$

On a alors

$$-P_1[S] + \sum_{n\geq 0} (-1)^n C_n(\Pi[S]) = \sum_{n\geq 0} (-1)^n \tilde{H}_n(\Pi[S])$$

Autrement dit,

$$\log(1+X)[S] = \sum_{n\geq 0}(-1)^n \tilde{H}_n(\Pi[S])$$

dès que Card S > 2. Pour terminer la démonstration on utilise le fait que le treillis Eq[S] est un treillis géométrique [F.1, S.1] et par suite que

$$\tilde{H}_n(\Pi[S]) = 0 \qquad \text{sauf si } n = (\text{Card } S) - 3.$$

Théorème 4.

$$\mathrm{Lie}(X) = \log 1 / 1-X$$

et

$$\mathrm{Lie}[1] = k.1$$

$$\mathrm{Lie}[2] = \Lambda[2]$$

$$\mathrm{Lie}[n] = \Lambda[n] \otimes \tilde{H}_{n-3}(\Pi[n]) \qquad (n\geq 3)$$

Preuve. La première égalité résulte de la proposition 1 du chapitre 3. On en déduit que

$$\mathrm{Lie}(X) = -\log(1-X)$$

Utilisant la Règle des signes et le théorème précédent, on obtient le résultat.

4.4. Pour terminer, nous allons calculer le caractère de l'espèce tensorielle Lie(X). Rappelons quelques résultats. A toute espèce tensorielle finitaire F(X) est associée une série *indicatrice*

$$Z_F = \sum_{n\geq 0} 1/n! \sum_{\sigma\in \mathfrak{S}_n} \mathrm{Tr}\, F[\sigma]\, I[\sigma]$$

où Tr F[σ] est la trace de l'opérateur linéaire $F[\sigma]: F[n] \longrightarrow F[n]$

et I(σ) est le monôme $x_1^{d_1} \ldots x_n^{d_n}$ où d_i est le nombre de cycles de longueur i dans σ.

Z_F est une série formelle d'une infinité de variables: $Z_F = Z_F(x_1, x_2, x_3, \ldots)$

On a les identités

$$Z_{F+G} = Z_F + Z_G$$

$$Z_{FG} = Z_F Z_G$$

$$Z_{F(G)} = Z_F(Z_G)$$

où le membre de droite de la dernière égalité désigne le *pléthisme* : si $f = f(x_1, x_2, \ldots)$ et $g = g(x_1, x_2, \ldots)$ alors

$$f(g) = f(g_1, g_2, \ldots)$$

où

$$g_i = g(x_i, x_{2i}, \ldots)$$

Proposition 3.

$$Z_{\log(1+X)} = \sum_{n\geq 1} \mu(n)/n \log(1+x_n)$$

$$= \sum_{n\geq 1} -1/n \sum_{d|n} \mu(d)(-1)^{n/d} x_d^{n/d}$$

Preuve. Inversant la relation $Y = \log(1+X)$, on obtient

$$1 + X = \exp(Y)$$

et par suite

$$1 + x_1 = \exp(\sum_{n\geq 1} y_n/n)$$

$$\log(1+x_1) = \sum_{n\geq 1} y_n/n$$

ce qui entraîne que pour tout $k\geq 1$

$$\log(1+x_k) = \sum_{n\geq 1} y_{nk}/n$$

Une inversion de Möbius nous donne finalement le résultat

$$y_1 = \sum_{k\geq 1} \mu(k)/_k \log(1+x_k)$$

Remarque. Si on combine ce résultat avec celui de la proposition on obtient le résultat de P. Hanlon sur la fonction de Möbius du treillis des partitions invariantes par une permutation donnée [H.1]

Proposition 4.

$$Z_{Lie(X)} = \sum_{n\geq 1} \mu(n)/_n \log 1/1-x_n$$

$$= \sum_{n\geq 1} 1/n \sum_{d|n} \mu(d)\, x_d^{\,n/d}$$

Preuve. Il suffit d'utiliser le théorème 4:

$$Lie(X) = \log 1/1-X$$

Remarque. La proposition permet de généraliser la formule de Witt. Soit Γ un groupe fini agissant sur un espace vectoriel V. L'action du groupe Γ se prolonge à l'algèbre de Lie libre sur V:

$$\Gamma \times Lie(V) \longrightarrow Lie(V)$$

Lie(V) est une algèbre graduée

$$Lie(V) = \sum_{n\geq 0} Lie^n(V)$$

et le groupe Γ agit sur la partie homogène de degré n

$$\Gamma \times Lie^n(V) \longrightarrow Lie^n(V)$$

Désignons par χ le caractère de la représentation V et par χ_n le caractère de la représentation $Lie^n(V)$.

Proposition 5. On a pour tout $\sigma \in \Gamma$

$$\chi_n(\sigma) = 1/_n \sum_{d|n} \chi(\sigma^d)^{\,n/d}$$

Appendice

Dans cet appendice, nous allons démontrer l'unicité de la série de Taylor d'un foncteur analytique et donner des conditions simples qui caractérisent les foncteurs analytiques.

Définition 1. Le *diagramme* DF d'un foncteur F: Ens $\longrightarrow$ Ens est la catégorie dont les objets sont les couples (x, A), où $x \in F(A)$, et dont les morphismes $(x, A) \longrightarrow (y, B)$ sont les fonctions $f: A \longrightarrow B$ telles que

$$F(f)(x) = y$$

Définition 2. Un élément $x \in F(A)$ est *générique* si pour tout

$$\begin{array}{ccc} & & (z, C) \\ & & \downarrow g \\ (x, A) & \xrightarrow{f} & (y, B) \end{array}$$

il existe $h: (x, A) \longrightarrow (z, C)$ tel que le triangle

$$\begin{array}{ccc} & & (z, C) \\ & \overset{h}{\nearrow} & \downarrow g \\ (x, A) & \xrightarrow{f} & (y, B) \end{array}$$

soit commutatif.

Lemme 1. Soit $u: (x, A) \longrightarrow (y, B)$. Si y est générique, alors u est surjective. Si x et y sont génériques, alors u est bijective.

Preuve. Si y est générique, il existe $h: (x,B) \longrightarrow (y, A)$ tel que le triangle

$$\begin{array}{ccc} & & (x, A) \\ & \overset{h}{\nearrow} & \downarrow u \\ (y, B) & \xrightarrow{1_B} & (y, B) \end{array}$$

soit commutatif. Ceci entraîne que u est surjective. Si de plus x est générique, alors h est surjective et, par suite, u est bijective.

Pour tout ensemble fini S posons

$$F^\circ[S] = \{x \in F(S) \mid x \text{ est générique}\}$$

Le foncteur $F^\circ[\ ]$ est défini sur la catégorie **B**; c'est donc une espèce de structures. Considérons le foncteur analytique $F^\circ(\ )$ dont les coefficients sont donnés par $F^\circ[\ ]$:

$$F^{\circ}(A) = \tilde{\sum_{S}} F^{\circ}[S] \times A^{S}$$

On a une transformation naturelle

$$i: F^{\circ} \longrightarrow F$$

qui envoie la classe de $(x, f) \in F^{\circ}[S] \times A^{S}$ dans l'élément $F(f)(x) \in F(A)$.

Lemme 2. La transformation $i: F^{\circ} \longrightarrow F$ est injective.

Preuve. Soit

$$(x, f) \in F^{\circ}[S] \times A^{S} \text{ et } (y, g) \in F^{\circ}[T] \times A^{T}$$

Supposons que $F(f)(x) = F(g)(y)$. Nous allons montrer que (x, f) et (y, g) sont *isomorphes* . On a

$$\begin{array}{ccc} & & (y, T) \\ & & \downarrow g \\ (x, S) & \xrightarrow{f} & (z, A) \end{array}$$

où $z = F(f)(x) = F(g)(y)$. Comme x est générique, il existe $h: (x, S) \longrightarrow (y, T)$ tel que le triangle

$$\begin{array}{ccc} & & (y, T) \\ & {}^{h}\nearrow & \downarrow g \\ (x, S) & \xrightarrow{f} & (z, A) \end{array}$$

soit commutatif. La bijectivité de h est conséquence du lemme 1.

C.Q.F.D.

Proposition 1. Soit F: Ens $\longrightarrow$ Ens un foncteur analytique. On a un isomorphisme canonique

$$F^{\circ}[\,] \simeq F[\,]$$

Preuve. Cet isomorphisme envoie $x \in F[S]$ dans la classe de

$$(x, 1_{S}) \in F[S] \times S^{S}$$

Le lecteur pourra compléter la démonstration.

Proposition 2. Un foncteur F: Ens $\longrightarrow$ Ens est analytique si et seulement si il est engendré par ses éléments génériques finis, c'est-à-dire si pour tout $x \in F(A)$, il existe un élément générique (y, S), avec S fini, et un morphisme $(y, S) \longrightarrow (x, A)$.

Preuve. En effet, $x \in F(A)$ appartient à l'image de $i: F^{\circ}(A) \longrightarrow F(A)$ si et seulement si il existe un élément générique (y, S) avec S fini et un morphisme $(y, S) \longrightarrow (x, A)$. La proposition 1 montre que la condition est nécessaire. La suffisance résulte du fait que si i est surjective, alors elle est bijective

(Lemme 2), ce qui montre que F est isomorphe au foncteur analytique F°.

Lemme 3. Soit α: F ⟶ G une transformation naturelle régulière entre les foncteurs F et G. Si $x \in F(A)$ est générique, alors $\alpha(x) \in G(A)$ est générique.

La démonstration est laissée au lecteur.

Proposition 3. Soit α: F ⟶ G une transformation naturelle régulière entre les foncteurs analytiques F et G. Alors il existe une et une seule transformation naturelle θ: F[ ] ⟶ G[ ] telle que $\alpha = \underline{\theta}$ (voir le chapitre 1).

Preuve. D'après le lemme précédent, α induit une transformation

$$\alpha^\circ: F^\circ[\ ] \longrightarrow G^\circ[\ ]$$

et par suite (prop. 1) on a une transformation naturelle

$$\theta: F[\ ] \longrightarrow G[\ ]$$

Le reste de la démonstration est laissée au lecteur.

Il reste à caractériser les foncteurs analytiques.

Définition 3. Un foncteur F: Ens ⟶ Ens est *continu* s'il préserve les limites inductives filtrantes. Si F préserve de plus les limites projectives filtrantes, nous dirons qu'il est *bicontinu.*

Définition 4. Un foncteur F: Ens ⟶ Ens est ***régulier*** s'il transforme les carrés commutatifs réguliers en carrés réguliers.

Théorème 1. Un foncteur F: Ens ⟶ Ens est analytique si et seulement si il est bicontinu et régulier.

Preuve. Pour démontrer la nécessité, remarquons que la classe C des foncteurs bicontinus et réguliers est close sous 1) les produits finis 2) les sommes quelconques 3) le quotient par l'action d'un groupe fini. Remarquons ensuite que le foncteur identité X: Ens ⟶ Ens appartient à C et que tout foncteur analytique F possède un développement en série

$$F = \sum_{n \geq 0} F[n] \times X^n / \mathfrak{S}_n$$

Pour démontrer la suffisance, nous allons vérifier qu'un foncteur bicontinu et régulier est engendré par ses éléments génériques (prop. 2)

Définition 5. Soit F: Ens ⟶ Ens un foncteur. Un élément $x \in F(A)$ est ***minimal*** si pour tout morphisme

$$(y, B) \xrightarrow{f} (x, A)$$

f est surjectif.

Lemme 4. Soit F: Ens $\longrightarrow$ Ens un foncteur continu. Alors F est engendré par ses éléments minimaux finis.

Preuve. Soit $x \in F(A)$. La continuité de F entraîne qu'il existe un sous-ensemble fini $S \subset A$ et un élément $y \in F(S)$ tel que $F(i)(y) = x$, où i dénote l'inclusion $S \subset A$. Choisissons un couple (y, S) de sorte que la cardinalité de S soit la plus petite possible. L'élément (y, S) est minimal. C.Q.F.D.

Lemme 5. Soit F: Ens $\longrightarrow$ Ens un foncteur continu régulier. Un élément $x \in F(A)$ est générique si et seulement si tout morphisme

$$(x, A) \longleftarrow (y, B)$$

où B est fini et y minimal, est un isomorphisme.

Preuve. Démontrons la suffisance, la nécessité étant laissée au lecteur. Soit

$$\begin{array}{ccc} & & (z, C) \\ & & \downarrow \\ (x, A) & \longrightarrow & (y, B) \end{array}$$

La régularité du foncteur F entraîne l'existence d'un carré commutatif (où $D = A \times_B C$):

$$\begin{array}{ccc} (w, D) & \longrightarrow & (z, C) \\ \downarrow & & \downarrow \\ (x, A) & \longrightarrow & (y, B) \end{array}$$

La continuité de F entraîne l'existence d'un morphisme (lemme 4)

$$(v, S) \longrightarrow (w, D)$$

où (v, S) est fini et minimal. Le composé

$$\sigma: (v, S) \longrightarrow (w, D) \longrightarrow (x, A)$$

est alors inversible si l'hypothèse est satisfaite. On a par suite un triangle commutatif

$$\begin{array}{ccccccc} (x, A) & \xrightarrow{\sigma^{-1}} & (v, S) & \longrightarrow & (w, D) & \longrightarrow & (z, C) \\ & \searrow & & & & & \downarrow \\ & & & & & & (y, B) \end{array}$$

C.Q.F.D.

Définition 6. Soit F: Ens $\longrightarrow$ Ens. Nous dirons que $x \in F(A)$ est *irrégulier* si

$$x \notin \operatorname{Im}(F^{\circ}(A) \xrightarrow{i} F(A))$$

Lemme 6. Soit F un foncteur continu régulier et soit $x_0 \in F(S_0)$ où S_0 est fini. Si x_0 est irrégulier et minimal, il existe un morphisme

$$(x_0, S_0) \xleftarrow{p} (x_1, S_1)$$

tel que

1) S_1 est fini et x_1 est irrégulier et minimal
2) p est surjective sans être bijective

Preuve. Si x_0 est irrégulier, à plus forte raison il n'est pas générique, ce qui entraîne, en vertu du lemme précédent, l'existence d'un morphisme

$$(x_0, S_0) \longleftarrow (x_1, S_1)$$

tel que

1) (x_1, S_1) est minimal
2) u n'est pas bijectif

La minimalité de (x_0, S_0) entraîne que u est surjective. De plus, si on avait $x_1 \in F^{\circ}(S_1)$, on aurait a fortiori $x_0 = F(f)(x_1) \in F^{\circ}(S_0)$ ce qui est absurde puisque x_0 est irrégulier. Donc x_1 est irrégulier.

C.Q.F.D.

Lemme 7. Soit F un foncteur continu et régulier. Alors les deux conditions suivantes sont incompatibles entre elles:

1) F préserve les limites projectives filtrantes
2) Il existe un élément irrégulier $x \in F(A)$

Preuve. Supposons les deux conditions satisfaites. La continuité entraîne l'existence d'un morphisme (lemme 4):

$$(x, A) \longleftarrow (x_0, S_0)$$

où S_0 est fini et x_0 est minimal. L'irrégularité de x entraîne celle de x_0. Une utilisation itérative du lemme précédent implique l'existence d'une chaîne infinie

$$(x_0, S_0) \overset{q_1}{\longleftarrow} (x_1, S_1) \overset{q_2}{\longleftarrow} (x_2, S_2) \longleftarrow \ldots$$

telle que pour tout $n \geq 0$

1) (x_n, S_n) est fini et minimal
2) q_n est surjective sans être bijective

Posons

$$S_\infty = \varprojlim S_n$$

et soient $p_n \colon S_\infty \longrightarrow S_n$ $(n \geq 0)$ les projections. L'hypothèse que F préserve les limites projectives entraîne l'existence d'un élément $z \in F(S_\infty)$ tel que

$$F(p_n)(z) = x_n \qquad \text{pour tout } n \geq 0$$

Il existe par continuité un morphisme

$$(z, S_\infty) \overset{i}{\longleftarrow} (w, S)$$

où S est fini. Comme

$$\lim_{n\to\infty} \text{Card } S_n = \infty$$

il existe un entier n tel que le composé

$$(x_n, S_n) \xleftarrow{p_n \circ i} (w, S)$$

n'est pas surjectif. Ceci contredit la minimalité de (x_n, S_n).

Bibliographie

[B.1] M.G. Barratt. *Twisted Lie Algebras*, Lecture Notes in Math., 658, Springer-Verlag, 1977.

[B.2] N. Bourbaki. *Groupes et algèbres de Lie*, Chap. 2. Actualités Scientifiques et Industrielles, Herman, Paris, 1972.

[F.1] J. Folkman. *The homology groups of a lattice*, J. Math. Mech. 15 (1966), 631-636.

[G.1] A.M. Garsia, S.C. Milne. *A Rogers-Ramanujan Bijection*, J. Comb. Th. (A) **31** (1981), 289-339.

[H.1] P. Hanlon. *The fixed-point partition lattices*, Pacific J. Math. 96 (1981), 319-341.

[J.1] A. Joyal. *Une théorie combinatoire des séries formelles*, Advances in Mathematics, Vol. 42 (1981), 1-82.

[J.2] A. Joyal. *Règle des signes en algèbre combinatoire*. C. R. Math. Acad. Sci. Soc. Royale Canada, Vol. VII (1985), 285-290.

[K.1] G.M. Kelly. *On clubs and doctrines*, in "Category Seminar" (G.M. Kelly, Ed.) Lecture Notes in Mathematics No 420, Springer Verlag, 1974.

[K.2] D. Knutson. *λ-Rings and the Representation theory of the Symmetric Group*. Lecture Notes in Mathematics No 308, Springer Verlag, 1973.

[L.1] G. Labelle. *Sur l'inversion et l'itération continue des séries formelles*, Europ. J. Combin. Vol. 1 (1980), 113-138.

[M.1] I.G. Macdonald. *Symmetric Functions and Hall Polynomials*, Clarendon Press, Oxford, 1979.

[M.2] S. MacLane. *Categories for the Working Mathematician*, Springer Verlag, New York, 1971.

[N.1] O. Nava, G.C. Rota. *Plethysm, Categories, and Combinatorics*, Advances in Mathematics 58 (1985), 61-88 .

[R.1] C. Reutenauer. *Theorem of Poincaré-Birkhoff-Witt, logarithm, and symmetric group representations of degrees equal to Stirling numbers*. Ce volume.

[S.1] R. Stanley. *Some aspects of groups acting on finite posets*, Journal of combinatorial theory, Series A, 32 (1982), 132-161.

[Y.1] Y.N. Yeh. *On the Combinatorial Species of Joyal*, Thèse, State University of New York at Buffalo, 1985.

ENUMERATION UNDER FINITE GROUP ACTION: SYMMETRY CLASSES OF MAPPINGS

Adalbert Kerber
Lehrstuhl II für Mathematik, Bayreuth, W.-Germany

It is the aim of this talk both to give a quick and easy *introduction* to as well as a *review* of part of the theory which is often called *Pólya's theory of enumeration*. I do not use this name for two reasons. First of all I want to embed it into the much more general theory of finite group action (which does not need more effort, in fact as I dare say, it makes things easier to understand), and secondly this theory is older than Pólya's famous paper, it is in fact due to J.H. Redfield.
The main point is that along these lines the main theorems of Pólya's theory of enumeration turn out to be beautiful but particular cases of lemmas on finite group actions.

1. Introduction

The theory in question is devoted to the *general problem of definition, enumeration, classification and construction of structures in mathematics and sciences* by way of considering them as orbits of finite groups on sets of mappings. The historic origin dates back to the middle of the nineteenth century when chemists discovered the phenomenon of isomerism, part of which was solved by representing molecules by graphs. The question arose for the number of graphs with a given edge degree sequence since it is the same as the number of connectivity isomers with a given gross formula (as long as the different atoms have pairwise different valencies).
But it needed another seventy years until J.H. Redfield (1927) and G. Pólya (1937) put this problem into a group theoretical context which uses the following quite general *Ansatz*:

(i) Choose suitable finite sets $X,Y \neq \emptyset$ and groups G,H acting on X,Y, which yield natural actions of $G,H,H\times G,H\wr_X G$ on $Y^X := \{f:X \to Y\}$.

(ii) Recognize the structure in question, say, a graph on v vertices as an orbit of such an action.

(iii) Use algebraic and combinatorial methods in order to get as much of information as possible on these orbits.

2. A list of problems

Let us consider an example, the graphs on 4 vertices, and list a few of the related problems:

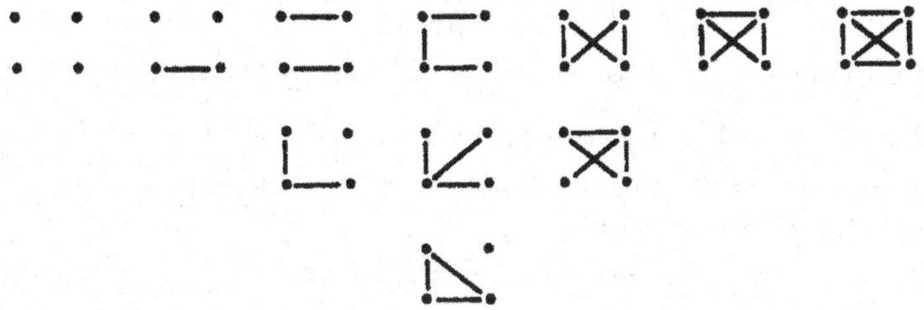

Some of the obvious questions are:

(i) Give a definition of "graph" which is flexible in the sense that it easily generalizes to multigraphs, directed graphs,....

(ii) Provide a formula for the total number of graphs on v vertices (which is 11 if v = 4, as you see).

(iii) Enumerate the graphs on v vertices by weight (:= no. of edges). In our example the answer is the following sequence of numbers: 1,1,2,3,2,1,1, which again gives rise to the question if all such sequences of numbers of graphs by weight are unimodal, i.e. weakly increasing to the middle term and then weakly decreasing.

(iv) Enumerate graphs by edge degree sequences.

(v) For which v do selfcomplementary graphs exist?

(vi) Count graphs by automorphism group. Characterize the subgroups of the symmetric group $S_{\underline{v}}$ which occur as automorphism groups of graphs on v vertices.

(vii) Enumerate graphs by weight and automorphism group.

(viii) Construct the graphs on v vertices exhaustively and redundancy--free.

(ix) Construct graphs on v vertices uniformly at random.

All these questions can be answered except question (iv), for which no satisfactory answer is known yet and which is, as it was mentioned in the introduction, in fact the question which gave rise to the whole theory. I shall describe some of the solutions in the next sections starting right now with question (i) which asks for a flexible definition of graphs on v vertices.

Let $\underline{v} := \{1,\ldots,v\}$ denote the set of vertices, so that the set of 2-subsets

$$\underline{v}^{[2]} := \{\{i,j\} \mid i,j \in \underline{v}, i \neq j\}$$

is the set of *pairs of vertices*. Hence putting $Y := 2 := \{0,1\}$ and $X := \underline{v}^{[2]}$, a *labelled graph* on v vertices can be considered as an $f \in Y^X = 2^{(\underline{v}^{[2]})}$, where

$$f(\{i,j\}) = \begin{cases} 0, & \text{if } i \text{ and } j \text{ are not connected} \\ 1, & \text{otherwise.} \end{cases}$$

The symmetric group $S_{\underline{v}}$ acts on $\underline{v}$ and hence on $\underline{v}^{[2]}$ and therefore also on $2^{(\underline{v}^{[2]})}$ in a natural way, namely by *renumbering* the vertices, so that an orbit is an isomorphism class of labelled graphs, which is exactly what we mean by a *graph*.

This definition is very flexible, for if we want to shift to k-multigraphs on v vertices, i.e. if we want to allow up to k-fold edges (but still no loops), then we need only replace $2 = \{0,1\}$ by $k+1=\{0,\dots,k\}$. If instead we want to consider directed graphs without loops, then we replace $\underline{v}^{[2]}$ by $\underline{v}^{<2>} := \{(i,j) \mid i,j \in \underline{v}, i \neq j\}$, and so on.

3. Finite group actions

Let G denote a finite group (written multiplicatively) and S a finite nonempty set. An *action* of G on S is a mapping

$$G \times S \to S : (g,s) \mapsto gs$$

subject to the conditions

$$g(g's) = (gg')s, \quad 1_G s = s.$$

The basic notions in connection with actions of finite groups are:

- the *orbit* of s: $G(s) := \{gs \mid g \in G\}$,
- the *stabilizer* of s: $G_s := \{g \mid gs = s\}$,
- the *fixed points* of g: $S_g := \{s \mid gs = s\}$.

We remark that for a *transversal* T of the orbits we have $S = \dot{\bigcup}_{t \in T} G(t)$.

Furthermore it is very easy to see that $G_{gs} = gG_sg^{-1}$, so that

$$\tilde{G}_s := \{G_{gs} = gG_sg^{-1} \mid g \in G\}$$

is a full class of conjugate subgroups, and that $G(s) \to G/G_s : gs \mapsto gG_s$ turns out to be a bijection between the orbit $G(s)$ and the set G/G_s of left cosets of G_s in G. This leads to the following very important result on the order of $G(s)$: $|G(s)| = |G/G_s|$.
Using this and denoting by

$$S/G := \{G(t) \mid t \in T\}$$

the *set* of orbits of G on S, one easily derives the following lemma, usually called Burnside's lemma (but in fact it is older, it is due to Cauchy and Frobenius):

3.1 $$|S/G| = \frac{1}{|G|} \sum_{g \in G} |S_g|.$$

There exist various refinements of this lemma. The most important ones are contained in the following result (due to Stockmeyer):

3.2 The Fundamental Lemma:

Let μ denote the Moebius-function on the lattice of subgroups of G. Let $w: S \to R$ denote a (weight-) function from S into a ring containing $\mathbb{Q}$, w being constant on the orbits ω_i of G on S, w_i the value on ω_i, $s_i \in \omega_i$. Then for each subgroup U of G and its class $\tilde{U}$ of conjugate subgroups we have

$$\sum_{G_{s_i} \in \tilde{U}} w_i = \frac{|\tilde{U}|}{|G/U|} \sum_V \mu(U,V) \sum_{V \leq G_s} w(s).$$

Let me show how easy it is to prove this lemma:

$$\sum_{U \leq G_s} w(s) = \sum_{U \leq V} \sum_{V = G_s} w(s)$$

$$= \sum_{U \leq V} \frac{1}{|\tilde{V}|} \sum_{G_s \in \tilde{V}} w(s)$$

$$= \sum_{U \leq V} \frac{|G/V|}{|\tilde{V}|} \sum_{G_{s_i} \in \tilde{V}} w_i,$$

Moebius-inversion now yields the statement.

□

3.3 Corollaries:

(i) *The generating function for the enumeration of the orbits of* G *by weight* w *is*

$$\sum_i w_i = \frac{1}{|G|} \sum_{g\in G} \sum_{s\in S_g} w(s).$$

(ii) *The number of orbits of length* k *of* G *on* S *ist*

$$\frac{1}{k} \sum_{|G/U|=k} \sum_{V} \mu(U,V)\,|S_V|$$

(if $S_V := \{s \mid \forall\, g\in V\ (gs=s)\}$*).*

(iii) *The number of orbits of type* $\tilde{U}$ *(i.e. the stabilizers of the elements are in* $\tilde{U}$*) is*

$$\frac{|\tilde{U}|}{|G/U|} \sum_V \mu(U,V)\,|S_V|.$$

It is most important to notice that in order to apply 3.2/3 *we need only to characterize the set* S_V *of fixed points of* V*, for each subgroup* V. The rest is done by an application of the Moebius-function μ which does *not* depend on the action of G on S which is examined.
Let us conclude this section by slightly simplifying the above equation by using that (as w is constant on the orbits):

$$\sum_{V\leq G_s} w(s) = \sum_{gVg^{-1}\leq G_s} w(s),$$

and introducing the abbreviation

$$\mu(U,\tilde{V}) := \sum_{W\in\tilde{V}} \mu(U,W).$$

Now if $\tilde{U}_1,\ldots,\tilde{U}_d$ are the classes of conjugate subgroups of G, $U_i \in \tilde{U}_i$, then we obtain the following equation

$$\sum_{G_{s_i}\in\tilde{U}_j} w_i = \frac{|\tilde{U}_j|}{|G/U_j|} \sum_k \mu(U_j,\tilde{U}_k) \sum_{U_k\leq G_s} w(s).$$

Putting

$$b_{jk} := \frac{|\tilde{U}_j|}{|G/U_j|}\,\mu(U_j,\tilde{U}_k),\quad B(G) := (b_{jk}),$$

a matrix which we call the *Burnsidematrix* of G, and which is the inverse of the *table of marks:*

$$B(G)^{-1} = M(G) = (m_{jk}),\quad m_{jk} = \frac{|G/U_k|}{|\tilde{U}_k|}\,\zeta(U_j,\tilde{U}_k),$$

ζ being the zeta function of the subgroup lattice. Thus:

3.4 Corollary:

The generating function for the enumeration of such orbits by weight w, the elements of which have stabilizers in $\tilde{U}_j$, is the element in the j-th row of the column vector

$$B(G)\cdot\begin{bmatrix}\vdots\\ \sum\limits_{U_k \leqslant G_s} w(s)\\ \vdots\end{bmatrix}.$$

This theorem has the advantage that it clearly seperates the things which do depend on the particular action of G from B(G) which does only depend on the isomorphism class of G. Unfortunately the evaluation of B(G) needs the knowledge of the subgroup lattice. A different approach which allows to avoid this can be found in the cited paper by Rota/Smith. I wanted to emphasize the finite group action aspect.

4. Symmetry classes of mappings

Let X,Y denote two finite nonempty sets on which finite groups G,H act. Then G,H,H×G and the wreath product $H\wr_X G := H^X \times G$ act on Y^X in a natural way:

$$\begin{aligned}
G\times Y^X &\to Y^X : (g,f) \mapsto f\circ g^{-1},\\
H\times Y^X &\to Y^X : (h,f) \mapsto h\circ f,\\
(H\times G)\times Y^X &\to Y^X : ((h,g),f) \mapsto h\circ f\circ g^{-1},\\
(H\wr_X G)\times Y^X &\to Y^X : ((\varphi,g),f) \mapsto \tilde{f}, \text{ where}\\
&\quad \tilde{f}(x) := \varphi(x)f(g^{-1}x).
\end{aligned}$$

The orbits under these actions are called *symmetry classes of mappings*. In order to apply the above mentioned results on finite group actions we first of all remark that we obviously can embed G,H and H×G into the wreath product $H\wr_X G$ so that all what we need to derive is a characterization of the $f\in Y^X$ which remain fixed under an element $(\varphi,g)\in H\wr_X G$.

4.1 Lemma

Let $g = \prod_{\nu=1}^{c(g)} (x_\nu g x_\nu \dots g^{l_\nu - 1} x_\nu)$ *denote a cycle decomposition of the permutation induced by g on X into disjoint cycles. For* $1 \leqslant \nu \leqslant c(g)$ *we put*

$$h_\nu(\varphi,g) := \varphi(x_\nu)\varphi(g^{-1}x_\nu)\dots\varphi(g^{-l_\nu+1}x_\nu).$$

Then $f \in Y^X$ *is left fixed by* (φ,g) *if and only if*

(i) *For each* ν $f(x_\nu)$ *is a fixed point of* $h_\nu(\varphi,g)$:

$$f(x_\nu) \in Y_{h_\nu(\varphi,g)},$$

and

(ii) *the other values of* f *can be obtained from these values* $f(x_\nu)$ *according to the equations*

$$f(x) = \varphi(x)f(g^{-1}x),\ x \in X.$$

Proof: Iteration of $f(x) = \varphi(x)f(g^{-1}x)$ yields

$$f(x) = \varphi(x)\varphi(g^{-1}x)f(g^{-2}x),\ldots\ f(x_\nu) = h_\nu(\varphi,g)f(x_\nu).$$ □

An application of 3.1 now yields for the total number of symmetry classes:

4.2 (i) $$|Y^X/G| = \frac{1}{|G|}\sum_g |Y|^{c(g)},$$

(ii) $$|Y^X/H| = \frac{1}{|H|}\sum_h |Y_h|^{|X|},$$

(iii) $$|Y^X/H\times G| = \frac{1}{|H||G|}\sum_{(h,g)}\prod_i |Y_{h^i}|^{a_i(g)}$$

($a_i(g)$:= no. of i-cycles induced by g on X).

(iv) $$|Y^X/H\wr_X G| = \frac{1}{|H|^{|X|}|G|}\sum_{(\varphi,g)}\prod_\nu |Y_{h_\nu(\varphi,g)}|.$$

This answers the second one of the questions put in section 2. The answer to the third one now follows by an application of 4.1 to 3.3 (i). The most general weight on Y^X which is constant on the orbits of G on Y^X is the mapping

4.3 $$w: Y^X \to \mathbb{Q}[Y]: f \to \prod_{x\in X} f(x),$$

where we took Y as set of (commuting) indeterminates over $\mathbb{Q}$. 4.1 shows that f is left fixed by $g \in G$ iff f is constant on the cycles of g (on X). If such a cycle is of length i, then it contributes to the weight of f the factor y^i, if y is the value of f on this cycle. Thus

$$\prod_i \left(\sum_{y\in Y} y^i\right)^{a_i(g)}$$

is the sum of the weights, of all the $f \in Y^X$ fixed under g. Hence 3.3 (i) yields Pólya's famous theorem:

4.4 *The sum of the values of the weight* w *on the orbits of* G *on* Y^X *is*

$$\sum_i w_i = \frac{1}{|G|} \sum_g \prod_i \left(\sum_y y^i\right)^{a_i(g)} .$$

In terms of the *cycle-indicator* for the action of G on X, i.e. the polynomial

$$Cyc(G,X) := \frac{1}{|G|} \sum_{g \in G} \prod_i z_i^{a_i(g)} \in \mathbb{Q}[z_1,\ldots,z_{|X|}]$$

and the so-called Pólya-insertion of polynomials $p(\ldots,y,\ldots) \in \mathbb{Q}[Y]$ into the cycle-indicator

$$Cyc(G,X \mid p(\ldots,y,\ldots)) := \frac{1}{|G|} \sum_{g \in G} \prod_i p(\ldots,y^i,\ldots)^{a_i(g)} ,$$

we can rephrase 4.4 as follows:

4.5 Pólya's Theorem:
The generating function for the enumeration of symmetry classes of G *on* Y^X *by weight (cf. 4.3) is*

$$Cyc(G, X \mid \sum_{y \in Y} y).$$

The other theorems on the enumeration by weight under the actions of $H, H \times G$ and $H \wr_X G$ can be derived similarly, since lack of space I have to leave that to the reader.

5. Enumeration of G-classes by weight and automorphism group

According to 3.4 we obtain for the enumeration of G-classes on Y^X by weight and autormorphism group the following matrix product

5.1
$$B(G) \begin{bmatrix} \vdots \\ \prod_{i=1}^{|X/U_k|} \left(\sum_{y \in Y} y^{l_i(U_k)}\right) \\ \vdots \end{bmatrix}$$

where $l_i(U_k)$ denotes the length of the i-th orbit of U_k on X. This column contains in its j-th row the generating function for the enumeration of such orbits by weight, that have their stabilizers in $\tilde{U}_j$.
If we are interested only in the enumeration by automorphism group we map each y onto 1, obtaining the matrix

$$\underline{5.2} \qquad B(G) \cdot \begin{bmatrix} \vdots \\ |Y|^{|X/U_k|} \\ \vdots \end{bmatrix} .$$

Let us consider an example: $G := S_{\underline{4}}$. The Aachen subgroup lattice program yields the following transversal of the classes of conjugate subgroups:

$U_1 = \langle 1\rangle$, $U_2 = \langle(24)\rangle$, $U_3 = \langle(12)(34)\rangle$, $U_4 = \langle(132)\rangle$, $U_5 = \langle(24),(13)\rangle$,
$U_6 = \langle(1324)\rangle$, $U_7 = \langle(12)(34),(14)(23)\rangle$, $U_8 = \langle(132),(13)\rangle$,
$U_9 = \langle(1234),(24)\rangle$, $U_{10} = \langle(132),(142)\rangle$, $U_{11} = \langle(1324),(1342)\rangle$.

It is easy to evaluate the sequence of numbers of orbits of the U_j on $\underline{4}^{[2]}$, which is

$$|\underline{4}^{[2]}/U_j| = 6,4,4,2,3,2,3,2,2,1,1.$$

Besides this we obtain from the Aachen subgroup lattice program the table $M(S_{\underline{4}})$ of marks which we have to invert (which is again easy since $M(S_{\underline{4}})$ is triangular. The resulting columns for k = 1,2,3,4,5 which give the numbers of (k-1)-multigraphs by automorphism group are (cf. 5.2):

$$B(S_{\underline{4}}) \cdot \begin{bmatrix} k^6 \\ k^4 \\ k^4 \\ k^2 \\ k^3 \\ k^2 \\ k^3 \\ k^2 \\ k^2 \\ k^1 \\ k^1 \end{bmatrix} = \overset{k:\ 1 \quad 2 \quad 3 \quad 4 \quad 5}{\begin{bmatrix} 0 & 0 & 11 & 100 & 465 \\ 0 & 2 & 21 & 84 & 230 \\ 0 & 1 & 9 & 36 & 100 \\ 0 & 0 & 0 & 0 & 0 \\ 0 & 2 & 9 & 24 & 50 \\ 0 & 0 & 0 & 0 & 0 \\ 0 & 0 & 1 & 4 & 10 \\ 0 & 2 & 6 & 12 & 20 \\ 0 & 2 & 6 & 12 & 20 \\ 0 & 0 & 0 & 0 & 0 \\ 1 & 2 & 3 & 4 & 5 \end{bmatrix}}$$

The last column (k=5) shows for example that there exist exactly 900 different 4-multigraphs on 4 vertices of which 465 have trivial automorphism group $U_1 = \langle 1\rangle$.
Notice that the fourth, sixth and tenth row of this matrix consists of zeros only. This means that neither $U_4 = \langle(132)\rangle = A_{\underline{3}}$, nor $U_6 = \langle(1324)\rangle$

$= C_4$ nor $U_{10} = \langle(132),(142)\rangle = A_4$ occur as automorphism groups of k-multigraphs on 4 vertices, $1 \leq k \leq 4$. Let me show how it can be proved that this is true also for bigger k and that A_v is, for $v \geq 3$, never the automorphism group of a graph on v vertices.
Choose an $f \in Y^X$. Its inverse images $X_y := f^{-1}[\{y\}]$ form a decomposition of X into disjoint subsets (some of which may of course be empty): $X = \dot{\cup} X_y$. The stabilizer of f in the *symmetric group* S_X is the direct sum of the symmetric groups on the X_y:

$$(S_X)_f = \bigoplus_{y \in Y} S_{X_y} .$$

These subgroups of S_X are called the *Young subgroups*. Thus the stabilizer of f in $\bar{G}$, the permutation group induced by G on X, must be the intersection of this Young subgroup $(S_X)_f$ with $\bar{G}$ and we have obtained

5.3 Corollary

A subgroup U of G occurs as the stabilizer of an $f \in Y^X$ if and only if $\bar{U}$ is the intersection of $\bar{G}$ with a suitable Young subgroup of S_X.

Thus in particular the alternating group A_v, $v \geq 3$, does never occur as automorphism group of a graph on v vertices. For $\bar{A}_v = W \cap \bar{S}_v$, W a Young subgroup of S_X, $X = \underline{v}^{[2]}$, yields $W = S_X$ as $\bar{A}_v$ acts transitively on $\underline{v}^{[2]}$ and hence $W \cap \bar{S}_v = \bar{S}_v$. Similarly it turns out that the cyclic group C_v cannot be the automorphism group of a graph on v vertices, if $v \geq 3$.
This can be made more explicit by considering the mapping

$$\varphi : S_X \to Y^X : \pi \mapsto f \circ \pi .$$

We call the $|Y|$-tuple of the orders of the inverse images the *content* of f:

$$C(f) := (\ldots, |f^{-1}[\{y\}]|, \ldots),$$

It is not difficult to check that φ has the following properties:

5.4 (i) *φ maps S_X onto the union of G-classes on Y^X that consist of mappings of content C(f).*

(ii) *The inverse image of the G-class of $f \circ \pi$ is the double coset $(S_X)_f \pi \bar{G}$.*

Thus in particular a transversal of the double cosets $(S_X)_f \pi \bar{G}$ in S_X is mapped under φ onto a transversal of the G-classes of content C(f). (cf. Hässelbarth/Ruch/Richter). This can be used for a redundancy-free

construction. One can take advantage of the fact that S_X can be linearly ordered if X has this property, so that there exists a canonic system of representatives, consisting of the least members of the double cosets (see Brown et. al.).
This can be considered as an answer to question (viii) put in section 2, but one should be aware of the fact that double coset calculations are very cumbersome. One gets a list of all the graphs with v vertices only up to v = 9 in a reasonable computer time.
An algorithm for generating orbit representatives uniformly at random was recently given by Dixon and Wilf. The method is: choose a conjugacy class C of G, which acts on S, with the probability

$$p(C) := \frac{|C|\,|S_g|}{|S/G|\,|G|}, \quad g \in C,$$

then pick a $g \in C$ and construct a fixed point s of g, uniformly at random. Then the probability that s belongs to the orbit ω of G on S is $|S/G|^{-1}$. One can use this for graphs with $v \leq 30$ vertices easily and for bigger v if a very good long integer arithmetic is at hand.

6. The Burnside ring and a generalization

The enumeration of the orbits of G on S by stabilizer class $\tilde{U}_i$ can be considered as the problem of identifying S with a certain element in a specific ring, the Burnside ring of G, as we shall see in a minute. This ring theoretic approach has a useful generalization (Plesken) which will be described next. Assume $(S,\leq)$ is a poset on which G acts as a group of automorphisms, i.e. we assume in addition that

$$\forall\, x,y \in S,\ g \in G\ (x < y \Leftrightarrow gx < gy).$$

If again both S and G are finite, then the following is easy to verify:

<u>6.1</u> (i) *Elements in the same orbit are incomparable.*

(ii) *The orbits B_i can be numbered in such a way that*

$$B_i \ni x \leq y \in B_k \Rightarrow i \leq k .$$

(iii) *If* $s < s'$ *then*

$$|G(s)|\,|\{y \in G(s') \mid s < y\}| = |G(s')|\,|\{x \in G(s) \mid x < s'\}|.$$

Lattices are posets. Assume that the finite group G acts on a finite lattice $(L,\wedge,\vee)$. Then the following properties are equivalent:

6.2 (i) $\forall\ x,y \in L,\ g \in G:\ x < y \Rightarrow gx < gy,$

(ii) $\forall\ x,y \in L,\ g \in G:\ g(x \wedge y) = gx \wedge gy,$

(iii) $\forall\ x,y \in L,\ g \in G:\ g(x \vee y) = gx \vee gy.$

The lattice $(L,\wedge,\vee)$ defines two semigroups: $(L,\wedge)$ and $(L,\vee)$. Hence let us assume that $(S,\cdot)$ is a finite semigroup and that the finite group G acts on it as a group of automorphisms:

$$\forall\ x,y \in S, g \in G\ (g(x\cdot y) = gx\cdot gy).$$

The *semigroup ring* $\mathbb{Z}[S,\cdot]$ consists of the $f:S \to \mathbb{Z}$ where addition $f+f'$ is pointwise and where the multiplication comes from that of S:

$$(f*f')(x) := \sum_{yz=x} f(y)\cdot f'(z).$$

As usual we write its elements as "formal sums"

$$f = \sum_{x\in S} f_x x,\ f_x := f(x).$$

It is very important to notice that we can put, for each i,j,k and $s \in B_k$:

$$a^{\cdot}_{ijk} := |\{(x,y) \in B_i \times B_j \mid x\cdot y = s\}|,$$

(the upper point in $a^{\cdot}_{ijk}$ indicates the semigroup multiplication) for this number does not depend on the chosen representative s of B_k. These numbers are the structure constants of the following subring.

$$\mathbb{Z}[S.\cdot]_G := \{f \in \mathbb{Z}[S,\cdot] \mid \forall\ g \in G\ (f = f \circ g^{-1})\},$$

consisting of the G-*invariant* elements of the semigroup ring.

6.3 $\mathbb{Z}[S,\cdot]_G$ has the orbit sums $b_i := \sum_{x\in B_i} x$ as $\mathbb{Z}$-basis. Their products satisfy the equations

$$b_i * b_j = \sum_k a^{\cdot}_{ijk}\ b_k .$$

The main theorem (Plesken) is

6.4 Theorem

Let $B_1,\ldots,B_d$ *denote the orbits of the finite group* G *acting on the finite lattice* $(L,\wedge,\vee)$ *as group of automorphism, and indicate by* b_k *the orbit sums. Then*

(i) *If, for any* $x \in B_i$

$$a^{\wedge}_{ik} := |\{y \in B_k \mid x \leqslant y\}|,$$

then the mapping

$$b_k \mapsto \begin{bmatrix} a^{\wedge}_{1k} \\ \vdots \\ a^{\wedge}_{dk} \end{bmatrix} =: \alpha^{\wedge}_k$$

defines a ringisomorphism between $\mathbb{Z}[L,\wedge]_G$ *and* $\mathbb{Z}^{\underline{d}}$.

(ii) *If, for any* $y \in B_i$

$$a^{\vee}_{ik} := |\{x \in B_k \mid x \leq y\}| \ ,$$

then the mapping

$$b_k \mapsto \begin{bmatrix} a^{\vee}_{1k} \\ \vdots \\ a^{\vee}_{dk} \end{bmatrix} =: \alpha^{\vee}_k$$

defines a ringisomorphism between $\mathbb{Z}[L,\vee]_G$ *and* $\mathbb{Z}^{\underline{d}}(+,*)$.

(iii) *These rings and their images are related by the identities*

$$a^{\wedge}_{ik}|B_i| = a^{\vee}_{ki}|B_k| \ .$$

(iv) *For the (coordinatewise) products* $\alpha^{\wedge}_i * \alpha^{\wedge}_j$ *of the columns of the matrix* $A^{\wedge} = (a^{\wedge}_{ik})$ *we have the unique linear combination*

$$\alpha^{\wedge}_i * \alpha^{\wedge}_j := \begin{bmatrix} a^{\wedge}_{1i} a^{\wedge}_{1j} \\ \vdots \\ a^{\wedge}_{di} a^{\wedge}_{dj} \end{bmatrix} = \sum_k a^{\wedge}_{ijk} \alpha^{\wedge}_k \ .$$

And correspondingly

$$\alpha^{\vee}_i * \alpha^{\vee}_j = \sum_k a^{\vee}_{ijk} \alpha^{\vee}_k \ .$$

(v) $a^{\wedge}_{ik} = a^{\wedge}_{kii}$, $a^{\vee}_{ik} = a^{\vee}_{kii}$.

The main point in the proof is to show that $A^{\wedge}$ and $A^{\vee}$ are triangular matrices with 1's along the main diagonal.

Beautiful examples are provided by the regular polyhedra and their symmetry groups. For example, the tetrahedron on which A_4 acts gives rise to the matrix

$$A^{\wedge} = \begin{bmatrix} 1 & 4 & 6 & 4 & 1 \\ & 1 & 3 & 3 & 1 \\ & & 1 & 2 & 1 \\ & & & 1 & 1 \\ & & & & 1 \end{bmatrix},$$

where $\alpha^{\wedge}_3 * \alpha^{\wedge}_4 = 2\alpha^{\wedge}_3 + 3\alpha^{\wedge}_2$. This means (6·4 (v)): $a^{\wedge}_{343} = 2$, $a^{\wedge}_{342} = 3$, so that each edge can be represented in exactly two ways as the infimum

of a face and an edge while a vertex can be represented in exactly three ways as such an infimum.
Now what about the Burnside ring?
From G and a transversal $\{U_1,\ldots,U_d\}$ of the classes of conjugate subgroups we obtain a complete set of pairwise nonisomorphic transitive G-sets ω_i as follows:

$$\omega_i := G/U_i = \{xU_i \mid x \in G\},\ 1 \leqslant i \leqslant d.$$

G acts on G/U_i by

$$G\times G/U_i \to G/U_i : (g,xU_i) \mapsto gxU_i.$$

We indicate the G-isomorphism class of ω_i by

$$\Omega_i := \{S \mid \text{G-set, G-isomorphic to } \omega_i\}\ .$$

Thus

$$\Omega := \{\Omega_1,\ldots,\Omega_d\}$$

is the complete set of G-isomorphism classes of transitive G-sets. The *Burnside ring* $\mathbb{Z}^{\Omega} := \{\psi:\Omega \to \mathbb{Z}\}$ consists of the "formal sums"

$$\psi = \sum_1^d z_i\Omega_i,\ z_i := \psi(\Omega_i),$$

which we can add:

$$\Omega_i+\Omega_j = \overline{\omega_i \dot{\cup} \omega_j}$$

(form the disjoint union $\omega_i \dot{\cup} \omega_j$ and then take its G-isomorphism class) and multiply

$$\Omega_i\cdot\Omega_j := \sum_k b_{ijk}\Omega_k,\ \text{if}\ \overline{\omega_i\times\omega_j} = \sum_k b_{ijk}\Omega_k.$$

6.5 Theorem:

(i) *The mapping*

$$\Omega_i \to |N_G(U_i):U_i|u_i,\ u_i := \sum_{U\in\tilde{U}_i} U \in \mathbb{Z}[L(G),\wedge]_G$$

defines an embedding

$$\mathbb{Z}^{\Omega} \to \mathbb{Z}[L(G),\wedge]_G$$

of the Burnside ring of G *into* $\mathbb{Z}[L(G),\wedge]_G$.

(ii) *For the matrix* $A^{\wedge}$ *of* $\mathbb{Z}[L(G),\wedge]_G$ *and the table* M(G) *of marks we have*

$$M(G) = A^{\wedge} \cdot \begin{bmatrix} \ddots & & 0 \\ & |N_G(U_i):U_i| & \\ 0 & & \ddots \end{bmatrix}$$

This finally explains the role of the table of marks.

References

A complete list of all the contributions to this theory would have to contain approximately 700 titles. I therefore have to restrict attention to particular topics and papers closely related to the present text.

(i) Papers of historical interest:

A. Cayley: On the Mathematical Theory of Isomers, Philosophical Magazine (4) 47, (1874), 444-446.

A. Crum Brown: On the theory of isomeric compounds, Trans. Roy. Soc. Edinb. 23 (1864), 707-719.

A.C. Lunn/J.K. Senior: Isomerism and configuration, J. Phys. Chem. 33 (1929), 1027-1079.

G. Pólya: Kombinatorische Anzahlbestimmungen für Gruppen, Graphen und chemische Verbindungen, Acta Math. 68 (1937), 145-254.

J.H. Redfield: The theory of group-reduced distributions. Amer. J. Math. 49 (1927), 433-455.

J.H. Redfield: Enumeration by frame group and range groups, J. Graph Theory 8 (1984), 205-224.

J.J. Sylvester: On an application of the new atomic theory to the graphical representation of the invariants and covariants of binary quantics, - with three appendices, Amer. J. Math. 1 (1878), 64-125.

(ii) On the history of this theory:

N.L. Biggs/E.K. Lloyd/R.J. Wilson: Graph Theory 1736-1936, Clarendon Press, 1977.

E.K. Lloyd: J. Howard Redfield (1879-1944), J. Graph Theory 8 (1984), 195-203.

P.M. Neumann: A lemma that is not Burnside's, Math. Scientist 4 (1979), 133-141.

(iii) On applications to chemistry:

A.T. Balaban: Chemical Applications of Graph Theory, Academic Press 1976, 389 pp.

R.A. Davidson: Unified Combinatorial Molecular Stereoanalysis, Ph.D. Thesis, The Pennsylvania State University, 1977.

R.K. Lindsay/B.G. Buchanan/E.A. Feigenbaum/J. Lederberg: Applications of Artificial Intelligence for Organic Chemistry, The DENDRAL Project, McGraw-Hill Book Company 1980.

W. Hässelbarth/B. Richter/E. Ruch: Doppelnebenklassen als Nomenklaturprinzip für Isomere und ihre Abzählung, Theoret. Chim. Acta 19 (1970), 288-300.

(iv) Introductions, longer manuscripts, books:

L. Beineke/Harary, F (ed.): A Seminar on Graph Theory, Holt, Rinehart and Winston, 1967.

F. Harary/E. Palmer: Graphical Enumeration, Academic Press, 1973.

N.G. de Bruijn: Pólya's theory of counting, in: Applied Combinatorial Mathematics, Ed. Beckenbach, Wiley, 1964, pp. 144-184.

N.G. de Bruijn: Pólyas Abzähl-Theorie: Muster für Graphen und chemische Verbindungen, in: Selecta Mathematica III, Ed. K. Jacobs, Springer, 1971.

A. Kerber/K.-J. Thürlings: Symmetrieklassen von Funktionen und ihre Abzählungstheorie (Teil I: Die Grundprobleme), Bayreuther Math. Schr. 12 (1983), 235 pp.

A. Kerber/K.-J. Thürlings: Symmetrieklassen von Funktionen und ihre Abzählungstheorie (Teil II: Hinzunahme darstellungstheoretischer Begriffsbildungen), Bayreuther Math. Schr. 16 (1983), 338 pp.

A. Kerber/K.-J. Thürlings: Symmetrieklassen von Funktionen und ihre Abzählungstheorie (Teil III: Der Burnsidering und Verallgemeinerungen, Unimodalitätsfragen), Bayreuther Math. Schr. 21 (1985).

(v) Enumeration by stabilizer class:

W. Burnside: The Theory of Groups of Finite Order, 2nd Ed. Cambridge 1911, reprint Dover Publ., 1955.

A. Kerber/K.-J. Thürlings: Counting symmetry classes of functions by weight and automorphism group, Proceedings, Lecture Notes in Mathematics 969, Springer 1982, 191-211.

W. Plesken: Counting with groups and rings, Journal für die reine und angewandte Mathematik 334 (1982), 40-68.

G.-C. Rota/D.A. Smith: Enumeration Under Group Action, Annali della Scuola Norm. Sup. di Pisa 4 (1977), 637-646.

J. Sheehan: The number of graphs with a given automorphism group, Can. J. Math. 20 (1968), 1068-1076.

J. Sheehan: On the number of graphs with a given automorphism group, Proc. Coll. Tihany 1966, 271-277, Academic Press 1968.

P.K. Stockmeyer: Enumeration of graphswith prescribed automorphism group, Ph.D. Thesis, Ann Arbor, Michigan, 1971.

D.E. White: Counting Patterns with a given Automorphism Group, Proceedings of the AMS, Vol. 47 (1975).

D.E. White: Classifying Patterns by Automorphism Group: an operator theoretic approach, Discrete Math. Vol. 13 (1975).

(vi) Construction of representatives:

H. Brown/L. Hjelmeland/L. Masinter: Constructive graph labeling using double cosets, Discrete Math. 7 (1974), 1-30.

J.D. Dixon/H.S. Wilf: The Random Selection of Unlabeled Graphs, Journal of Algorithms 4 (1983), 205-213

(vii) Papers on the unimodality (problem (iii)):

A. Kerber/K.-J. Thürlings: Symmetrieklassen von Funktionen und ihre Abzählungstheorie, Teil III, see references (iv).

R.P. Stanley: Unimodal sequences arising from Lie algebra, Combinatorics, representation theory and statistical methods in groups, Young Day Proc., Lect. Notes pure appl. Math., Vol. 57 (1980), 127-136.

R.P. Stanley: Unimodality and Lie superalgebras, Studies Appl. Math. 72 (1985), 263-281.

D. White: Monotonicity and unimodality of the Pattern Inventory, Advances in Math. Vol. 38 (1980).

See also the book

I.G. Macdonald: Symmetric Functions and Hall Polynomials, Clarendon Press, Oxford, 1979, 180 pp.

(viii) Pólya-Redfield theory and linear representation theory:

H.O. Foulkes: On Redfield's Group Reduction Functions, Canad. J. Math. 15, 272-284 (1963).

H.O. Foulkes: On Redfield's Range-Correspondences, Canad. J. Math. 18, 1060-1071 (1966).

A. Kerber/K.-J. Thürlings: Symmetrieklassen von Funktionen und ihre Abzählungstheorie, Teil II, see references (iv).

W. Lehmann: Ein vereinheitlichender Ansatz für die Redfield-Pólya-de Bruijnsche Abzähltheorie, Dissertation, RWTH Aachen, 1976.

R.C. Read: The use of S-functions in combinatorial analysis, Canad. J. Math. 20 (1968), 808-841.

J. Sheehan: On Pólya's Theorem, Can. J. Math. 19 (1967), 792-799.

K.J. Thürlings: Eine Verallgemeinerung des Lemmas von Cauchy-Frobenius, kombinatorische und algebraische Zusammenhänge, Dissertation 1981, Bayreuther Math. Schr. 8 (1981), 39-131.

JOINT DISTRIBUTIONS
OF THREE DESCRIPTIVE PARAMETERS OF BRIDGES

Germain Kreweras
Université Pierre et Marie Curie (Paris VI)
4 place Jussieu 75005 Paris

§1. Introduction

A *bridge* is a finite word written with a's (ascents) and b's (descents) in equal number, such that, for whatever i , the i-th descent never appears before the i-th ascent. If the total length of the word is 2n , we shall call it a bridge of *span* n , or n-bridge. It is well known that the total number of n-bridges is the Catalan number $\frac{1}{n+1}\binom{2n}{n} = c_n$.

Bridges have been investigated extensively, especially from the following viewpoint : if some descriptive parameter of an n-bridge is subject to take a prescribed value z , the corresponding number of n-bridges is an integer function f(n,z), with the property $\sum_z f(n,z) = c_n$.

One of the most familiar distributions f(n,z) is the so-called Narayana distribution

$$N_0(n,z) = \frac{1}{n}\binom{n}{z}\binom{n}{z-1}, \qquad N_0(n,0) = \begin{cases} 0 \text{ if } n \geq 1 \\ 1 \text{ if } n = 0 \end{cases},$$

which enumerates the n-bridges with z "peaks" (occurrences of ab).

It will be convenient for the aims of this paper to consider, instead of the number z of peaks, the number of occurrences of aa (double ascents), which is of course n-z since any of the n ascents a is followed either by another a or by a b . In all the sequel we shall denote the latter parameter n-z with the letter h. Thus the number of n-bridges with a prescribed number h of double ascents will be

$$N(n,h) = \frac{1}{n}\binom{n}{h}\binom{n}{h+1} \qquad h \in \{0,1,\dots,n-1\} \qquad (1)$$

The discovery of this distribution essentially goes back to a paper by Narayana [6].

More recently it was discovered that two other descriptive parameters, the consideration of which may appear as slightly less natural, also follow the same Narayana distribution. Namely :

(i) the number d of *ascents in even position*, with the obvious extreme values

$d = 0$ for the bridge abab...ab = $(ab)^n$

$d = n-1$ for the bridge aabab...abb = $a(ab)^{n-1}b$

The general result is implied by the set of papers [1], [2] and [3].

(ii) The number k of *long non final sequences*. By "sequences" we mean here maximal subwords made either only of ascents, which we call *jumps*, or only of descents, which we call *landings* (if there are z peaks, there are z jumps and z landings) ; we call a sequence *long* if it is made of at least two letters, and we exclude the final sequence, necessarily a landing, from the enumeration, irrespectively of whether it is long or not. Example : the 9-bridge

a a a b a a b b a b b a b a a b b

has 3 long jumps and 2 long non-final landings, so that the total number of *long non-final sequences* (l.n.f.s.) is k = 5 . Again it is obvious that the extreme values of k are

$k = 0$ for the bridge $(ab)^n$

$k = n-1$ for the bridge $(aabb)^{\frac{n}{2}}$ if n is even

or $(aabb)^{\frac{n-1}{2}}$ if n is odd .

The general result has been proved by the author and P. Moszkowski [4].

Once ascertained that the three parameters h (number of double ascents), d (number of ascents in even position) and k (number of l.n.f.s.) follow the same distribution, it is natural to wonder what the joint distributions are, i.e. to calculate the number of n-bridges for which *two* of the three parameters have prescribed values.

This was first done for the parameters h and k by the author and Y. Poupard [5], who were naturally led to a problem in which not only k is specified at the same time as h , but even the number i of long jumps and the number j of long non-final landings are specified separately (of course i+j=k). The noteworthy result can be stated in the following way : the number of n-bridges with h double ascents, i long jumps and j long non-final landings is given by the monomial expression

$$E(n,h,i,j) = \frac{1}{h}\binom{h}{i}\binom{h}{j}\binom{n-h-1}{i-1}\binom{n-h}{j+1} \tag{2}$$

The treatment of the particular cases i=0 or h=0, which both lead to the only bridge $(ab)^n$, is trivial.

We shall have to remember in §2 that the way to prove (2) in [5] leans on the combinatorial proof of an identity equivalent to

$$E(n,h,i,j) = \sum_{(u,v)} \binom{h-1+u}{i+j-1}\binom{n-h+v}{i+j} E(i+j,j,u,v) \tag{3}$$

As far as only the joint distribution of h and k is needed, it can be easily calculated by

$$B(n,h,k) = \sum_{j=0}^{n-1} E(n,h,k-j,j) \quad .$$

Tables of B(n,h,k) for $n \leq 8$ are given in §4 , illustrating the fact that

$$\sum_{k=0}^{n-1} B(n,h,k) = N(n,h)$$

and

$$\sum_{h=0}^{n-1} B(n,h,k) = N(n,k) \quad .$$

In the following sections, we proceed in a similar way to investigate what happens if we prescribe d and h (§2), or d and k (§3). In both cases we shall have first to solve a more refined problem than simply finding the joint distribution of two parameters.

§2. Prescribed d and h

2.1 The double ascents, i.e. the occurences of aa, can have two sorts of positions : odd-even (OE) or even-odd (EO) ; the first double ascent encountered is necessarily in OE position. Let us call r the number of double ascents in OE position (DAOE) and s the number of double ascents in

EO position (DAEO), so that $r+s=h$. In the following example with $n=9$, where the word is sliced into 9 pairs for clarity, we have $r=2$ (underlined double ascents) ans $s=3$ (overlined) :

$$P = \underline{a\overline{a}}|\overline{a}b|\underline{aa}|b\overline{a}|\overline{a}b|b\overline{a}|\overline{a}b|bb|bb \quad .$$

In the same 9-bridge we have $d=4$ (pointed ascents).

Our goal in this section is to determine the number $E^*(n,d,r,s)$ of n-bridges with d ascents in even position, r DAOE ' s and s DAEO ' s ; we shall in fact prove that $E^*(n,d,r,s)$ depends on its four arguments exactly in the same way as $E(n,h,i,j)$ depends on its own.

2.2 We shall need the concept of "squeezed bridge" as introduced by Y. Poupard [7].

Squeezing a bridge P means spotting all the occurrences of ab , deleting them and joining the remaining subwords together. The example P above, once squeezed, yields the bridge

$$P' = a\ a\ a\ a\ b\ a\ b\ b\ b\ b$$

(the "scars" of the deletions are marked).

We shall first prove the following

<u>Lemma</u> : If an n-bridge P has d ascents in even position, r double ascents in "odd-even" position (DAOE) and s double ascents in "even-odd" position (DAEO), its squeezed bridge P' is an (r+s)-bridge with r ascents in odd position (AO) and s ascents in even position (AE).

The lemma is easily verified for small values of n . Assume that it is true for any span $< n-1$, and consider an n-bridge P with r DAOE ' s and s DAEO ' s .

If P is connected (i.e. is not the concatenation of two non-empty bridges), the deletion of the initial a and the final b yields a (n-1)-bridge Q ; this can be written $P = aQb$. Clearly enough $P' = aQ'b$, where Q' is the bridge obtained by squeezing Q . Obviously

$$r(P) = s(Q) + 1$$
and $s(P) = r(Q)$,

so that, when passing from Q to P, the sum r+s is increased by 1 unit.

Furthermore any odd position in Q becomes even in P and conversely, so that DAOE's and DAEO's are exchanged between P and Q. Thus the proof of the lemma is straightforward if P is connected.

If P is not connected, it is a concatenation of several connected components (or "arches"), each of which has a span $\leq n-1$. Since the lemma is supposed true for each arch separately, it is true for P because each arch is of even length by definition.

2.3 Consider now a given (r+s)-bridge P', with r AO's and s AE's. P' has a certain number of double ascents, say u DAOE's and v DAEO's. It follows easily that the occurrences of ab in P' are r-u times in OE position and s-v times in EO position.

In order to go back from P' to the n-bridge P whose squeezed bridge is P', we have to *insert*, in each of the 2r + 2s + 1 possible positions, a certain number (possibly 0) of occurrences of ab, i.e. $(ab)^z$ with a certain *non-negative* exponent. The exponent must in fact be *positive* for the positions between an ascent a and a descent b (else this a and this b would have been deleted by squeezing P into P'). The sum of all these exponents must be n-r-s.

Instead of calling the successive exponents

$$z_0\ z_1\ z_2\ z_3\ z_4\ \cdots\cdots\ z_{2r+2s-1}\ \ z_{2r+2s}$$

it will be convenient to call them

$$y_0\ x_1\ y_1\ x_2\ y_2\ \cdots\cdots\ x_{r+s}\ y_{r+s}\quad .$$

$(ab)^{x_\lambda}$ will be inserted between the positions $2\lambda-1$ and 2λ, thus r-u of the exponents x_λ must be positive since in P' ab appears r-u times in position $(2\lambda-1,2\lambda)$. In the same way (ab^{y_λ}) will be inserted between the positions 2λ and $2\lambda+1$, thus s-v of the exponents y_λ must be positive since in P' ab appears s-v times in position $(2\lambda,2\lambda+1)$. In P the inserted $(ab)^{x_\lambda}$ will all be in EO position, while the inserted $(ab)^{y_\lambda}$ will all be in OE position ; the former alone will increase (by x_λ units) the number of ascents in even position (AE's).

It follows that the exponents $x_1\ x_2\ \ldots\ x_{r+s}$ will be a sequence of r+s integers, r-u of which must be positive and s+u non-negative ; their sum must be d-s since the number of AE ' s must be increased from s (in P') to d (in P). The number of such sequences is easily proved to be the binomial $\binom{d-1+u}{r+s-1}$.

In a similar way the exponents $y_0\ y_1\ \ldots\ y_{r+s}$ will be a sequence of r+s+1 integers, s-v of which must be positive and r+v+1 non-negative ; as for their sum $y_0+y_1+\ldots+y_{r+s}$, it must be such that

$$\sum_{\lambda=1}^{r+s} x_\lambda + \sum_{\lambda=0}^{r+s} y_\lambda = n-r-s \quad ,$$

span increase from P' to P ; whence

$$\sum_{\lambda=0}^{r+s} y_\lambda = n-r-s-(d-s) = n-d-r \quad .$$

Again the number of such sequences is easily proved to be a binomial, viz. $\binom{n-d+v}{r+s}$.

Finally, given a squeezed (r+s)-bridge P' with s AE ' s , u DAOE ' s and v DAEO ' s, the number of possible n-bridges P having d AE ' s , r DAOE ' s and s DAEO ' s , whose squeezing generates P' , is the product

$$\binom{d-1+u}{r+s-1}\binom{n-d+v}{r+s} \quad .$$

The immediate consequence is that the numbers E*(n,d,r,s) satisfy the identity

$$E^*(n,d,r,s) = \sum_{(u,v)} \binom{d-1+u}{r+s-1}\binom{n-d+v}{r+s} E^*(r+s,s,u,v) \tag{4}$$

We observe here that (4) is identical to (3) in §1., except for the notations. Since (3) was the central tool to prove (2), we can now state the result corresponding to (2), viz.

$$E^*(n,d,r,s) = \frac{1}{d}\binom{d}{r}\binom{d}{s}\binom{n-d-1}{r-1}\binom{n-d}{s+1} \tag{5}$$

§3. Prescribed d and k

3.1 Again we shall proceed to enumerate n-bridges with d AE ' s, and prescribe additionally not only the total number k of long non-final sequences, but separately the number of long jumps and the number of non-final landings.

For this purpose we shall need a "symmetry lemma" which will be introduced by the following remarks.

Let us start from the expression (2) of §1, which enumerates the n-bridges with h double ascents, i long jumps and j long non-final landings. Among these bridges, some begin with aa... ; their number is

$$E_0(n,h,i,j) = E(n,h,i,j) - E(n-1,h,i,j) \quad , \tag{6}$$

since the subtracted term counts those which begin with ab . Let us now find an expression F(n,h,i,j) for the number of n-bridges with h double ascents, i long *non-initial* jumps and j long non-final landings. Among such bridges, F_1 begin with ab... and F_2 begin with aa... Clearly

$$F_1 = E(n-1,h,i,j)$$

and $F_2 = E_0(n,h,i+1,j)$,

so that, with use of the suitable form of (6) ,

$$\begin{aligned} F(n,h,i,j) &= F_1 + F_2 \\ &= E(n-1,h,i,j) + E(n,h,i+1,j) - E(n-1,h,i+1,j) \ . \end{aligned}$$

Since the expression of E is known by (2), it is elementary (although not so short) to derive the result

$$F(n,h,i,j) = \binom{h}{i}\binom{h}{j}\binom{n-h-1}{i}\binom{n-h-1}{j} - \binom{h+1}{i+1}\binom{h-1}{j-1}\binom{n-h-2}{i-1}\binom{n-h}{j+1} \tag{7}$$

Any of the n-bridges counted by (7), if preceded by an additional ascent a and followed by an additional descent b , is bijectively transformed into a connected (n+1)-bridge with h+1 double ascents, i+1 long jumps (initial or not) and j+1 long landings (final or not). Thus the number $H_{n,i,j}(h)$ of connected n-bridges with i long jumps, j long landings and h double ascents, is given by F(n-1,h-1,i-1,j-1), whence

$$H_{n,i,j}(h) = \binom{h-1}{i-1}\binom{h-1}{j-1}\binom{n-h-1}{i-1}\binom{n-h-1}{j-1} - \binom{h}{i}\binom{h-2}{j-2}\binom{n-h-2}{i-2}\binom{n-h}{j} \ ,$$

expression which makes it easy to check that

$$H_{n,i,j}(h) = H_{n,i,j}(n-h) \ .$$

This result, which we call the "symmetry lemma", is interesting by itself because it extends to the connected n-bridges *with prescribed numbers of long jumps and long landings* a remark which is a straightforward consequence of the symmetry of the Narayana distribution as far as *all* the connected n-bridges are concerned.

3.2 In this section we shall consider the set $\mathcal{P}(n,i,j)$ of the *connected* n-bridges with prescribed numbers i and j of long jumps and long landings.

The following remarks are obvious :

(i) $\mathcal{P}(1,i,j)$ is non-empty only if $i = j = 0$
(bridge ab or "micro-arch")

(ii) $\mathcal{P}(2,i,j)$ is non-empty only if $i = j = 1$

(iii) For any $n > 3$, $\mathcal{P}(n,i,j)$ is non-empty if and only if $i > 1$, $j > 1$ and $i+j < n-1$.

Our main statement will be the following : in $\mathcal{P}(n,i,j)$ the parameters "total number of landings (or jumps, or occurrences of ab)" and "number of ascents in odd position (AO 's)" have the same distribution.

The assertion is trivially checked for $n = 1$ and $n = 2$. $\mathcal{P}(3,i,j)$ reduces to the two bridges aaabbb and aababb , the first of which has 1 landing and 2 AO ' s , the second 2 landings and 1 AO .

Let us assume that our assertion is true for the spans $1,2,\ldots,n-1$ and for whatever i and j (recurrence assumption), and consider a bridge P belonging to $\mathcal{P}(n,i,j)$.

Since P is connected, it may be written

$$P = a\,Q\,b\ ,$$

where Q is an (n-1)-bridge. The latter is either connected or decomposable into m arches $Q_1\ Q_2 \ldots Q_m$, the t-th of which is an n_t-bridge with i_t long jumps and j_t long landings ; $Q_t \in \mathcal{P}(n_t, i_t, j_t)$ and $n_t < n-1$, so that we can use the recurrence assumption.

Note that $N = (n_1\ n_2 \ldots n_m)$ is a sequence of *positive* integers (with sum n-1), but not necessarily $I = (i_1, i_2 \dots i_m)$ or $J = (j_1 j_2 \dots j_n)$, which can have vanishing terms because of "micro-arches" ; it is easy to make sure that

$$i_1 + i_2 + \dots + i_m = \begin{cases} i & \text{if } n_1 \geq 2 \\ i-1 & \text{if } n_1 = 1 \end{cases}$$

and
$$j_1 + j_2 + \dots + j_m = \begin{cases} j & \text{if } n_m \geq 2 \\ j-1 & \text{if } n_m = 1 \end{cases} .$$

Let the sequence N be specified and consider the subset of $\mathcal{P}(n,i,j)$ for which the successive arches of the (n-1)-bridge Q have spans $n_1 \, n_2 \dots n_m$. This subset can in turn be partitioned into classes $C = (N,I,J)$, each of which corresponds to specified N, I and J (general class $\mathcal{P}(n,i,j;C)$).

Example : $(n,i,j) = (12,4,3)$

$N = (5,6)$ $(m = 2,\ n_1 = 5,\ n_2 = 6,\ n_1+n_2 = 12-1)$.

There are four possible systems (I,J) ; one of them is defined by

$I = (2,2)$ $(i_1 = 2,\ i_2 = 2)$ and $J = (2,1)$ $(j_1 = 2,\ j_2 = 1)$

$|\mathcal{P}(n_1,i_1,j_1)| = |\mathcal{P}(5,2,2)| = 2$

Bridges :			
	aabaabbabb	3	landings, 2 AO ' s
	aaabbaabbb	2	" 3 " .

$|\mathcal{P}(n_2,i_2,j_2)| = |\mathcal{P}(6,2,1)| = 10$

Bridges :			
	aaaabaabbbbb	2	landings, 3 AO ' s
	aaabaaabbbbb	2	" 4 "
	aaabaababbbb	3	" 3 "
	aaababaabbbb	3	" 4 "
	aabaaaabbbbb	2	" 3 "
	aabaaababbbb	3	" 2 "
	aabaababab bb	4	" 4 "
	aababaaabbbb	3	" 2 "
	aababaababbb	4	" 3 "
	aabababaabbb	4	" 2 " .

(In both lists above, the AO ' s are pointed).

The equal distribution of landings and AO ' s on each of the sets $\mathcal{P}(n_t, i_t, j_t)$ is achieved according to the recurrence assumption (and immédiately checked on the above example).

More precisely, if we call $A_t(z_t)$ (resp. $B_t(z_t)$) the number of bridges in $\mathcal{P}(n_t, i_t, j_t)$ that have z_t AO ' s (resp. z_t landings), the recurrence assumption means that

$$A_t(z_t) = B_t(z_t) \qquad (8)$$

As a consequence, in the class of (n-1)-bridges Q corresponding to $\mathcal{P}(n,i,j;C)$, i.e. the cartesian product

$$\prod_{t=1}^{m} \mathcal{P}(n_t, i_t, j_t) \quad ,$$

the number $A_C(z)$ of bridges having z AO ' s is equal to the number $B_C(z)$ of bridges having z landings, since both are obtained by multiplying together the corresponding number $A_t(z_t)$ and $B_t(z_t)$ and summing up with respect to all the admissible sequences $(z_1\ z_2\ \ldots\ z_p)$ with sum z .

In the above example, the class C is composed of 20 (=2 x 10) bridges Q ; the number of landings, as well as the number of AO ' s, is equal

to 4	:	3 times	(1 x 3 ,	since 4 = 2 + 2)
to 5	:	7 times	(1 x 4 + 1 x 3 ,	" 5 = 2 + 3 = 3 + 2)
to 6	:	7 times	(1 x 3 + 1 x 4 ,	" 6 = 2 + 4 = 3 + 3)
to 7	:	3 times	(1 x 3 ,	" 7 = 3 + 4) .

Once proved that in any class C we have

$$A_C(z) = B_C(z) \qquad (9)$$

it is convenient to consider a (vertical) list of the bridges P in $\mathcal{P}(n,i,j)$ grouped by classes $C_1\ C_2\ \ldots$ and to write in front of each row in a column (a), the number of AO ' s in P : say y

in a column (a'), the number of AE ' s in P (or of AO's in Q) : n-y

in a column (b), the number of landings (common to P and Q) .

Let y be a given integer and x its complement to n ; x = n-y . The number of occurrences of y in column (a) is equal to the number of occurrences of x in column (a'), i.e. to the number of bridges Q having x AO's. By virtue of (9), the number of occurrences of x in (a') is the same as in column (b), since the equality of these numbers is achieved *in each of the classes C*. But by virtue of the "symmetry lemma" proved above, this x appears in (b) the same number of times as its complement $y = n - x$. Finally y appears the same number of times in (a) and (b), which means that in $\mathcal{P}(n,i,j)$ the number of bridges with y AO's is equal to the number of bridges with y landings ; which completes the proof of the assertion announced in this section.

The list $\mathcal{P}(6,2,1)$ taken from the above example illustrates the argument ; we indicate here the classes C_1 C_2 C_3 C_4 and in each bridge P the decomposition into aQb and the decomposition of Q into its arches.

		P		(a)	(a')	(b)
C_1	a	aaabaabbbb	b	3	3	2
	a	aabaaabbbb	b	4	2	2
	a	aabaaabbbb	b	3	3	3
	a	aababaabbb	b	4	2	3
C_2	a	ab|aaaabbbb	b	3	3	2
	a	ab|aaababbb	b	2	4	3
	a	ab|aabababb	b	4	2	4
C_3	a	ab|ab|aaabbb	b	2	4	3
	a	ab|ab|aababb	b	3	3	4
C_4	a	ab|ab|ab|aabb (Q)	b	2	4	4

In the sequel we shall use the result in the following equivalent form : in the set $\mathcal{P}(n,i,j)$ of connected n-bridges with i long jumps and j long landings, there are as many bridges with x double ascents as with x ascents in even position (AE's).

<u>3.3</u> It is now easy to come back to the original problem, which concerns the set $\mathcal{R}(n,i,j)$ of n-bridges (no longer necessarily connected) having i long jumps and j long *non-final* landings, and to show that in this set the parameters "number h of double ascents" and "number d of ascents in even position" are also equally distributed.

For this purpose, consider separately two types of bridges in

$\mathcal{R}(n,i,j)$. The first type consists of the bridges ending with ...ab ; for these bridges, prescribing the span n and j long non-final landings goes back to prescribing the span n-1 and simply j long landings. For the second type, i.e. for all the bridges of $\mathcal{R}(n,i,j)$ ending with ...bb, prescribing j long non-final landings goes back to prescribing simply j+1 long landings.

To whichever type a bridge of $\mathcal{R}(n,i,j)$ belongs, it can be decomposed into its successive arches $P_1, P_2, \ldots$, where P_t belongs to some $\mathcal{P}(n_t, i_t, j_t)$; moreover $\mathcal{R}(n,i,j)$ can be partitioned into classes $C = (N,I,J)$, each of which is included in one of the two types and corresponds to given sequences of n_t , i_t and j_t . The same argument as in § 3.2 proves that within each class the number h of double ascents and the number d of AE's are equally distributed. The result obviously extends to either "type" , thus to $\mathcal{R}(n,i,j)$, as is illustrated by the following example, with n = 5, i = 2, j = 1 :

		h	d
type I	aabaabbbab	2	2
type II	aaabbaabbb	3	2
	aabaabbabb	2	3
	aaabbb|aabb	3	2
	aababb|aabb	2	3
	aabb|aaabbb	3	2
	aabb|aababb	2	3
	aabb|ab|aabb	2	2
	ab|aabb|aabb	2	2

The final conclusion is that if, among the n-bridges with i long jumps and j long non-final landings, we have to enumerate those with d ascents in even position, we find the same number as the one that enumerates those with d double ascents. The latter, as was proved in [5], is given by (2), with replacement of h by d ; thus the number looked for is

$$\frac{1}{d} \binom{d}{i} \binom{d}{j} \binom{n-d-1}{i-1} \binom{n-d}{j+1} ,$$

a third interpretation of the same function of four arguments as met in (2) and (5).

§4. Summing up the last two arguments

In order not to privilege any of the three problems solved by the same four-argument function, we shall now adopt Greek letters and write

$$E(n,\lambda,\rho,\sigma) = \frac{1}{\lambda}\binom{\lambda}{\rho}\binom{\lambda}{\sigma}\binom{n-\lambda-1}{\rho-1}\binom{n-\lambda}{\sigma+1} \qquad (10)$$

To recapitulate, we know now that (10) enumerates the n-bridges with

(a) λ ascents in even position, ρ double ascents in "odd-even" position, σ double ascents in "even-odd" position (§2)

(b) λ ascents in even position, ρ long jumps and σ long non-final landings (§3)

(c) λ double ascents, ρ long jumps and σ long non-final landings ([5]).

Moreover, if we set $\rho+\sigma=\mu$, the function of three arguments defined by

$$B(n,\lambda,\mu) = \sum_{\sigma=0}^{n-1} E(n,\lambda,\mu-\sigma,\sigma) \qquad (11)$$

gives explicitly the three joint distributions of the parameters d (ascents in even position), h (double ascents) and k (number of long non-final "sequences", i.e. jumps or landings indistinctly) in the cases

(a) $\lambda = d$ $\mu = h$
(b) $\lambda = d$ $\mu = k$
(c) $\lambda = h$ $\mu = k$.

The function E would not be easy to tabulate because of its four arguments. Its values are readily calculated ; they are, in a way, generalizations of the Narayana numbers defined by (1).

On the other hand, B is susceptible of a tabulation, which is given below for $n < 8$; it illustrates in particular the announced property of the row sums and column sums, viz. that both yield the Narayana numbers.

n = 2	μ = 0	1	
λ = 0	1		1
1		1	1
	1	1	

n = 3	μ = 0	1	2	
λ = 0	1			1
1		2	1	3
2		1		1
	1	3	1	

n = 4	μ = 0	1	2	3	
λ = 0	1				1
1		3	3		6
2		2	3	1	6
3		1			1
	1	6	6	1	

n = 5	μ = 0	1	2	3	4	
λ = 0	1					1
1		4	6			10
2		3	9	7	1	20
3		2	5	3		10
4		1				1
	1	10	20	10	1	

n = 6	μ = 0	1	2	3	4	5	
λ = 0	1						1
1		5	10				15
2		4	18	22	6		50
3		3	15	22	9	1	50
4		2	7	6			15
5		1					1
	1	15	50	50	15	1	

n = 7	μ = 0	1	2	3	4	5	6	
λ = 0	1							1
1		6	15					21
2		5	30	50	20			105
3		4	30	70	55	15	1	175
4		3	21	45	30	6		105
5		2	9	10				21
6		1						1
	1	21	105	175	105	21	1	

n = 8	μ = 0	1	2	3	4	5	6	7	
λ = 0	1								1
1		7	21						28
2		6	45	95	50				196
3		5	50	160	185	80	10		490
4		4	42	144	185	96	18		490
5		3	27	76	70	20			196
6		2	11	15					28
7		1							1
	1	28	196	490	490	196	28	1	

These numbers B deserve to be studied for their own sake. They exhibit various properties which may be more or less easy to prove ; we only mention one of them as an example, leaving the proof up to the reader :

If n is odd (=2m+1), the differences between row m and row m+1 are positive and form a *symmetric* sequence (taking the same values for μ and n-μ).

If n is even (=2m), the differences between row m-1 and row m form a *skew symmetric* sequence (taking opposite values for μ and $n-\mu$, thus value 0 for $\mu=0$, which means $B(2m,m-1,m)=B(2m,m,m)$).

REFERENCES

[1] M. DELEST & G. VIENNOT, Algebraic languages and Polyominoes enumeration, Theor. Comp. Sc., 34 (1984), 169-206.

[2] P. FLAJOLET, Combinatorial aspects of continued fractions, Discrete Math., 32 (1980), 125-161.

[3] I. GESSEL, A non-commutative generalization of the Lagrange inversion formula, Trans. Amer. Math. Soc., 257 (1980), 455-481.

[4] G. KREWERAS & P. MOSZKOWSKI, A new enumerative property of the Narayana numbers, Journal of Statistical Planning and Inference, 14 (1986), 63-67.

[5] G. KREWERAS & Y. POUPARD, Subdivision des nombres de Narayana suivant deux paramètres supplémentaires, Europ. J. of Combinatorics, 7 (1986), 141-149.

[6] T.V. NARAYANA, Sur les treillis formés par les partitions d'un entier, C.R. Ac. Sci. Paris, 240-I (1955), 1188.

[7] Y. POUPARD, Sur les quasi-ponts, Cahiers du B.U.R.O., 32 (1980), 3-20.

SOME NEW COMPUTATIONAL METHODS IN THE THEORY OF SPECIES

Gilbert Labelle*
Université du Québec à Montréal

0. Introduction.

The strong interactions between Classical Analysis and Enumerative Combinatorics are reflected by the fact that different kinds of series -- including the "calculus" of the operations between them -- can be associated with each given combinatorial species [J2,J5]. The purpose of the present paper is to describe some computational techniques, involving inversion and Newton-Raphson iteration, that can be applied in a "uniform manner" to such series.

For simplicity of presentation, we shall state and prove our results within the context of **unisorted unweigthed** combinatorial species . Their generalizations to **multisorted** and/or **weighted** species (see [J2], also [L7]) can be obtained in a straigthforward manner. It is also possible to further extend the results to the case of **virtual** species (i.e. formal differences of species) using the various operations between them (including substitution) introduced in [J3] and [Y1-2].

Let F be a species. We shall be concerned with the following series :

- The **generating series** of F, given by $F(x) = \Sigma_n f_n x^n/n!$ where f_n is the number of (labeled) F-structures on n distinct points.

- The **type-generating series** of F, given by $\widetilde{F}(x) = \Sigma_n \widetilde{f}_n x^n$ where $\widetilde{f}_n$ is the number of isomorphism types of F-structures on n points (i.e. unlabeled F-structures).

- The **cycle indicator series** of F, given by

$$Z_F(x_1,x_2,x_3,\ldots) = \Sigma_{\beta_1+2\beta_2+3\beta_3+\ldots<\infty} f_{\beta_1,\beta_2,\beta_3,\ldots} x_1^{\beta_1} x_2^{\beta_2} x_3^{\beta_3}\ldots / 1^{\beta_1}\beta_1!\, 2^{\beta_2}\beta_2!\, 3^{\beta_3}\beta_3!\ldots$$

where $f_{\beta_1,\beta_2,\beta_3,\ldots} = \text{fix}\, F[\beta_1,\beta_2,\beta_3\ldots]$ is the number of F-structures on n distinct points ($n = \beta_1+2\beta_2+3\beta_3+\ldots$) which are invariant under the action of any permutation β, of these n points, of type $\beta_1,\beta_2,\beta_3\ldots$ (here β_k denotes the number of cycles of length k in β).

---

* Avec l'appui financier du programme FCAR (Québec, EQ1608) et du CRSNG (Canada, A5660).

Cycle indicator series are also called **indicatrix series**, for short. Standard references about them are [BBN;B1,B6-7;H1-2;HP;HR;J2;JK;L2,L5-6,L9;P;R1,R2,R3;W].

• The **molecular series** of F, given by $F(\mathbf{X}) = \sum_{n,H} f_H \mathbf{X}^n/H$ where f_H denotes the multiplicity (up to isomorphism) of the molecular component $\mathbf{X}^n/H$, of F. Here, for each n, the variable H runs through a (fixed) set of representatives of the conjugacy classes of subgroups of the symmetric group $\mathfrak{S}_n$ and $\mathbf{X}$ denotes, as usual, the species of all singletons. See [J3-5], [L4,L8-9] and [Y1-2] for more informations (and various tables) concerning the calculus of molecular series (see [L10] for related series).

Consider now any two species F and G. It turns out that the equality $F(\mathbf{X}) = G(\mathbf{X})$ of their respective molecular series is a much stronger condition than the mere equality $Z_F = Z_G$ of their corresponding cycle indicator series (see [L8] for the "simplest" explicit combinatorial illustration of this phenomenon). This can be explained as follows : Each species F determines, by functoriality, a sequence $(F[n])_{n\geq 0}$ of **permutation - representations** of the symmetric groups $\mathfrak{S}_n$ and, by linearization, a corresponding sequence $(\mathfrak{Lin}\, F[n])_{n\geq 0}$ of **linear representations** of each $\mathfrak{S}_n$ (these two sequences are arising from the sets F[n] of all F-structures on {1, 2, 3, ..., n}, n=0, 1, 2, ...). One can check that

$$F(\mathbf{X}) = G(\mathbf{X}) \quad \text{iff} \quad \text{each } F[n] \text{ is isomorphic to } G[n]$$

while

$$Z_F = Z_G \quad \text{iff} \quad \text{each } \mathfrak{Lin}\, F[n] \text{ is isomorphic to } \mathfrak{Lin}\, G[n],$$

which is a weaker condition. In the first situation, we say that F and G are **isomorphic species** (and usually write F = G); in the second, we say that they are **linearly isomorphic species**. Still weaker conditions are

$$F(x) = G(x) \quad \text{(i.e. F and G are \textbf{equipotent species})}$$

and

$$\tilde{F}(x) = \tilde{G}(x) \quad \text{(i.e. F and G are \textbf{type-equipotent species})}.$$

Moreover [J2], $F(x) = Z_F(x,0,0,\ldots)$ and $\tilde{F}(x) = Z_F(x,x^2,x^3,\ldots)$.

The whole situation is conveniently summarized by Figure 1 which displays the 3 main levels of structural information in the theory of species: the "**combinatorial**" level, the "**linear**" level and the "**analytical**" level. "Going down" in this figure means "forgetting structural information".

In §1 we present a general principle by means of which one can "lift" formulas from classical (multidimensional) Analysis up to the linear level. We apply it to obtain explicit expressions for the cycle indicator series of certain species of "enriched" rooted trees (in the sense of [J2, L1]), including, as a special case, the species of "ordinary" rooted trees. This is done by lifting to the linear level, the classical multidimensional inversion formulas due to I.J.Good [G2] and S.A.Joni [J1]. In §2, we show how a certain combinatorial approximation problem concerning recursively defined species gives rise, in a natural manner, to an efficient Newton - Raphson iterative scheme that can be applied to all 3 levels. This generalizes the combinatorial approach to Newton - Raphson iteration given in [DLL]. The standard references about combinatorial species are [B2-5; D; DLL; FL1-2; J2-5; L1-10; LS; LV; Y1-2]. For the representation theory of the symmetric group see [JK] for instance.

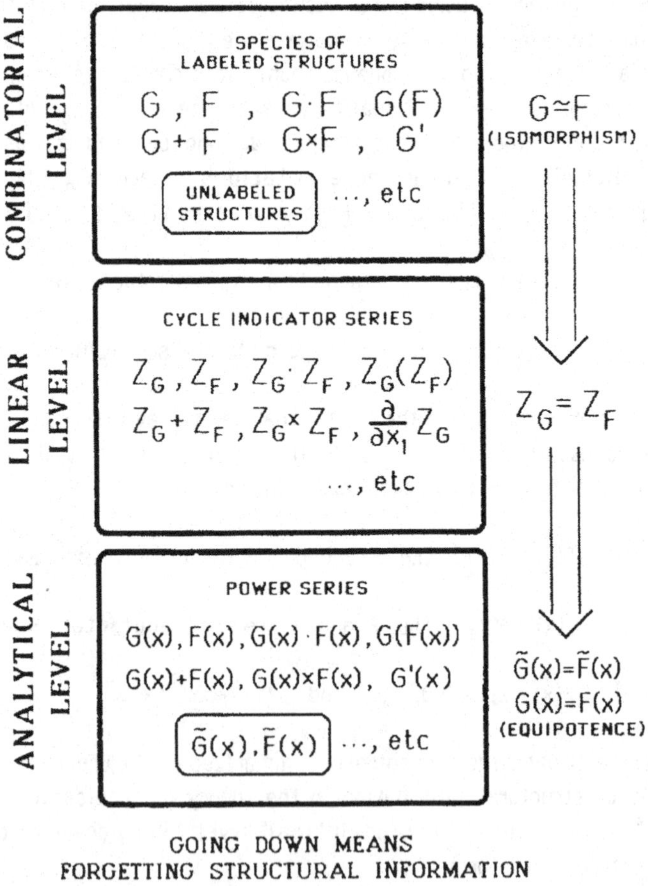

Figure 1

1. A "lifting" Principle from classical Analysis.

Let $\mathbb{C}[[x_1,x_2,x_3, \ldots]]$ be the ring of (formal) indicatrix series

$$f(x_1,x_2,x_3, \ldots) = \Sigma f_{\beta_1,\beta_2,\beta_3,\ldots} x_1^{\beta_1} x_2^{\beta_2} x_3^{\beta_3} \ldots / 1^{\beta_1}\beta_1 !\, 2^{\beta_2}\beta_2 !\, 3^{\beta_3}\beta_3 ! \ldots$$

whose coefficients $f_{\beta_1,\beta_2,\beta_3,\ldots}$ belong to the field $\mathbb{C}$ of all complex numbers. Many supplementary operations can be added to that ring structure, including a **derivation** and a **substitution** (plethysm) defined by (see [J2]):

$$f' = \partial f/\partial x_1 \qquad \text{and} \qquad g(f) = g(f_1,f_2,f_3, \ldots)$$

where the following convention is used

$$f_k = f(x_k, x_{2k}, x_{3k}, \ldots), \quad k=1, 2, 3, \ldots .$$

Each of these operations has its counterpart in both the Combinatorial and the Analytical levels. Moreover, the "chain rule" takes its usual form:

$$h = g(f) \Rightarrow h' = g'(f)f' .$$

This analogy with classical Analysis goes very far in view of the following observations (see [L2]): Define, for each indicatrix series f, the **infinite vector** $\underline{f}$ and the **infinite "jacobian" matrix** $\underline{f}'$ by

$$\underline{f} = (f_i)_{1 \leq i < \infty} = (f(x_i, x_{2i}, x_{3i}, \ldots))_{1 \leq i < \infty} \qquad \text{and} \qquad \underline{f}' = ((\partial/\partial x_j)f_i)_{1 \leq i,j < \infty}$$

respectively, then

Observation 1. $h = g(f) \Longleftrightarrow \underline{h} = \underline{g}(\underline{f})$ *where the right member denotes the* *__usual__* *(infinite dimensional) vectorial substitution.*

Observation 2. $h' = g'(f)f' \Longleftrightarrow \underline{h}' = \underline{g}'(\underline{f})\underline{f}'$ *where the right member denotes the* *__usual__* *(infinite dimensional) matricial multiplication.*

Observation 3. *The jacobian matrix* $\underline{f}' = ((\partial/\partial x_j)f_i)_{1 \leq i,j < \infty}$ *is always upper triangular and we have, for each* $i,j \geq 1$,

$$(\partial/\partial x_j)f_i = (\partial f/\partial x_{j/i})_i \quad \textit{if } i \textit{ divides } j, \qquad \textit{and } = 0 \textit{ otherwise.}$$

Moreover, if $f'(0,0,0, \ldots) = 1$ *then* $\det \underline{f}' = \prod_{i \geq 1} f'(x_i, x_{2i}, x_{3i}, \ldots)$.

Hence, we can state the following "heuristic" principle:

Lifting Principle : *"Every" formula from classical m-dimensional power series Analysis gives rise, letting* $m \longrightarrow \infty$, *to a corresponding formula at the level of indicatrix series.*

We shall illustrate the use of this principle by lifting the classical multi-dimensional inversion formulas due to I.J.Good [G2], and S.A.Joni [J1] to compute the indicatrix series of the species $A = A(\mathbf{X})$ and $B = B(\mathbf{X})$ which are respectively characterized by the following two combinatorial equations

$$A(\mathbf{X}) = \mathbf{X}R(A(\mathbf{X})) \qquad \text{and} \qquad B(\mathbf{X}) = \mathbf{X} + G(B(\mathbf{X})),$$

where R and G are any two given species satisfying $R[\emptyset] \neq \emptyset$ and $G[\emptyset] = G'[\emptyset] = \emptyset$. A structure of species A can be described[J2,L1] as an **R-enriched rooted tree** (French: arborescence R-enrichie) and a generic example is given by Figure 2a. Similarly, a B-structure can be described[L2] as a **G-Catalan rooted tree** (French: G-arborescence de Catalan) and is given by Figure 2b.

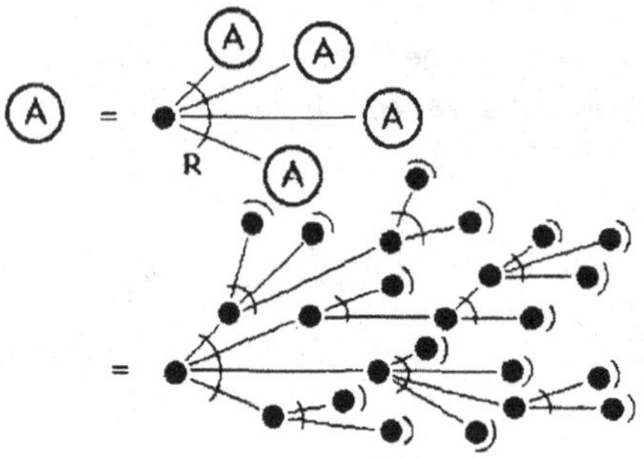

Figure 2a

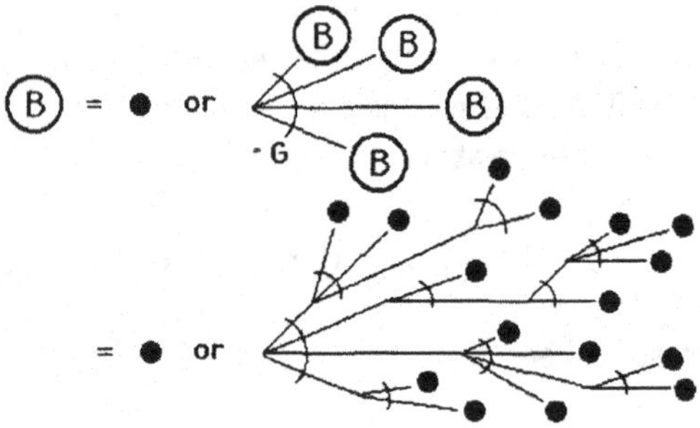

Figure 2b

Theorem A. *Let* $A = \mathbf{X}R(A)$ *be the species of* R*-enriched rooted trees and let* F *be an arbitrary species. Then the indicatrix series* $Z_{F(A)}$ *of the species* $F(A)$ *is given by*

$$\text{fix}\, F(A)[\beta_1, \beta_2, \beta_3, \dots] = \textit{coeff. of}\ (x_1^{\beta_1} x_2^{\beta_2} x_3^{\beta_3} \dots / 1^{\beta_1}\beta_1!\, 2^{\beta_2}\beta_2!\, 3^{\beta_3}\beta_3! \dots) \quad \textit{in}$$

$$Z_F(x_1, x_2, x_3, \dots) \prod_{i \geq 1} \rho(x_i, x_{2i}, x_{3i}, \dots)\, Z_R(x_i, x_{2i}, x_{3i}, \dots)^{\beta_i}$$

where

$$\rho(x_1, x_2, x_3, \dots) = 1 - x_1 Z_R'(x_1, x_2, x_3, \dots)/Z_R(x_1, x_2, x_3, \dots).$$

Proof. We have $Z_{F(A)} = Z_F(Z_A)$ where $Z_A = Z_X Z_R(Z_A)$. To simplify the notation, let us write f, a and r for Z_F, Z_A and Z_R respectively. We must compute $f(a) = f(a_1,a_2,a_3, \ldots)$ given that $a = x_1 r(a)$. By Observation 1 above, this last equation is equivalent to the infinite system

$$a_i = x_i r_i(a_1,a_2,a_3, \ldots), \quad i = 1,2,3,\ldots .$$

Now, Good's multidimensional inversion formula, which is valid for finite systems $a_i = x_i r_i(a_1,a_2,\ldots,a_m)$, $1 \le i \le m$, with $a_i = a_i(x_1,x_2,\ldots,x_m)$, can be written as

$$[x_1^{\beta_1}\ldots x_m^{\beta_m}]\, f(a_1,\ldots,a_m) = [x_1^{\beta_1}\ldots x_m^{\beta_m}]\, f(x_1,\ldots,x_m)\, |K(x_1,\ldots,x_m)| \prod_{1\le i\le m} r_i(x_1,\ldots,x_m)^{\beta_i}$$

where $|K(x_1,\ldots,x_m)| = \det(\delta_{i,j} - x_i(\partial r_i/\partial x_j)/r_i)$. Hence, the result follows by letting $m \longrightarrow \infty$ and using the fact, due to Observation 3, that the resulting determinant involves an upper triangular infinite matrix. ∎

Theorem B. *Let* $B = X + G(B)$ *be the species of* G - *Catalan trees and let* F *be an arbitrary species. Then the indicatrix series* $Z_{F(B)}$ *of the species* F(B) *is given by*

$$Z_{F(B)} = \sum_{\underline{k}} \underline{D}^{\underline{k}}\, Z_F(x_1,x_2,x_3,\ldots) \prod_{i\ge 1} \gamma(x_i,x_{2i},x_{3i},\ldots)\, Z_G(x_i,x_{2i},x_{3i},\ldots)^{k_i}/k_i!$$

where

$$\gamma(x_1,x_2,\ldots) = 1 - Z_G'(x_1,x_2,\ldots) \quad \textit{and} \quad \underline{D}^{\underline{k}} = (\partial/\partial x_1)^{k_1}(\partial/\partial x_2)^{k_2}\ldots .$$

Proof. We have $Z_{F(B)} = Z_F(Z_B)$ where $Z_B = Z_X + Z_G(Z_B)$. Again, to simplify, write f, b and g for Z_F, Z_B and Z_G respectively. We must compute $f(b) = f(b_1,b_2,b_3,\ldots)$ given that $b = x_1 + g(b)$. By Observation 1 above, this last equation is equivalent to the infinite system

$$b_i = x_i + g_i(b_1,b_2,b_3,\ldots), \qquad i = 1, 2, 3, \ldots .$$

Now, Joni's multidimensional inversion formula, which is valid for finite systems $b_i = x_i + g_i(b_1,b_2,\ldots,b_m)$, $1 \le i \le m$, with $b_i = b_i(x_1,x_2,\ldots,x_m)$, can be written as

$$f(b_1,\ldots,b_m) = \sum_{\underline{k}} D_1^{k_1}\ldots D_m^{k_m} f(x_1,\ldots,x_m)\, |M(x_1,\ldots,x_m)| \prod_{1\le i\le m} g_i(x_1,\ldots,x_m)^{k_i}/k_i!$$

where $|M(x_1,\ldots,x_m)| = \det(\delta_{i,j} - \partial g_i/\partial x_j)$. Hence, the result follows by letting $m \longrightarrow \infty$ and using the fact, due to Observation 3, that the resulting determinant involves an upper triangular infinite matrix. ∎

Other multidimensional inversion formulas (e.g. Abhyankar's[BCW], Gessel's[G1], Labelle's[L2], Stieltjes'[S]) can be lifted using the same technique. Each particular choices of the species F, R or G in the above theorems gives an explicit formula for corresponding series Z_A or Z_B :

Corollary A1. *Let* $\mathfrak{End}$ *be the species of all endofunctions, then*

$$\text{fix}\,\mathfrak{End}[\beta_1,\beta_2,\beta_3,\ldots] \;=\; \beta^{\beta_1}(\text{fix}\,\beta^2)^{\beta_2}(\text{fix}\,\beta^3)^{\beta_3}\cdots$$

where β_k *is the number of cycles of length* k *in the permutation* β *and* $\text{fix}\,\beta^k = \Sigma_{d|k}\, d\beta_d$ *denotes the number of fixed points of the* k^{th} *iterate* β^k *of* β, *(i.e.* k*-recurrent points of* β*).*

Proof. Take, in Theorem A, R = the species E of all sets and F = the species S of all permutations. Then $A = X\exp(A)$ = the species $\mathfrak{T}$ of all ordinary rooted trees and $F(A) = S(\mathfrak{T})$ = the species $\mathfrak{End}$ of all endofunctions. But $R' = E' = E = R$, so that $\rho(x_1,x_2,\ldots) = 1-x_1Z_R'/Z_R = 1-x_1$. Moreover, it is well-known [see BK1;J2;R1;R2;R3] that $Z_S = \Pi_i\,(1-x_i)^{-1}$ and $Z_E = \exp \Sigma_j\, x_j/j$. Hence,

$$Z_F(x_1,x_2,x_3,\ldots)\,\Pi_{i\geq 1}\;\rho(x_i,x_{2i},x_{3i},\ldots)\,Z_R(x_i,x_{2i},x_{3i},\ldots)^{\beta_i}$$

$$=\;\Pi_i\,(1-x_i)^{-1}\,\Pi_i\,(1-x_i)\exp(\beta_i\,\Sigma_j\,x_{ij}/j) \;=\; \Pi_k \exp[(\text{fix}\,\beta^k)x_k/k].$$

Taking the coefficient of $(x_1^{\beta_1}x_2^{\beta_2}x_3^{\beta_3}\ldots/1^{\beta_1}\beta_1!\,2^{\beta_2}\beta_2!\,3^{\beta_3}\beta_3!\ldots)$ in this last expression gives the desired result. ∎

Corollary A2. *Let* $\mathfrak{T}$ *be the species of all (ordinary) rooted trees, then*

$$\text{fix}\,\mathfrak{T}[\beta_1,\beta_2,\beta_3,\ldots] \;=\; \beta^{\beta_1-1}\Pi_{k\geq 2}\,[(\text{fix}\,\beta^k)^{\beta_k} - k\beta_k(\text{fix}\,\beta^k)^{\beta_k-1}], \quad \textit{if}\;\; \beta_1 \geq 1;$$

$$=\;0, \quad \textit{if}\;\; \beta_1 = 0.$$

Proof. This time, take R = E and F = X (the species of all singletons). Then $F(A) = X(\mathfrak{T}) = \mathfrak{T}$ and $Z_X = x_1$. So, proceeding as above, one gets

$$Z_F(x_1,x_2,x_3,\ldots)\,\Pi_{i\geq 1}\;\rho(x_i,x_{2i},x_{3i},\ldots)\,Z_R(x_i,x_{2i},x_{3i},\ldots)^{\beta_i}$$

$$=\;x_1\,\Pi_i\,(1-x_i)\exp(\beta_i\,\Sigma_j\,x_{ij}/j) \;=\; x_1\,\Pi_k\,(1-x_k)\exp[(\text{fix}\,\beta^k)x_k/k]$$

$$=\;x_1\,\Pi_k\,\Sigma_i\;[(\text{fix}\,\beta^k)^i - ki(\text{fix}\,\beta^k)^{i-1}]\;x_k{}^i/k^i i!\,,$$

which gives the desired result. ∎

Corollary B. *Let* $\mathfrak{Bcp}$ *be the species of all binary commutative parenthetisations* [J2, L2], *then*

$$Z_{\mathfrak{Bcp}} \;=\; \Sigma_{\underline{k}}\,(\partial/\partial x_1)^{k_1}(\partial/\partial x_2)^{k_2}\ldots\,x_1\Pi_{i\geq 1}\,(1-x_i)[(x_i^2+x_{2i})/2]^{k_i}/k_i!\,.$$

Proof. Use Theorem B with G = the species E_2 of all unordered pairs of (distincts) points and $F = \mathbf{X}$. Here, $Z_G = (x_1^2 + x_2)/2$ and $Z_G' = Z_{\mathbf{X}} = x_1$. ■

Notes. An alternate proof of Corollary A2, involving the matrix tree theorem, has been recently suggested by Doron Zeilberger [Z]. Also, Corollary A1 can be easily verified by a direct combinatorial argument.

Another computational tool consists of the following "convolution" operation between sequences of (formal) indicatrix series.

Definition. *The convolution* $a * b$ *of any two sequences*

$$a = (a(n))_{n\geq 1} = (a(n;\, x_1, x_2, x_3, \dots))_{n\geq 1} \quad \textit{and} \quad b = (b(n))_{n\geq 1} = (b(n;\, x_1, x_2, x_3, \dots))_{n\geq 1}$$

of indicatrix series is the sequence $c = (c(n))_{n\geq 1} = (c(n;\, x_1, x_2, x_3, \dots))_{n\geq 1}$ *given by the formula*

$$c(n;\, x_1, x_2, x_3, \dots) = \sum_{d|n} a(d;\, x_1, x_2, x_3, \dots)\, b(n/d;\, x_d, x_{2d}, x_{3d}, \dots), \quad n\geq 1.$$

In short:

$$c = a * b \quad \textit{iff} \quad c(n) = \sum_{d|n} a(d)\, [b(n/d)]_d .$$

The next proposition shows, among other things, that this convolution is closely related to the chain rule for indicatrix series.

Proposition C. a) *The set* $\mathbb{C}[[x_1, x_2, x_3, \dots]]^{\mathbb{N}^*}$ *of all sequences of indicatrix series is a non-commutative ring, with unity, under termwise addition and the above convolution. The unity is the sequence* $\varepsilon = (\varepsilon(n))_{n\geq 1} = (1, 0, 0, 0, \dots)$ *of constant indicatrix series.*

b) *Denote by* $\nabla f = (\partial f/\partial x_n)_{n\geq 1}$ *the "gradient" of the indicatrix series* f. *For any three indicatrix series* f, g, h *such that* $h = g(f)$, *the chain rule* $h' = g'(f) f'$ *takes the following form:* $\nabla h = [(\nabla g)(f)] * \nabla f$.

c) *Let* $a, b : \mathbb{N}^* \longrightarrow \mathbb{C}[[x_1, x_2, x_3, \dots]]$ *be two sequences of indicatrix series and* f, g, h *be indicatrix series such that*

$$h = \sum_{n\geq 1} a(n)\, g_n \ , \quad g = \sum_{n\geq 1} b(n)\, f_n$$

then

$$h = \sum_{n\geq 1} c(n)\, f_n \quad \textit{where} \quad c = a * b \quad (\textit{here } f_n \textit{ means } f(x_n, x_{2n}, \dots)).$$

Proof. a) Straightforward verification.
b) It is easy to see that the jacobian matrix $\underline{f}'$ of any indicatrix series f is completely determined by its first line ∇f. Now, the matrix version $\underline{h}' = \underline{g}'(\underline{f})\, \underline{f}'$ of the chain rule

gives, for any i, j (where i divides j):

$$(\partial h/\partial x_{j/i})_i = \sum_{i|k|j} [(\partial g/\partial x_{k/i})(f)]_i \cdot (\partial f/\partial x_{j/k})_k .$$

Hence, the result follows by putting $i = 1$ in this last equality.

c) We have successively,

$$h = \sum_{i\geq 1} a(i) g_i = \sum_{i\geq 1} a(i)(\sum_{j\geq 1} b(j) f_j)_i = \sum_{i,j\geq 1} a(i)(b(j))_i f_{ji} = \sum_{n\geq 1} c(n) f_n.$$

■

Remark. Note that the "usual" commutative convolution $c = a * b$,

$$c(n) = \sum_{d|n} a(d)\, b(n/d), \quad n \geq 1$$

of multiplicative number theory [HW], for complex-valued arithmetical functions $a, b, c : \mathbb{N}^* \longrightarrow \mathbb{C}$, is a special case of the above convolution. This is easily seen by viewing $a(n)$, $b(n)$ and $c(n)$ as "constant" indicatrix series and using the fact that $(b(n/d))_d = b(n/d)$ in this case. In the general noncommutative case (of arbitrary sequences of indicatrix series), it is easily seen that the classical Möbius inversion splits into two different forms: $[\, b = u * a \Longleftrightarrow \mu * b = a \,]$, $[\, b = a * u \Longleftrightarrow b * \mu = a \,]$. Here $\mu(n)$ = Möbius function evaluated at n and $u(n) = 1$, $n \geq 1$.

From a practical point of view, the "convolution of gradients" in part b) of Proposition C can be used to simplify the computation of certain classes of indicatrix series. For example, it can help in the manipulation of the Lie-Gröbner indicatrix series introduced by the author in [L2]. Explicit formulas for indicatrix series can also be obtained by direct applications of part c) of Proposition C, as the following corollary shows.

Corollary C. [C1;L5;R3] *Let* E, S *and* C *respectively denote the species of all sets, all permutations and all cyclic permutations. Let* M *be a species for which* Z_M *is already known. The following formulas hold:*

a) *If* N *is a species such that* $M = E(N)$ *then*

$$Z_N = \sum_{k\geq 1} (\, \mu(k)/k \,) \log Z_M(\, x_k, x_{2k}, x_{3k}, \ldots \,).$$

b) *If* N *is a species such that* $M = S(N)$ *then*

$$Z_N = 1 - \prod_{k\geq 1} \{\, Z_M(\, x_k, x_{2k}, x_{3k}, \ldots \,) \,\}^{-\mu(k)}.$$

c) *If* N *is such that* $M = C(N)$ *then*

$$Z_N = 1 - \exp[-\Sigma_{k\geq 1}\, \omega(k)\, Z_M(x_k, x_{2k}, x_{3k}, \ldots)]$$

where $\omega(k) = k^{-1}\prod_{p|k}(1-p)$, p *prime*

Proof. Let $a = (a(n))_{n\geq 1}$ and $b = (b(n))_{n\geq 1}$ be any two sequences such that $a * b = \varepsilon$. Then by part c) of Proposition C, we have for any two indicatrix series f and g: $[g = \sum_{k\geq 1} a(k) f_k \iff f = \sum_{k\geq 1} b(k) g_k]$. The corollary will follow from special choices of a, b, f, g.
a) Let $M = E(N)$ and write, for simplicity, $m = Z_M$ and $n = Z_N$. Then we have $m = \exp(n_1 + n_2/2 + n_3/3 + \cdots)$ and so $\log m = n_1 + n_2/2 + n_3/3 + \cdots$. Putting $g = \log m$, $f = n$, $a(k) = 1/k$, we get $b(k) = \mu(k)/k$. Hence the formula $n = \sum_{k\geq 1}(\mu(k)/k)\log m_k$, as desired.
b) Let $M = S(N)$, $m = Z_M$, $n = Z_N$. Then $m = (1-n_1)^{-1}(1-n_2)^{-1}(1-n_3)^{-1}\cdots$ and so $(-\log m) = \sum_{k\geq 1}(\log(1-n))_k$. Putting $g = -\log m$, $f = \log(1-n)$, $a(k) = u(k) = 1$, $(k\geq 1)$, we get $b(k) = \mu(k)$. Hence $\log(1-n) = \sum_{k\geq 1}\mu(k)(-\log m)_k$, which gives the desired result.
c) Let $M = C(N)$, $m = Z_M$, $n = Z_N$. Then (see [J2] example 22) using the Euler totient function φ we can write $m = \sum_{k\geq 1}(\varphi(k)/k)(-\log(1-n_k))$. Putting $g = m$, $f = -\log(1-n)$, $a(k) = \varphi(k)/k$, we get after short manipulations, $b(k) = \omega(k) = k^{-1}\prod_{p|k}(1-p)$. Hence $\log(1/(1-n)) = \sum_{k\geq 1}\omega(k)\, m_k$, as desired. ■

Referee's notes. The idea of using the Möbius function for the sort of inversion given in part a) was propounded by Cadogan [C1] in a fairly general setting, and applied specifically in the cycle index sum setting by Robinson [R3]. An alternate proof of the result in part c) can be obtained by applying part b) to $E(M) = E(C(N)) = S(N)$.

2. Combinatorial Newton-Raphson iteration.

A computational iterative scheme for successive approximations to an object is often better than an explicit expression -- unless the object is given in "closed form" (as in Corollaries A1 and A2 above, for example). With this in mind, we shall solve, at the combinatorial level, a problem involving successive approximations of certain very general families of species, including all the above "enriched" ones. Down, at the analytical level, the solution of the problem will correspond to the classical Newton-Raphson iteration (the "tangent" method) frequently used in Numerical Analysis.

Definitions. *Given any species* F *and any* $k \geq 0$, *denote by* F_k *(respectively,* $F_{\leq k}$ *) the species of all* F*-structures built on exactly* k *(respectively, at most* k*) points.*

a) *The species* $F_{\leq k} = F_0 + F_1 + \cdots + F_k$ *is called an* ***approximation to*** F ***of order*** k.

b) *For any two species* F *and* G, *a natural isomorphism* $F_{\leq k} \xrightarrow{\sim} G_{\leq k}$ *is called a* ***contact of order*** k ***between*** F ***and*** G *and is denoted by* $F =_k G$.

Let now $M = M(X,Y)$ be a given (bi-sorted) species such that $M[\emptyset, \emptyset] = (\partial M/\partial Y)[\emptyset, \emptyset] = \emptyset$. It is known [J2] that the combinatorial equation

$$A = M(X, A),$$

with the initial condition $A[\emptyset] = \emptyset$, determines, up to isomorphism, a unique species $A = A(X)$. So that it can be taken as a "recursive definition" of the species $A = A(X)$. For convenience, an arbitrary A-structure shall be called a **M-tree.** Figure 3 shows a typical M-tree.

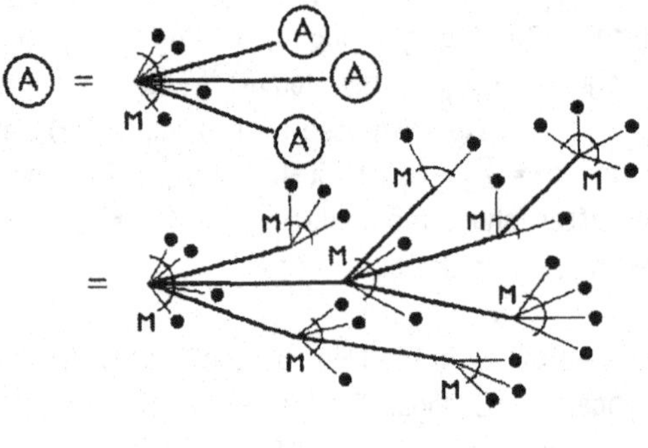

Figure 3

Of course, R-enriched rooted trees and G-Catalan rooted trees are particular instances of M-trees (take $M(X,Y) = XR(Y)$ and $M(X,Y) = X + G(Y)$ respectively).

Problem. *Let* $A = M(X, A)$ *be the species of all* M*-trees and let* $k \geq 0$ *be a given, fixed integer. Consider the approximation* $\mathcal{A}$ *to* A *defined by*

$$\mathcal{A} = A_0 + A_1 + \cdots + A_k =_k A.$$

Is it possible to compute, in a natural way, a new species $\mathcal{A}^+$ *such that*

$$\mathcal{A}^+ =_{2k+1} A,$$

using combinatorial constructions depending only on $\mathcal{A}$ *and* M *?*

Here, "naturality" means that if $\mathcal{Q}$ is isomorphic with another species $\mathcal{Q}^*$ then $\mathcal{Q}^+$ must be isomorphic with the corresponding species $\mathcal{Q}^{*+}$.

Note that the contact of $\mathcal{Q}^+$ with A is required to be more than twice that of $\mathcal{Q}$. Here is the (positive) solution to the problem:

Solution. Define a species $\mathcal{B}$ by $\mathcal{B} = A_{k+1} + A_{k+2} + \cdots + A_{2k+1}$ and let us agree with the following suggestive terminology :

An $\mathcal{Q}$-structure is a **light** tree, a $\mathcal{B}$-structure is a **heavy** tree.

Analyzing any given heavy tree (see Figure 4a) we find that at most one heavy tree can be attached to its "root".

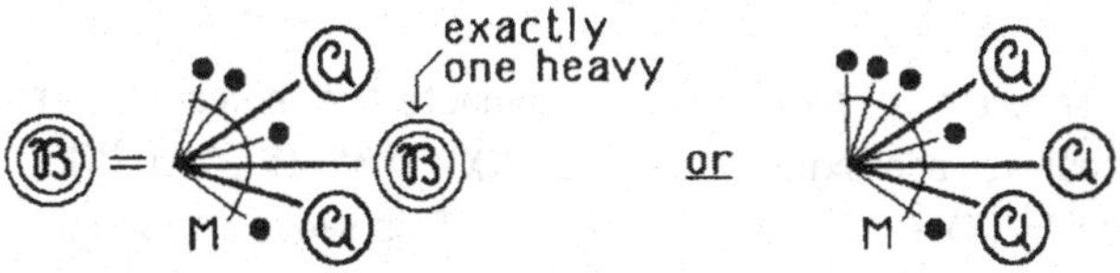

Figure 4a

This is due to the fact that if at least two heavy trees were attached to the root, then the underlying set of the given heavy tree would be of cardinality at least $(k+1)+(k+1) > 2k+1$, a contradiction. Iterating this decomposition until there remain only **light** trees we find Figure 4b :

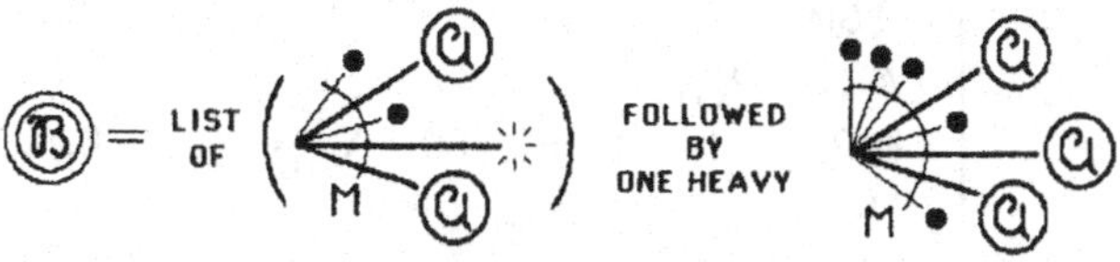

Figure 4b

Of course, the right member in this figure depends, in a natural way, only on $\mathcal{Q}$ and M. Hence, using classical operations among species, we have established the following contact:

$$\mathcal{B} =_{2k+1} L(M_Y(X,\mathcal{Q}))\cdot(M(X,\mathcal{Q}) - \mathcal{Q})$$

where $L = L(X)$ is the species of all linear orders and $M_Y = \partial M/\partial Y$. The "combinatorial substraction" in the right-hand side is due to the fact that the last tree in Figure 4b is a $M(X,\mathcal{Q})$-structure which is heavy, i.e. not an $\mathcal{Q}$-structure (it can be checked that $\mathcal{Q}$ is a subspecies of $M(X,\mathcal{Q})$). Summarizing, we have established the following theorem describing $\mathcal{Q}^+$:

Theorem D. (A Newton-Raphson scheme for species)
Let $\mathcal{Q} = A_0 + A_1 + \cdots + A_k =\!=_k A$ *be a given approximation to the species* A *of all* M*-trees. Then the new species* $\mathcal{Q}^+$ *defined by the formula*

$$\mathcal{Q}^+ = \mathcal{Q} + L(M_Y(X,\mathcal{Q}))\cdot(M(X,\mathcal{Q}) - \mathcal{Q})|_{\leq 2k+1}$$

coincides (up to isomorphism) with the better approximation

$$\mathcal{Q} + \mathbb{B} = A_0 + A_1 + \cdots + A_k + A_{k+1} + \cdots + A_{2k+1} =\!=_{2k+1} A.$$

■

In practise, Theorem D can be used to compute the molecular series $A(X)$ via a sequence of successive approximations $\mathcal{Q}^{\langle 0\rangle}(X)$, $\mathcal{Q}^{\langle 1\rangle}(X)$, $\mathcal{Q}^{\langle 2\rangle}(X), \ldots$ defined by $\mathcal{Q}^{\langle 0\rangle}(X) = 0$ and $\mathcal{Q}^{\langle n+1\rangle}(X) = \mathcal{Q}^{\langle n\rangle +}(X)$, $n=0,1,2,\ldots$. This sequence is highly convergent to $A(X)$. In fact, it is a "quadratically convergent" sequence and we have $\mathcal{Q}^{\langle n\rangle} =\!=_{2^n-1} A$. The following algorithm can be used to generate $\mathcal{Q}^+$ from $\mathcal{Q}$ in a very efficient manner, without having to compute first the "auxiliary" species

$$L(M_Y(X,\mathcal{Q})) = 1 + M_Y(X,\mathcal{Q}) + M_Y(X,\mathcal{Q})^2 + M_Y(X,\mathcal{Q})^3 + \cdots.$$

Let $P := M(X,\mathcal{Q})$ and $Q := M_Y(X,\mathcal{Q})$.

Algorithm D. (INPUT : $A_0, A_1, \ldots, A_k$; OUTPUT : $A_{k+1}, \ldots, A_{2k+1}$)

```
FOR i := 1 TO k+1 DO
    BEGIN
    A_{k+i} := P_{k+i} ;
    FOR j := 1 TO i-1 DO A_{k+i} := A_{k+i} + Q_j A_{k+i-j}
    END.
```

Proof. The $\mathbb{B}$-structure in the left-hand side of Figure 4a is, in fact, an A_{k+i}-structure, for a certain value of i, $1 \leq i \leq k+1$ (the cardinality of the underlying set being $k+1$). For this value of i, the right-hand side displays <u>either</u> a $Q_j\mathbb{B}$-structure (for a certain value of j) <u>or</u> a P_{k+i}-structure. The condition $(\partial M/\partial Y)[\emptyset,\emptyset] = \emptyset$ implies that $j \geq 1$, so that the $Q_j\mathbb{B}$-structure (if it exists) must be a Q_jA_{k+i-j}-structure with $1 \leq j \leq i-1$. Hence,

$$A_{k+i} = P_{k+i} + Q_1A_{k+i-1} + \cdots + Q_jA_{k+i-j} + \cdots + Q_{i-1}A_{k+1}.$$

■

Example D. Let $A = A(X)$ be recursively defined by

$$A \;=\; X + C(X)A + E_{\geq 2}(X + A), \qquad A[\emptyset] = \emptyset$$

where C denotes the species of all oriented cycles and $E_{\geq 2}$ that of all sets having at least 2 points (see Figure 5). Here $M(X,Y) = X + C(X)Y + E_{\geq 2}(X+Y)$. Put $\mathcal{Q}^{\langle 0\rangle} = 0$, $\mathcal{Q}^{\langle n+1\rangle} = \mathcal{Q}^{\langle n\rangle +}$, $n=0,1,2,\ldots$; then $\mathcal{Q}^{\langle 1\rangle} = A_0 + A_1$, $\mathcal{Q}^{\langle 2\rangle} = A_0 + A_1 + A_2 + A_3$ where

$$A_0 = 0, \qquad A_1 = X, \qquad A_2 = 2X^2 + 2E_2, \qquad A_3 = 6X^3 + 9XE_2 + 2E_3,$$

Moreover, $\mathcal{Q}^{\langle 3\rangle} = A_0 + A_1 + A_2 + \cdots + A_7 \equiv_7 A$ contains dozens of terms involving products and compositions of species such as X, E_k, C_k for small values of k.

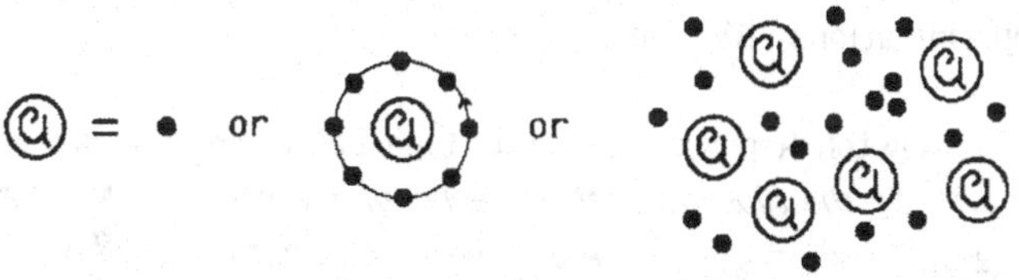

Figure 5

The computations involved in the above example have been done with the help of Algorithm D and various elementary identities about species such as: $E = \sum_{i\geq 0} E_i$, $E_0 = 1$, $C_0 = 0$, $E_1 = C_1 = X$, $C_2 = E_2$, $E_{\geq 2}(M+N+P+\cdots) = E(M+N+P+\cdots) - 1 - (M+N+P+\cdots) = E(M)E(N)E(P)\cdots - 1 - (M+N+P+\cdots)$. A computer program dealing directly with molecular series would be of great use to carry on similar computations, say up to degree 15 (or more).

Corollary D1. (A Newton-Raphson scheme for indicatrix series)
Let $\alpha = \alpha(x_1, x_2, x_3, \ldots) \equiv_k Z_A$ *be a given approximation to the indicatrix series* Z_A *of the species* A *of all* M-*trees. Let* $Z_M(x_1, x_2, \ldots; y_1, y_2, \ldots)$ *be the indicatrix series of the (bi-sorted) species* $M = M(X,Y)$.
Then the new indicatrix series $\alpha^+ = \alpha^+(x_1, x_2, x_3, \ldots)$, *defined by the formula*

$$\alpha^+ = \alpha + (1 - (\partial Z_M/\partial y_1)(x_1, x_2 \ldots; \alpha_1, \alpha_2 \ldots))^{-1} \cdot (Z_M(x_1, x_2 \ldots; \alpha_1, \alpha_2 \ldots) - \alpha)\big|_{\leq 2k+1}$$

where $\alpha_k = \alpha(x_k, x_{2k}, x_{3k}, \ldots)$, *satisfies* $\alpha^+ \equiv_{2k+1} Z_A$.

Proof. Simply apply the operator Z to every formula in Theorem D and use the fact that $Z_L = (1-x_1)^{-1}$. For any two indicatrix series u and v, the notation $u =_k v$ means that every monomial $x_1^{k_1} x_2^{k_2} x_3^{k_3} \ldots$ satisfying $\Sigma\, i k_i \leq k$, appears with the same coefficient in both u and v. ■

Of course, an adaptation of Algorithm D to the context of indicatrix series can also be made.

Example D1. Let $B = X + E_2(B)$ be the species of all binary commutative parenthetisations considered in Corollary B, section 1, above.

Let $\beta = \beta(x_1, x_2, x_3, \ldots) =_k Z_B$ and define β^+ by

$$\beta^+ = \beta + (1-\beta)^{-1} \cdot (x_1 + \tfrac{1}{2}\beta^2 + \tfrac{1}{2}\beta_2 - \beta)\big|_{\leq 2k+1}$$

then $\beta^+ =_{2k+1} Z_B$. This leads to a very efficient iterative method for the computation of successive approximations $\beta^{\langle n \rangle}$ to Z_B, $n \geq 0$.

Corollary D2. (A Newton-Raphson scheme for type-generating series).
Let $\tilde{A}(x) = \Sigma_{i \geq 0} \tilde{a}_i x^i$ *be the type-generating series of the species* A *of all* M*-trees and let* $\tilde{\mathfrak{a}}(x) = \Sigma_{0 \leq i \leq k} \tilde{a}_i x^i =_k \tilde{A}(x)$ *be a given approximation to* $\tilde{A}(x)$. *Then the new series* $\tilde{\mathfrak{a}}^+(x)$ *defined by*

$$\tilde{\mathfrak{a}}(x) + (1 - (\partial Z_M/\partial y_1)(x, x^2, \ldots; \tilde{\mathfrak{a}}(x), \tilde{\mathfrak{a}}(x^2), \ldots))^{-1} \cdot (Z_M(x, x^2, \ldots; \tilde{\mathfrak{a}}(x), \tilde{\mathfrak{a}}(x^2), \ldots) - \tilde{\mathfrak{a}}(x))\big|_{\leq 2k+1}$$

satisfies $\tilde{\mathfrak{a}}^+(x) =_{2k+1} \tilde{A}(x)$.

Proof. Simply replace $(x_1, x_2, x_3, \ldots)$ in Corollary D1 by $(x, x^2, x^3, \ldots)$. ■

Example D2. When applied to the species B of all binary commutative parenthetisations, Corollary D2 takes the following form

$$\tilde{\mathfrak{B}}(x) =_k \tilde{B}(x) \Rightarrow \tilde{\mathfrak{B}}^+(x) = \tilde{\mathfrak{B}}(x) + (1 - \tilde{\mathfrak{B}}(x))^{-1} \cdot (x + \tfrac{1}{2}\tilde{\mathfrak{B}}^2(x) + \tfrac{1}{2}\tilde{\mathfrak{B}}(x^2) - \tilde{\mathfrak{B}}(x)) =_{2k+1} \tilde{B}(x)$$

It is interesting to compare the resulting computational scheme with the (linearly convergent) one given in [C2, vol.1, p.67].

Corollary D3. (The usual Newton-Raphson scheme for generating series).
Let $A(x) = \Sigma_{i \geq 0} a_i x^i / i!$ *be the generating series of the species* A *of all* M*- trees and* $M(x,y) = \Sigma_{i,j \geq 0} m_{i,j} x^i y^j / i! j!$ *be that of the (bi-sorted) species* $M = M(X,Y)$. *Let* $\mathfrak{a}(x) = \Sigma_{0 \leq i \leq k} a_i x^i / i! =_k A(x)$ *be a given approximation to* $A(x)$. *Then the new series* $\mathfrak{a}^+(x)$ *defined by*

$$Q^{+}(x) = Q(x) + \left(1 - M_y(x, Q(x))\right)^{-1} \cdot \left(M(x, Q(x)) - Q(x)\right)\Big|_{\leq 2k+1}$$

satisfies $Q^{+}(x) \equiv_{2k+1} A(x)$. (*Here,* $M_y(x,y)$ *stands for* $\partial M(x,y)/\partial y$). ■

This time, the "iteration step" corresponds to the classical "newtonian step" $\eta^{+} = \eta - \Psi(\eta)/\Psi'(\eta)$, currently used in the computation of truncated power series (see [BK2-3]), when applied to the equation $\Psi(y) = y - M(x,y) = 0$ to obtain successive approximations $\eta^{\langle 0\rangle}(x)$, $\eta^{\langle 1\rangle}(x)$, $\eta^{\langle 2\rangle}(x)$, ..., to the "unknown" series $y = y(x)$, $y(0) = 0$.

Example D3. The first few terms of the generating series $A(x)$ of the species $A = X + C(X)A + E_{22}(X + A)$ (considered in example D above) are given by

$$x + 6x^2/2! + 65x^3/3! + 1092x^4/4! + 25272x^5/5! + 749034x^6/6! + 27108440x^7/7! + \cdots$$

Aknowledgements. The author wishes to thank F. Bergeron, H. Décoste, A. Joyal, J. Labelle, P. Leroux, and the referee for their helpful suggestions.

REFERENCES

[BBN] **M. Barnabei, A. Brini and G. Nicoletti.** A General Umbral Calculus in infinitely many Variables, Adv. in Mathematics, 50, **1983**, 49-93.

[BCW] **H. Bass, E. H. Connel and D. Wright.** The Jacobian conjecture: Reduction of degree and formal expansion of the inverse, Bull. Amer. Math. Soc. (N.S.) 7, **1982**, 287-330.

[B1] **M. Barnabei.** Lagrange Inversion in infinitely many Variables, J. of Mathematical Analysis and Applications, Vol.108, No.1, **1985**, 198-210.

[B2] **F. Bergeron.** Une systématique de la combinatoire énumérative, Ph.D. Thesis, **U. de Montréal-UQAM**, **1986**.

[B3] **F. Bergeron.** Modèles combinatoires de familles de polynômes orthogonaux, Rapp. Tech. no.3, Dép. Math. et Info., **UQAM**, **1984**, (submitted for publication).

[B4] **F. Bergeron.** Une approche combinatoire de la méthode de Weisner, in Polynômes orthogonaux et applications, Proc. Bar-le-Duc (1984), ed. C. Brézinski, A. Draux, A.P. Magnus, P. Maroni, A. Ronveaux, Springer Lecture Notes in Mathematics., no. 1171, **1985**, 111-119.

[B5] **F. Bergeron.** Combinatorial Representations of some Lie Groups and Lie Algebras, (this volume).

[B6] **N. G. de Bruijn.** Enumerative combinatorial problems concerning structures. Nieuw Arch. Wisk. (3), 11, **1963**, 142-161.

[B7] **N. G. de Bruijn.** Enumeration of tree-shaped molecules. Recent Progress in Combinatorics, (Proc. 3rd Waterloo Conf. on Combinatorics, 1968), Academic, New York, **1969**, 59-68.

[BK1] **N. G. de Bruijn and D.A.Klarner.** Multisets of aperiodic cycles. SIAM J. Algebraic Discrete Methods, 3, **1982**, 359-368.

[BK2] **R. P. Brent and H. T. Kung.** $O((n \log n)^{3/2})$ algorithms for composition and reversion of power series in "Analytic Computational Complexity", Proceedings of the Symposium on Analytic Computational Complexity, Carnegie-Mellon Univ., Pittsburgh, Penn., 1975, Acad. Press, New York, **1976**, 217-225.

[BK3] **R. P. Brent and H. T. Kung.** Fast algorithms for manipulating formal power series. J. Assoc. Comput. Mach. 25 , **1978**, 581-595.

[C1] **C. C. Cadogan.** The Möbius function and connected graphs, J. Comb. Theory, Ser. B, 11, **1971**, 193-200.

[C2] **L. Comtet.** Analyse combinatoire, (2 volumes), Collection SUP, Presses Universitaires de France, **1970**.

[D] **H. Décoste.** Séries indicatrices d'espèces pondérées et q-analogues. Ph.D. Thesis, **U. de Montréal-UQAM**, (under preparation).

[DLL] **H. Décoste, G. Labelle et P. Leroux.** Une approche combinatoire pour l'itération de Newton-Raphson, Adv. in Applied Math. 3, **1982**, 407-416.

[FL1] **D. Foata (U. de Strasbourg) et J. Labelle.** Modèles combinatoires pour les polynômes de Meixner, J. Européen de Combinatoire, **1983**, 305-311.

[FL2] **D. Foata (U. de Strasbourg) et P. Leroux.** Polynômes de Jacobi, interprétation combinatoire et fonction génératrice, Proc. Amer. Math. Soc. 87, **1983**, 47-53.

[G1] **I. Gessel.** A Combinatorial Proof of the Multivariable Lagrange Inversion Formula , (manuscript).

[G2] **I. J. Good.** Generalization to several variables of Lagrange's expansion with applications to stochastic processes, Proc. Cambridge Phil. Soc. 56, **1960**, 367-380.

[H1] **P. Hanlon.** A Cycle Index Sum Inversion Theorem , J. Comb. Theory, Series A,30, **1981**, 248-269.

[H2] **P. Hanlon.** The Characters of the Wreath Product Group Acting on the Homology Groups of the Dowling Lattices, J. of Algebra 91, No.2, **1984**, 430-463.

[HP] **F. Harary, E. M. Palmer.** Graphical Enumeration , Acad. Press, N.Y., **1973**.

[HR] **P. Hanlon and R. W. Robinson.** Counting Bridgeless Graphs , J. Comb. Theory, Series B,33, **1982**, 276-305.

[HW] **G. H. Hardy, E. M. Wright.** An Introduction to the Theory of Numbers 4th ed., Oxford: Clarendon Press, **1960**.

[J1] **S. A. Joni.** Lagrange inversion in higher dimensions and umbral operators , Linear and Multilinear Algebra 6, **1978**, 111-121.

[J2] **A. Joyal.** Une théorie combinatoire des séries formelles, Adv. in Math. 42, **1981**, 1-82.

[J3] **A. Joyal.** Règle des signes en algèbre combinatoire, C.R. Math. Rep. Acad. Sci. Canada, VII, 5, **1985**, 285-290.

[J4] **A. Joyal.** Calcul intégral combinatoire et homologie des groupes symétriques, C.R. Math. Rep. Acad. Sci. Canada, VII, 6, **1985**, 337-342.

[J5] **A. Joyal.** Foncteurs analytiques et espèces de structures, (this volume).

[JK] **G. James and A. Kerber.** The Representation Theory of the Symmetric Group , Encycl. of Maths, Vol.16, Addison-Wesley, N.Y., **1981**.

[L1] **G. Labelle.** Une nouvelle démonstration combinatoire des formules d'inversion de Lagrange , Adv. in Math. 42, **1981**, 217-247.

[L2] **G. Labelle.** Éclosions combinatoires appliquées à l'inversion multidimensionnelle des séries formelles, J. of Comb. Theory Ser. A, 39, No.1, **1985**, 52-82.

[L3] **G. Labelle.** Une combinatoire sous-jacente au théorème des fonctions implicites , J. of Combinatorial Theory, Ser. A, 40, No.2, **1985**, 377-393.

[L4] **G. Labelle.** On Combinatorial Differential Equations, J. of Math. Anal. and Appl., 113, No.2, Feb. **1986**, 344-381.

[L5] **G. Labelle.** The computation of the cycle index series of some combinatorial species, Lecture notes, Combinatorial Year, **M.I.T.**, Nov. **1984**.

[L6] **G. Labelle.** The Cyclic Type of Combinatorial Species, Lecture notes, Special Session on Enumerative Combinatorics, 819th Meeting of the **A.M.S.**, April **1985**.

[L7] **J. Labelle.** Applications diverses de la théorie combinatoire des espèces de structures , Ann. Sc. Math. du Québec, 7, no.1, **1983**, 59-94.

[L8] **J. Labelle.** Quelques espèces sur les ensembles de petite cardinalité , Ann. Sc. Math. du Québec, 9, no.1, **1985**, 31-58.

[L9] **J. Labelle.** On the Decomposition of Species, Rapport de recherche no.14, UQAM, mai **1986**.

[L10] **A. Longtin.** Une combinatoire non-commutative pour l'étude des nombres sécants, (this volume).

[LS] **P. Leroux et V. Strehl.** Jacobi polynomials: combinatorics of the basic identities, Discrete Math., 57, **1985**, 167-187.

[LV] **P. Leroux et G. Viennot.** Combinatorial Resolution of Systems of differential Equations, I. Ordinary differential Equations, (this volume).

[P] **G. Pólya.** Kombinatorische Anzahbestimmungen fur Gruppen, Graphen und chemische Verbindungen , Acta Math. 68, **1937**, 145-254.

[R1] **R. C. Read.** A note on the number of functional digraphs. Math. Ann., 143, **1961**, 109-110.

[R2] **J. H. Redfield.** The theory of group-reduced distributions. Amer. J. Math., 49, **1927**, 443-455.

[R3] **R. W. Robinson.** Enumeration of nonseparable Graphs, J. Comb. Theory, Series B, 9, **1970**, 327-356.

[S] **T. J. Stieltjes.** Sur une généralisation de la série de Lagrange, Oeuvre Complète, vol.1, P.Noordhoff, Groningen, **1914**, 445-450.

[W] **T. R. S. Walsh.** Counting unlabelled three-connected and homeomorphically irreducible two-connected Graphs, J. Comb. Theory, Series B, 32, **1982**, 12-32.

[Y1] **Y. N. Yeh.** On the Combinatorial Species of Joyal , Ph.D. Thesis, State University of New York at **Buffalo, 1985**.

[Y2] **Y. N. Yeh.** The calculus of virtual species and $\mathbb{K}$-species, (this volume).

[Z] **D. Zeilberger.** Personal communication (June 10, **1985**).

COMBINATORIAL RESOLUTION OF SYSTEMS OF DIFFERENTIAL EQUATIONS, I. ORDINARY DIFFERENTIAL EQUATIONS.

by

Pierre Leroux*, Université du Québec à Montréal
Gérard X. Viennot, Université de Bordeaux I

§1. Introduction

There is a strong interplay between differential equations and enumerative combinatorics. A simple and classical example goes back to 1879 with the enumeration by André [AN] of **alternating permutations** $\sigma = \sigma(1) > \sigma(2) < \sigma(3) > \sigma(4) \ldots$ of $[n] = \{1, 2, \ldots, n\}$. The sequence of numbers E_n of such permutations (known as the **Euler numbers**) satisfies a certain recurrence relation which is equivalent to the following system of differential equations

$$\begin{aligned} y' &= 1 + y^2, \quad & y(0) = 0 \\ z' &= yz, \quad & z(0) = 1 \end{aligned} \tag{1.1}$$

where $y = y(t) = \sum_{n\geq 0} E_{2n+1}\, t^{2n+1}/(2n+1)!$ and $z = z(t) = \sum_{n\geq 0} E_{2n}\, t^{2n}/(2n)!$, whose unique solution is given by the trigonometric (generating) functions $y = \tan t$ and $z = \sec t$.

One direction is the use of differential equations in order to obtain some information about the generating functions of finite combinatorial structures. For example, Collins et al. [CGJ] consider the enumeration of permutations of $[n]$ having a "periodic up-down sequence". They show that the corresponding generating functions satisfy a differential system of Riccati type. A similar enumeration problem involving the resolution of systems of differential equations can be found in [CA1], [CA2] where the so-called Ollivier functions appear. As another example, see [TU] where Tutte establishes a second order differential equation for the generating function of map colourings.

A second direction is to start from a (system of) differential equation(s) satisfied by a given function $f(t) = \sum_{n\geq 0} a_n t^n/n!$. Now the problem is to find finite combinatorial structures enumerated by the integers a_n. This gives a so-called "combinatorial interpretation" of the numbers a_n, or equivalently of the function $f(t)$. The combinatorialists

* Avec l'appui financier du programme FCAR (Québec, EQ 1608) et du CRSNG (Canada, A5660).

want to derive the identities satisfied by these functions f(t) from bijections and correspondences between the combinatorial objects themselves. These constructions constitute a "combinatorial" or "geometric" theory of the functions, giving a new insight into their analytic properties. Such a methodology can be applied to a wide class of functions, including, for example, elementary trigonometric functions or generating functions of orthogonal polynomials. A sample is given by the following trigonometric identities which can be proven directly from correspondences between finite combinatorial structures:

$$\sec^2 t = 1 + \tan^2 t \qquad (1.2) \qquad\qquad \tan(u+v) = \frac{\tan u + \tan v}{1 - \tan u \tan v} \qquad (1.3)$$

$$\int_0^t \tan x \, dx = \log \sec t \qquad (1.4) \qquad\qquad \int_0^t \sec x \, dx = \log(\tan t + \sec t) \qquad (1.5)$$

See [LO] in this volume for a combinatorial proof of (1.3) and §6 for (1.5). Another example is the combinatorial theory of Jacobi elliptic functions that has been initiated by Dumont [D1], [D2], Flajolet [FLA] and Viennot [V2] and where many questions remain open.

The first purpose of this paper is to give a systematic way to find combinatorial interpretations of the solutions of (systems of) differential equations of the form

$$y_i' = f_i(t, y_1, \dots, y_k) \quad , \qquad y_i(0) = \alpha_i \quad , \qquad i = 1, \dots, k \qquad (1.6)$$

where $f_1, \dots, f_k$ are generating functions (in k+1 variables) of some combinatorial finite structures. The combinatorial objects may be weighted by some formal parameters. We will give a "**canonical**" **combinatorial interpretation** of the solution of such systems (see §3 and §6). In order to work at this level of generality, without expliciting the combinatorial structures having $f_1, \dots, f_k$ as generating functions, we need an abstract formalization of what we mean by "combinatorial structures", together with some basic operations on these structures.

The fundamental notion used here is that of $\mathbb{L}$-species, a variant in the theory of combinatorial species of structures ([BL], [J1], [LA]). Intuitively, a $\mathbb{L}$-species corresponds to a certain kind of combinatorial objects constructed on totally ordered sets. An example of $\mathbb{L}$-species is the "alternating permutations" or, more generally, "permutations having a given up-down sequence". "Young tableaux" is another example. In comparison, the species of structures originally introduced by André Joyal (herein called $\mathbb{B}$-species, see §2) do not make use of a total order on the underlying sets. Examples of $\mathbb{B}$-species are "endofunctions", "permutations", "involutions", "linear lists", "graphs", "trees", "binary trees", etc.

In fact the differential equations themselves can be lifted to the combinatorial level and written, in the case of one equation, in the form

$$Y' = M(T,Y) \quad , \qquad Y'(0) = Z \quad , \tag{1.7}$$

where Y, Y' and M are certain 2-sorted $\mathbb{L}$-species and T and Z are variables corresponding to two distinct sorts. The standard operations on $\mathbb{L}$-species, such as derivation and substitution alluded to in (1.7), are defined in §2.

Our main result is that any differential equation (1.7) has a "canonical" solution which is given in terms of **M-enriched increasing arborescence**, as displayed in figure 3.2. , and that this solution is **unique**, up to a canonical isomorphism. A similar result also holds for combinatorial systems of the form (1.6). The unicity is in contrast with the case of $\mathbb{B}$-species: G. Labelle [L4] has shown that the same equation can have several non-isomorphic solutions (as $\mathbb{B}$-species).

We now illustrate the general method by giving the "canonical" combinatorial solution of the system (1.1). Here y and z will be denoted by Y and Z and considered as $\mathbb{L}$-species, that is, as constructions which can be performed on linearly ordered sets. Starting with the initial conditions, Y and Z are built up recursively using (1.1). The equations $Y' = 1 + Y^2$, $Y(0) = 0$, and $Z' = YZ$, $Z(0) = 1$, are respectively visualized in figures 1.1 and 1.2, where "min." designates the minimum element of the underlying sets.

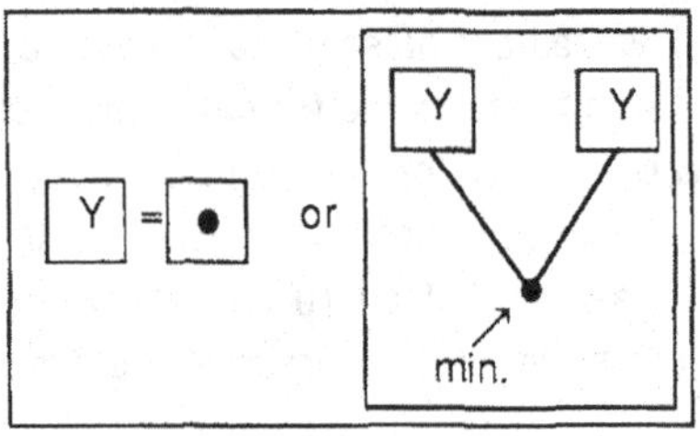

Figure 1.1

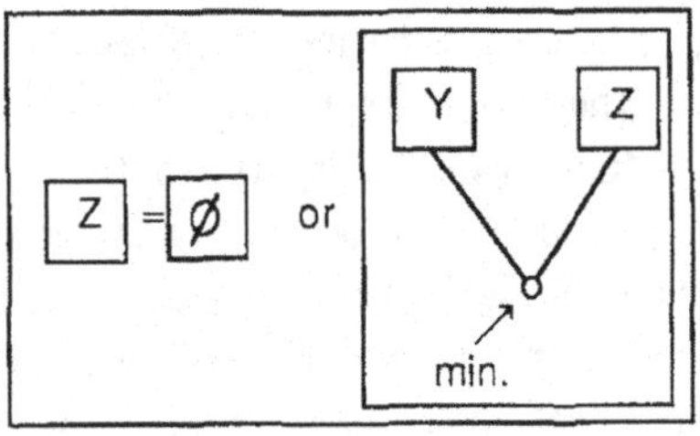

Figure 1.2

Iterating these procedures, one finds that Y can be canonically identified with the construction of so-called **complete increasing binary trees**. The same is true for Z, except that there remains an empty leaf at the extreme right end (see fig. 1.3 and 1.4 respectively). The sets of vertices of the trees are linearly ordered and "increasing" means that their elements are increasing when going away from the root.

This canonical solution, the $\mathbb{L}$-species of "increasing binary trees" should, by unicity, be isomorphic to that of "alternating permutations". In this case, the canonical isomorphism is the well known bijection due to Foata and Schützenberger (see [FS2, 2nd part] or [FST], [FR], [V1], [V3], [GJ, 5.2.14]) which associates to any increasing binary tree (not necessarily complete) a permutation, by "**projection**". For example the trees displayed on figure 1.3 (resp figure 1.4) give the permutation of an odd (resp. even) number of elements σ = 8 3 5 1 7 2 6 4 9 (respect. σ = 4 1 8 3 5 2 7 6). The alternating property comes from the fact that the binary tree is complete (resp. "almost" complete).

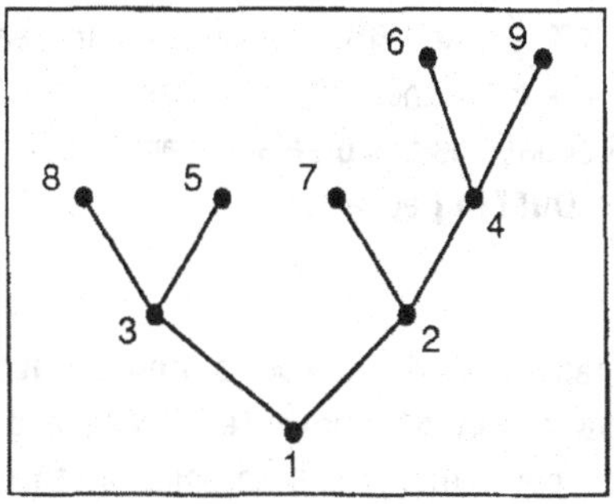

Figure 1.3

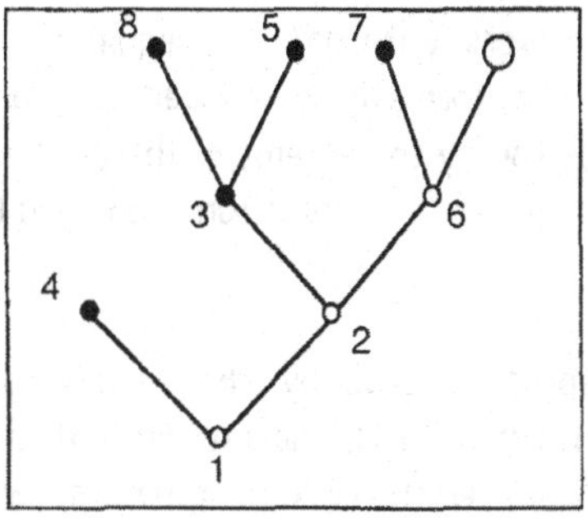

Figure 1.4

In fact, at least three different classes of permutations can be used to give a combinatorial interpretation of the Euler numbers: the above alternating permutations, the **André permutations** introduced by Foata and Schützenberger [FS1], [FST], and the **Jacobi permutations** defined by Viennot [V2]. These classes are naturally associated with the "canonical" interpretations of three different systems of differential equations (see §6 below).

Note that the canonical solutions of systems of differential equations given in terms of M-enriched increasing arborescences, remain at a certain "recursive level". This is the price to pay for a general method that works for any system of differential equations. Usually more combinatorial work needs to be done to go to other levels of combinatorics, as above in going back to the alternating, André or Jacobi permutations. This last step does not involve differential equations any more. It is taking place only at the "bijective combinatorics" level and is somewhat analoguous to the transformation of recursive programs into non-recursive ones in computer science. An advantage of our method is to bring some order inside the "zoo" of the many known different combinatorial interpretations of classical functions.

This paper is the first of a series devoted to a combinatorial theory of differential equations. Our objective is to lift as much as possible of the classical theory to the combinatorial or discrete level, thus obtaining a completely new insight and also new tools to effectively handle or solve differential equations.

In this paper we use the **M-enriched increasing arborescences** to give combinatorial ("bijective") proofs of some classical results such as, for example, the method of "variation of the constant" for linear differential equations. Using the beautiful idea of **eclosions** of G. Labelle [L2], we give in §5 a combinatorial proof of the classical **Lie-Gröbner formulas**. We recall basic concepts about $\mathbb{L}$-species in §2. For simplicity we work with only one equation in §3, §4 and §5. M-enriched increasing arborescences are formally introduced in §3 and the main theorem is proven there. In §6 we extend this theory to systems of differential equations and give some applications, in particular to the Jacobi elliptic functions and the **Duffing** equation.

The topics covered by the following papers will include a **combinatorial integral calculus** [LV1], the **method of separation of variables** [LV2], and **non linear systems with forced entries,** with particular combinatorial methods of resolution [LV3]. An interesting fact is that the combinatorial interpretation of systems with forced entries is contained in the interpretation of the corresponding ordinary system with no entries. Using Fliess theory of such systems (see for example [F1], [FL1], [FL2], [LL]) we will be able to find an explicit combinatorial interpretation of the coefficients of the so-called **Volterra kernels**. Then, using other combinatorial tools, such as the notion of "**histories**", we will give explicit new formulas for these coefficients in certain special cases, such as the one given in [FL2, p. 566]. The cases of the Duffing equation and of elliptic functions and integrals will also be treated in more detail in other papers.

ACKNOWLEDGMENTS

Much of this paper has been inspired by recent work of Gilbert Labelle on functional equations, differential equations, Lie Gröbner series and Newton-Raphson iteration, in the context of species of structures (see [L2], [L3] and [L4]), and also of Michel Fliess and his school on the resolution of non-linear differential equations with forced entries in the context of non-commuting formal power series (see [F1], [FL1], [FL2] and [LL]), together with some previous work of M.P. Schützenberger about the non-commutative solution of the differential equation related to André permutations [SC].

We would like to thank many colleagues for discussions and helpful suggestions, in particular François Bergeron, Michel Fliess, Dominique Foata, Gilbert Labelle, Françoise Lamnabhi-Lagarrigue, and Don Rawlings, and also Nantel Bergeron who MACDREW the figures of this paper.

§2. $\mathbb{L}$-species

The $\mathbb{L}$-species are introduced as a variant of the theory of species of structures to account for the combinatorial constructions that make use of a linear order on the underlying set, for example "alternating permutations" or "increasing binary trees". To give precise and concise definitions, we introduce the following categories:

- $\mathbb{E}$ = the category of finite sets and functions
- $\mathbb{B}$ = the category of finite sets and bijections
- $\mathbb{L}$ = the category of finite linearly ordered sets and order preserving bijections.

Recall then (see [J1], [LA], or [L5] and [YE] in this volume) that a **species of structures** F (herein called $\mathbb{B}$-species to emphasize the distinction with $\mathbb{L}$-species) is a functor

$$F : \mathbb{B} \longrightarrow \mathbb{E}. \tag{2.1}$$

By contrast, an **$\mathbb{L}$-species** M is defined to be a functor

$$M : \mathbb{L} \longrightarrow \mathbb{E}. \tag{2.2}$$

This means that to each linearly ordered finite set ℓ, M associates a finite set, denoted by $M[\ell]$, whose elements are called **M-structures on** ℓ, and to each order preserving bijection $\varphi : \ell_1 \longrightarrow \ell_2$, a function

$$M[\varphi] : M[\ell_1] \longrightarrow M[\ell_2], \tag{2.3}$$

called the **transport of structures**, in a functorial way, that is such that

$$M[\varphi \circ \psi] = M[\varphi] \circ M[\psi] \quad \text{and} \quad M[1_\ell] = 1_{M[\ell]}. \tag{2.4}$$

A convenient and useful graphical representation of a generic or typical M-structure on a linearly ordered set is given by figure 2.1, where the curved arrow indicates the linear order on the set of points and the label M represents the M-structure.

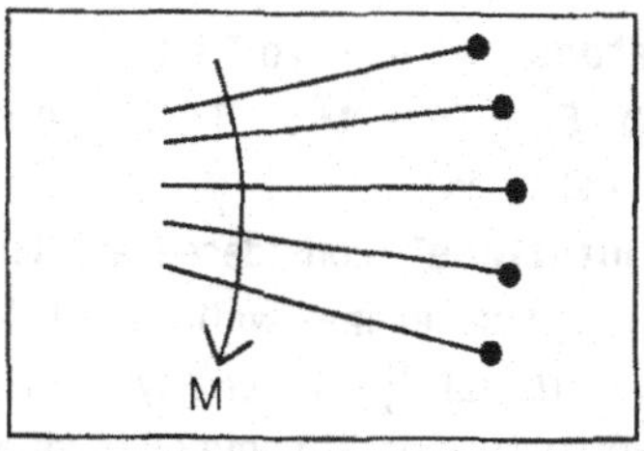

Figure 2.1. Generic M-structure

Two $\mathbb{L}$-species M and N are **isomorphic** if, by definition, there exists a natural isomorphism of functors

$$\alpha : M \xrightarrow{\sim} N. \tag{2.5}$$

In other words, there should exist a bijection $\alpha_\ell : M[\ell] \longrightarrow N[\ell]$, for each $\ell \in \mathbb{L}$, such that for any increasing bijection $\varphi : \ell \longrightarrow h$, the following diagram commutes:

$$\begin{array}{ccc} M[\ell] & \xrightarrow{\alpha_\ell} & N[\ell] \\ M[\varphi] \downarrow & & \downarrow N[\varphi] \\ M[h] & \xrightarrow{\alpha_h} & N[h] \end{array} \tag{2.6}$$

Given a $\mathbb{L}$-species M, it follows from the definition that the cardinality card $M[\ell]$ of $M[\ell]$ depends only on the cardinality of the linearly ordered set ℓ. The **cardinality** of M, or the (exponential) **generating function** of M, denoted by Card M or M(t) is defined to be the following formal power series of Hürwitz type

$$\text{Card}\, M = M(t) = \sum_{n\geq 0} \text{Card}\, M[n]\, t^n/n!, \tag{2.7}$$

where, for any $n\geq 0$, we set $[n] = \{1,2,..,n\}$, including $[0] = \emptyset$, and we write $M[n]$ instead of $M[[n]]$. Of course, $[n]$ is equipped with its usual order $1 < 2 < ... < n$.

The following are examples of $\mathbb{L}$-species:

1. "Alternating permutations"; and also other classes of permutations determined by their up-down sequences.

2. "Increasing binary trees"; and other classes of rooted increasingly labelled trees. As we have seen, the $\mathbb{L}$-species of "complete increasing binary trees" and of "alternating descending odd permutations" are isomorphic.

3. If F is any $\mathbb{B}$-species, then F can be considered as an $\mathbb{L}$-species, after composition with the forgetful functor $\mathbb{L} \longrightarrow \mathbb{B}$; an F-structure on ℓ simply ignores its linear order. This class of species includes some important ones that we now point out:

   0 : the "**empty**" species, with Card 0 = 0 ;
   1 : the "**empty set**" species, with Card 1 = 1 ;
   T : the species of "**singletons**", with Card T = t ;
   T^2 : "**ordered pairs**", with Card $T^2 = t^2$; $T^2/2!$: "**doubletons**" ;
   T^3 : "**ordered triples**". And so on...
   L : the species of "**permutations**", considered as "**lists**" or "**linear orders**". We have $L(t) = 1/(1-t)$ and sometimes write $L = 1/(1-T)$.
   E : the **uniform** species, with $|E[\ell]| = 1$ for any ℓ. There are several ways to describe the unique E-structure on ℓ; some of them, such as "the increasing (or decreasing) list of elements of ℓ", actualy use the linear order of ℓ. We have $E = \sum_{n\geq 0} T^n/n!$, $E(t) = \exp(t)$, and sometimes write $E = \exp(T)$.

Note that two non isomorphic $\mathbb{B}$-species can become isomorphic, when considered as $\mathbb{L}$-species. This is the case for the two species of permutations: L, for "linear orders", and P, for "bijective endofunctions", for which $L(t) = 1/(1-t) = P(t)$. The following proposition shows that $\mathbb{L}$-species constitute a more rigid combinatorial lifting of formal power series than $\mathbb{B}$-species.

Proposition 2.1. Two $\mathbb{L}$-species M and N are isomorphic if and only if $M(t) = N(t)$.

Proof. The condition certainly is necessary. To show that it is sufficient, assume $M(t) = N(t)$ and select for each $n \geq 0$ a bijection $\alpha_n : M[n] \longrightarrow N[n]$. Now for any linearly ordered set ℓ, say of cardinality n, there is a unique order preserving bijection $\varphi : [n] \longrightarrow \ell$. It is then easy to see that the family $\alpha = \{\alpha_\ell\}_{\ell \in \mathbb{L}}$, with α_ℓ defined as the composite

$$M[\ell] \xrightarrow{M[\varphi]^{-1}} M[n] \xrightarrow{\alpha_n} N[n] \xrightarrow{N[\varphi]} N[\ell], \tag{2.8}$$

is a natural isomorphism between M and N. □

Note. We often write $M = N$ to denote isomorphism of the $\mathbb{L}$-species M and N.

Operations are defined on $\mathbb{L}$-species, with the purpose of reflecting the usual operations on functions or formal power series. The definitions are similar to those of $\mathbb{B}$-species, with the additional following observations on linearly ordered sets ℓ.

We will denote by

$$\ell_1 + \ell_2 = \ell , \tag{2.9}$$

the situation where ℓ_1 and ℓ_2 are subsets of ℓ, with their linear order induced from that of ℓ, such that

$$\ell_1 \cap \ell_2 = \emptyset \quad \text{and} \quad \ell_1 \cup \ell_2 = \ell . \tag{2.10}$$

This supplies the category $\mathbb{L}$ with a decomposition law (see [J1] or [B1]) which determines the theory of $\mathbb{L}$-species.

Also, if $p \in R[\ell]$ = the set of all equivalence relations on ℓ, then each equivalence class $c \in \ell/p$ has an induced linear order and the factor set ℓ/p is itself linearly ordered according to the order of the smallest elements in each equivalence class. As usual, equivalence classes are required to be non empty. The empty set admits one equivalence relation with an empty set of equivalence classes.

In the following definitions, the operations + (and $\sum$) and × (and $\prod$) on sets are the disjoint union and cartesian product, respectively. We let $\min(\ell)$ denote the minimum element of ℓ and $1+\ell$ denote the ordered set obtained by adjunction of a new minimum element. Recall that $[0] = \emptyset$.

Let M and N be $\mathbb{L}$-species, and ℓ be a linearly ordered set. The following operations are defined:

The **addition**, $M + N$, by

$$(M + N)[\ell] = M[\ell] + N[\ell] . \tag{2.11}$$

The **product**, $M \cdot N$, (see fig. 2.2) by

$$(M \cdot N)[\ell] = \sum_{\ell_1 + \ell_2 = \ell} M[\ell_1] \times M[\ell_2] . \tag{2.12}$$

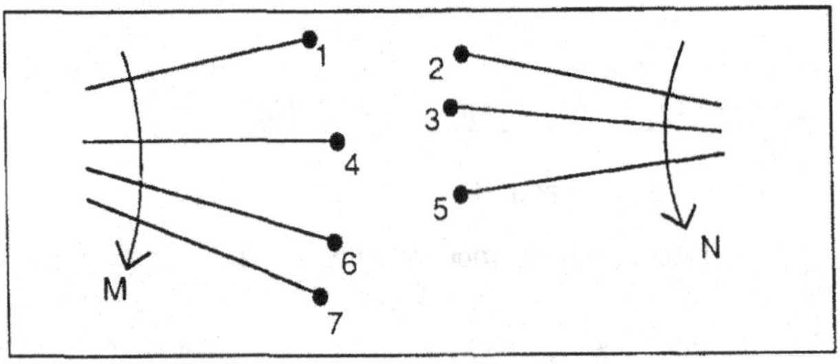

Figure 2.2. Generic M·N-structure

The **substitution**, $M(N)$, when $N[0] = \emptyset$, by

$$(M(N))[\ell] = \sum_{\rho \in R[\ell]} M[\ell/\rho] \times \prod_{c \in \ell/\rho} N[c] . \tag{2.13}$$

M(N)-structures are often called M-**assemblies of** N-**structures** (see fig. 2.3). If $M = E$, the uniform species, we simply speak of **assemblies** or **sets** of N-structures.

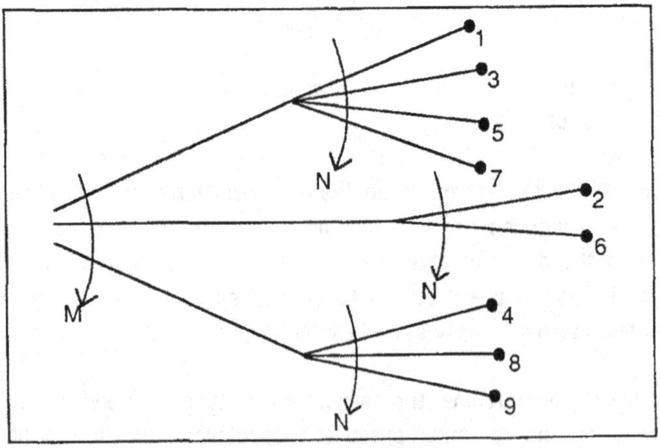

Figure 2.3. Generic M(N)-structure

The operation of substitution for $\mathbb{L}$-species is closely related to the concept of **composé partitionnel** of Foata and Schützenberger [FS1]. Note that substitution of the singleton species T is neutral so that we can write $M = M(T)$.

The **derivative**, M', also denoted by (dM/dT), (see fig. 2.4), by

$$M'[\ell] = M[1+\ell] \qquad (2.14)$$

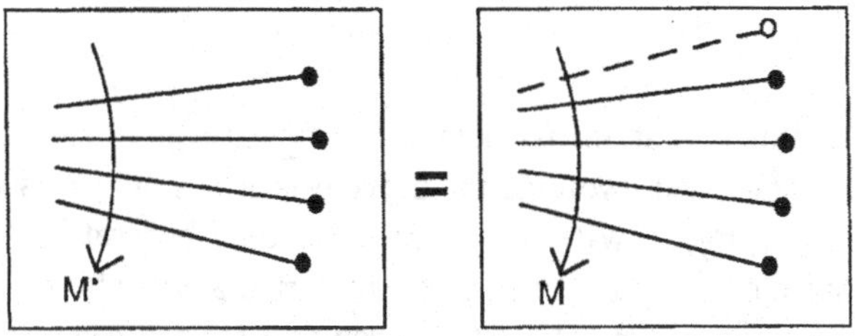

Figure 2.4. Generic M'-structure.

The **integral**, $F = F(T) = \int_0^T M(X)\,dX$, (see fig. 2.5), by

$$F[0] = \emptyset, \quad F[\ell] = M[\ell\backslash\{\min(\ell)\}], \quad \text{for } \ell \neq \emptyset. \qquad (2.15)$$

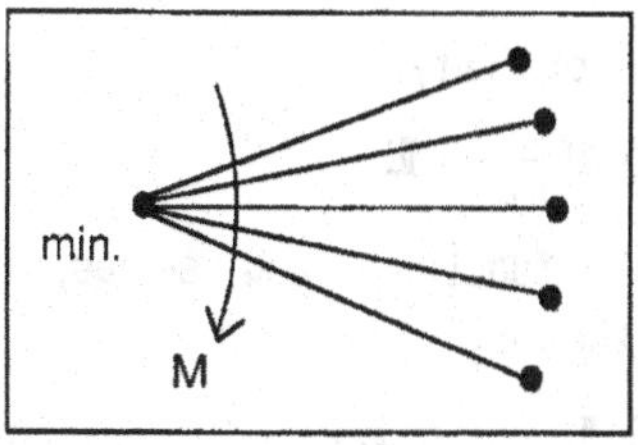

Figure 2.5. Generic $\int_0^T M(X)\,dX$-structure.

All the elementary properties, associativity, commutativity, distributivity, linearity, etc., of the operations are true at the combinatorial level, including the **Leibnitz rule** and the **chain rule** for the derivative:

$$(M{\cdot}N)' = M{\cdot}N' + M'{\cdot}N \quad , \qquad (2.16)$$

$$M(N)' = M'(N){\cdot}N' \quad , \qquad (2.17)$$

with equality meaning isomorphism of $\mathbb{L}$-species (see also [J1] and [LA]). Properties particular to combinatorial integration (2.15) will be studied in a forthcoming paper [LV2]. The following proposition is now easily verified.

Proposition 2.2. For any $\mathbb{L}$-species M and N, we have

$$(M+N)(t) = M(t)+N(t) \quad , \tag{2.18}$$

$$(M\cdot N)(t) = M(t)\cdot N(t) \quad , \tag{2.19}$$

$$M(N)(t) = M(N(t)) \quad , \quad (N[0] = \emptyset) \tag{2.20}$$

$$M'(t) = dM(t)/dt \quad , \tag{2.21}$$

$$\mathrm{Card} \int_0^T M(X)dX = \int_0^t M(x)dx \quad . \tag{2.22}$$

□

A consequence of this is that any formula $\mathcal{F}(M_1, M_2, \ldots) = \mathcal{G}(M_1, M_2, \ldots)$, expressing an isomorphism of $\mathbb{L}$-species and involving the above operations, which is valid for any sequence of $\mathbb{L}$-species $M_1, M_2, \ldots$, will also be valid for their generating functions. By the principle of extension of algebraic identities, the formula will also be valid for any sequence of formal power series $m_1(t), m_2(t), \ldots$ and also for any sequence of analytic functions. This is the basic principle at the root of combinatorial proofs of analytic formulas such as the Leibnitz rule (2.16) or the chain rule (2.17) for derivation, the Lagrange inversion formulas [L1], etc...

We will also consider k-**sorted** $\mathbb{L}$-**species** ($k \geq 2$), that is functors of the type

$$\mathbb{L}^k = \mathbb{L} \times \mathbb{L} \times \ldots \times \mathbb{L} \longrightarrow \mathbb{E}, \tag{2.23}$$

and **mixed species**, for example of the type

$$\mathbb{L} \times \mathbb{B} \longrightarrow \mathbb{E}, \tag{2.24}$$

corresponding to multivariable functions and series, as well as $\mathbb{K}$-**weighted** $\mathbb{L}$-**species**, that is functors of the type

$$M : \mathbb{L} \longrightarrow \mathbb{E}_{\mathbb{K}} , \tag{2.25}$$

where $\mathbb{K}$ is a commutative ring with unity and $\mathbb{E}_{\mathbb{K}}$ denotes the category of **finite** $\mathbb{K}$-**weighted sets**, i.e. pairs (A,v) where A is a set and $v : A \longrightarrow \mathbb{K}$ is a **weight function**; a **morphism** between two $\mathbb{K}$-weighted sets (A, v_A) and (B, v_B) is a function $f : A \longrightarrow B$ such that $v_A = v_B \circ f$.

The reader is refered to [J1, §5, §6], [LA, §3] and [YE], for a discussion of k-sorted and $\mathbb{K}$-weigted $\mathbb{B}$-species, in particular for a definition of their operations and generating functions, which can easily be adapted to the case of $\mathbb{L}$-species. Perhaps is it worthwhile to give an explicit definition of partial derivatives: let $M = M(S,T)$ be a 2-sorted $\mathbb{L}$-species; then we set

$$(\partial M/\partial S)[h,\ell] = M[1+h, \ell] \quad , \tag{2.26}$$

$$(\partial M/\partial T)[h,\ell] = M[h, 1+\ell] \quad . \tag{2.27}$$

§3. Case of one differential equation

In this section we examine the case of one differential equation of the form

$$(dY/dT) = Y' = M(T,Y) \quad , \quad Y(0) = Z \tag{3.1}$$

where $M(T,Y)$ is a given 2-sorted $\mathbb{L}$-species, Z is an indeterminate which will correspond to an extra sort of points, and $Y(0)$ is to be interpreted as the $\mathbb{L}$-species obtained from Y by substitution of the empty species 0. Specific examples, with, for instance, $M(T,Y) = 1 + Y^2$, or $a_0 + a_1Y + a_2Y^2 + \ldots + a_nY^n$ (autonomous), and $M(T,Y) = G(T)Y + F(T)$ (linear), will be considered in §4.

Formally, a solution of (3.1) is defined to be a pair (A, ψ) where $A = A(T)$ ($=Y$) is a $\mathbb{L}$-species such that $A(0) = Z$, and ψ is an isomorphism of $\mathbb{L}$-species

$$A'(T) \xrightarrow[\sim]{\psi} M(T, A(T)) . \tag{3.2}$$

Note that in fact A also depends on the initial condition Z and that we should write $A = A(T,Z)$ and

$$\partial A(T,Z)/\partial T \xrightarrow[\sim]{\psi} M(T, A(T,Z)) \quad , \quad A(0,Z) = Z . \tag{3.3}$$

Now keeping Z fixed, (3.3) is equivalent to the integral equation

$$A(T,Z) = Z + \int_0^T M(X, A(X,Z))dV . \tag{3.4}$$

By virtue of the definition (2.15) of the integral (see fig. 2.5) and of the definition of the substitution in a 2-sorted species (see [J1, Def. 19, p. 46] or [LA, p. 89]), this integral equation can be visualized as in figure 3.1, where:

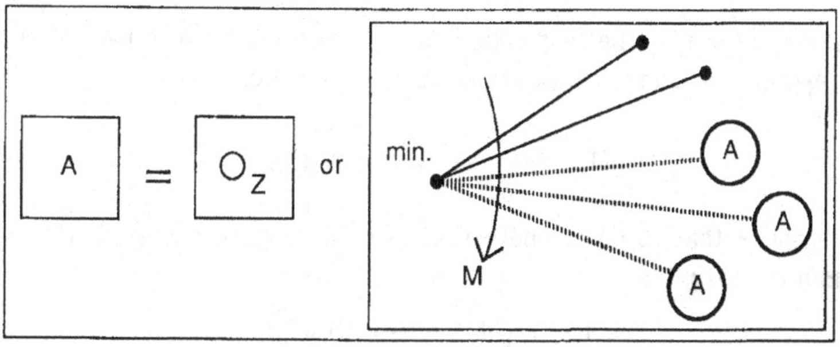

Figure 3.1

- the dots ■ and circles ○ represent singletons of the sorts T (and X) and Z, respectively, and will be called "points" and "buds", respectively,

- the circled A's represent A-structures on equivalence classes of the underlying set,
- the two sorts of elements on which the M-structure is constructed are symbolized by continuous and mashed lines respectively.

It now suffices to iterate this process to obtain a canonical combinatorial solution of (3.1), that is the $\mathbb{L}$-species $A = A_M(T,Z)$ of so-called M-**enriched increasing arborescences**, generically described by Figure 3.2. An $A_M(T,Z)$-structure lies over a couple (ℓ,s) of linearly ordered sets (in fig. 3.2, $\ell = \{1, 2, ..., 18\}$ and $s = \{a, b, ..., e\}$). Elements of ℓ and s will be called **points** (T-singletons) and **buds** (Z-singletons) respectively. We make the convention that "all points are smaller than all buds". In such an M-enriched increasing arborescence, a point is called **fertile** if it is the root of some A_M-substructure and **sterile** otherwise. Note that the buds, like the sterile points, do not have any sons.

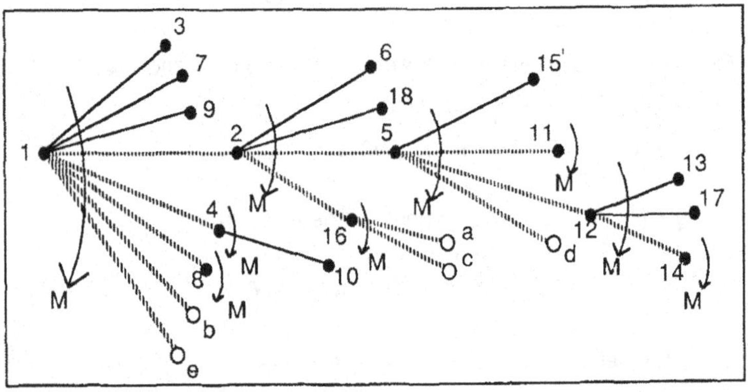

Figure 3.2. Generic $A_M(T,Z)$-structure

It should be clear to the reader that the $\mathbb{L}$-species $Y = A_M$ is indeed a solution of (3.1) since $A_M(0,Z) = Z$ and there exists an obvious isomorphism

$$\partial A_M(T,Z)/\partial T \xrightarrow[\sim]{\psi} M(T, A_M(T,Z)) . \tag{3.5}$$

Now suppose that (B,φ) is another solution of 3.1, then there will be a unique **isomorphism of solutions**

$$\Phi : (A_M, \psi) \xrightarrow{\sim} (B, \varphi) , \tag{3.6}$$

that is an isomorphism of $\mathbb{L}$-species $\Phi : A_M \xrightarrow{\sim} B$ such that the following diagram commutes

$$\begin{array}{ccc} (\partial A_M/\partial T) & \xrightarrow[\sim]{\psi} & M(T,A_m) \\ (\partial\Phi/\partial T)\ \big\downarrow\wr & & \wr\big\downarrow\ M(T,\Phi) \\ (\partial B/\partial T) & \xrightarrow[\sim]{\varphi} & M(T,B) \end{array} \tag{3.7}$$

where the natural transformations $\partial\Phi/\partial T$ and $M(T,\Phi)$ are defined in an obvious fashion. This is shown by induction on the cardinality of ℓ, where (ℓ, s) is the couple of linearly ordered sets on which the A_M- and B-structures are taken: to start with, we have $A_M(0,Z) = Z = B(0,Z)$ and the unique choice for Φ_0 is the identity $Z \xrightarrow{\sim} Z$. Now, for $n \geq 0$, suppose that the natural bijection

$$\Phi_{(h,r)} : A[h,r] \longrightarrow B[h,r] \tag{3.8}$$

has been uniquely defined for all linearly ordered sets h of cardinality $\leq n$ and for all r, and let ℓ be of cardinality n. Then we have, for any s,

$$A_M[1+\ell,s] = (\partial A_M/\partial T)[\ell,s] \xrightarrow{\psi(\ell,s)} M(T,A_M)[\ell,s] \tag{3.9}$$

and also, by hypothesis,

$$B[1+\ell,s] = (\partial A_M/\partial T)[\ell,s] \xrightarrow{\varphi(\ell,s)} M(T,B)[\ell,s] \tag{3.10}$$

In other words, the equivalent of Figure 3.1 with A replaced by either A_M or B is valid. But each A_M- and B-substructures that appear in this decomposition (the circled A-structures in Fig. 3.1) lies over a couple (h,r) with $|h| \leq n$ and hence, by the induction hypothesis, correspond isomorphically to each other, using $\Phi_{(h,r)}$. Consequently the bijection $M(T,\Phi)_{(\ell,s)}$ can be constructed and afterwards, also $\Phi'_{(\ell,s)}$ by asking that the diagram (3.7), applied to (ℓ,s) commutes. This determines $\Phi_{(1+\ell,s)} = (\partial\Phi/\partial T)_{(\ell,s)}$ uniquely. We thus have proved the following:

Theorem 3.1. For any 2-sorted $\mathbb{L}$-species M, the 2-sorted $\mathbb{L}$-species $Y = A_M(T,Z)$ of M-enriched increasing arborescences with buds, described above, together with the natural isomorphism $\psi : (\partial A_M/\partial T)(T,Z) \xrightarrow{\sim} M(T,A_M(T,Z))$, is a (canonical) solution of the differential equation (3.1). Moreover, for any other solution (B,φ) of 3.1, there is a unique isomorphism of solutions $\Phi : (A_M,\psi) \xrightarrow{\sim} (B,\varphi)$.

□

We conclude this section by noting that, as shown by G. Labelle in [L4, theorem B], the combinatorial Newton-Raphson iteration scheme, first introduced in [DLL], can be applied in the resolution of a differential equation and gives a sequence of approximations with quadratic convergence. More precisely, for any $\mathbb{L}$-species F, we introduce the $\mathbb{L}$-species $F_{\leq n}$, the "truncation of F to sets of cardinality at most n", by

$$F_{\leq n}[\ell] = F[\ell], \text{ if } |\ell| \leq n, \text{ and } F_{\leq n}[\ell] = \emptyset, \text{ otherwise.}$$

We then have the following:

Theorem 3.2. Let $Y = A = A(T)$ be the solution of the equation $Y' = M(T,Y)$, $Y(0) = 0$ and set $\mathcal{Q} = A_{\leq n}$. Let $Y = \mathbb{B}$ be the solution of the 1^{rst} order linear differential equation

$$Y' = F(T)Y + G(T), \quad Y(0) = 0, \tag{3.11}$$

where

$$F(T) = \partial M(T,Y)/\partial Y \big|_{Y=\mathcal{Q}(T)} \quad \text{and} \quad G(T) = M(T,\mathcal{Q}(T)) - \mathcal{Q}'(T).$$

Then the $\mathbb{L}$-species $\mathcal{Q}^+ := \mathcal{Q} + \mathbb{B}$ has a contact of order $2n + 2$ with A, i.e. there exists a canonical isomorphism

$$\mathcal{Q}^+_{\leq 2n+2} \xrightarrow{\sim} A_{\leq 2n+2}. \tag{3.12}$$

Proof. See [LA, §3]. □

We will see in the next section how to deal combinatorialy with 1^{rst} order linear differential equations.

§4. Examples

In this section, we consider special cases of first order differential equations of the form (3.1), including autonomous and linear equations.

Note first that the initial condition $Y(0) = Z$, can take special forms, by substitution into Z, some of which actually make the solution independant of Z. In particular, $Y(0) = 0$, the **empty** species, is to be interpreted as "no buds are allowed", and $Y(0) = 1$, the **empty set** species, as "buds are unlabelled, indistinguishable and not accounted for". This last case however, the substitution of 1 for Z, is not always possible or legal. In particular, the generation of an infinite number of structures on any given ℓ should be avoided. More precisely, writing

$$Y(T,Z) = \sum_{k\geq 0} Y_k(T)\, Z^k/k!\,, \tag{4.1}$$

then each $Y_k(T)$ should be combinatorialy divisible by $k!$ and the family $\{Y_k(T)/k!\}_{k\geq 0}$ of $\mathbb{L}$-species should be summable.

A first order differential equation is called **autonomous** if $M(T,Y) = G(Y)$ does not depend on T, i.e. if it is of the form

$$Y' = G(Y) \quad , \quad Y(0) = Z. \tag{4.2}$$

In this case, the M-enriched (or rather G-enriched) increasing arborescences will have no sterile points and, equivalently, only mashed edges will appear (these are then **unmashed** for simplicity of representation). See figure 4.1 for an illustration of this canonical solution. The following four examples are special cases of autonomous differential equations.

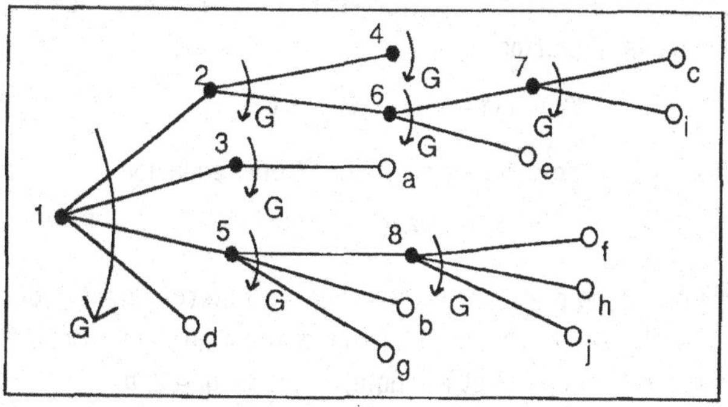

Figure 4.1

Example 4.1. Consider the autonomous differential equation

$$Y' = 1 + Y^2 \quad , \qquad Y(0) = 0 \tag{4.3}$$

where $M(T,Y) = G(Y) = 1 + Y^2$ and $Z = 0$. Since there will be no buds, the canonical solution is that of "complete increasing binary trees", as we saw in §1 (see fig. 1.3). Moreover the unique isomorphism of solutions between this and the other solution of (4.1), that of "alternating descending odd permutations" was described in §1 as the "projection along the x-axis".

Example 4.2. The generating function $y = \tan t + \sec t$ of alternating permutations (without distinction between the odd and even case) is the solution of the following differential equation

$$y' = (1 + y^2)/2 \quad , \qquad y(0) = 1 \ . \tag{4.4}$$

Defining $z = y - 1$, this is equivalent to

$$z' = 1 + z + z^2/2 \quad , \qquad z(0) = 0 \ . \tag{4.5}$$

Similarly to example 4.1, we see that the canonical solution of (4.5) at the species level is that of the so-called **increasing 1-2 arborescences**, i.e. arborescences such that every vertex has at most two sons. Note that in the case of two sons, no distinction is made between left and right, contrarily to the case of binary trees. Adapting the bijection between increasing binary trees and permutations mentionned earlier, one can easily give a bijection between increasing 1-2 arborescences and "André permutations"; see [FS1], [FS2], [V3].

It is also possible to construct directly the canonical combinatorial solution of (4.4), where $(1 + Y^2)/2$ is to be interpreted as the $\mathbb{L}$-species $1 + Y^2$ weigted by $1/2$. The reader will easily show the equivalence between the corresponding weigted arborescences and the 1-2 arborescences.

Example 4.3. **Planar trees** are, by definition, L-enriched arborescences (see [L1]), where $L(T) = 1/(1 - T)$ is the $\mathbb{L}$-species of permutations, considered as lists. Now, from theorem 3.1, the $\mathbb{L}$-species $Y = Pla(T)$ of **increasing** planar trees (see Fig. 4.2,a)) is the solution of the differential equation

$$Y' = L(Y) \quad , \quad Y(0) = 0 . \tag{4.6}$$

This $\mathbb{L}$-species is also solution of the **functional** equation

$$Y = T + Y^2/2! \tag{4.7}$$

which says that an increasing planar tree is either a singleton or a set of two increasing planar trees. One way to realize this fact is to cut the right-most branch at the root of any increasing planar tree which is not a singleton (see fig. 4.2, b)).

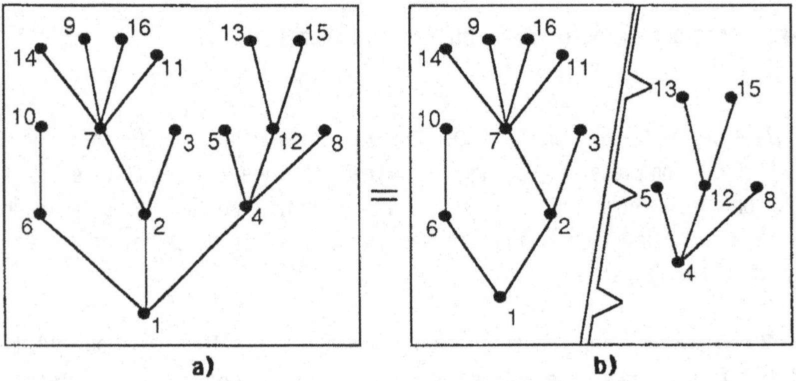

Figure 4.2

It follows from (4.7) that the generating function $y = Pla(t)$ satisfies the quadratic equation

$$y^2 - 2y + 2t = 0 \tag{4.8}$$

which can be solved to give

$$Pla(t) = 1 - \sqrt{1-2t} \tag{4.9}$$

Example 4.4. Let $a_0, a_1, \ldots, a_n$ be scalar parameters. From the previous example, the solution of the equation

$$Y' = a_0 1 + a_1 Y + \ldots + a_n Y^n \quad , \quad Y(0) = Z \tag{4.10}$$

is seen to be the $\mathbb{L}$-species of weighted increasing planar trees, such that each vertex has at most n sons, the weight of a vertex having i sons being a_i. We can also consider the infinite case

$$Y' = G(Y) = \sum_{i\geq 0} a_i Y^i \quad , \quad Y(0) = Z. \tag{4.11}$$

The solution is that of "weighted increasing planar trees". This point of view is different from the one adopted throughout this paper which considers $G = G(T)$ as an abstract $\mathbb{L}$-species rather than as the species of "weighted lists". It would be possible to develop the theory using these weighted increasing planar trees (see [BR1], [BR2]). Another option would be to start with $G(T) = \sum_{i\geq 0} a_i T^i/i!$ considered as the species of "weigted sets".

Example 4.5. The linear equation. The general first order linear differential equation can be expressed, at the combinatorial level, as follows:

$$Y' = F(T)Y + G(T) \quad , \quad Y(0) = Z \tag{4.12}$$

where F and G are given $\mathbb{L}$-species.

a) **The homogeneous case**. If $G(T) = 0$, we have the homogeneous equation

$$Y' = F(T)Y \quad , \quad Y(0) = Z\ . \tag{4.13}$$

Its canonical combinatorial solution, denoted by $Y = A_F(T)$, is given by enriched increasing arborescences of a special form, as illustrated in figure 4.3, with $m_1 < m_2 < m_3 < m_4$.

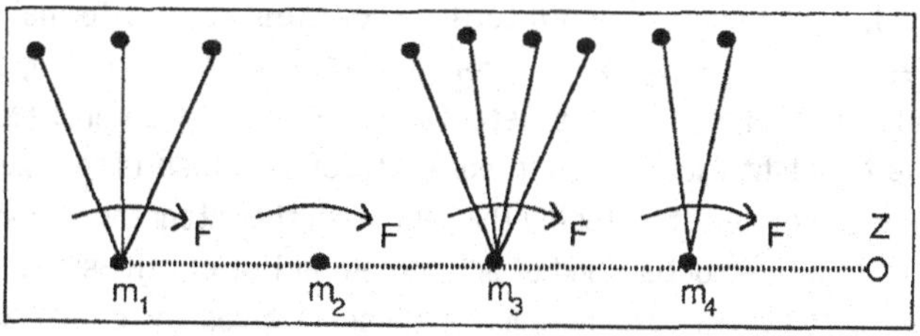

Figure 4.3. Generic A_F-structure

This is simply an increasing sequence of fertile points to which are attached F-structures on bigger elements, followed by a bud. Remembering how the integral $\int_0^T F(X)dX$ (or $\int F$ for short) was defined (see (2.15) and fig. 2.5), we see that we essentially have an assembly of $\int F$-structures, naturally written in the increasing order of the smallest elements (**increasing list**), multiplied by Z. In other words, the solution can be expressed as

$$A_F = \exp\left(\int_0^T F(X)dX\right)\cdot Z\ . \tag{4.14}$$

b) **The general case**. If $Y = B_{F,G}(T)$ is the solution of

$$Y' = F(T)Y + G(T) \quad , \quad Y(0) = 0\ , \tag{4.15}$$

that is with $Z = 0$, then it is easily checked that the solution of the general equation (4.12) is simply given by the sum

$$Y = A_F(T) + B_{F,G}(T)\ . \tag{4.16}$$

The solution $Y = B_{F,G}(T)$ of (4.15) also has a simple combinatorial description, given generically by figure 4.4. We see that a $B_{F,G}$-structure can be described as an increasing list of $\int F$-structures ended by an $\int G$-structure. Unfortunately the combinatorial operation "to end by" is not easily expressed in terms of the other operations.

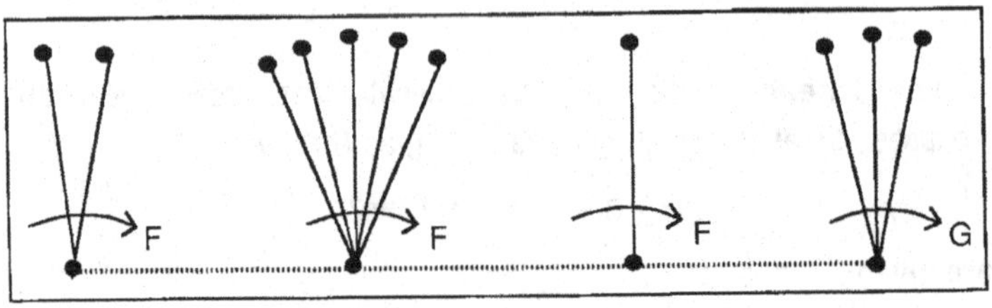

Figure 4.4 Generic $B_{F,G}$-structure.

Analytically, the classical method of "variation of the constant" gives the solution

$$Y = C_{F,G}(T) = \exp\left(\int_0^T F(X)dX\right) \cdot \int_0^T G(X) \cdot \exp\left(-\int_0^X F(U)dU\right) dX \tag{4.17}$$

for equation (4.15). This expression can be interpreted combinatorially and shown to be equivalent to $B_{F,G}(T)$, in the context of **virtual $\mathbb{L}$-species**, that is formal differences $M - N$ of $\mathbb{L}$-species (with $M_1 - N_1 \cong M_2 - N_2 \iff M_1 + N_2 \cong M_2 + N_1$), as follows. Notice that $C_{F,G}(T)$ is a product. The left factor is simply an increasing list of $\int F$-structures while the right factor is an integral structure which is realized by setting aside the minimum element and structuring the rest with the integrand $G \cdot \exp(-\int F)$. This is itself the product of a G-structure and of an increasing list of $-\int F$-structures, that is of $\int F$-structures weighted by -1. The product of these -1 gives the global weight of the $C_{F,G}$-structure. See figure 4.5 for a generic $C_{F,G}$-structure, where the G-structure has been attached to the right-hand side minimum thus giving an $\int G$-structure denoted by g. Some of the labels are explicitely given and mashed edges are omitted in this figure.

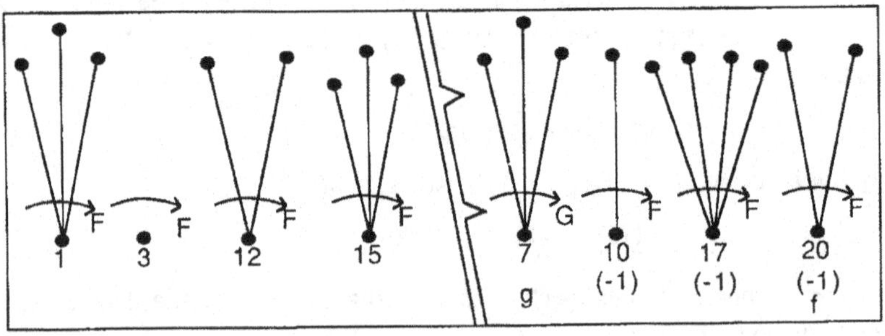

Figure 4.5. Generic $C_{F,G}$-structure.

There is a sign reversing involution Φ on $C_{F,G}$ defined as follows: let f denote the maximum $\int F$-structure (according to the roots, of course). If f is bigger than g then the involution Φ simply switches f from the right list to the left list or conversely and changes its sign accordingly. This cancels all $C_{F,G}$-structures except the fixed points of Φ which occur when g is actually the maximum of all integral structures of both sides. These are obviously isomorphic to $B_{F,G}$-structures. In other words, we have, as desired,

$$B_{F,G} \cong \text{Fix}\,\Phi \cong C_{F,G}\,. \tag{4.18}$$

§ 5. Lie-Gröbner formulas

In this section we adapt to $\mathbb{L}$-species the method of "combinatorial eclosions" first introduced by Gilbert Labelle in his work on multidimensional Lagrange inversion, the implicit function theorem and differential equations in the context of $\mathbb{B}$-species (see [L2], [L3] and [L4]. More precisely we will lift to the combinatorial level, the Lie-Gröbner type formulas (see [G1], [G2]) for the solution of the differential equation

$$Y' = M(T,Y) \quad , \quad Y(0) = Z\,. \tag{5.1}$$

The case of a system of equations will be considered in the following section.

Theorem 5.1 Let $M(T,Y)$ be a 2-sorted $\mathbb{L}$-species. The canonical solution of (5.1), the $\mathbb{L}$-species $Y = A_M(T)$ of M-enriched increasing arborescences, can be expressed as

$$A_M(T) = e^{\Gamma}Z\,\Big|_{X=0} \tag{5.2}$$

where $\Gamma = T\mathcal{D}$ and $\mathcal{D}$ is the differential operator defined by

$$\mathcal{D} = \partial/\partial X + M(X,Z)\,\partial/\partial Z\,. \tag{5.3}$$

Before proving the theorem, we have to understand the action of the operators $T\mathcal{D}$, $(T\mathcal{D})^n/n!$ $(n\geq 0)$, and $e^{T\mathcal{D}}$, on $\mathbb{L}$-species. Actually, these operators act on 3-sorted $\mathbb{L}$-species $\Psi = \Psi(T,X,Z)$. The 3 sorts of elements will be named as follows:

points, ●, minibuds, ○, maxibuds, ◯, corresponding to T, X, Z respectively.

Let $\Psi = \Psi(T,X,Z)$ be a 3-sorted $\mathbb{L}$-species and $\mathcal{D}$ be defined by (5.3) and set $\Gamma = T\mathcal{D}$. Then the 3-sorted $\mathbb{L}$-species $\Gamma\Psi(T,X,Z)$ is the sum

$$\Gamma\Psi(T,X,Z) = (T\partial/\partial X)\Psi(T,X,Z) + (TM(X,Z)\partial/\partial Z)\Psi(T,X,Z)\,. \tag{5.4}$$

Hence, using a proper combinatorial interpretation of the partial derivatives of $\mathbb{L}$-species (see 2.26, 2.27) and the idea of Gilbert Labelle, a $\Gamma\Psi$-structure can be described as a Ψ-structure which has undergone a **combinatorial eclosion** of one of the following two types:

Type I eclosion, $T\partial/\partial X$: "Replace the (phantomatic) minimum minibud by an arbitrary point". This point comes from the multiplication by T and is called the **eclosion point**. See figure 5.1 for a generic $(T\partial/\partial X\ \Psi)$-structure.

Type II eclosion, $TM(X,Z)\partial/\partial Z$: "Replace the (phantomatic) minimum maxibud by a $TM(X,Z)$-structure, that is an arbitrary (eclosion) point with an M-structure of minibuds and maxibuds attached". See figure 5.2 for a generic $(T{\cdot}M(X,Z)\partial/\partial Z\ \Psi)$-structure.

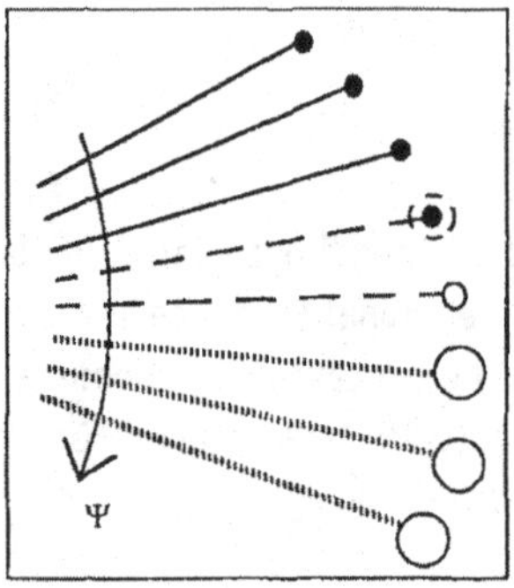

Figure 5.1.

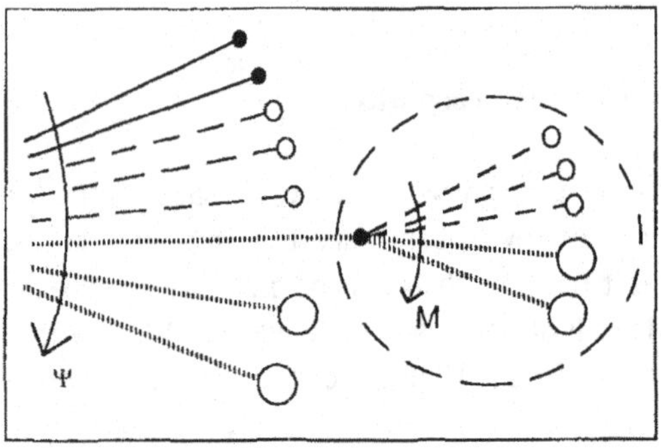

Figure 5.2.

For $n \geq 0$, $\Gamma^n\Psi$ is obviously obtained by n successive applications of $\Gamma = T\mathcal{D}$, that is by n successive eclosions of either type. The order of eclosion is important and can be recorded by noting it above the unique "eclosion point" that appears at each step (see fig. 5.3). This numbering is not to be confused with the labels of these points. Indeed, for a given set of n eclosion points there are n! different labellings of these in an otherwise fixed $\Gamma^n\Psi$-structure, since their occurence comes from n multiplications by T. In each case, this gives n! distinct structures; however there is only one of these whose labels are in the same order as the eclosions; we call this an **orderly labelled** $\Gamma^n\Psi$-structure. See for example the $\Gamma^n\Psi$-structure of figure 5.3 where $n = 13$ and the set of points is $\{a, b, ..., o\}$, in alphabetical order.

Now $\Gamma^n/n!$ can be interpreted as follows: a $(\Gamma^n/n!)\Psi$-structure is identified with an orderly labelled $\Gamma^n\Psi$-structure. Finally e^Γ is easily interpreted since it is defined as

$$e^\Gamma = \sum_{n\geq 0} \Gamma^n/n! \qquad (5.5)$$

so that a $e^\Gamma\Psi$-structure is an orderly labelled $\Gamma^n\Psi$-structure for some $n \geq 0$.

Proof of Theorem 5.1. If we take $\Psi(T,X,Z) = Z$, the starting structure is simply a maxibud. After one eclosion, we get $\Gamma(Z) = T\,M(X,Z)$ and after a few more eclosions we get an M-enriched arborescence on points or minibuds, and maxibuds. Moreover all the points are eclosion points so that in the properly labelled case, that is for $e^\Gamma Z$, the point labels can be identified with their order of apparition. Hence they should be increasing as we go away from the root since the eclosions occur in this way.

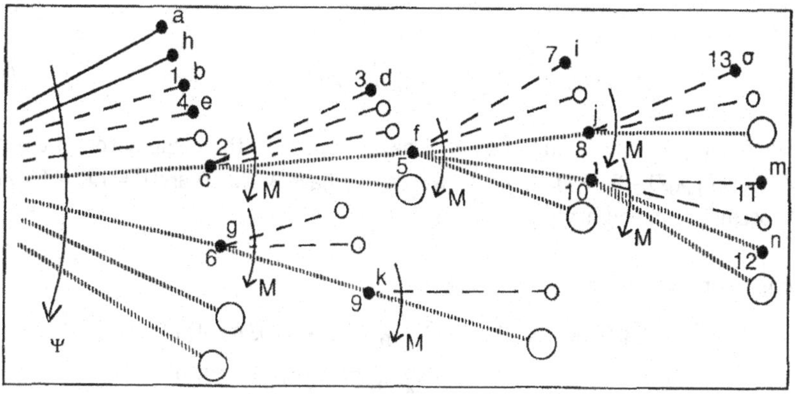

Figure 5.3. Generic orderly labelled $\Gamma^n\Psi$-structure, n = 13.

Finally, setting $X = 0$ in $e^{\Gamma}Z$ simply imposes the restriction that there should be no minibuds remaining in the structures. These can obviously be identified with the increasing M-enriched arborescences defined in § 3 (see fig. 3.2). In other words we have

$$A_M \cong e^{T\mathcal{D}} Z \Big|_{X=0} . \qquad \square$$

If we start with $\Psi(T,X,Z) = F(Z)$, that is with an F-assembly of maxibuds, where F is any given $\mathbb{L}$-species, and if we apply e^{Γ} and set $X = 0$, we obviously obtain F-assemblies of A_M-structures; in other words we have the slightly more general result:

Proposition 5.2. For any $\mathbb{L}$-species F, we have

$$F(A_M) = e^{\Gamma}F(Z) \,\big|_{X=0} \tag{5.6}$$

where A_M and Γ are defined as in theorem 5.1.

□

In the case of the autonomous equation

$$Y' = G(Y), \quad Y(0) = Z, \tag{5.7}$$

where G is a given $\mathbb{L}$-species, we have $M(T,Y) = G(Y)$ and, as seen in § 4, the canonical solution $Y = A(T,Z)$ is given by "G-enriched increasing rooted trees, with buds" (see fig. 4.1). We then have the following:

Corollary 5.3. The solution $Y = A(T,Z)$ of the autonomous equation (5.7) can be expressed as

$$A(T,Z) = \exp(TG(Z)\partial/\partial Z)\, Z \tag{5.8}$$

and

$$A(T,Z) = \textstyle\sum_{n\geq 0} (G(Z)\partial/\partial Z)^n Z\, T^n/n!\ . \tag{5.9}$$

Moreover, for any $\mathbb{L}$-species F, we have

$$F(A(T,Z)) = \textstyle\sum_{n\geq 0} (G(Z)\partial/\partial Z)^n F(Z)\, T^n/n!\ . \tag{5.10}$$

Proof. Since $M(X,Z) = G(Z)$, minibuds will never appear in the combinatorial eclosions applied to any $\Psi(T,X,Z) = \Psi(T,Z)$. In other words, we have $\partial/\partial X = 0$ and $\Gamma = T\mathcal{D} = TG(Z)\partial/\partial Z$ so that theorem 5.1 immediately gives (5.8).

Furthermore we also have (5.9) since

$$\begin{aligned}\exp(TG(Z)\partial/\partial Z)\, Z &= \textstyle\sum_{n\geq 0} ((TG(Z)\partial/\partial Z)^n/n!)\, Z \\ &= \textstyle\sum_{n\geq 0} (G(Z)\partial/\partial Z)^n Z\ T^n/n!\ .\end{aligned}$$

Formula (5.10) is obtained similarly from proposition 5.2. □

The expression (5.9) is interpreted as follows: First perform all eclosions with buds only, keeping track of the order of eclosions (see fig. 5.4), and, when this is completed, put the points into their unique position (compare with fig. 4.1).

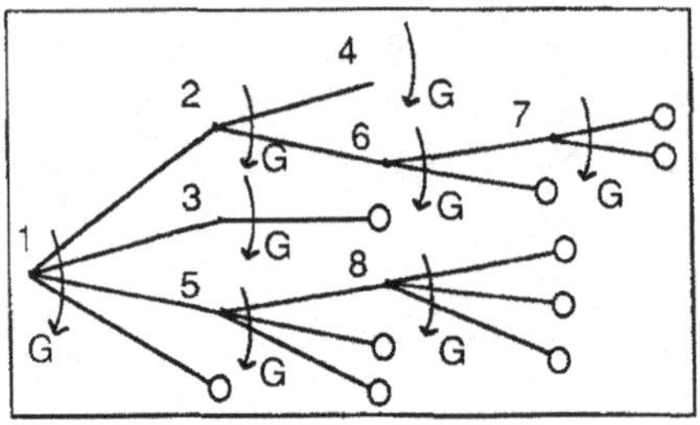

Figure 5.4. $(G(Z)\,\partial/\partial Z)^n Z$ $(n = 8)$

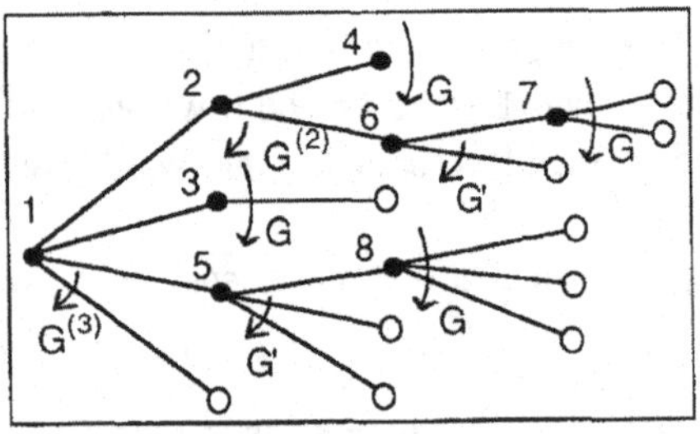

Figure 5.5

Moreover each $(G(Z)\,\partial/\partial Z)^n Z$-structure such as that of figure 5.4 can be transformed isomorphically into a structure given generically by figure 5.5. From this observation results the following proposition.

Proposition 5.4. [BR1] The $\mathbb{L}$-species $H_n(Z) = (G(Z)\,\partial/\partial Z)^n Z$, for $n \geq 1$, can be written in the form

$$H_n(Z) = \textstyle\sum_{\alpha} C_n(\alpha)\, G^{\alpha_0} (G')^{\alpha_1} (G'')^{\alpha_2} \ldots (G^{(p)})^{\alpha_p} \tag{5.11}$$

where α ranges over all sequences $\alpha = (\alpha_0, \alpha_1 \ldots)$ of non negative integers and $C_n(\alpha)$ is the number of increasing arborescences on $[n]$ having, for $p = 0, 1, \ldots, \alpha_p$ vertices with p sons. □

Note that the conditions

$$n = \sum_{p\geq 0} \alpha_p \qquad \text{and} \qquad \alpha_0 = 1 + \sum_{p\geq 1} (p-1)\alpha_p \tag{5.12}$$

are necessary for having $C_n(\boldsymbol{\alpha}) \neq 0$, so that the expression (5.11) is a finite sum. The first values of $H_n(Z)$ are

$$\begin{aligned} H_1(Z) &= G(Z) \\ H_2(Z) &= G(Z)\, G'(Z) \\ H_3(Z) &= G^2(Z)\, G''(Z) + G(Z)\,(G'(Z))^2 \\ H_4(Z) &= G^3(Z)\, G^{(3)}(Z) + 4\, G^2(Z)\, G'(Z)\, G''(Z) + G(Z)\,(G'(Z))^3 . \end{aligned} \tag{5.13}$$

Another observation which follows from (5.9), or from figure 5.4, is that, for $n \geq 0$,

$$(G(Z)\, \partial/\partial Z)^n Z = (\partial/\partial T)^n A(T,Z)\big|_{T=0} . \tag{5.14}$$

Similar interpretations can be given for the general expansion of the differential operator $(G(Z)\, \partial/\partial Z)^n$ applied to any $F(Z)$, using equation (5.10). See [CO] and [BR1].

Example 5.5. We now consider a simple example to illustrate the use of combinatorial eclosions. Let $Y = Ter(T,Z)$ be the solution of the autonomous differential equation

$$Y' = Y^3 \quad , \quad Y(0) = Z . \tag{5.15}$$

It can be viewed as the $\mathbb{L}$-species of increasing ternary trees, with buds as leaves (see fig. 5.6). Setting $Z = 1$, we get

$$Ter(T,Z)\big|_{Z=1} = Ter(T), \tag{5.16}$$

the $\mathbb{L}$-species of **ternary trees**.

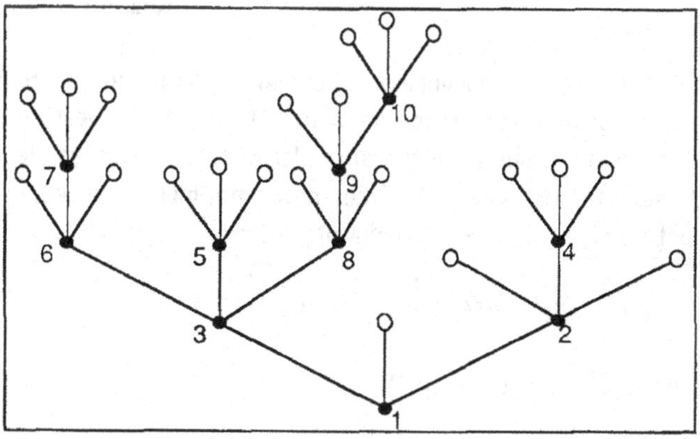

Figure 5.6

Now corollary 5.3 can be applied, with $G(Z) = Z^3$, to get

$$\mathrm{Ter}(T,Z) = e^{TZ^3\, \partial/\partial Z}\, Z. \tag{5.17}$$

Hence, to get an increasing ternary tree with buds, we can do the following:

1. Start with a single bud: O

2. Apply a certain number of times the combinatorial eclosion $TZ^3\, \partial/\partial Z$ which can be interpreted as "replace a bud by a TZ^3-structure", as in Figure 5.7.

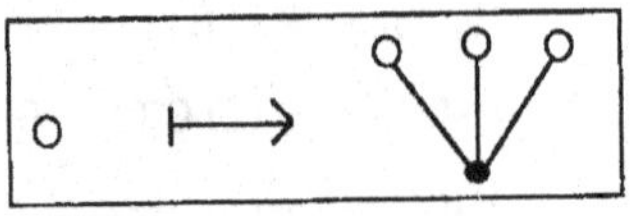

Figure 5.7. $TZ^3\, \partial/\partial Z$

3. Label the eclosion points according to their order of apparition.

However formula (5.17) can be given a different interpretation by changing slightly the first two rules given above, as follows:

1 bis. Start with a single bud, with a stem planted in a flower box labelled by zero (Fig. 5.8)

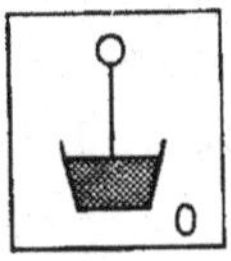

Figure 5.8

2 bis. The combinatorial eclosion $TZ^3\partial/\partial Z$ is to be interpreted as "replace a stemmed bud by a TZ^3-structure as in Figure 5.9".

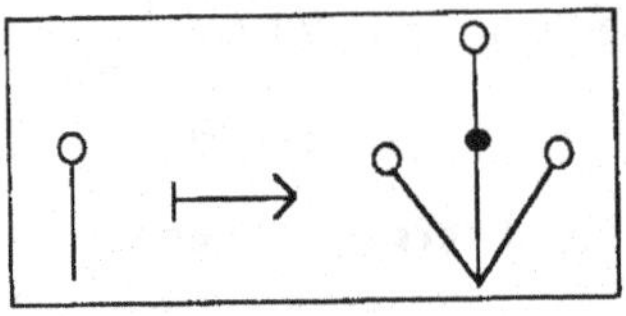

Figure 5.9

For example, if we apply a sequence of 10 eclosions, as in figure 5.6, but with this new interpretation, we get a planar tree like that of figure 5.10. Setting $Z = 1$ gives figure 5.11 which is nothing but an increasing planar tree (see example 4.3) on the linearly ordered set $1+[10]$ (zero is the new minimum element), that is a $(d/dT)\mathrm{Pla}(T)$-structure on $[10]$. Thus, in general, we have

$$\mathrm{Ter}(T) = e^{(TZ\partial/\partial Z)}\, Z \Big|_{Z=1} = (d/dT)\, \mathrm{Pla}\,(T) \tag{5.18}$$

and, for the generating function, using (4.9),

$$\mathrm{Ter}(t) = (d/dt)\, \mathrm{Pla}(t) = 1/\sqrt{1-2t}. \tag{5.19}$$

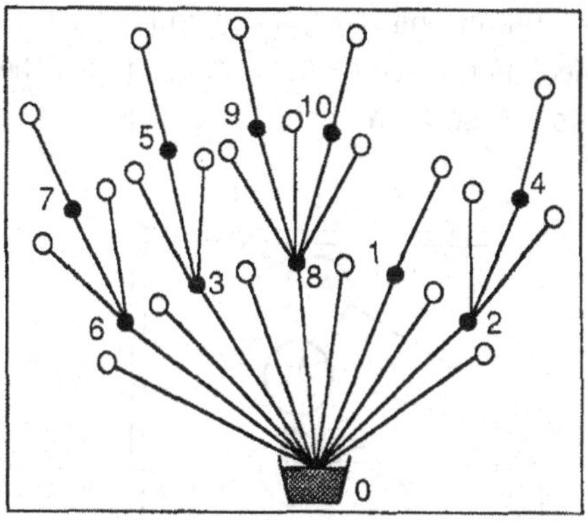

Figure 5.10

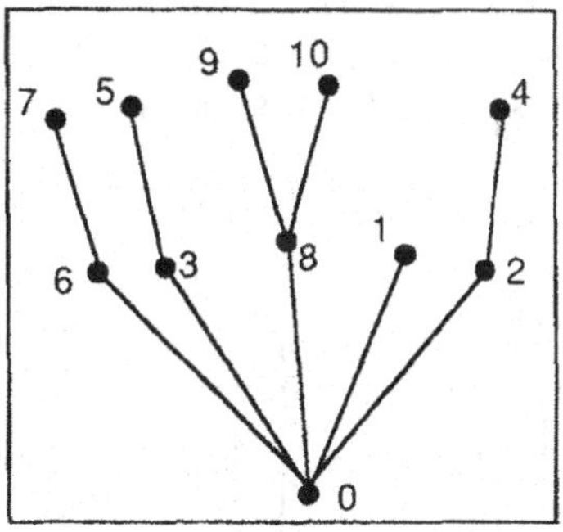

Figure 5.11

§ 6. Systems of differential equations

Most of the theory developed so far can be easily extended to systems of differential equations. We now sketch briefly the main results in this case. Let $M_1,\ldots,M_p$ be given p-sorted $\mathbb{L}$-species. We will consider the following system of first order autonomous differential equations:

$$dY_i/dT = Y_i' = M_i(Y_1,\ldots,Y_p)\ , \qquad Y_i(0) = Z_i \quad , \quad i = 1, 2, \ldots, p\,. \tag{6.1}$$

There is no loss of generality in assuming, as we do, that the equations are autonomous since we can always reduce ourselves to this case at the cost of the supplementary equation (6.2), if necessary.

$$Y_0' = 1\ , \qquad Y_0(0) = 0 \tag{6.2}$$

The initial conditions $Z_1, \ldots, Z_p$ are considered as p extra sorts of linearly ordered sets (of buds) and the solutions will in fact depend also on them, altough it will not be explicitely written, to lighten the notations.

A solution of (6.1) is a family (A_i, ψ_i) where for $i = 1,\ldots,p$, A_i is a $\mathbb{L}$-species such that $A_i(0) = Z_i$ and ψ_i is an isomorphism

$$dA_i/dT \xrightarrow[\sim]{\psi_i} M_i(A_i(T), \ldots, A_p(T))\,. \tag{6.3}$$

As in the case of one equation, the system (6.1) is equivalent to the integral equations

$$A_i(T) = Z_i + \int_0^T M_i(A_1(X),\ldots,A_p(X))\, dX \ , \qquad i = 1, \ldots, p\,. \tag{6.4}$$

For each i, this equation can be represented combinatorially as in Figure 6.1, where in the non-empty case, the minimum element is marked by the number i and the M_i-structure lies over a p-tuple of sets of lines of p sorts: a line of sort j is used to attach an A_j-structure to the minimum element.

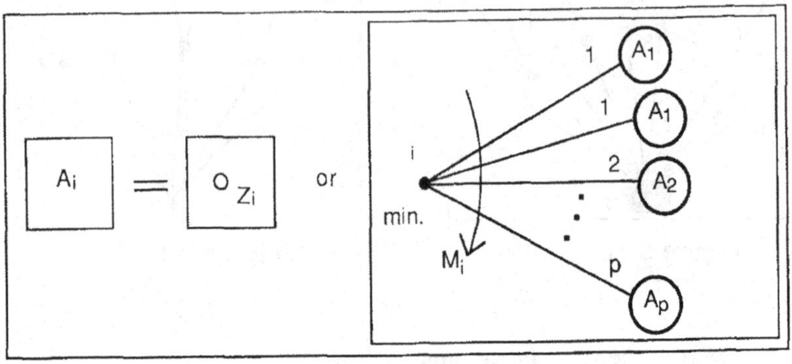

Figure 6.1

Since this process can be iterated,we obtain a canonical solution $(A_{\mathbf{M},1},\ldots,A_{\mathbf{M},p})$ where $A_{\mathbf{M},i}(T)$ is the $\mathbb{L}$-species of p-colored $\mathbf{M}$-enriched increasing arborescences as represented in figure 6.2 (where $p = 2$), where each point is marked with some color j, $1\leq j\leq p$, and is the root of an $A_{\mathbf{M},j}$-structure; the buds of sort i are empty $A_{\mathbf{M},i}$-structures and are marked by Z_i.

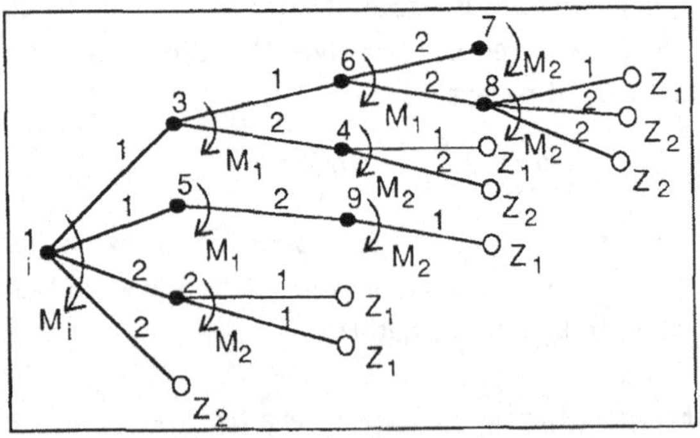

Figure 6.2 $A_{\mathbf{M},i}$-structure

Theorem 6.1. The $\mathbb{L}$-species of p-colored $\mathbf{M}$-enriched increasing arborescences is, up to isomorphism, the unique solution $A_{\mathbf{M},1},\ldots,A_{\mathbf{M},p}$ of the system (6.1) Moreover this solution can be expressed as, for $i = 1,\ldots,p$,

$$A_{\mathbf{M},i}(T) \;=\; e^{T\mathcal{D}}\, Z_i \qquad (6.5)$$

where $\mathcal{D}$ is the combinatorial differential operator

$$\mathcal{D} = \sum_{j=1}^{p} M_j(Z_1,\ldots,Z_p)\, \partial/\partial Z_j \tag{6.6}$$

Proof. The expression (6.5) is very similar to (5.8), in the case of one autonomous equation (see cor. 5.3). The only difference here is that the eclosions can be of p different sorts: for $j = 1,\ldots,p$, the operator $TM_j(Z_1,\ldots,Z_p)\partial/\partial Z_j$ is interpreted as "replace the (phantom) minimum bud of sort j by a point (marked j) with an M_j-assembly of buds attached". □

Note also that (6.5) can be written as

$$A_{\mathbf{M},i}(T) = \sum_{n\geq 0} \mathcal{D}^n Z_i\, T^n/n! \tag{6.7}$$

which says that pointless eclosions can be performed first and then the points placed in position according to the order of eclosions as in corollary 5.3.

The system of differential equations that we considered in the introduction (see (1.1)) can be written as

$$\begin{aligned} Y_1' &= 1 + Y_1^2 , & Y_1(0) &= 0 \\ Y_2' &= Y_1 Y_2 , & Y_2(0) &= 1 \end{aligned} \tag{6.8}$$

Its canonical solution, as given then, is that of complete increasing binary trees for Y_1 and complete (except for an empty bud at the far right) increasing binary trees for Y_2. See figures 1.3 and 1.4 where colors 1 and 2 are modestly realized as black and white respectively.

The tangent and secant functions are also solutions of the following slightly different system of equations:

Example 6.2. Consider the system of equations

$$\begin{aligned} Y_1' &= Y_2^2 , & Y_1(0) &= 0 \\ Y_2' &= Y_1 Y_2 , & Y_2(0) &= 1. \end{aligned} \tag{6.9}$$

Here the two sorts of eclosions are as follows:

1. $TZ_2^2\, \partial/\partial Z_1$ (see fig. 6.3).
2. $TZ_1Z_2\, \partial/\partial Z_2$ (see fig. 6.4).

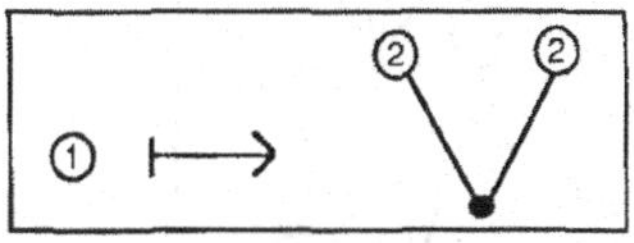

Figure 6.3

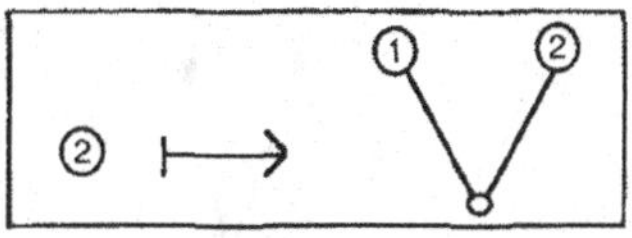

Figure 6.4

In order to get the canonical solution $Y_1 = A_1(T)$, $Y_2 = A_2(T)$, we can then start with a bud Z_1 or Z_2 and iterate the eclosion operators a certain number of times. If we set $Z_1 = 0$ and $Z_2 = 1$, we get, for $A_1(T)$, increasing binary trees of the form illustrated by Figure 6.8 and for $A_2(T)$, increasing binary trees like those rooted at white points in Figure 6.8 (the labels a, c and d on points, and C_0 and D_0 on buds should be disregarded in Figure 6.8 for the time being. Define a **left branch** to be a maximal subtree having only left edges. These increasing binary trees, called **Jacobi arborescences**, are characterized by the following properties:

A_1: All the left branches are even (number of points, not counting buds), except the leftmost one which is odd.

A_2: All the left branches are even.

This combinatorial model of the tangent and secant functions is appropriate for establishing the integral (1.5) which can be equivalently written as

$$\tan T + \sec T = \exp\left(\int_0^T \sec X \, dX\right). \tag{6.10}$$

Indeed, if you cut the edges of the leftmost left branch of an A_1- or A_2-structure, what you get is an assembly (in the form of a decreasing list), of $\int A_2$-structures.

Note also that this model can take other interesting forms. For example, when the bijection between increasing binary trees and permutations (projection on the x-axis) is applied to A_1 and A_2, it gives the class of **Jacobi permutations** introduced by one of the authors in [V2] in order to obtain a combinatorial interpretation for the Jacobi elliptic functions (see below).

Another possibility is to apply the bijection between permutations, or increasing binary trees, and forests of increasing arborescences (see [BU], [V1] or [V3]). This gives forests, that is assemblies (odd or even), of increasing arborescences where all points have an even number of sons. For example, the forest of figure 6.5 corresponds to the binary tree of figure 6.8 under this bijection.

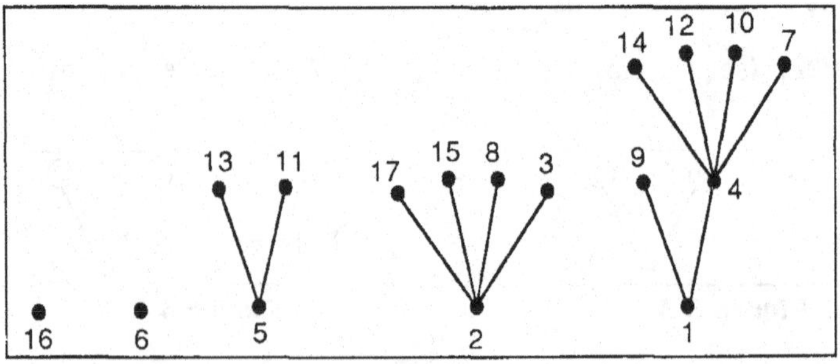

Figure 6.5

Example 6.3. Elliptic functions. We consider the following system

$$\begin{aligned} S' &= aCD\ , & S(0) &= S_0\ , \\ C' &= cDS\ , & C(0) &= C_0\ , \\ D' &= dSC\ , & D(0) &= D_0\ . \end{aligned} \tag{6.11}$$

where a, c, d are some scalar parameters. The classical Jacobi elliptic functions correspond to this system (sn, cn, dn, respectively, for S, C, D) with $a = 1$, $c = -1$, $d = -k^2$, $S_0 = 0$, $C_0 = D_0 = 1$. They can be expanded in the form

$$\begin{aligned} sn(t, k) &= \textstyle\sum_{n\geq 0} (-1)^n J_{2n+1}(k)\, t^{2n+1}/(2n+1)!\ , \\ cn(t, k) &= 1 + \textstyle\sum_{n\geq 1} (-1)^n J_{2n}(k)\, t^{2n}/(2n)!, \\ dn(t, k) &= 1 + \textstyle\sum_{n\geq 1} (-1)^n k^{2n} J_{2n}(1/k)\, t^{2n}/(2n)!. \end{aligned} \tag{6.12}$$

Here $J_{2n+1}(k)$ and $J_{2n}(k)$ are even polynomials with non negative integer coefficients, of degree respectively $2n$ and $2n-2$.

We will briefly show how the general theory gives back the interpretations of Viennot [V2] and of Schett's polynomials [SCH]. There are three types of eclosions, as displayed on figure 6.6.

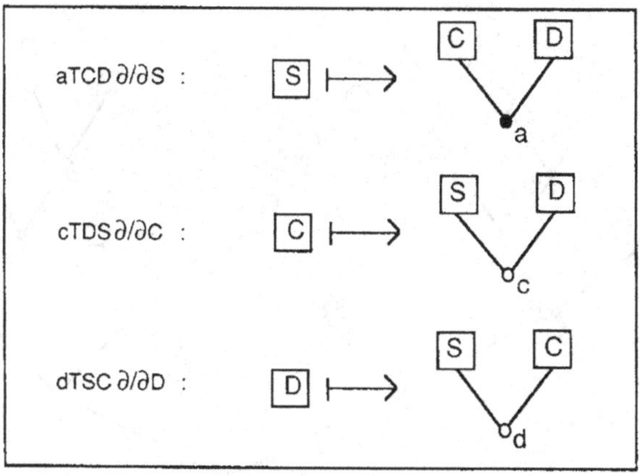

Figure 6.6

Let us assume that $S_0 = 0$. The underlying binary trees of the solutions S(T), C(T), D(T) of (6.11) are of the same type as in the previous example (see fig. 6.8)). Moreover, we have to consider the weight of these trees in terms of the parameters a , c and d . In fact, once the underlying binary tree is known, the weight $a^i c^j d^h$ of an S- (or C- or D-) structure can easily be determined recursively, starting at the root.

As pointed out in the introduction a little more work has to be done in order to define this weight in a "global" form, without recursivity. Define the **right-height** of a vertex in a binary tree as the number of right edges of the path going from the root to that vertex. We redefine the eclosions of the first type as follows:

If the eclosion appears at an even right-height vertex, use fig 6.7, a):

If the eclosion appears at an odd right-height vertex, use fig 6.7, b):

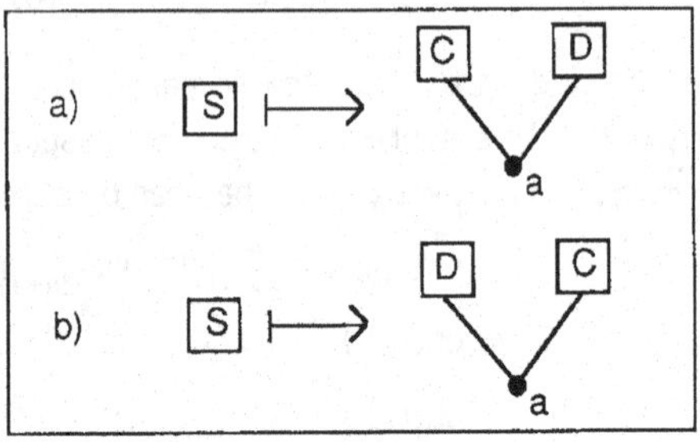

Figure 6.7

The binary trees produced are the same as before. In particular all the left branches are even, except the leftmost in the case of S(T). See the figure 6.8, where the weights a, c and d as well as the bud labels S_0, C_0, D_0, have been displayed according to these new eclosion rules.

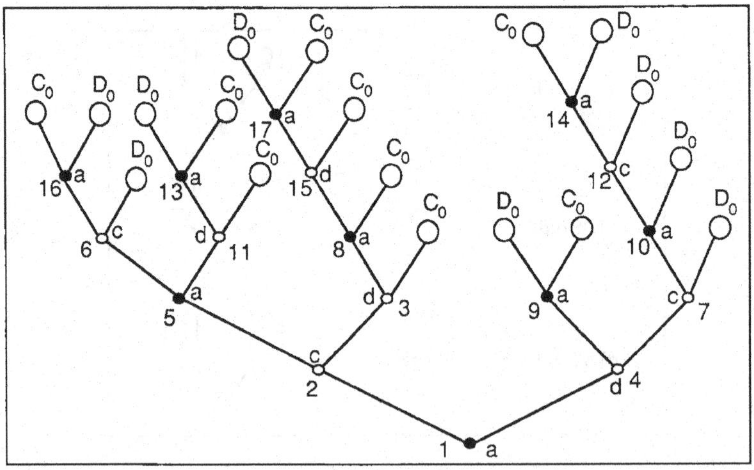

Figure 6.8

It is easy to see that all the eclosions of the third type appear at an odd right-height in the binary tree (vertices labelled d). Also, on each left branch containing such vertices, exactly half of them are labelled d (the others being labelled a). Thus the weight of a S-binary tree of size 2n+1 (buds are not counted) is $a^{n+1}c^{n-j}d^{j}$, where j is equal to half the number of nodes having an odd right-height. Setting $a = 1$, $c = -1$ and

$d=-k^2$ and taking generating functions, we obtain from (6.12) that, for example,

$$J_{2n+1}(k) = \sum_{j=0}^{n} J_{2n+1,2j}\, k^{2j}, \qquad (6.13)$$

where $J_{2n+1,2j}$ is the number of Jacobi arborescences with $2n+1$ vertices among which exactly $2j$ have an odd right height. By projection of these binary trees, we get the Jacobi permutations and the interpretation of Viennot [V2]. Also, as in the previous example, using the bijection between increasing binary trees and assemblies of increasing arborescences (see fig. 6.5), one gets another result of [V2]: $J_{2n+1,2j}$ is equal to the number of forests of increasing arborescences of size $2n+1$, with all vertices of even degree and having $2j$ vertices at an odd height.

The functions cn and dn can be deduced similarly.

Dumont's interpretation of the Jacobi elliptic functions is based on Schett's method [SCH]. Schett introduced (in a slightly different form) the following polynomials:

$$S_m(x,y,z) = \mathcal{D}^m x, \qquad m \geq 1, \qquad (6.14)$$

where

$$\mathcal{D} = yz\partial/\partial x + zx\partial/\partial y + xy\partial/\partial z. \qquad (6.15)$$

These polynomials can be written explicitly as

$$\begin{aligned} S_{2n+1}(x,y,z) &= \sum_{i,j\geq 0} a_{2n+1,i,j}\, x^{2i} y^{2j+1} z^{2n-2i-2j+1}, \\ S_{2n}(x,y,z) &= \sum_{i,j\geq 0} a_{2n,i,j}\, x^{2i+1} y^{2j} z^{2n-2i-2j}. \end{aligned} \qquad (6.16)$$

Schett's result is that the coefficients $a_{2n+1,0,j}$ (resp. $a_{2n,i,0}$) are precisely the coefficients $J_{2n+1,2j}$ of the polynomials $J_{2n+1}(k)$ (resp. $J_{2n,2i}$ of $J_{2n}(k)$).

This can be shown as follows. The operator $\mathcal{D}$ is the operator of theorem 6.1. Thus $S_m(x,y,z)$ is the polynomial enumerating the $m!$ increasing binary trees according to the number of buds labelled S_0, C_0 and D_0. Let $a^e c^g d^j$ be the weight of such a tree; i.e. there have been e eclosions of first type, g of second type, and j of third type and $e+g+j=m$. Let α (resp. γ, resp. δ) be the number of buds labelled S_0 (resp. C_0, resp D_0). We have the equations

$$\alpha = 1-e+g+j, \qquad \gamma = e-g+j, \qquad \delta = e+g-j. \qquad (6.17)$$

The case of sn is given by putting $\alpha = 0$ (no buds labelled S_0) and $m = 2n+1$. Eliminating e gives $\gamma = 2j+1$ and $\delta = 2n-2j+1$ and we get Schett's result. The functions cn and dn can be treated similarly. We will give more results, in particular the relationship between the present general theory and Dumont's work [D1], [D2] in another paper.

Example 6.4 Duffing's equation

Higher order differential equation are generally reduced to systems of first order differential equations. We give here the classical example of the cubic anharmonic oscillator, commonly known as Duffing's equation

$$y'' = ay' + by + cy^3 , \qquad y(0) = \alpha . \tag{6.18}$$

Denoting $\beta = y'(0)$, this is equivalent to the system

$$\begin{aligned} y' &= u , & y(0) &= \alpha , \\ u' &= au + by + cy^3 , & u(0) &= \beta . \end{aligned} \tag{6.19}$$

There are four types of eclosions giving birth to the weighted increasing rooted planar trees as shown in figure 6.9 :

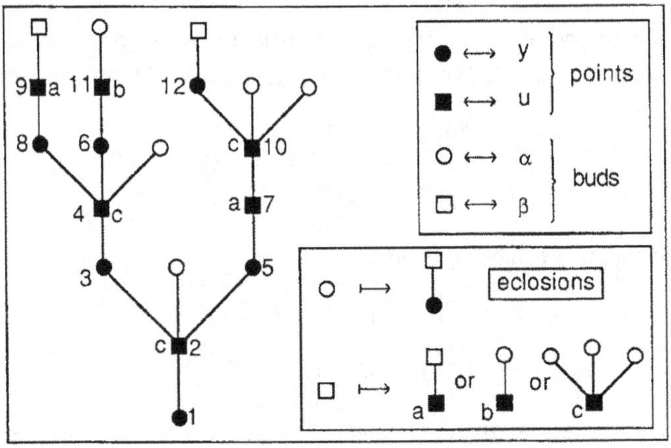

Figure 6.9.

The tree of figure 6.9 gives the contribution $a^2bc^3\alpha^4\beta^2$ to the coefficient of $t^{12}/12!$ of the Taylor expansion of the solution. This kind of planar trees can be drawn in a more compact form, as shown in figure 6.10.

There are three types of nodes:

1) some nodes contain two numbers (unordered),

2) some contain one number and a "□",

3) some contain a single number.

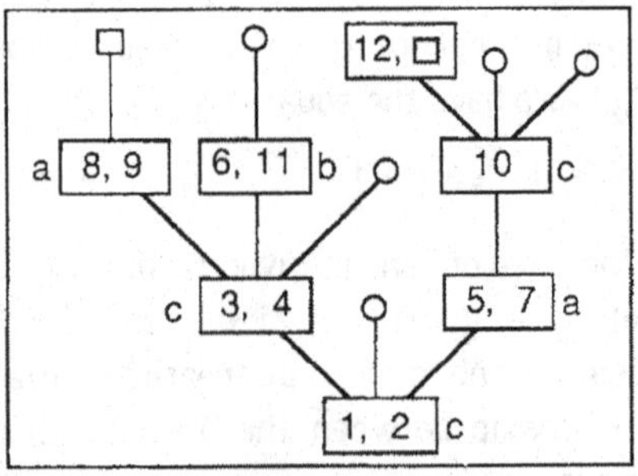

Figure 6.10

In case 1 or 2, the father is weighted b or c (depending if there is one or three sons). In case 3, the father is weighted a. These planar rooted trees are 1-3 (one or three sons) and still have the increasing property.

The case $a=0$ is particularly important since it contains that of Jacobi elliptic functions sn, cn and dn. The equation $y'' = 2y^3$, $y(0) = 1$, $y'(0) = 1$, gives birth to increasing ternary trees with nodes formed by pairs (unordered) of integers or, possibly for external nodes, by a single integer. Such trees can be put in bijection with permutations. Increasing binary trees correspond to binary search trees in Computer Science (see Françon [FR]). The above ternary trees also have a corresponding concept in Computer Science, as shown by Jonassen & Knuth [JK]. The analysis of the average cost of comparisons involves elliptic functions.

We will give more details in another paper, combining the general concepts of this paper with the concept of "histories". This will be particularily important when the Duffing equation has forcing terms. Starting from Fliess & Lamnabhi-Lagarrigue's paper [FL1] we will give more explicit computations of the coefficients of the functional expansion of the solution, in terms of certain weighted paths.

Bibliography

[A2] D. André, Développement en séries des fonctions elliptiques et de leurs puissances, Ann. Ecole Norm. Sup., (2) 6 (1877), 265-328.

[AN] D. André, Sur les permutations alternées, J. Math. Pures Appl. 5 (1879), 31-46.

[B1] F. Bergeron, Une systématique de la combinatoire énumérative, Ph.D. Thesis, Un. de Montréal - UQAM, 1986.

[B2] F. Bergeron, Modèles combinatoires de familles de polynômes orthogonaux, UQAM 1984, submitted.

[B3] F. Bergeron, Combinatorial representations of some Lie groups and Lie algebras, This volume.

[BR1] F. Bergeron, C. Reutenauer, Combinatorial interpretation of the powers of a linear differential operator, submitted.

[BR2] F. Bergeron, C. Reutenauer, Combinatorial resolution of systems of differential equations, III Differentiably algebraic functions, submitted.

[BL] C. Blais, Espèces de structures et polynômes eulériens, Mémoire de maîtrise, UQAM, 1982.

[BU] W.H. Burge, An analysis of a tree sorting method and some properties of a set of trees, First USA-Japan Computer Conference, 1972, 372-379.

[CA1] L. Carlitz, Generating functions for a special class of permutations, Proc. Amer. Math. Soc., 47 (1975), 251-256.

[CA2] L. Carlitz, R. Scoville, Generating functions for certain types of permutations, J. of Comb. Th. A, 18 (1975), 262-275.

[CGJ] C.B. Collins, I.P. Goulden, D.M. Jackson, D.M. Neirstrasz, A combinatorial application of matrix Riccati equations and their q-analogue, Discrete Math., 36 (1981), 139-153.

[CO] L. Comtet, Une formule explicite pour les puissances successives de l'opérateur de dérivation de Lie. C. R. Acad. Sc. Paris, t. 267 (1973), Série A, 165-168.

[DLL] H. Décoste, G. Labelle, P. Leroux, Une approche combinatoire pour l'itération de Newton-Raphson, Adv. Applied Math. 3 (1982), 407-416.

[D1] D. Dumont, A combinatorial interpretation for the Schett recurrence of the Jacobian elliptic functions, Math. Comp., 33 (1979), 1293-1297.

[D2] D. Dumont, Une approche combinatoire des fonctions elliptiques de Jacobi, Adv. in Math. 41 (1981), 1-39.

[FLA] P. Flajolet, Combinatorial aspects of continued fractions, Discrete Math. 32(1980), 125-161

[F1] M. Fliess, Fonctionnelles causales non linéaires et indéterminées non commutatives, Bull. Soc. math. France, 109 (1981), 3-40.

[FL1] M. Fliess, F. Lamnabhi-Lagarrigue, Application of a new functional expansion to the cubic anharmonic oscillator, J. Math. Phys. 23 (1982), 495-502.

[FL2] M. Fliess, M. Lamnabhi, F. Lamnabhi-Lagarrigue, An algebraic approach to nonlinear functional expansions, I.E.E.E. Trans. Circuits and Systems, 30 (1983), 554-570.

[FO1] D. Foata, La série génératrice exponentielle dans les problèmes d'énumération, Séminaire de Math. Sup. Presses de l'Université de Montréal, Montréal, 1974.

[FO2] D. Foata, Combinatoire des identités sur les polynômes orthogonaux, International congress of Mathematicians, Warsaw, 1983.

[FS1] D. Foata, M.-P. Schutzenberger, Théorie géométrique des polynômes eulériens, Lecture Notes in Math., 138, Springer-Verlag, Berlin 1970.

[FS2] D. Foata, M.-P. Schutzenberger, Nombres d'Euler et permutations alternantes, manuscrit, University of Florida, Gainesville, 1971. Première partie publiée dans: A Survey of Combinatorial Theory, J.N. Srivastava et als. eds., North Holland, 1973, 173-187.

[FST] D. Foata, F. Strehl, Rearrangement of the symmetric group and enumerative properties of the tangent and secant numbers, Math. Z., 137 (1974), 257-264.

[FR] J. Françon, Arbres binaires de recherche: propriétés combinatoires et applications, R.A.I.R.O. Informatique Théorique, 10 (1976), 35-50.

[FV] J. Françon, G. Viennot, Permutations selon les pics, creux, doubles descentes, nombres d'Euler et nombres de Genocchi, Discrete Math. 28 (1979), 21-35.

[GA] J.F. Gagné, Rapport existant entre la théorie des espèces et les équations différentielles, Mémoire de maîtrise, UQAM, 1985.

[GJ] I.P. Goulden, D.M. Jackson, Combinatorial Enumeration, John Wiley and Sons, New York, Toronto, 1983.

[G1] W. Gröbner, Die Lie-Reihen und ihre Anwendungen, VEB Deutscher Verlag der Wissenschaften, Berlin, 1967.

[G2] W. Gröbner, H. Knapp, Contributions to the theory of Lie series, Bibliographisches Institut-Mannheim Hochschultaschenbücher Verlag. Mannheim, 1967.,

[JK] A. Jonassen, D.C. Knuth, A trivial algorithm whose analysis isn't, J. of Computer and System Sc., 16 (1978), 301-322.

[J1] A. Joyal, Une théorie combinatoire des séries formelles, Adv. in Math., 42 (1981), 1-82.

[J2] A. Joyal, Calcul intégral combinatoire et homologie du groupe symétrique, C.-R. Math. Acad. Sci., Soc. Roy. Canada, Vol. VII, No. 6 (1985), 337-342.

[J3] A. Joyal, Foncteurs analytiques et espèces de structures, This volume.

[LA] J. Labelle, Applications diverses de la théorie combinatoire des espèces de structures. Ann. Sci. Math. du Québec, 7 (1983), 59-94.

[L1] G. Labelle, Une nouvelle démonstration combinatoire des formules d'inversion de Lagrange, Adv. in Math. 42 (1981), 217-247).

[L2] G. Labelle, Eclosions combinatoires appliquées à l'inversion multidimensionnelle des séries formelles, J. Combin. Theory (A), 39 (1985), 52-82.

[L3] G. Labelle, Une combinatoire sous-jacente au théorème des fonctions implicites, J. Combin. Theory (A), 40 (1985), 377-393.

[L4] G. Labelle, On combinatorial differential equations, J. Math. Analysis and Applic. (to appear)

[L5] G. Labelle, Méthodes de calcul en théorie des espèces, Communication au Colloque de combinatoire énumérative, UQAM 1985, This Volume.

[LL] F. Lamnabhi-Lagarrigue, M. Lamnabhi, Algebraic computation of the solution of some nonlinear forced differential equations, in Computer algebra (J. Calmet, ed.), Lecture Notes in Computer Science, Vol. 144, Springer Verlag, 1982, 204-211.

[LV1] P. Leroux, G. Viennot, Résolution combinatoire des systèmes d'équations diffèrentielles, II. Calcul intégral combinatoire, en préparation.

[LV2] P. Leroux, G. Viennot, Combinatorial resolution of systems of differential equations, IV. Separation of variables. En préparation.

[LV2] P. Leroux, G. Viennot, Combinatorial resolution of systems of differential equations, V. Non linear systems with forcing terms. En préparation.

[LO] A. Longtin, Une combinatoire non-commutative pour l'étude des nombres sécants. This volume.

[SCH] A. Schett, Recurrence formula of the Taylor series expansion coefficients of the Jacobian elliptic functions, Math. Comp., 31 (1977), 1003-1005.

[SC] M. P.Schützenberger, Solution non commutative d'une équation différentielle classique, in New concepts and technologies in parallel information processing, M. Caianiello Ed., NATO Advanced studies Institute Series, (E: Applied Sc.), Vol. 9 (1975), 381-401, Noordhoff, Leyden.

[ST] V. Strehl, Combinatorics of Jacobi configurations, I. Complete oriented matchings. This volume.

[TU] W. T. Tutte, Map colourings and differential equations, in Progress in Graph Theory, Bondy & Murty, ed., Academic Press, 1984.

[V1] G. Viennot, Quelques algorithmes de permutations, in Journées algorithmiques, Astérisque 38-39, S. M. F. Paris, 1976, 275-293.

[V2] G. Viennot, Une interprétation combinatoire des coefficients de développements en série entière des fonctions elliptiques de Jacobi, J. Combin. Th. (A), 29 (1980), 121-133.

[V3] G. Viennot, Interprétations combinatoires des nombres d'Euler et de Genocchi, Séminaire de théorie des nombres, Université de Bordeaux, 1980-1981, exposé no 11.

[YE] Y.N. Yeh, The calculus of virtual species and $\mathbb{K}$-species, This volume.

UNE COMBINATOIRE NON-COMMUTATIVE POUR L'ETUDE DES NOMBRES SECANTS

André Longtin
Département de mathématiques et d'informatique
Université du Québec à Trois-Rivières
C.P. 500, Trois-Rivières, G9A 5H7

There is a non-commutative combinatorial setting in which the up-down structure of permutations can most naturally be studied. In this setting the definition of various differential and integral operators and different types of substitution operations provides us with the tools and language needed to derive and express many of the laws governing these combinatorial structures.

Introduction. Si σ est une permutation de l'ensemble $\{1,2,..,n\}$ nous disons qu'elle est de spécification ascendante ou de type $\tau = (m_1,\dots,m_\ell)$ si $m_1 + \dots + m_\ell = n$ et $\sigma(1) < \dots < \sigma(m_1) > \sigma(m_1 + 1) < \dots < \sigma(m_1 + m_2) > \sigma(m_1 + m_2 + 1) < \dots < \sigma(m_1 + \dots + m_{\ell-1}) > \sigma(m_1 + \dots + m_{\ell-1} + 1) < \dots < \sigma(m_1 + \dots + m_\ell)$; le nombre de permutations de ce type est dénoté par $S((m_1, \dots, m_\ell))$. Il revient semble-t-il à Désiré André en 1879 d'avoir le premier démontré que la série génératrice exponentielle des nombres $S((2,\dots,2))$ est $\sec(x)$ et celle des nombres $S((2,\dots,2,1))$, $\tan(x)$. Depuis ce temps les nombres $S(\tau)$ ont été étudiés par plusieurs et de diverses façons (voir [5], [1], [3]) et nous pouvons dire qu'ils sont généralement bien connus. Dans cette étude nous adoptons cependant le point de vue que le contexte combinatoire naturel dans lequel ces objets doivent être étudiés est essentiellement non-commutatif et que le langage dans lequel leurs lois doivent être formulées est celui des séries $a(X,x)=\sum_{\tau,n} a_{\tau,n}X^\tau x^n/n!$ Le problème nous sert donc à la fois de raison et de guide pour jeter les premiers jalons d'une théorie de ces séries formelles.

Après avoir défini les opérations d'addition et de multiplication nous posons $\cos(X,x) = \sum_\tau (-1)^{\ell(\tau)} X^\tau x^{|\tau|}/|\tau|!$, où $\ell(\tau)$ dénote la longueur de τ et $|\tau|$ sa cardinalité, et nous démontrons que $\sec(X,x) = 1/\cos(X,x)$ est la série génératrice des permutations classées selon leur spécification ascendante. Nous développons ensuite le calcul différentiel et intégral des séries $a(X, x)$ et obtenons diverses formules de récurrence pour les nombres $S(\tau)$. A la section 6 nous étudions les séries de forme $a(X*Y,x)$, $a(X+Y,x)$ et $a(X-Y,x)$ afin d'obtenir des formules additives pour les nombres $S(\tau_1*\tau_2)$, $S(\tau_1+\tau_2)$ et $S(\tau_1-\tau_2)$, c'est-à-dire des égalités décrivant les relations existant entre eux et les nombres $S(\tau_1)$ et $S(\tau_2)$. Puis à la section 7 nous définissons une opération de substitution multivariée et l'utilisons pour sommer les nombres $S(\tau)$ selon plusieurs paramètres. Enfin cette étude ayant eu pour origine la recherche d'une preuve combinatoire pour les formules classiques $\sec(x+y) = \frac{\sec(x)\sec(y)}{1-\tan(x)\tan(y)}$ et $\tan(x+y) = \frac{\tan(x) + \tan(y)}{1-\tan(x)\tan(y)}$, nous démontrons à la section 8 que certaines identités trigonométriques ne sont en fait que la restriction aux permutations de type

(2,2,...,2) d'identités combinatoires générales, valides pour les permutations d'un type τ quelconque. Ainsi nous établissons l'identité

$$\sec(X, x + y) = \sec(X, y) \left(\frac{1}{1 - \sum_{r\geq 1} \Delta_{*r} \sec(X, x)_{-r}\Delta \sec(X, y)}\right) \sec(X, x)$$

et montrons qu'elle peut se prouver combinatoirement en interprétant $\sec(X, x + y)$ comme la série génératrice des permutations d'un ensemble bicoloré d'éléments classées selon leur spécification ascendante.

1. Les permutations de type $(m_1, \ldots, m_\ell)$. Soient $(E, \leq) = \{\ell_1 < \ldots < \ell_n\}$ un ensemble linéairement ordonné et σ une permutation de E; Si $m_1 + \ldots + m_\ell = n$ et $\{i : \sigma(\ell_i) > \sigma(\ell_{i+1})\} = \{m_1, m_1 + m_2, \ldots, m_1 + \ldots + m_{\ell-1}\}$, nous dirons que σ est de spécification ascendante $\tau = (m_1, \ldots, m_\ell)$ et nous écrirons $\sigma \in \mathcal{S}_E(\tau)$. Dans le cas particulier où $(E, \leq) = \{1 < 2 < \ldots < n\}$ nous écrirons simplement $\sigma \in \mathcal{S}(\tau)$ et $S(\tau)$ sera la cardinalité de $\mathcal{S}(\tau)$.

2. L'espace $(A, N^* \times N^+)$. N^* sera la monoïde libre engendré par $N = \{1,2,3,\ldots\}$, $\emptyset$ sera son élément neutre et ses autres éléments seront dénotés par $\tau = (m_1, \ldots, m_\ell)$ où $m_i \geq 1$ et $\ell \geq 1$. La longueur $\ell(\tau)$ de $\tau = (m_1, \ldots, m_\ell)$ sera ℓ et sa cardinalité $|\tau|$ sera $m_1 + \ldots + m_\ell$. Nous définissons aussi $\ell(\emptyset) = 0 = |\emptyset|$. Si $\tau_1 = (m_1, \ldots, m_k)$ et $\tau_2 = (m_1', \ldots, m_\ell')$ alors $\tau_1 * \tau_2$ représentera la concaténation $(m_1, \ldots, m_k, m_1', \ldots, m_\ell')$ de τ_1 et τ_2. N^+ pour sa part sera le monoïde des entiers $\{0,1,2,\ldots\}$ sous l'addition.

Si A est un anneau commutatif unitaire de caractéristique o, $(A, N^* \times N^+)$ sera l'ensemble des fonctions de $N^* \times N^+$ dans A et ses éléments seront représentés sous forme de suites par $a = (a_{\tau,n})_{\tau\in N^*, n\in N^+}$ et en séries formelles par

$a(X, x) = \sum_{\tau\in N^*} \sum_{n\in N^+} a_{\tau,n} X^\tau x^n/n!$ Géométriquement l'élément $\tau\in N^*$ sera décrit par une chaîne  et l'entier $n\in N$ par un ensemble linéairement ordonné (E) de cardinalité n. Une suite $a \in (A, N^* \times N^+)$ sera alors considérée comme étant une "sorte de structures" dont le poids sur

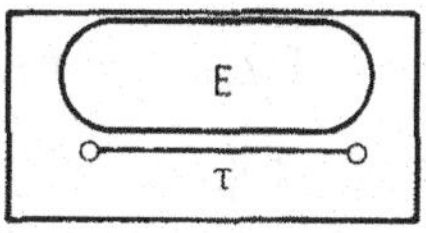

est $a_{\tau,|E|}$

3. L'addition et la multiplication dans $(A, N^* \times N^+)$. $\forall a,b \in (A, N^* \times N^+)$ on pose:

$$(a + b)_{\tau,n} = a_{\tau,n} + b_{\tau,n}$$

$$(a \cdot b)_{\tau,n} = \sum_{\substack{(n_1,n_2)\\ n_1+n_2=n}} \sum_{\substack{(\tau_1,\tau_2)\\ \tau_1 * \tau_2=\tau}} \binom{n}{n_1,n_2} a_{\tau_1,n_1} b_{\tau_2,n_2}, \quad \text{où} \binom{n}{n_1,n_2} = \frac{n!}{n_1!n_2!}$$

Proposition. $(A, N^* \times N^+)$ est un anneau unitaire non-commutatif. Il est intègre $\Leftrightarrow$ A est intègre et $a = (a_{\tau,n})$ est inversible $\Leftrightarrow a_{\emptyset,0}$ est inversible dans A.

Notation: 1) Les éléments neutres additif et multiplicatif de $(A, N^* \times N^+)$ seront dénotés par 0 et 1. 2) $\sum_{(n_1,n_2)=n}$ et $\sum_{(\tau_1,\tau_2)=\tau}$ remplaceront respectivement $\sum_{\substack{(n_1,n_2)\\ n_1+n_2=n}}$ et $\sum_{\substack{(\tau_1,\tau_2)\\ \tau_1*\tau_2=\tau}}$.

Ces opérations d'addition et de multiplication dans $(A, N^* \times N^+)$ s'appliquent naturellement aux représentations en séries formelles et nous avons

$a(X, x) + b(X, x) = (a + b)(X, x)$

$a(X, x)\, b(X, x) = (a \cdot b)(X, x)$

Ainsi l'addition s'effectue termes à termes et la multiplication s'obtient de la façon habituelle à la différence que $(a_{\tau_1,n_1} X^{\tau_1} x^{n_1}/n_1!)\,(b_{\tau_2,n_2} X^{\tau_2} x^{n_2}/n_2!) = a_{\tau_1,n_1} b_{\tau_2,n_2} X^{\tau_1*\tau_2} x^{n_1+n_2}/n_1!\, n_2!$.

Attention. $X^{\tau_1} X^{\tau_2} \neq X^{\tau_2} X^{\tau_1}$

Géométriquement nous écrirons

$(a+b)_{\tau,E} =$ 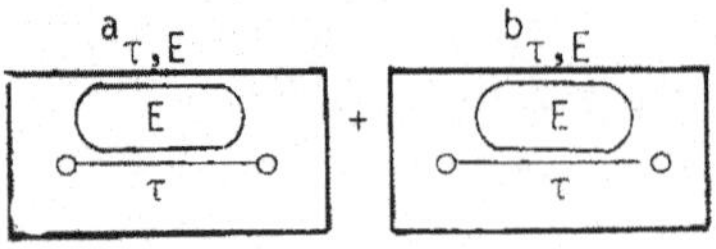 et

$(a\cdot b)_{\tau,E} = \sum$ 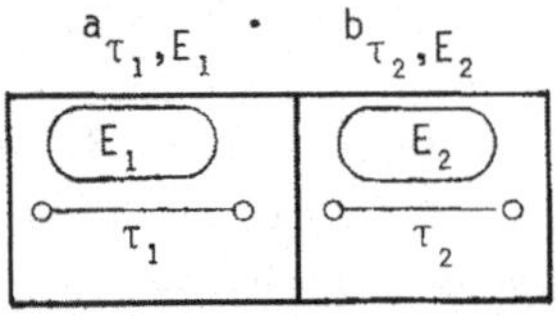 , où la somme s'effectue sur tous les par-

tages de E en deux sous-ensembles disjoints E_1 et E_2 que l'on munit des ordres linéaires induits par celui de E, et sur toutes les façons de séparer $\circ\!\!\underset{\tau=(m_1,\ldots,m_\ell)}{\text{———}}\!\!\circ$ en deux sous-chaînes $\circ\!\!\underset{\tau_1=(m_1,\ldots,m_i)}{\text{———}}\!\!\circ$ $\circ\!\!\underset{(m_{i+1},\ldots,m_\ell)=\tau_2}{\text{———}}\!\!\circ$.

4. Les séries cos(X, x) et sec(X, x). On définit $(\cos)_{\tau,n} = \begin{cases} (-1)^{\ell(\tau)}, & \text{si } |\tau| = n \\ 0, & \text{sinon} \end{cases}$

Donc $\cos(X, x) = \sum_{\tau} (-1)^{\ell(\tau)} X^{\tau} x^{|\tau|}/|\tau|!$. Puisque $(\cos)_{\emptyset,0} = 1$, cette suite est inversible dans $(A, N^* \times N^+)$ et nous posons $\sec = 1/\cos$, i.e. $\sec(X, x) = 1/\cos(X, x)$.

Théorème. sec(X, x) est la série génératrice des permutations classées selon leur spécification ascendante

i.e. $\mathcal{S}(X, x) = \sum_{\tau,n} |\mathcal{S}_{[n]}(\tau)|\, X^{\tau} x^{n}/n! = \sec(X, x)$.

Preuve: $\forall a \in (A, N^* \times N^+)$, si $a_{\emptyset,n} = \begin{cases} 1 & \text{si } n = 0 \\ 0 & \text{sinon} \end{cases}$ alors $1/a = \sum_{k\geq 0} (-1)^k (a-1)^k$ de sorte que la valeur de $\sec = 1/\cos = \sum_{k\geq 0} (-1)^k (\cos - 1)^k$ sur 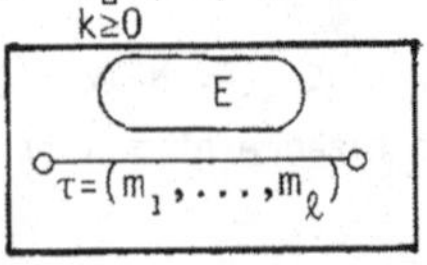 est

$\sum_{k\geq 0}(-1)^k \sum$ 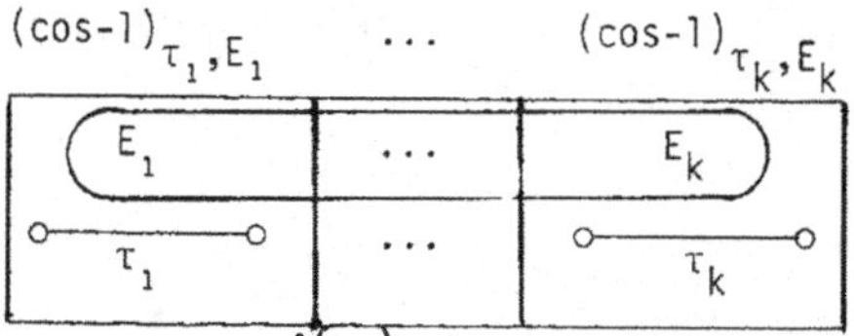

Puisque $(\cos - 1)_{\tau_i,E_i} = \begin{cases}(-1)^{\ell(\tau_i)}, & \text{si } |\tau_i| = |E_i| \neq o\\ 0, \text{ sinon}\end{cases}$

nous avons $(1/\cos)_{\tau,E} = 0$ si $|\tau| \neq |E|$, et si $|\tau| = |E|$ $(1/\cos)_{\tau,E} =$

$\sum_{k\geq 0}(-1)^k \sum_{|E_i|=|\tau_i|\neq 0}$ 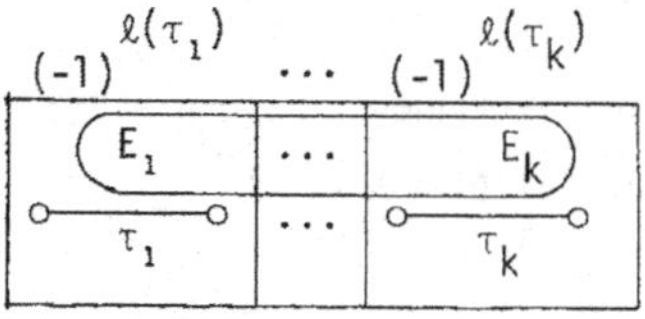 =

$(-1)^{\ell(\tau)} \sum_{k\geq 0}(-1)^k \left(\sum_{|E_i|=|\tau_i|\neq 0} \right.$ 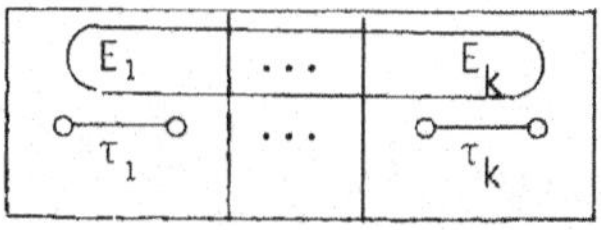 $\left. \right)$.

Or nous allons montrer que la plupart des termes de cette somme s'annulent deux à deux. En effet si

S = 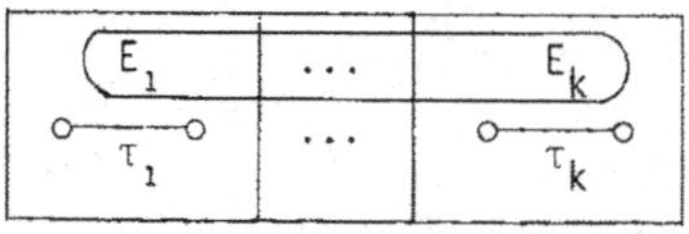

est une structure telle que

$|E_i| = |\tau_i| \neq 0\ \forall_i$ et $\ell(\tau_1) > 1$, alors on pose

$\Phi(S) =$ 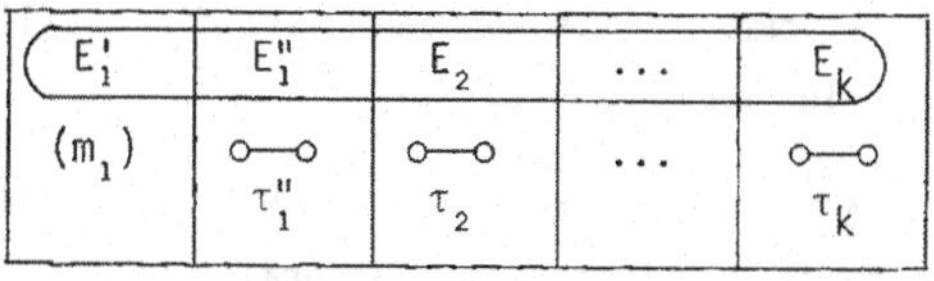

où $\tau_1 = (m_1) * \tau_1''$, E_1' est l'ensemble des <u>m_1 plus petits éléments de E_1</u> et E_1'' est son complément dans E_1. Nous obtenons de cette façon une bijection entre les structures satisfaisant $|E_i| = |\tau_i| \neq 0$ et $\ell(\tau_1) > 1$ d'une part et $|E_i| = |\tau_i| \neq 0$, $\ell(\tau_1) = 1$ et $\max(E_1) < \min(E_2)$ d'autre part. Cependant puisque dans la somme considérée S et $\Phi(S)$ seront respectivement affectées des coefficients $(-1)^k$ et $(-1)^{k+1}$ les poids de ces structures s'annulent mutuellement. Nous n'avons donc qu'à évaluer le poids des structures restantes, c'est-à-dire celles pour lesquelles $|E_i| = |\tau_i| \neq 0$, $\ell(\tau_1) = 1$ et $\max(E_1) > \min(E_2)$ que nous représenterons par

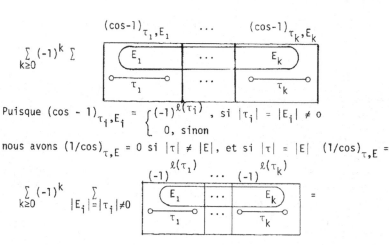

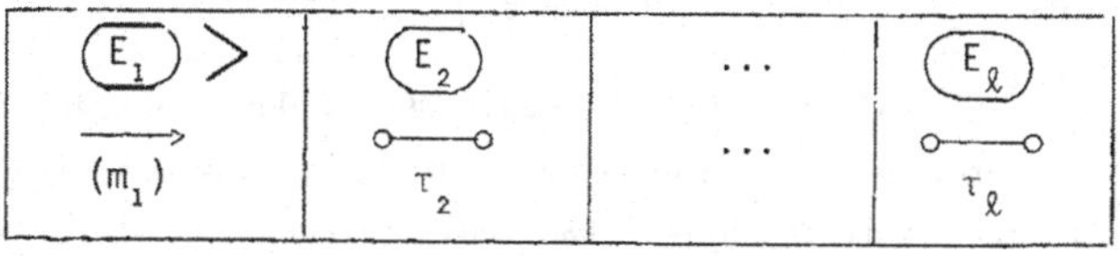

En appliquant

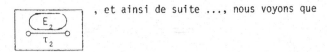

le même procédé au facteur 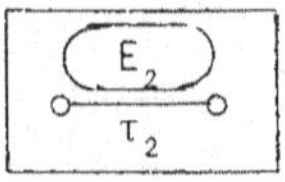, et ainsi de suite ..., nous voyons que

seules les structures de forme

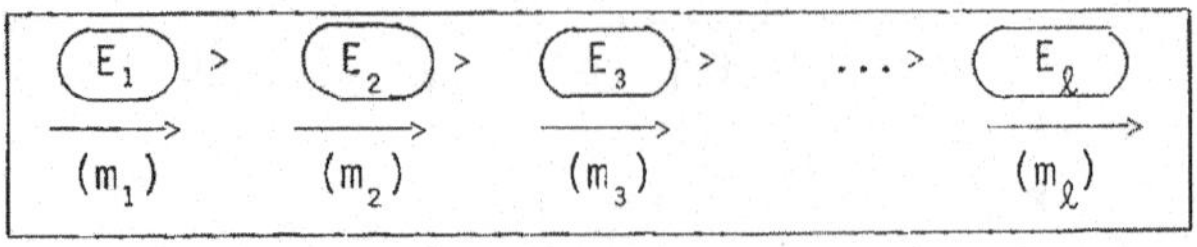

c'est-à-dire celles pour lesquelles $|E_i| = |\tau_i| \neq 0$, $\ell(\tau_i) = 1$ et $\max(E_i) > \min(E_{i+1})$, $\forall i$ sont effectivement comptées dans la somme. Puisqu'elles sont toutes des partages de E de longueur $\ell(\tau)$, chacune sera affectée du signe $(-1)^{\ell(\tau)}$ $(-1)^{\ell(\tau)}$ dans $(1/\cos)_{\tau,E}$ de sorte que $(\sec)_{\tau,E} = |S_E(\tau)|$.□

Corollaire. $\forall\ \tau \neq \emptyset$

$$S(\tau) = \sum_{k\geq 1} (-1)^{\ell(\tau)-k} \sum_{\substack{(\tau_1,\ldots,\tau_k)=\tau \\ \tau_i \neq \emptyset}} \binom{|\tau|}{|\tau_1|,\ldots,|\tau_k|}$$

où $\displaystyle\binom{|\tau|}{|\tau_1|,\ldots,|\tau_k|} = \frac{|\tau|!}{|\tau_1|!\ \ldots\ |\tau_k|!}$

Cette formule découverte et redécouverte par plusieurs au cours des années apparaît notamment dans [5]. Le langage que nous utilisons nous permet cependant de la traduire en une identité "analytique" simple, à savoir $\mathcal{S}(X,x) = 1/\cos(X,x)$. Puisque nous avons posé $\sec(X,x) = 1/\cos(X,x)$ il semblerait dorénavant justifié d'utiliser l'appellation "nombres sécants" pour tous les nombres $S(\tau)$.

5. Les opérateurs différentiels et intégraux dans $(A, N^* \times N^+)$. Dans l'espace $(A, N^* \times N^+)$ plusieurs opérateurs de type différentiel ou intégral peuvent être définis. Nous ne nous attarderons cependant qu'à ceux qui nous seront les plus utiles dans les sections suivantes pour écrire sous forme d'identités certaines lois auxquelles obéissent les nombres $S(\tau)$

Les dérivées par rapport à la variable x: $\forall r \geq 0$, $\forall a \in (A, N^* \times N^+)$ la dérivée r-ième de a par rapport à la variable x est la suite $d_r(a)$ définie par $(d_r a)_{\tau,n} = a_{\tau,n+r}$. Géométriquement $d_r(a)$ doit s'interpréter comme étant la suite dont le poids sur le couple 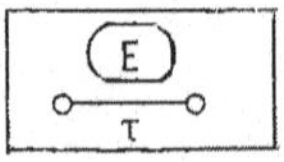 est celui de a sur le couple  où $E \oplus R$ est la réunion disjointe de E avec un ensemble linéairement ordonné R ayant r éléments, ces éléments étant tous réputés plus grands que ceux de E. Les opérateurs d_r sont essentiellement les opérateurs différentiels habituels. Cependant ils s'appliquent à des éléments d'un espace non-commutatif de sorte que certaines des formules usuelles s'en trouvent modifiées.

Proposition. $\forall a, a_1, \ldots, a_k$ et c inversible $\in (A, N^* \times N^+)$.

1) $d_{(r_1+r_2)}(a) = d_{r_1}\left(d_{r_2}(a)\right) = d_{r_2}\left(d_{r_1}(a)\right)$

2) $d_r(a_1 + a_2) = d_r(a_1) + d_r(a_2)$

3) $d_r(a_1 \cdots a_k) = \sum_{(r_1,\ldots,r_k)=r} \binom{r}{r_1,\ldots,r_k} d_{r_1}(a_1)\cdots d_{r_k}(a_k)$, $\forall r \geq 0$

4) $d_r(1/c) = \left[\sum_{k>0} (-1)^k \sum_{\substack{(r_1,\ldots,r_k)=r \\ r_i \neq 0}} \binom{r}{r_1,\ldots,r_k} \left(1/c \cdot d_{r_1}(c)\right) \ldots \left(1/c \cdot d_{r_k}(c)\right)\right] \cdot 1/c$, $\forall r > 0$

Preuve: Puisque $(d_r\, a)_{\tau,n} = a_{\tau,n+r}$ alors $\sum_{r\geq 0} (d_r\, a)\,(X,x)\, y^r/r! = a(X,x+y)$; on vérifie facilement que $(a\cdot b)(X,x+y) = a(X,x+y)b(X,x+y)$ de sorte que $d_r(a_1\ldots a_k)$ et $d_r(1/c)$ sont respectivement les coefficients de $y^r/r!$ dans $a_1(X,x+y)\ldots a_k(X,x+y)$ et $1/c(X,x+y) = \sum_{k\geq 0} (-1)^k \left[1/c(X,x) \cdot \left(c(X,x+y) - c(X,x)\right)\right]^k \cdot \left(1/c(X,x)\right)$ □

Corollaire. 1) $d_1(a\cdot b) = a \cdot d_1(b) + d_1(a) \cdot b$, 2) $d_1(a^k) = \sum_{(i,j)=k-1} a^i \cdot d_1(a) \cdot a^j$, 3) $d_1(1/c) = -1/c \cdot d_1(c) \cdot 1/c$.

Corollaire. $d_1 \sec(X,x) = -\sec(X,x)\, d_1\cos(X,x) \sec(X,x)$ d'où la récurrence

$$S(\tau) = -\sum_{\substack{(\tau_1,\tau_2,\tau_3)=\tau \\ \tau_2 \neq \emptyset}} \binom{|\tau| - 1}{|\tau_1|,\ |\tau_2| - 1,\ |\tau_3|} S(\tau_1)\, (-1)^{\ell(\tau_2)}\, S(\tau_3)\ .$$

Les dérivées et intégrales par rapport à la variable X: De façon générale toute fonction partiellement définie $N^* \xrightarrow{\delta} N^*$ se prolonge de façon naturelle en une transformation $(A, N^* \times N^+) \xrightarrow{\Delta_\delta} (A, N^* \times N^+)$ en posant:

$$(\Delta_\delta\, a)_{\tau,n} = \begin{cases} a_{\delta(\tau),n}\ , & \text{si } \delta(\tau) \text{ est définie} \\ 0\ , & \text{sinon} \end{cases}$$

C'est ainsi que $\forall r \geq 1$, en posant:

— $(*r)(\tau) = \tau * (r)$, $\forall \tau \in N^*$

— $(+r)(\tau) = \begin{cases} (m_1,\ldots,m_\ell + r), & \text{si } \tau = (m_1,\ldots,m_\ell),\ \ell \geq 1 \\ \text{Non définie si } \tau = \emptyset \end{cases}$

— $(-r)(\tau) = \begin{cases} (m_1,\ldots,m_\ell - r), & \text{si } \tau = (m_1,\ldots,m_\ell),\ \ell \geq 1 \text{ et } m_\ell > r \\ \text{Non définie ailleurs} \end{cases}$

— $(-\infty)(\tau) = \begin{cases} (m_1,\ldots,m_{\ell-1}), & \text{si } \tau = (m_1,\ldots,m_\ell),\ \ell > 1 \\ \emptyset \text{ si } \ell(\tau) = 1 \\ \text{Non définie ailleurs.} \end{cases}$

Nous obtenons quatre familles d'opérateurs que nous dénoterons respectivement par Δ_{*r} , Δ_{+r} , Δ_{-r} et $\Delta_{-\infty}$. Ces opérateurs obéissent aux lois suivantes.

Proposition. $\forall r \geq 1$, si $\{r\}$ dénote $*r$, $+r$, ou $-r$, alors $\forall$ a, b, c inversible $\in$ $(A, N^* \times N^+)$ nous avons

1) $\Delta_{\{r\}} (a + b) = \Delta_{\{r\}} (a) + \Delta_{\{r\}} (b)$.

2) $\Delta_{\{r\}} (a \cdot b) = a \cdot \Delta_{\{r\}}(b) + \Delta_{\{r\}} (a) \cdot b_\emptyset$

3) $\Delta_{\{r\}} (1/c) = - (1/c) \cdot \Delta_{\{r\}} (c) \cdot (1/c)_\emptyset$

4) $\Delta_{-\infty} (a + b) = \Delta_{-\infty} (a) + \Delta_{-\infty} (b)$

5) $\Delta_{-\infty} (a \cdot b) = a \cdot \Delta_{-\infty} (b)$

6) $\Delta_{-\infty} (1/c) = (1/c)^2 \cdot \Delta_{-\infty} (c)$

où en général $a_\emptyset$ dénote la suite définie par $(a_\emptyset)_{\tau,n} = \begin{cases} a_{\emptyset,n} & \text{si } \tau = \emptyset \\ 0 , & \text{si } \tau \neq \emptyset \end{cases}$

Remarque. Les produits apparaissant dans ces formules doivent être effectués dans l'ordre indiqué puisque la multiplication est non commutative. On peut cependant faire exception pour les suites "constantes" de type $a_\emptyset$ puisque $\forall a, b \in (A, N^* \times N^+)$, $a_\emptyset \cdot b = b \cdot a_\emptyset$

Corollaire. $\forall r \geq 1$, si $\{r\}$ dénote $*r$, $+r$ ou $-r$ alors:

$$\Delta_{\{r\}} \sec(X,x) = -\sec(X,x) \; \Delta_{\{r\}} \cos(X,x)$$

$$\Delta_{-\infty} \sec(X,x) = (\sec(X,x))^2 \; \Delta_{-\infty} \cos(X,x)$$

d'où les récurrences suivantes:

$$S(\tau * r) = \sum_{(\tau_1,\tau_2)=\tau} \binom{|\tau| + r}{|\tau_1|, \; |\tau_2| + r} S(\tau_1) \; (-1)^{\ell(\tau_2)} ,$$

$$S(\tau) = - \sum_{\substack{(\tau_1,\tau_2,\tau_3)=\tau \\ \tau_1\neq\tau,\tau_2\neq\tau}} \binom{|\tau|}{|\tau_1|, \; |\tau_2|, \; |\tau_3|} S(\tau_1) \; S(\tau_2) \; (-1)^{\ell(\tau_3)}$$

Preuve. La première récurrence peut s'obtenir de la première identité en prenant $\{r\} = *r$, $+r$ ou $-r$; la seconde s'obtient en évaluant $\Delta_{-\infty} \sec(X,x)$ en $(\tau * (m), |\tau|)$, où $m \neq 0$ □

Il est évidemment possible de combiner les divers opérateurs que nous avons définis pour analyser plus en détails la structure des nombres $S(\tau)$. Les deux applications suivantes serviront à l'illustrer mais aussi à faire valoir la précision du langage.

① Si l'on veut comparer les nombres $S(\tau * r)$ et $S(\tau + r)$ il suffit d'évaluer $\Delta_{*r} \sec(X,x)$ et $\Delta_{+r} \sec(X,x)$; sur la base de nos calculs on obtient immédiatement

$$\Delta_{*r} \sec(X,x) + \Delta_{+r} \sec(X,x) = \sec(X,x) \cdot x^r/r!$$

d'où l'égalité

$$S(\tau * r) + S(\tau + r) = \binom{|\tau| + r}{|\tau| \; , \; r} S(\tau)$$

② Proposition. $d_1 \sec(X,x) = (\Delta_{-1} \sec(X,x) + X^{(1)}) \sec(X,x)$.

<u>Preuve</u>. Nous savons déjà que $d_1\sec(X,x) = -\sec(X,x)\ d_1\cos(X,x)\ \sec(X,x)$; on vérifie directement que

$d_1\cos(X,x) = \Delta_{-1}\cos(X,x) - \cos(X,x)\ X^{(1)}$ de sorte que l'identité s'obtient en substituant cette formule dans la précédente et en se rappelant que $-\sec(X,x)\ \Delta_{-1}\cos(X,x) = \Delta_{-1}\sec(X,x)$ □

<u>Corollaire</u>. $\forall\ \ell \geq 1$

$$S(m_1,\ldots,m_\ell) = \begin{cases} \displaystyle\sum_{m_i>1} \binom{m_1+\ldots+m_\ell-1}{m_1+\ldots+m_i-1,\ m_{i+1}+\ldots+m_\ell} S(m_1,\ldots,m_i-1)\ S(m_{i+1},\ldots,m_\ell), \\ \text{si } m_1 \neq 1 \\ (\text{Expression précédente}) + S(m_2,\ldots,m_\ell)\ ,\ \text{si } m_1 = 1 \end{cases}$$

<u>Remarques</u>. 1) Il est facile d'interpréter géométriquement cette identité; la valeur de $d_1\sec(X,x)$ sur 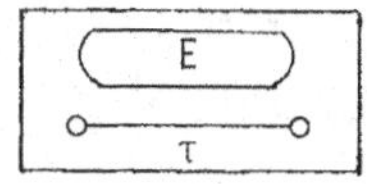

est celle de $\sec(X,x)$ sur 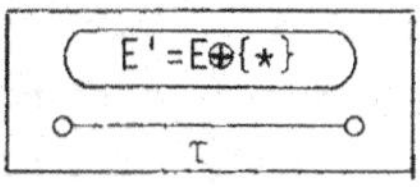

où * est le plus grand élément de E'. Or si σ est une permutation de type τ sur E' l'élément * sera ou bien le premier élément, et alors on aura

$$\begin{cases} \sigma = (\ * > \ .\ .\ .\ ) \\ \tau = (1)\ \ \circ\!\overset{}{\underset{\tau'}{\text{———}}}\!\circ \end{cases}$$

et σ sera de type $X^{(1)}\sec(X,x)$, ou bien il sera situé au sommet d'une montée

$$\begin{cases} \sigma = (\ .\ .\ .\ < * >\ .\ .\ .\ ) \\ \tau = \circ\!\underset{\tau_1}{\text{————}}\!\circ \quad \circ\!\underset{\tau_2}{\text{——}}\!\circ \end{cases},$$

dans lequel cas nous aurons une $\Delta_{-1}\sec(X,x)\ \sec(X,x)$ structure. Ceci montre que lorsque l'on a suffisamment apprivoisé le langage il est souvent possible d'écrire directement l'identité recherchée. Ainsi, obtenir $d_{+2}\sec(X,x)$ signifie déterminer où peuvent être situés les deux plus grands éléments $*_1 < *_2$ de E" dans une permutation de type τ ; en y réfléchissant quelques instants on trouve: $d_{+2}\sec(X,x) = 2\big(\Delta_{-1}\sec(X,x)\big)^2\sec(X,x) + \Delta_{-1}\sec(X,x)\ X^{(1)}\sec(X,x) + 2\ X^{(1)}\ \Delta_{-1}\sec(X,x)\ \sec(X,x) + X^{(1,1)}\sec(X,x) + X^{(2)}\sec(X,x) + \Delta_{-2}\sec(X,x)\ \sec(X,x)$ identité que l'on peut vérifier en calculant $d_1\big(d_1\sec(X,x)\big)$ au moyen de notre calcul différentiel.

2) La formule que nous avons obtenue pour $d_1\sec(X,x)$ nous procure une récurrence <u>non alternée</u> nous permettant de décrire $S(\tau)$ en termes des nombres $S(\tau')$ où $|\tau'| < |\tau|$. Mais elle est en fait à l'origine d'une famille complète de récurrences du même type que l'on obtient par spécialisation ou par sommation à travers une opération de substitution dont nous parlerons à la section 7. Ainsi par exemple on pourra vérifier qu'en substituant $(0,1,0,0,\ldots)$ à la variable X , l'identité précédente prend la forme bien connue $\frac{d}{dx}\sec(x) = \tan(x)\sec(x)$ et la récurrence devient

$E_{2n} = \sum_{(k,\ell)=n} \binom{2n-1}{2k,\,2\ell-1} E_{2k}\, E_{2\ell-1}$, où les nombres E_n sont maintenant les nombres d'Euler. De même en remplaçant X par (y,y,y,...) l'identité devient $\partial/\partial x\, A(x,y) = (A(x,y) - 1 + y)\, A(x,y)$, où $A(x,y) = (1 - y)/1 - y\, \ell^{x(1-y)}$ est la série génératrice exponentielle des polynômes eulériens et la récurrence devient alors

$$A_{n+1}(y) = (1 + y)\, A_n(y) + \sum_{\substack{(k,\ell)=n \\ k,\ell\neq 0}} \binom{n}{k,\ell} A_k(y)\, A_\ell(y)$$

On pourra voir en [6] jusqu'à quel point une récurrence relativement simple de forme semblable à celle obtenue ici peut engendrer des familles importantes de nombres, polynômes et q-analogues.

<u>Les opérateurs duaux</u>. Si nous revenons maintenant aux définitions des fonctions partielles $*r$, $+r$, $-r$ et $-\infty$ sur N^*, il est clair que nous pourrions tout aussi bien agir avec r à gauche sur les éléments $\tau \in N^*$ plutôt qu'à droite et obtenir ainsi des opérateurs duaux ${}_{r*}\Delta$, ${}_{r+}\Delta$, ${}_{-r}\Delta$ et ${}_{-\infty}\Delta$. En fait pour toute fonction partiellement définie $N^* \xrightarrow{\delta} N^*$ nous pouvons définir la <u>fonction duale</u> $N^* \xrightarrow{\overline{\delta}} N^*$ de δ en posant $\overline{\delta} = \sigma \circ \delta \circ \sigma$, où $N^* \xrightarrow{\sigma} N^*$ est définie par:

$$\sigma(\tau) = \begin{cases} (m_\ell,\ldots,m_1), \text{ si } \tau = (m_1,\ldots,m_\ell),\ \ell \geq 1 \\ \emptyset \ , \text{ si } \tau = \emptyset \end{cases}$$

La transformation $(A, N^* \times N^+) \xrightarrow{\Delta_{\overline{\delta}}} (A, N^* \times N^+)$ ainsi obtenue est appelée <u>duale</u> de Δ_δ et dénotée par ${}_\delta\Delta$.

<u>Lemme</u>. $\forall\ N^* \xrightarrow{\delta} N^*$ et $\forall\ a \in (A, N^* \times N^+)$ on a ${}_\delta\Delta(a) = \Delta_\sigma\big(\Delta_\delta\,(\Delta_\sigma(a)\big)$

<u>Proposition</u>. $\forall r \geq 1$, si $\{r\}$ dénote $r*$, $r+$ ou $-r$, alors $\forall a$, b , c inversible $\in (A, N^* \times N^+)$ on a:

1) ${}_{\{r\}}\Delta\,(a + b) = {}_{\{r\}}\Delta\,(a) + {}_{\{r\}}\Delta\,(b)$

2) ${}_{\{r\}}\Delta\,(a \cdot b) = {}_{\{r\}}\Delta\,(a) \cdot b + a_\emptyset \cdot {}_{\{r\}}\Delta\,(b)$

3) ${}_{\{r\}}\Delta\,(1/c) = -\,(1/c)_\emptyset \cdot {}_{\{r\}}\Delta\,(c) \cdot (1/c)$

4) ${}_{-\infty}\Delta\,(a + b) = {}_{-\infty}\Delta\,(a) + {}_{-\infty}\Delta\,(b)$

5) ${}_{-\infty}\Delta\,(a \cdot b) = {}_{-\infty}\Delta\,(a) \cdot b$

6) ${}_{-\infty}\Delta\,(1/c) = {}_{-\infty}\Delta\,(c) \cdot (1/c)^2$

<u>Preuve</u>. $N^* \xrightarrow{\sigma} N^*$ étant un anti-isomorphisme involutif, il en est de même de $(A, N^* \times N^+) \xrightarrow{\Delta_\sigma} (A, N^* \times N^+)$ relativement à la multiplication. Il suffit donc d'appliquer le lemme aux formules déjà obtenues pour les opérateurs Δ_δ . □

Ces opérateurs duaux seront utiles à la section 8 car ils apparaissent naturellement à l'intérieur de formules qui généralisent des identités trigonométriques bien connues.

6. Les séries $a(X * Y, x)$, $a(X + Y, x)$ et $a(X - Y, x)$. Afin d'être en mesure de traduire en identités certaines récurrences simples mais importantes concernant les nombres $S(\tau)$ il sera utile de considérer des séries à deux variables non-commutatives c'est à dire de type $a(X, Y, x) = \sum_{\tau_1,\tau_2,n} a_{\tau_1,\tau_2,n} X^{\tau_1} Y^{\tau_2} x^n/n!$. Les opérations d'addition et de multiplication correspondent évidemment à celles des suites $(a_{\tau_1,\tau_2,n})$ appartenant à l'espace $(A, N^* \times N^* \times N^+)$ et sont définies par

$$(a + b)_{\tau_1,\tau_2,n} = a_{\tau_1,\tau_2,n} + b_{\tau_1,\tau_2,n}$$

$$(a \cdot b)_{\tau_1,\tau_2,n} = \sum_{\substack{(\tau_1',\tau_1'') = \tau_1 \\ (\tau_2',\tau_2'') = \tau_2 \\ (n_1,n_2) = n}} \binom{n}{n_1,n_2} a_{\tau_1',\tau_2',n_1} b_{\tau_1'',\tau_2'',n_2}$$

Pour chaque suite $a \in (A, N^* \times N^+)$, les séries $\sum_{\tau,n} a_{\tau,n} X^\tau Y^\varnothing x^n/n!$ et $\sum_{\tau,n} a_{\tau,n} X^\varnothing Y^\tau x^n/n!$ seront comme à l'habitude dénotée simplement par $a(X,x)$ et $a(Y, x)$. Pour toute fonction partielle $N^* \xrightarrow{\delta} N^*$, les suites $\Delta_{(\delta,-)}(a)$ et $\Delta_{(-,\delta)}(a)$ seront respectivement définies par $\left(\Delta_{(\delta,-)}(a)\right)_{\tau_1,\tau_2,n} = \begin{cases} a_{\delta(\tau_1),\tau_2,n} , & \text{si } \delta(\tau_1) \text{ est définie} \\ 0 , & \text{sinon} \end{cases}$

et $\left(\Delta_{(-,\delta)}(a)\right)_{\tau_1,\tau_2,n} = \begin{cases} a_{\tau_1,\delta(\tau_2),n} , & \text{si } \delta(\tau_2) \text{ est définie} \\ 0 & \text{sinon} \end{cases}$, et $\Delta_{(\delta_1,\delta_2)}(a)$ dénotera $\Delta_{(\delta_1,-)}\left(\Delta_{(-,\delta_2)}(a)\right) = \Delta_{(-,\delta_2)}\left(\Delta_{(\delta_1,-)}(a)\right)$. Par contre si $N^* \times N^* \xrightarrow{\delta} N^*$ est une fonction partiellement définie nous utiliserons la notion $(A,N^* \times N^+) \xrightarrow{\nabla_\delta} (A, N^* \times N^* \times N^+)$ pour désigner la transformation définie par

$$(\nabla_\delta a)_{\tau_1,\tau_2,n} = \begin{cases} a_{\delta(\tau_1,\tau_2),n} , & \text{si } \delta(\tau_1,\tau_2) \text{ est définie} \\ 0 , & \text{sinon} \end{cases}$$

C'est le cas en particulier pour la concaténation $N^* \times N^* \xrightarrow{*} N^*$, et la suite $\nabla_*(a)$ alors obtenue s'écrit en série formelle sous la forme $\nabla_*(a)(X,Y,x) = \sum_{\tau_1,\tau_2,n} a_{\tau_1 * \tau_2,n} X^{\tau_1} Y^{\tau_2} x^n/n!$. Nous utiliserons cependant une notation plus suggestive pour la désigner; nous écrirons $a(X * Y, x)$.

Relativement aux opérations et opérateurs déjà définis cette transformation possède les propriétés suivantes:

Proposition. $\forall a, b, c$ inversible $\in (A, N^* \times N^+)$

1) $(a + b)(X * Y, x) = a(X * Y, x) + b(X * Y, x)$
2) $(a \cdot b)(X * Y, x) = a(X, x)\, b(X * Y, x) + a(X * Y, x)\, b(Y, x) - a(X, x)\, b(Y, x)$
3) $(1/c)(X * Y, x) = 1/c(X,x) + 1/c(Y,x) - \left(1/c(X,x)\right) c(X * Y, x) \left(1/c(Y, x)\right)$

4) $\forall\delta = +r, -r$ et $-\infty$

$$(\Delta_\delta\, a)\,(X * Y,\ x) = \Delta_{(-,\delta)}\, a(X*Y,\ x) + \Delta_\delta\, a(X,x)$$

5) $\forall r \geq 1, (\Delta_{*r}\, a)(X*Y,\ x) = \Delta_{(-,*r)}\, a(X*Y,\ x)$

Corollaire. $\sec(X*Y,\ x) = \sec(X,x) + \sec(Y,x) - \sec(X,x)\cos(X*Y,\ x)\sec(Y,x)$

d'où $\forall\ \tau_1, \tau_2 \neq \emptyset$

$$S(\tau_1 * \tau_2) = -\sum_{\substack{(\tau_1',\tau_1'') = \tau_1 \\ (\tau_2',\tau_2'') = \tau_2}} \binom{|\tau_1 * \tau_2|}{|\tau_1'|,\ |\tau_1'' * \tau_2'|,\ |\tau_2''|} S(\tau_1')\ (-1)^{\ell(\tau_1'' * \tau_2')} S(\tau_2'')$$

Applications.

① $\Delta_{(*r,-)} \sec(X*Y,\ x) + \Delta_{(-,r+)} \sec(X*Y,\ x) = \Delta_{*r} \sec(X,x)\sec(Y,x)$

d'où $\forall r \geq 1$

$$S(\tau_1 * r * \tau_2) + S(\tau_1 * (r + \tau_2) = \binom{|\tau_1| + r + |\tau_2|}{|\tau_1| + r,\ |\tau_2|} S(\tau_1 * r)\ S(\tau_2)$$

Ceci est essentiellement la récurrence que MacMahon nomme son <u>théorème de multiplication</u> [5] .

② Définissons $\Delta_{*\tau}(a) = \begin{cases} a\ , \text{ si } \tau = \emptyset \\ \Delta_{*m_\ell}\left(\cdots\left(\Delta_{*m_1}(a)\right)\right), \text{ si } \tau = (m_1,\ldots,m_\ell) \end{cases}$

alors

Proposition $\forall a, b, c$ inversible $\in (A, N^* \times N^+)$

1) $$\Delta_{*\tau}(a \cdot b) = a \cdot \Delta_{*\tau}(b) + \sum_{\substack{(\tau_1,\tau_2)=\tau \\ \tau_1 \neq \emptyset}} \Delta_{*\tau_1}(a) \cdot \left(\Delta_{*\tau_2}(b)\right)_\emptyset$$

2) $$\Delta_{*\tau}(1/c) = -\sum_{\substack{(\tau_1,\tau_2)=\tau \\ \tau_1 \neq \emptyset}} (1/c) \cdot \Delta_{*\tau_1}(c) \cdot \left(\Delta_{*\tau_2}(1/c)\right)_\emptyset$$

Preuve. Puisque $a(X*Y,\ x) = \sum_\tau \Delta_{*\tau} a(X,x)\, Y^\tau$, $\Delta_{*\tau}(a \cdot b)$ est donc le coefficient de Y^τ dans $(a \cdot b)\,(X*Y,\ x)$. □

Les séries $a(X+Y,x)$ et $a(X-Y,x)$ sont obtenues de façon tout-à-fait analogue à celle utilisée pour la série $a(X*Y,x)$, à partir des fonctions $N^* \times N^* \xrightarrow{+} N^*$ et $N^* \times N^* \xrightarrow{-} N^*$ définies par:

$$\tau_1 + \tau_2 = \begin{cases} (m_1,\ldots,m_k + m_1',\ m_2',\ldots,m_\ell')\ , \text{ si } \tau_1 = (m_1,\ldots,m_k),\ \tau_2 = (m_1',\ldots,m_\ell'), \\ \quad k \geq 1,\ \ell \geq 1 \\ \text{Non définie, ailleurs} \end{cases}$$

$$\tau_1 - \tau_2 = \begin{cases} (m_1,\ldots,m_k - m_1', m_2', \ldots, m_\ell'), \text{ si } \tau_1 = (m_1,\ldots,m_k),\ \tau_2 = (m_1',\ldots,m_\ell'), \\ \quad m_k > m_1' \text{ et } k \geq 1,\ \ell \geq 1 \\ \text{Non définie ailleurs.} \end{cases}$$

On montre:

Proposition. $\forall a, b, c$ inversible $\in (A, N^* \times N^+)$

1) $(a + b)(X \pm Y, x) = a(X \pm Y, x) + b(X \pm Y, x)$

2) $(a \cdot b)(X \pm Y, x) = a(X,x)\, b(X \pm Y, x) + a(X \pm Y, x)\, b(Y, x)$

3) $(1/c)(X \pm Y, x) = -\big(1/c(X,x)\big) c(X \pm Y, x) \big(1/c(Y, x)\big)$

Corollaire. $\sec(X \pm Y, x) = -\sec(X, x) \cos(X \pm Y, x) \sec(Y, x)$

Application.

Proposition. $\sec(X + Y, x) = \sec(X, x) \sec(Y, x) - \sec(X \star Y, x)$

d'où $\forall \tau_1, \tau_2 \neq \emptyset$

$$S(\tau_1 + \tau_2) = \binom{|\tau_1| + |\tau_2|}{|\tau_1|, |\tau_2|} S(\tau_1)\, S(\tau_2) - S(\tau_1 \star \tau_2)$$

Preuve. On vérifie directement que $\cos(X + Y, x) = \cos(X, x) + \cos(Y, x) - \cos(X \star Y, x) - 1$. □

Nous retrouvons donc à nouveau le théorème de multiplication de MacMahon. La définition des opérateurs $\nabla_\star$ et ∇_+ nous aura cependant permis d'en obtenir une formulation analytique à la fois simple, précise et élégante.

7. La substitution multivariée. Sur la base de l'égalité $X^{(m_1,\ldots,m_\ell)} = X^{(m_1)} \ldots X^{(m_\ell)}$ les séries $a(X, x)$ peuvent être considérées comme étant des séries à une infinité de variables indépendantes $X^{\emptyset}, X^{(1)}, X^{(2)}, \ldots$ et une opération de substitution consistant essentiellement à remplacer les variables $X^{(m)}$ par des séries $b_m(Y)$ et $X^{\emptyset}$ par Y peut être définie. De façon plus précise si I est un monoïde pour lequel $\{(i_1, i_2): i_1 i_2 = i\}$ est fini $\forall i \in I$ et (A, I) dénote l'ensemble des fonctions de I dans l'anneau commutatif unitaire A muni des opérations d'addition et de multiplication

$$(a + b)_i = a_i + b_i$$

$$(a \cdot b)_i = \sum_{(i_1, i_2) = i} a_{i_1} b_{i_2}$$

alors les éléments $c \in (A, I)$ peuvent être formellement représentés par $c(Y) = \sum_{i \in I} c_i\, Y^i$. Si $a \in (A, N^* \times N^+)$, et si $\{c_m\}_{m=1,2,\ldots}$ est une suite d'éléments de (A, I) pour laquelle la série

$$\sum_n a_{\emptyset,n}\, Y^{i_0}\, x^n/n! + \sum_n \left(\sum_{\ell \geq 1} \sum_{(m_1,\ldots,m_\ell)} a_{(m_1,\ldots,m_\ell)} c_{m_1}(Y) \ldots c_{m_\ell}(Y) \right) x^n/n!$$

est sommable, i_0 étant l'élément neutre de I, alors on dira qu'elle a été obtenue par substitution de $(c_1, c_2, \ldots)$ à X dans $a(X, x)$ et on la dénotera par $a(c_1, c_2, \ldots; x)$.

Proposision. $\forall a, b \in (A, N^* \times N^+)$, $\forall\{c_m\} \subseteq (A, I)$, sous les conditions de sommabilité appropriées, nous avons

1) $(a + b)(c_1, c_2, \ldots; x) = a(c_1, c_2, \ldots; x) + b(c_1, c_2, \ldots; x)$
2) $(a \cdot b)(c_1, c_2, \ldots; x) = a(c_1, c_2, \ldots; x)\, b(c_1, c_2, \ldots; x)$

Corollaire. Sous les conditions de sommabilité appropriées

$$\sec(c_1, c_2, \ldots; x) = 1/\cos(c_1, c_2, \ldots; x).$$

Cette opération de substitution est fort utile car elle nous permet de sommer les permutations selon divers paramètres. En voici quelques exemples.

Applications.

① Sur la base de nos définitions, $\sec(0, 0, \ldots, 0, \underset{k}{1}, 0, 0, \ldots; x) = \sum_{n\geq 0} S(\underbrace{k, k, \ldots, k}_{n})\, x^{nk}/(nk)!$ est la série génératrice des permutations de type $(k, k, \ldots, k)$. Puisque $\cos(0, \ldots, 0, \underset{k}{1}, 0, 0, \ldots; x) = \sum_{n\geq 0} (-1)^k\, x^{nk}/(nk)!$ nous écrirons $\cos(0, \ldots, 0, \underset{k}{1}, 0, 0, \ldots; x) = \cos_k(x)$ et $\sec(0, \ldots, 0, \underset{k}{1}, 0, 0, \ldots; x) = \sec_k(x) = 1/\cos_k(x)$. En particulier lorsque $k = 2$ nous retrouvons les fonctions trigonométriques usuelles $\cos(x)$ et $\sec(x)$ et, par le fait même, le résultat bien connu que $\sec(x)$ est la série génératrice exponentielle des permutations de type $(2, 2, \ldots, 2)$, dites alternées commençant par une montée.

② Si nous voulons sommer les permutations dont la spécification ascendante ne contient que des multiples de k, nous devons calculer $\sec(0, \ldots, 0, \underset{k}{1}, 0, \ldots, 0, \underset{2k}{1}, 0, \ldots, 0, \underset{3k}{1}, 0, \ldots; x)$; or on établit facilement que $\cos(0, \ldots, 0, \underset{k}{1}, 0, \ldots, 0, \underset{2k}{1}, 0, \ldots; x) = 1 - x^k/k!$ de sorte que $\sum_n \left(\sum_{\ell\geq 0} \sum_{(n_1,\ldots,n_\ell)=n} S(n_1 k, \ldots, n_\ell k) \right) x^{nk}/(nk)! = 1/(1 - x^k/k!)$ et $\sum_{\ell\geq 1} \sum_{(n_1,\ldots,n_\ell)=n} S(n_1 k, \ldots, n_\ell k) = \binom{nk}{k, k, \ldots, k}$

③ Si nous voulons sommer les permutations dont la spécification ascendante ne contient aucun entier égal à 1, on doit calculer $\sec(0, 1, 1, 1, \ldots; x)$; nous obtenons

$$\sum_{n\geq 0} \left(\sum_{\substack{\tau, |\tau|=n \\ m_i\neq 1}} |S(\tau)| \right) x^n/n! = \sec_3(x)/\left(1 - \tan_3^{(1)}(x)\right)$$

où $\tan_3^{(1)}(x) = -\frac{d}{dx}\cos_3(x)/\cos_3(x)$

Preuve. $\cos(0, 1, 1, 1, \ldots; x) = \sum_n \left(\sum_{\substack{\tau, |\tau|=n \\ m_i\neq 1}} (-1)^{\ell(\tau)} \right) x^n/n!$.

Or si $\tau \neq (3, 3, \ldots, 3)$ alors il existe un i tel que $\tau = (3, \ldots, 3, m_i, \ldots, m_\ell)$ où $m_i = 2$ ou $m_i \geq 4$. Dans le premier cas, si m_{i+1} existe on pose $\Phi(\tau) = (3, \ldots,$

$\underset{i-1}{3}, \underset{i}{2+m_{i+1}}, m_{i+2}, \dots, m_\ell)$ et dans le second on définit $\Phi(\tau) = (3, \dots, 3, \underset{i}{2}, m_i-2, m_{i+1}, \dots, m_\ell)$, de sorte que dans les deux cas $|\ell(\tau) - \ell(\Phi(\tau))| = 1$. Leurs effets s'annulent donc dans cos $(0,1,1,1,\dots; x)$ d'où $\cos(0,1,1,1,\dots;x) = \sum_{n\geq 0} (-1)^n x^{3n}/(3n)! + \sum_{n\geq 0} (-1)^{n+1} x^{3n+2}/(3n+2)! = \cos_3(x) + \frac{d}{dx}\cos_3(x)$. De là on obtient l'égalité

$$\sum_{\substack{\tau=(m_1,\dots,m_\ell)\\ m_i\neq 1, |\tau|=n}} S(\tau) = \sum_{r\geq 0} \sum_{\substack{3k_0+(3k_1+2)+\dots+(3k_r+2)=n\\ k_i\geq 0}} \binom{n}{3k_0, 3k_1+2, \dots, 3k_r+2} \cdot S(\underbrace{3,\dots,3}_{k_0})\; S(\underbrace{3,\dots,3,2}_{k_1}) \dots S(\underbrace{3,\dots,3,2}_{k_r}) \quad \square$$

④ Pour sommer les permutations dont la longueur des "ascensions" est inférieure à k nous devons évaluer sec(1, 1, ..., 1, 0, 0, ...; x). Or

$$\cos(1, 1, \dots, 1, \underset{k}{0}, 0, \dots; x) = \sum_{n\geq 0} x^{nk}/(nk)! - \sum_{n\geq 0} x^{nk+1}/(nk+1)!$$

de sorte que $\sum_{n} \sum_{\substack{\tau\\ m_i<k\\ |\tau|=n}} S(\tau)\, x^n/n! = \text{sech}_k(x)/1 + \tanh_k^{(k-1)}(x)$

où $\cosh_k(x) = \sum_{n\geq 0} x^{nk}/(nk)!$, $\text{sech}_k(x) = 1/\cosh_k(x)$

et $\tanh_k^{(k-1)}(x) = \frac{d^{(k-1)}}{dx^{(k-1)}} \cosh_k(x)/\cosh_k(x)$

<u>Preuve</u>. Si $\tau \neq (1, k-1, 1, k-1, \dots, 1, k-1)$ et $\tau \neq (1, k-1, \dots, 1, (k-1), 1)$ alors ou bien $\tau = (1, k-1, \dots, 1, k-1, m_i, \dots, m_\ell)$ où $m_i \neq 1$ et $m_j < k$ ou bien $\tau = (1, k-1, \dots, 1, k-1, 1, m_i, \dots, m_\ell)$ où $o < m_i < k-1$ et $m_j < k$. Dans le premier cas on pose $\Phi(\tau) = (1, k-1, \dots, 1, k-1, 1, m_i-1, m_{i+1}, \dots, m_\ell)$ et dans le second $\Phi(\tau) = (1, k-1, \dots, 1, k-1, 1+m_i, m_{i+1}, \dots, m_\ell)$. Puisque $|\ell(\Phi(\tau)) - \ell(\tau)| = 1$ leurs effets s'annulent dans cos$(1, 1, \dots, 1, \underset{k}{0}, 0, \dots; x)$. □

⑤ sec(y, y, y, ...; x) est évidemment la série génératrice des permutations classées selon la longueur de leur spécification ascendante. Nous retrouvons ainsi un résultat bien connu (voir [1]):

$$\sum_n \sum_\ell \sum_{\substack{|\tau|=n\\ \ell(\tau)=\ell}} S(\tau)\; y^\ell x^n/n! = (1-y)/1-ye^{x(1-y)}$$ = la série génératrice exponentielle des polynômes eulériens.

<u>Preuve</u>. Puisque $\sum_{\substack{|\tau|=n\\ \ell(\tau)=\ell}} (-1)^{\ell(\tau)} = \binom{n-1}{\ell-1} (-1)^\ell$ on obtient cos(y, y, ...; x) = $1-ye^{x(1-y)}/1-y$. □

(6) Si nous voulons sommer les permutations σ selon leur nombre $m(\sigma)$ de montées (i.e. le nombre d'éléments i tels que $\sigma(i) < \sigma(i+1)$). Nous devons calculer $\sec(1, y^1, y^2, y^3, \ldots; x)$. Nous obtenons:

$$\sum_n \sum_m \Big(\sum_{\substack{|\tau|=n \\ m(\tau)=m}} S(\tau) \Big) y^m x^n/n! = (y-1)/y-e^{(y-1)x}$$

Preuve. $\cos(1, y^1, y^2, y^3, \ldots; x) = \sum_n \sum_m \Big(\sum_{\substack{|\tau|=n \\ \ell(\tau)=n-m}} (-1)^{\ell(\tau)} \Big) y^m x^n/n! = \Big(\cos(y, y, y, \ldots; x)\Big)(1/y, yx)$. □

(7) Parmi les transformations qui nous permettent de passer d'un contexte non-commutatif à un contexte commutatif il y a celle qui consiste à remplacer la variable X par $(x_1, x_2, x_3, \ldots)$ où $\{x_i\}_{i=1,2,\ldots}$ est une suite de variables commutatives indépendantes. Or si nous définissons le genre d'un type τ comme étant $g(\tau) = (\alpha_1, \alpha_2, \alpha_3, \ldots)$ où α_m = nombre de fois que m apparaît dans τ, nous aurons

$$\sec(x_1, x_2, x_3, \ldots; 1) = \sum_{\alpha=(\alpha_i)} \Bigg(\sum_{\substack{\tau \\ g(\tau)=\alpha}} S(\tau) \Bigg) x_1^{\alpha_1} x_2^{\alpha_2} \ldots/(\alpha_1+2\alpha_2+\ldots)!$$

et ainsi $\sec(x_1, x_2, \ldots; 1)$ est la série génératrice des permutations classées par genre d'ascendance. Puisque $\sec(x_1, x_2, x_3, \ldots; 1) = 1/\cos(x_1, x_2, \ldots; 1)$, nous obtenons l'identité suivante:

$$\sum_{\substack{\tau \\ g(\tau)=\alpha}} S(\tau) = (-1)^{|\alpha|} \sum_{k\geq 0} (-1)^k \sum_{(\beta_1,\ldots,\beta_k)=\alpha} \binom{|\beta_1|}{(\beta_1)} \cdots \binom{|\beta_k|}{(\beta_k)} \binom{\|\alpha\|}{\|\beta_1\|,\ldots,\|\beta_k\|}$$

où $|\alpha| = \alpha_1 + \alpha_2 + \alpha_3 + \ldots$, $\|\alpha\| = \alpha_1 + 2\alpha_2 + 3\alpha_3 + \ldots$ et $\binom{|\beta|}{(\beta)} = \binom{\beta_1 + \beta_2 + \ldots}{\beta_1, \beta_2, \ldots}$

(8) Nous pouvons évidemment utiliser la substitution conjointement avec les opérateurs que nous avons définis dans $(A, N^* \times N^+)$. Ainsi, sommer les permutations dont la spécification ascendante commence par τ_0, c'est à dire calculer $\sum_{\substack{\tau \\ |\tau_0*\tau|=n}} S(\tau_0*\tau)$ revient à évaluer la série $\sec(X*Y,x)\,\big(X, (1,1,1,\ldots)\big)$ en (τ_0,n). Or

Proposition. $\sec(X*(1,1,\ldots), x) = \sec(X,x) + \Delta_{*1}\sec(X,x)/(1-x)$.

Preuve. Nous savons que $\sec(X*Y, x) = \sec(X,x) + \sec(Y,x) - \sec(X,x)\cos(X*Y,x)\sec(Y,x)$. En substituant $(1,1,1,\ldots)$ à Y de chaque côté de l'égalité et en vérifiant que $\cos(Y,x)(1,1,\ldots) = 1-x$ et $\cos(X*(1,1,\ldots), x) = \Delta_{*1} \cos(X,x) + \cos(X,x)$ nous obtenons l'identité ci-haut . □

Corollaire. $$\sum_{\substack{\tau \\ |\tau_0*\tau|=n}} S(\tau_0*\tau) = \begin{cases} S\big(\tau_0*(1)\big).\, n!/(|\tau_0|+1)!\ , & \text{si } |\tau_0| < n \\ S(\tau_0)\ , & \text{si } |\tau_0|=n \end{cases}$$

8. Quelques identités trigonométriques. Nous avons vu à la section précédente que les fonctions trigonométriques usuelles cos(x) et sec(x) s'obtiennent des séries cos(X, x) et sec(X, x) par substitution de (0,1,0,0,...) à X. Il était donc normal de chercher à savoir si certaines identités trigonométriques classiques proviennent d'identités plus généralement valides pour les séries cos(X, x) et sec(X, x). Dans cette section nous verrons que la réponse est affirmative.

De façon générale nous poserons $a(X, -x) = \sum_{\tau,n} a_{\tau,n} (-1)^n X^\tau x^n/n!$ et $a(X, x+y) = \sum_{\tau,n_1,n_2} a_{\tau,n_1+n_2} X^\tau x^{n_1} y^{n_2}/n_1!n_2!$. Cette notation se justifie évidemment par le fait que les séries en question sont celles que l'on obtient en substituant -x et x + y à la variable x dans a(X, x)

Théorème. $\cos(X, x) \cos(X, -x) - \sum_{r\geq 1} \Delta_{*r} \cos(X, x) \; {}_{-r}\Delta \cos(X, -x) = 1$

Preuve. L'égalité est facilement vérifiée pour $\tau = \emptyset$. Si $\tau \neq \emptyset$, les structures S dont nous devons calculer le poids dans le terme de gauche de l'égalité ont la forme générale suivante

S =

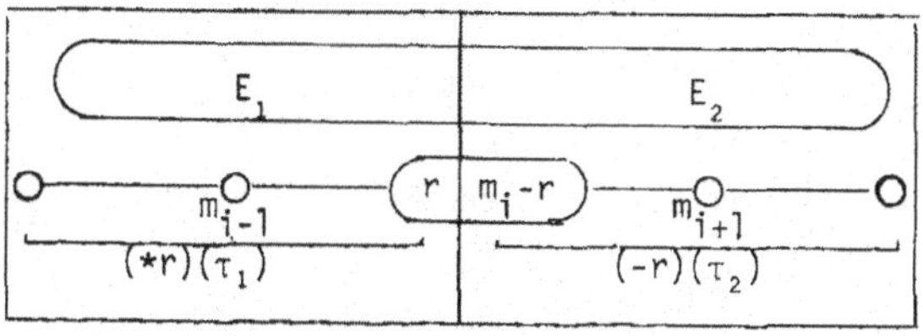

$$\mathrm{pds}(S) = \alpha \cdot \big(\cos(X, x)\big)_{\tau_1 * r, |E_1|} \cdot \big(\cos(X, -x)\big)_{(-r)+\tau_2, |E_2|}$$

où $\alpha = +1$ si $r = 0$ et -1 si $r \geq 1$. En raison de nos définitions, les structures S dont le poids est non nul sont celles pour lesquelles

$\begin{cases} 1 \leq i \leq \ell \\ 0 < r < m_i \end{cases}$ ou $\begin{cases} 0 \leq i \leq \ell \\ r = 0 \end{cases}$, et $\begin{cases} |E_1| = |\tau_1| + r \\ |E_2| = |\tau_2| - r \end{cases}$

Etant donné que $(1)_{\tau,n} = 0$ lorsque $\tau \neq \emptyset$, l'égalité sera établie si nous montrons que le poids total de ces structures est 0. Pour ce faire il suffit de vérifier que l'opération par laquelle on transfère le plus grand élément $*$ de E_1 vers E_2 ou de E_2 vers E_1 selon qu'il se trouve dans E_1 ou E_2, définit une involution Φ sur l'ensemble de ces structures pour laquelle on a $\mathrm{pds}\, \Phi(S) = -\mathrm{pds}(S)$. □

Corollaire. $\sec(X, x) \sec(X, -x) + \sum_{r\geq 1} \Delta_{*r} \sec(X, x) \; {}_{-r}\Delta \sec(X, -x) = 1$

Preuve. Il suffit de multiplier les termes apparaissant de chaque côté de l'égalité précédente, à gauche par sec(X, x) et à droite par sec(X, -x), et utiliser les formules déjà obtenues pour $\Delta_{*r} \sec(X, x)$ et ${}_{-r}\Delta \sec(X, x)$. □

Pour établir la prochaine identité nous devons considérer les séries de type

$$a(X, x, y) = \sum_{\tau,n_1,n_2} a_{\tau,n_1,n_2} X^{\tau} \frac{x^{n_1}}{n_1!} \frac{y^{n_2}}{n_2!}$$

Nous nous situons donc dans l'espace $(A, N^* \times N^+ \times N^+)$ où l'addition est définie termes à termes et la multiplication par

$$(a \cdot b)_{\tau,n_1,n_2} = \sum_{\substack{(\tau_1,\tau_2)=\tau \\ (n_1',n_1'')=n_1 \\ (n_2',n_2'')=n_2}} \binom{n_1}{n_1',n_1''} \binom{n_2}{n_2',n_2''} a_{\tau_1 n_1' n_2'} b_{\tau_2 n_1'' n_2''} .$$

A nouveau, dans ce contexte, les séries correspondant respectivement aux suites qui satisfont $a_{\tau,n_1,n_2} = 0$ si $n_2 \neq 0$ et $a_{\tau,n_1,n_2} = 0$ si $n_1 \neq 0$ seront dénotées simplement par $a(X, x)$ et $a(X, y)$.

<u>Théorème</u>. $\cos(X, x + y) = \cos(X, x) \cos(X, y) - \sum_{r\geq 1} \Delta_{*r} \cos(X, x) \;{}_{-r}\Delta \cos(X, y)$.

<u>Preuve</u>. $\forall \tau$, n_1, n_2 nous avons

$$\big(\cos(X, x + y)\big)_{\tau,n_1,n_2} = \begin{cases} (-1)^{\ell(\tau)} & \text{si } |\tau| = n_1 + n_2 \\ 0 & \text{sinon} \end{cases}$$

D'autre part

$$\big(\cos(X, x) \cdot \cos(X, y)\big)_{\tau,n_1,n_2} = \sum_{(\tau_1,\tau_2)=\tau} \big(\cos(X, x)\big)_{\tau_1,n_1,0} (\cos X, y)_{\tau_2,0,n_2}$$

$$= \begin{cases} (-1)^{\ell(\tau_1)}(-1)^{\ell(\tau_2)} = (-1)^{\ell(\tau)} & \text{s'il existe un couple } (\tau_1,\tau_2) \text{ tel que } \tau_1 * \tau_2 = \tau \\ & \text{et } |\tau_1| = n_1 \;,\; |\tau_2| = n_2 \\ 0 \;, & \text{sinon} \end{cases}$$

et $\forall r \geq 1$ $\big(\Delta_{*r} \cos(X, x) \cdot {}_{-r}\Delta \cos(X, y)\big)_{\tau,n_1,n_2} = \sum_{(\tau_1,\tau_2)=\tau} (\cos)_{\tau_1 * r,n_1} (\cos)_{-r+\tau_2,n_2}$

$$= \begin{cases} (-1)^{\ell(\tau_1)+1} (-1)^{\ell(\tau_2)} = (-1)^{\ell(\tau)+1} & \text{s'il existe un couple } (\tau_1,\tau_2) \text{ tel que} \\ & \tau_1 * \tau_2 = \tau \text{ et } |\tau_1| + r = n_1 \;,\; |\tau_2| - r = n_2 \\ 0 \;, \text{sinon} \end{cases}$$

Par conséquent le terme de droite est non nul en $(\tau,n_1,n_2) \Leftrightarrow |\tau| = n_1 + n_2$, et alors il existe un et un seul $r \geq 0$ pour lequel $\Delta_{*r} \cos(X, x) \;{}_{-r}\Delta \cos(X, y)$ est non nul. Puisque sa valeur est alors égale à

$(-1)^{\ell(\tau)}$ si $r = 0$ et $(-1)^{\ell(\tau)+1}$ si $r \geq 1$, l'égalité est démontrée.□

<u>Remarque</u>. La première identité peut évidemment s'obtenir de celle-ci par substitution de $-x$ à y.

<u>Corollaire</u>. $\sec(X, x + y) = \sec(X, y) \left(\frac{1}{1 - \sum_{r\geq 1} \Delta_{*r} \sec(X, x) \;{}_{-r}\Delta \sec(X,y)}\right) \sec(X, x)$

Preuve. $\sec(X, x + y) = 1/\cos(X, x + y) =$

$$\sec(X, y)\left(\frac{1}{\sec(X, x)\cos(X, x+y)\sec(X, y)}\right)\sec(X, x)\ . \ \square$$

Attention. L'ordre de multiplication ne peut être changé ni dans la formule ni dans la preuve.

Une des qualités des identités que nous venons d'établir réside dans leur généralité. En effet puisque leur formulation ne fait appel qu'à l'addition et la multiplication elles demeurent valides lorsque nous substituons à X toute suite $(c_1, c_2, \ldots)$ qui rend ses termes sommables. En particulier, nous avons vu à la section précédente que $\cos(0, 0, 0, \underset{k}{1}, 0, 0, \ldots, x) = \sum_{k\geq 0} (-1)^k x^{nk}/(nk)! = \cos_k(x)$ et que $\sec_k(x) = 1/\cos_k(x)$ est la série génératrice des permutations de type $(k, k, \ldots, k)$. Or la fonction $\cos_k(x)$ possède essentiellement k-1 dérivées différentes de sorte que par analogie avec les définitions usuelles on peut considérer les fonctions $-\frac{d^{(r)}}{dx^{(r)}}\cos_k(x)$, $0 < r < k$ et $-\frac{d^{(r)}}{dx^{(r)}}\cos_k(x)/\cos_k(x)$, $0 < r < k$, comme étant respectivement k-1 fonctions "sinus" et "tangentes". Si l'on dénote ces dernières par $\tan_k^{(r)}(x)$ nous pouvons vérifier que la substitution de $(0, 0, \ldots, 0, \underset{k}{1}, 0, \ldots)$ à X dans les identités que nous venons d'établir mène aux formules suivantes:

$$\sec_k(x)\ \sec_k(-x) + \sum_{r=1}^{k-1} \tan_k^{(k-r)}(x)\ \tan_k^{(r)}(-x) = 1$$

$$\sec_k(x + y) = \frac{\sec_k(x)\ \sec_k(y)}{1 - \sum_{r=1}^{k-1} \tan_k^{(k-r)}(x)\ \tan_k^{(r)}(y)}$$

Ces identités furent à l'origine établies de façon purement algébrique à partir de la représentation complexe $\cos_k(x) = \frac{e^{\xi_1 x} + \ldots + e^{\xi_k x}}{k}$, où ξ_i sont les racines de $x^k + 1$, dans le but de généraliser à la fonction $\sec_k(x)$ une preuve combinatoire de la formule $\sec(x + y) = \sec(x)\cdot\sec(y)/1 - \tan(x)\cdot\tan y$ obtenue en 1983 en utilisant les permutations alternées comme modèle des nombres sécants et tangents. Alors que cette démonstration s'avéra relativement difficile, nous verrons dans les lignes qui suivent qu'il en est autrement de la formule générale obtenue pour la série $\sec(X, x + y)$ puisqu'une interprétation combinatoire simple peut en être donnée. Nous pourrons par la suite expliquer cet apparent paradoxe.

Une interprétation combinatoire de la formule de la $\sec(X, x + y)$.

Remarque. Cette démonstration s'inspire des méthodes de preuve utilisées en théorie des espèces de structures, plus précisément en théorie des espèces linéaires (voir [4]).

Observons d'abord qu'un élément $a \in (A, N^* \times N^+ \times N^+)$ peut se décrire comme auparavant dans le langage ensembliste en posant $a_{\tau,E_1,E_2} = a_{\tau,|E_1|,|E_2|}$, E_1 et E_2 étant des ensembles linéairement ordonnés. Puisque $\left(a(X, x+y)\right)_{\tau,n_1,n_2} = a_{\tau,n_1+n_2}$, l'interprétation ensembliste de $\left(a(X, x+y)\right)_{\tau,E_1,E_2}$ doit être définie par $a_{\tau,E_1 \oplus E_2}$ où $E_1 \oplus E_2$ dénote la réunion disjointe de E_1 et E_2 munie de l'ordre linéaire

$\ell < \ell' \Leftrightarrow \ell,\ell' \in E_1$ et $\ell < \ell'$ dans E_1, ou $\ell,\ell' \in E_2$ et $\ell < \ell'$ dans E_2, ou $\ell \in E_1$ et $\ell' \in E_2$, c'est à dire en considérant tous les éléments de E_1 comme étant plus petits que ceux de E_2. Dans ce contexte $a(X, x+y)$ peut être considérée comme étant la série des a-structures existant sur un ensemble E linéairement ordonné bicoloré dans lequel les éléments de E_1, disons les éléments verts, sont plus petits que ceux de E_2, disons les rouges. Ainsi $\sec(X, x+y)$ apparaît comme étant la série génératrice des permutations d'un ensemble bicoloré vert-rouge avec $V < R$, classées selon leur spécification ascendante.

Ceci dit, l'identité trigonométrique précédente peut alors se lire ainsi: une permutation de spécification ascendante τ sur un ensemble bicoloré $V < R$ est composée dans l'ordre suivant: a) d'une permutation sur un ensemble (possiblement vide) d'éléments rouges (i.e. $\sec(X, y)$), b) d'une structure de type

$\circledast = \left[\dfrac{1}{1 - \sum_{r\geq 1} \Delta_{*r} \sec(X, x)\ {}_{-r}\Delta \sec(X, y)}\right]$ et c) d'une permutation sur un ensemble (possiblement vide) d'éléments verts (i.e. $\sec(X, x)$). Or comme $1/(1-a) = \sum_{k\geq 0} a^k$, une $\circledast$ -structure est tout simplement un produit fini de $\sum_{r\geq 1} \Delta_{*r} \sec(X, x)\ {}_{-r}\Delta\sec(X,y)$ -structures, et une $\Delta_{*r}\sec(X, x)\ {}_{-r}\Delta\sec(X, y)$-structure n'est rien d'autre qu'une permutation bicolorée composée d'une permutation sur un ensemble non vide d'éléments verts suivie d'une permutation sur un ensemble non vide d'éléments rouges; la présence de Δ_{*r} et ${}_{-r}\Delta$ exprime simplement le fait que le changement du vert au rouge doit s'effectuer à une distance $r \geq 1$ d'un point de séparation de τ. L'identité obtenue pour la série $\sec(X, x+y)$ décrit donc dans notre langage le théorème de (dé)composition des permutations bicolorées suivant:

Théorème. Toute permutation d'un ensemble bicoloré linéairement ordonné avec $V < R$ se compose (dans le sens de la composition des spécifications ascendantes) de façon unique d'une permutation sur un ensemble (possiblement vide) d'éléments rouges, suivie d'un nombre fini (peut-être 0) de permutations vert-rouge (dans cet ordre), suivies en dernier d'une permutation sur un ensemble (possiblement vide) d'éléments verts.

Illustration du théorème. Prenons $V = \{1<2<..\ 8\} < R = \{①< ② <...< ⑫\}$ et considérons la permutation $\sigma = (①, ⑧, ④, 1, 4, 5, 3, 8, ⑨, ②, ⑫, 7, ⑤,$

⑥, ⑩, ⑪, 2, ③, ⑦, 6). Elle est de spécification ascendante $\tau = (2, 1, 3, 3, 2, 5, 3, 1)$ et la décomposition de σ décrite par la formule de la $\sec(X, x + y)$ est la suivante:

	(r=2 sur 3 8)	(r=1 sur 7)	(r=1 sur 2 ③)	
σ = (① ⑧ ④	1 4 5 3 8 ⑨ ② ⑫	7 ⑤ ⑥ ⑩ ⑪	2 ③ ⑦	6)
τ = (2 , 1)	(3 , 3 , 2)	(5)	(3)	(1)
R	V-R	V-R	V-R	V
sec(X, y)				sec(X,x)

$$\sec(X,y)\left[\frac{1}{1-\sum_{r\geq 1}\Delta_{\star r}\sec(X,x)\;{}_{-r}\Delta\sec(X,y)}\right]\sec(X,x)$$

Ainsi l'identité donnée par la formule de la $\sec(X, x + y)$ peut s'interpréter facilement dans notre contexte combinatoire, par simple lecture. Comment expliquer alors que la recherche d'une démonstration combinatoire des formules obtenues pour la fonction $\sec_k(x)$ soit plus difficile? C'est essentiellement pour deux raisons: Premièrement, en passant, au moyen de la substitution de $(0,1,0,0 \ldots)$ à X, du contexte général à un contexte particulier, nous avons perdu la <u>non-commutativité initiale et fondamentale</u> de la multiplication combinatoire des objets considérés. Deuxièmement en passant des séries $\Delta_{\star r}\sec(X, x)$ et ${}_{-r}\Delta\sec(X, y)$ aux fonctions $\tan_k^{(k-r)}(x)$ et $\tan_k^{(r)}(y)$ nous avons en partie perdu la <u>signification combinatoire</u> de ces séries.

Il faut donc reconnaître que les identités que nous avons établies, en plus d'être générales, ont la qualité d'être précises, significatives. C'est là à mon avis un des précieux avantages que l'on peut obtenir à étudier les objets combinatoires dans le contexte et le langage qui leur sont propres.

9. <u>Extensions et applications</u>. Il est clair que le contexte combinatoire que nous avons utilisé peut servir à l'étude de plusieurs autres types d'objets combinatoires. Nous ne parlerons cependant ici que de certaines extensions et applications du problème que nous avons abordé dans ce texte et des techniques que nous avons employées pour en faire l'étude.

1) Nous avons étudié les permutations σ de $\{1 < 2 < 3 < \quad < n\}$ sous l'angle des relations d'ordre prévalant entre les éléments $\sigma(i)$ et $\sigma(i+1)$; mais nous pouvons tout aussi bien nous intéresser à elles sous l'angle des relations d'ordre existant entre les éléments i et $\sigma(i)$. A cet effet, il est bon de mentionner qu'à chaque permutation σ on peut associer de façon bi-univoque une permutation τ telle que $\forall i = 1, \ldots, n$, $\begin{pmatrix} \cdots & i & i+1 & \cdots \\ \cdots & \sigma(i) < & \sigma(i+1) & \cdots \end{pmatrix} \Leftrightarrow \begin{pmatrix} \cdots & \sigma(i) & \cdots \\ \cdots & \tau(\mathring{\sigma}(i)) & \cdots \end{pmatrix}$. En effet, si $\sigma = \big(\sigma(1), \ldots, \sigma(i_1), \ldots, \sigma(i_2), \ldots, \sigma(i_k), \ldots, \sigma(n)\big)$ où $\forall j$, $\sigma(i_{j+1})$ est le premier élément à

droite de $\sigma(i_j)$ tel que $\sigma(i_j) > \sigma(i_{j+1})$ et si τ est la permutation dont la décomposition en cycles disjoints est $\big(\sigma(1) \ldots \sigma(i_1-1)\big)\big(\sigma(i_1) \ldots \sigma(i_2-1)\big) \ldots \big(\sigma(i_k) \ldots \sigma(n)\big)$ alors on a $\sigma(i) < \sigma(i+1) \Leftrightarrow \sigma(i) < \tau\big(\sigma(i)\big) \forall i$. A travers cette bijection qui est essentiellement la transformation fondamentale de Foata [2] il est donc possible d'appliquer nos méthodes et résultats à l'étude de ce problème.

2) Cette théorie s'étend par ailleurs aux ensembles avec répétitions que l'on nomme aussi tas. Pour ce faire il suffit d'ajuster notre définition de spécification ascendante en conséquence et de remplacer la combinatoire des ensembles, c'est à dire celle associée à la variable x, par celle des tas, c'est à dire celle des fonctions symétriques.

3) Déterminer les permutations de $\{1 < 2 < \ldots < n\}$ dont la spécification ascendante est $(m_1, \ldots, m_\ell)$ revient à déterminer tous les étiquetages de l'arborescence linéaire orientée suivante

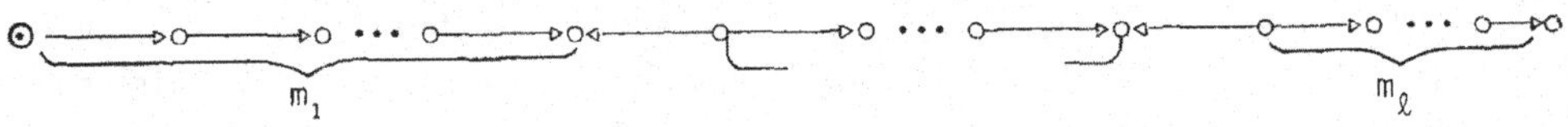

au moyen des entiers 1, 2, ..., n de façon à ce que l'ordre prévalant entre les étiquettes soit compatible avec l'orientation des arêtes. Or nous pouvons montrer que les méthodes et résultats décrits dans ce texte se généralisent pour l'essentiel au problème du calcul des étiquetages d'une arborescence orientée quelconque compatibles avec son orientation. Même les identités trigonométriques demeurent valides pour ces nombres. Ceci fera l'objet d'une publication subséquente.

4) Mentionnons enfin qu'il est aussi possible d'utiliser cette théorie pour obtenir et faire l'étude de nombreux q-analogues. Par exemple, dans le contexte des séries de type
$a_q(X, x) = \sum_{\tau,n} a_{\tau,n} X^\tau x^n/(n!)_q$ on vérifie aisément que $\sec_q(X,x)$ est la série génératrice des permutations classées selon leur spécification ascendante et leur nombre d'inversions.

Références bibliographiques

[1] Carlitz L. - Permutations with prescribed pattern I., Math. Nachr. (58), 1973, pp 31-53.
- Permutations with prescribed pattern II. Applications, Math. Nachr. (83), 1978, pp 101-126.

[2] Foata D., Schutzenberger M.-P., Théorie géométrique des polynômes eulériens, Lecture Notes in Mathematics #138, Springer-Verlag, 1970.

[3] Goulden I.P., Jackson D.M., Combinatorial enumeration, John Wiley & Sons, 1983, pp 230-260.

[4] Joyal A., Une théorie combinatoire des séries formelles, Advances in mathematics, vol. 42 no. 1, oct. 81.

[5] MacMahon P.A., Combinatory analysis, Vol. I, II, Chelsea Publishing company, N.Y., 1960, Vol. I, pp 187-196.

[6] Rawlings D., The ABC's of classical enumeration, 1985, (soumis pour publication).

THEOREM OF POINCARE-BIRKHOFF-WITT, LOGARITHM AND SYMMETRIC GROUP REPRESENTATIONS OF DEGREES EQUAL TO STIRLING NUMBERS.

Christophe Reutenauer

Université du Québec à Montréal
et
CNRS (Paris)

à S.P.

0. Introduction

The (historical) starting point of the present work is a paper of Ree [10], where an extension of a formula of Baker-Campbell-Hausdorff is given. We give it here in a slightly different formulation.

Let A be an alphabet and $Q\langle\langle A\rangle\rangle$ the algebra of noncommutative formal power series over Q, equipped with the usual noncommutative product ("concatenation algebra") The space $Q\langle\langle A\rangle\rangle$ possesses another product, the shuffle product: the shuffle algebra is commutative; see [10], [8] for the definition of the shuffle.

Let $\mathcal{T}$ be the complete tensor product

$$\mathcal{T} = Q\langle\langle A\rangle\rangle \otimes Q\langle\langle A\rangle\rangle$$

where the left factor is the shuffle algebra and the right factor the concatenation algebra. For instance, if a,b are in A, then

$$(a\otimes b)(b\otimes a) = (ab+ba) \otimes ba$$

Each element of $\mathcal{T}$ is an infinite linear combination of couple of words over A

$$\sum_{u,v\in A^*} \alpha_{u,v}\, u\otimes v \qquad (\alpha_{u,v}\in Q)$$

where A^* denotes the free monoid generated by A. In $\mathcal{T}$, consider the particular element

$$S = \sum_{\omega\in A^*} \omega\otimes\omega$$

(a kind of diagonal). As

$$S = 1\otimes 1 + T$$

(1 is the empty word), as $1\otimes 1$ is the neutral element of $\mathcal{T}$ and as

$$\lim_{n\to 0} T^n = 0$$

(in the usual topology of $\mathcal{T}$), one may define log S by the usual formula

$$(0.1) \qquad \log S = \sum_{n\geq 1} \frac{(-1)^{n-1}}{n} T^n$$

Note that, because of the special form of S, one has

$$(0.2) \qquad \log S = \sum_{u\in A^*} u\otimes P_u$$

where P_u is an homogeneous polynomial of degree length(u).

<u>Theorem</u> (Ree, see [10] th. 2.5)

<u>The polynomials P_u are Lie polynomials, that is, they belong to the Lie algebra generated by the elements of A</u>.

One has, for instance

$$(0.3) \qquad P_{abc} = \frac{1}{6}(2abc - acb - bac - bca - cab + 2cba) = \frac{1}{3}[[a,b],c] - \frac{1}{6}[[a,c],b]$$

<u>Remarks</u>

1. One may recover the Baker-Campbell-Hausdorf formula in the following way. Let A = {a,b} and define a linear mapping f

$$Q\langle\langle A\rangle\rangle \to Q$$

by

$$\omega \to \frac{1}{i!j!} \text{ if } \omega = a^i b^j$$

$$0 \text{ otherwise}$$

for any word ω. Then it is easily verified that this mapping is a homomorphism with respect to the shuffle product (or use the fact that a^ib^j are Lyndon words, and that the latter form a transcendance basis for the shuffle [8] ex. 5.3.6, [9]).

Now, apply the homomorphism f ⊗ id to (0.2), obtaining that

$$\log\left(\sum_{i,j\geq 0} \frac{a^ib^j}{i!j!}\right) = \log(e^ae^b)$$

is a Lie element c, hence $e^ae^b = e^c$, which is the B. - C. - H. formula.

2. Ree uses his formula to prove a theorem of Chen [2]: the (non-commutative) formal series defined by iterated integrals of a given path in $R^{|A|}$ is a Lie element; this formula may also be applied to nonlinear system theory [4], [5], to algebraic groups [11] and to combinatorics on words [12].

The aim of the present work is to compute explicitly the coefficients of the series (0.2) (section 1). Moreover, we show that this series corresponds to the canonical projection π_1 of the free associative algebra onto the free Lie Algebra; in terms of Hopf algebras, this would be written $id = \exp(\pi_1)$. The free associative algebra U is, by the theorem of Poincaré-Birkhoff-Witt, the direct sum of its subspaces U^q, $q \geq 0$, where U^q is the linear span of the q-th powers of Lie elements. We show that the corresponding projections π_q satisfy $\pi_q = \pi_1^q$, in the algebra structure of End (U) deduced from the Hopf algebra structure of U. This may be interpreted by the fact that certain polynomials P_u of (0.2), viewed as elements of the algebra of the

symmetric group, are idempotents (section 2). Finally, we study the action of the symmetric group S_n on the spaces U^q. This leads to representations of S_n whose orders are the Stirling numbers, and whose corresponding idempotents are given by the series (0.2).

1. Coefficients

We want to compute explicitly the polynomials P_u of formula (0.2). By (0.1), we have

$$(1.1) \qquad \sum_{u\in A^*} u \otimes P_u = \sum_{k\geq 1} \frac{(-1)^{k-1}}{k} \left(\sum_{w\in A^+} w\otimes w\right)^k$$

where $A^+ = A^*\backslash 1$ (the semigroup of nonempty words).

By definition of the product in $\mathcal{T}$,

$$\left(\sum_{w\in A^+} w\otimes w\right)^k = \sum_{u_1,\dots,u_k\in A^+} (u_1 \circ \dots \circ u_k) \otimes (u_1 \dots u_k)$$

where o denotes the shuffle product. Denote by (P,v) the coefficient of the word v in the polynomial P. Then we obtain

$$(1.3) \qquad P_u = \sum_{k\geq 1} \frac{(-1)^{k-1}}{k} \sum_{u_1,\dots,u_k\in A^+} (u_1 \circ \dots \circ u_k, u) u_1 \dots u_k$$

We shall compute explicitly the coefficients of P_u in the case where u is <u>multilinear</u> (i.e. no letter occurs more than once in u). In this case, we may write $u = a_1 \dots a_n$, where the a_i's are distinct letters.

<u>Theorem 1.1.</u> <u>Let</u> $u = a_1 \dots a_n$ <u>be a multilinear word.</u> <u>Then</u>

$$(1.4) \qquad P_u = \sum_{\sigma\in S_n} \frac{(-1)^{d_\sigma - 1}}{d_\sigma} \binom{n}{d_\sigma}^{-1} a_{\sigma(1)} \dots a_{\sigma(n)}$$

<u>where</u> d_σ <u>is the number of ascending runs of</u> σ (= 1+ <u>number of descents of</u> σ)

We prove first a lemma.

<u>Lemma 1.2</u>: Let $a_1, \ldots, a_n$ be distinct letters. Then

(1.5) $$\sum_{u_1,\ldots,u_k \in A^+} (u_1 \circ \ldots \circ u_k,\ a_1 \ldots a_n)\ u_1 \ldots u_k$$

$$= \sum_{\sigma \in S_n} \binom{n-d_\sigma}{k-d_\sigma} a_{\sigma(1)} \ldots a_{\sigma(n)}$$

<u>Proof</u> Denote by Q_k the polynomial on the left-hand side. Let $w = a_{\sigma(1)} \ldots a_{\sigma(n)}$. Then the coefficient (Q_k, w) of w in Q_k is

(1.6) $$(Q_k, w) = \sum_{\substack{u_1,\ldots,u_k \in A^+ \\ w = u_1 \ldots u_k}} (u_1 \circ \ldots \circ u_k,\ a_1 \ldots a_n)$$

Let us say that a word is <u>growing</u> if it may be written as $a_{i_1} \ldots a_{i_\ell}$, with $i_1 < \ldots < i_\ell$.

Then, by definition of the shuffle, the coefficient $(u_1 \circ \ldots \circ u_k, a_1 \ldots a_n)$ vanishes, unless each u_i is growing, in which case this coefficient is 1. Hence, the value of (1.6) is the number of factorizations of w in k growing nonempty words. Let

(1.7) $$w = v_1 \ldots v_d$$

where each v_i is growing and where d is the number of ascending runs of σ.

Schematically

w = 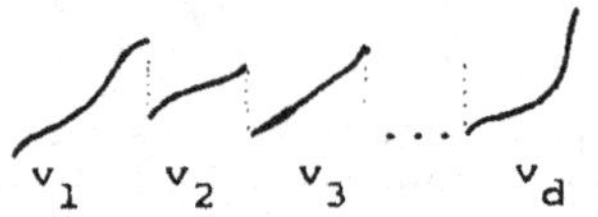

A factorization of w into growing words is necessarily a subfactorization of (1.7). Thus we have to factorize each v_i in s_i nonempty words, where the s_i's are such that $\Sigma s_i = k$. If p_i = length (v_i), then there are $\binom{p_i - 1}{s_i - 1}$ factorizations of v_i in s_i nonempty words. Thus

$$(Q_k, w) = \sum_{s_1 + \ldots + s_d = k} \binom{p_1 - 1}{s_1 - 1} \ldots \binom{p_d - 1}{s_d - 1}$$

Now, note that

$$(1 + x)^{p-1} = \sum_{1 \le s \le p} \binom{p-1}{s-1} x^{s-1}$$

hence

$$\prod_{1\le i\le d} (1+x)^{p_i-1} = \prod_i \Big(\sum_{1\le s_i\le p} \binom{p_i-1}{s_i-1} x^{s_i-1} \Big),$$

$$= \sum_\ell x^\ell \Big(\sum_{s_1+\ldots+s_d-d=\ell} \binom{p_1-1}{s_1-1}\cdots\binom{p_d-1}{s_d-1} \Big)$$

But it is also equal to

$$(1+x)^{p_1+\ldots+p_d-d} = (1+x)^{n-d} = \sum_\ell \binom{n-d}{\ell} x^\ell$$

which shows that

$$\sum_{s_1+\ldots+s_d = \ell+d} \binom{p_1-1}{s_1-1}\cdots\binom{p_d-1}{s_d-1} = \binom{n-d}{\ell}$$

This shows that

$$(Q_k, w) = \binom{n-d}{k-d}$$

as desired. □

We still need another lemma

<u>Lemma 1.3</u> <u>The following relation holds</u>

$$\sum_{k\ge 1} \Big(\frac{(-1)^{k-1}}{k} \binom{n-d}{k-d} = \frac{(-1)^{d-1}}{d} \binom{n}{d}^{-1}$$

<u>Proof</u> Denote by $a(n,d)$ the left-hand side. We prove the lemma by induction on $n-d$. If $n-d = 0$, it is obvious. Let $n-d \ge 1$. Using the usual relation for binomial coefficients, we obtain

$$a(n,d) = a(n-1,d) + a(n,d+1)$$

This is equal, by induction, to

$$\frac{(-1)^{d-1}}{d} \binom{n-1}{d}^{-1} + \frac{(-1)^d}{d+1} \binom{n}{d+1}^{-1}$$

$$= \frac{(-1)^{d-1}}{d} \frac{d!\ (n-1-d)!}{(n-1)!} + \frac{(-1)^d}{d+1} \frac{(d+1)!\ (n-d-1)!}{n!}$$

$$= \frac{(-1)^{d-1}\ (d-1)!\ (n-d-1)!}{n!} (n-d)$$

$$= \frac{(-1)^{d-1}}{d} \binom{n}{d}^{-1}$$

as desired. □

Now, the theorem follows, using (1.3) and the two lemmas.

<u>Remark</u> If $u = a_1 \ldots a_n$ is any word, not necessarily multilinear, then formula (1.4) still holds (but the words $a_{\sigma(1)} \ldots a_{\sigma(n)}$ are not all distinct). This is a consequence of (1.1) (or see lemma 2.4). For instance, one obtains P_{aba} from P_{abc} by replacing c by a in (0.3):

$$P_{aba} = \frac{1}{6} (2\ aba - aab - baa - baa - aab + 2\ aba)$$

$$= \frac{1}{6} (4\ aba - 2\ aab - 2\ baa)$$

$$= \frac{1}{3} (2\ aba - aab - baa)$$

$$= \frac{1}{3} [[a,b], a]$$

2. <u>The canonical projection</u>

Consider an alphabet $A = \{a_1, \ldots, a_n\}$. Then each multilinear word of length n may be viewed as a permutation, element of S_n. For instance, if $A = \{a_1, a_2, a_3\}$. the word $a_1 a_2 a_3$ will represent 1 (the identity permutation), and $a_3 a_2 a_1$ will be the transposition (13).

Consequently, a polynomial which is a linear combination of such words may be viewed as an element of the group algebra $Q\{S_n\}$. For instance, the polynomial P_{abc} of (0.3) will be

$$\frac{1}{6} (2 - (23) - (12) - (123) - (132) + 2\ (13))$$

A direct calculation shows that this element of $Q[S_3]$ is idempotent!

We shall now explain this fact.

Let Q<A> be the algebra of noncommutative polynomials (it is a subalgebra of the concatenation algebra Q<<A>>). Let $\mathcal{L}$<A> denote the Lie algebra generated, in Q<A>, by the <u>letters</u> (= elements of A); an element in $\mathcal{L}$ <A> is called a <u>Lie polynomial</u>. It is known that $\mathcal{L}$<A> is the free Lie algebra generated by A, with Q<A> as an enveloping algebra (see [8]). Let U^q denote the subspace of Q<A> generated by the polynomials P^q, where P ranges over $\mathcal{L}$<A>. Then, by the theorem of Poincaré-Birkhoff-Witt

$$(2.1) \qquad Q\langle A\rangle = \bigoplus_{q \geq 0} U^q$$

see [3], 2.4.6.

The direct sum decomposition (2.1) defines a family of linear projections

$$(2.2) \qquad \pi_q : Q\langle A\rangle \to Q\langle A\rangle$$

defined by $\pi_q \mid U^q = id$, $\pi_q \mid U^{q'} = 0$ if $q' \neq q$. Note that $U^1 = \mathcal{L}\langle A\rangle$. The projection π_1 is called the canonical projection of $Q\langle A\rangle$ onto $\mathcal{L}\langle A\rangle$.

Theorem 2.1 The canonical projection $\pi_1 : Q\langle A\rangle \to \mathcal{L}\langle A\rangle$ is also defined by the condition

$$\pi_1(u) = P_u$$

for any word u (where P_u is defined by (0.2)). In particular, if $u = a_1 \ldots a_n$ is a multilinear word and if P_u is considered as an element of $Q[S_n]$, then P_u is idempotent in $Q[S_n]$.

We need some lemmas. The first one is well-known.

Lemma 2.2 Define the concatenation homomorphism

$$c_k : Q\langle A\rangle \to Q\langle A\rangle^{\otimes k}$$

by

$$c_k(a) = a\otimes 1\otimes \ldots \otimes 1 + 1\otimes a\otimes \ldots \otimes 1 + \ldots + 1\otimes \ldots \otimes 1\otimes a$$

for any letter a. Then, for any w one has

$$(2.3) \quad c_k(w) = \sum_{u_1,\ldots,u_k \in A^*} (u_1 \circ \ldots \circ u_k, w)\, u_1\otimes \ldots \otimes u_k$$

Furthermore if P is a Lie polynomial, then

$$c_k(P) = P\otimes 1\otimes \ldots \otimes 1 + 1\otimes P\otimes \ldots \otimes 1 + \ldots + 1\otimes \ldots \otimes 1\otimes P$$

Proof The first relation is a simple consequence of the definition of the shuffle product. For the second, it is enough to note that it is true when P is a letter, and then check that if it is true for P, Q, then also for their Lie bracket $[P,Q] = PQ - QP$. □

Lemma 2.3 Let $f_1,\ldots,f_k,g$ be linear endomorphisms $Q\langle A\rangle \to Q\langle A\rangle$ such that the following equality holds in the algebra $\mathcal{T}$ (see section 0):

$$\prod_{1\leq i\leq k} \Big(\sum_{u\in A^*} u\otimes f_i(u)\Big) = \sum_{w\in A^*} w\otimes g(w)$$

Then, for any polynomial P

$$g(P) = p_k \circ (f_1\otimes \ldots \otimes f_k) \circ c_k(P)$$

where $p_k : Q\langle A\rangle^{\otimes k} \to Q\langle A\rangle$ is the concatenation product:

$$p_k(u_1\otimes \ldots \otimes u_k) = u_1 \ldots u_k$$

Proof By definition of the product in $\mathcal{T}$, one has

$$\sum_{w\in A^*} w\otimes g(w) = \sum_{u_1,\dots,u_k} (u_1 o\dots o u_k)\otimes(f_1(u_1)\dots f_k(u_k))$$

This shows that

$$g(w) = \sum_{u_1,\dots,u_k} (u_1 o\dots o u_k, w) f_1(u_1)\dots f_k(u_k)$$

Now, by (2.3), we obtain

$$p_k o(f_1\otimes\dots\otimes f_k)\ o c_k(w)$$

$$= p_k\Big(\sum_{u_1,\dots,u_k} (u_1 o\dots o u_k, w)\ f_1(u_1)\otimes\dots\otimes f_k(u_k)\Big)$$

$$= \sum_{u_1,\dots,u_k} (u_1 o\dots o u_k, w)\ f_1(u_1)\dots f_k(u_k)$$

which proves the lemma, because A^* is a basis of Q<A>. □

Remark The product $(f,g) \to f*g$ in End (Q<A>) defined by

$$(f*g)(P) = p_2 o(f\otimes g)o\ c_2(P)$$

is a classical one in the context of bialgebras, see [7] p. 5. It is associative, and

$$f_1*\dots*f_k = p_k o(f_1\otimes\dots\otimes f_k)o\ c_k$$

Lemma 2.4 *Let* I *be the endomorphism of* Q<A> *defined by* $I(w) = w$ *if* $w \neq 1$ *and* $I(1) = 0$. *Let* α *be the endomorphism defined by*

$$\alpha = p_k o\ I^{\otimes k} o\ c_k$$

Let f *be an algebra homomorphism* Q<A> → Q<A> *such that* f(a) *is a Lie polynomial for any letter* a. *Then*

$$\alpha\ o\ f = f\ o\ \alpha$$

Proof If $w = a_1\dots a_n$ $(a_i\in A)$, then

$$c_k(w) = \prod_{1\le i\le n} (a_i\otimes\dots\otimes 1 +\dots+ 1\otimes\dots\otimes a_i)$$

Moreover, $c_k\ o\ f(w) = c_k\Big(\prod_{1\le i\le n} f(a_i)\Big)$

$$= \prod_{1\le i\le n} c_k(f(a_i))$$

Because each $f(a_i)$ is a Lie polynomial, this is equal, by lemma 2.2, to

$$\prod_{1\le i\le n} (f(a_i)\otimes\dots\otimes 1 +\dots+ 1\otimes\dots\otimes f(a_i))$$

This shows that we have

$$c_k \circ f = f^{\otimes k} \circ c_k$$

Now, because any Lie polynomial has zero constant term, we have

$$I \circ f = f \circ I$$

Moreover, as f is an algebra homomorphism, we have

$$p_k \circ f^{\otimes k} = f \circ p_k$$

Taking these relations together, we obtain

$$\begin{aligned} \alpha \circ f &= p_k \circ I^{\otimes k} \circ c_k \circ f \\ &= p_k \circ I^{\otimes k} \circ f^{\otimes k} \circ c_k \\ &= p_k \circ (I \circ f)^{\otimes k} \circ c_k \\ &= p_k \circ (f \circ I)^{\otimes k} \circ c_k \\ &= p_k \circ f^{\otimes k} \circ I^{\otimes k} \circ c_k \\ &= f \circ p_k \circ I^{\otimes k} \circ c_k = f \circ \alpha \end{aligned}$$

what was to be shown. □

<u>Proof of theorem 2.1</u>

Let π be the endomorphism of Q<A> defined by

$$\pi(u) = P_u$$

for any word u. We have to show that $\pi = \pi_1$. For this, it is enough to show that for any Lie polynomial P, one has $\pi(P) = P$ and $\pi(P^q) = 0$ if $q \neq 1$.
Let I : Q<A> → Q<A> the endomorphism defined by $I(1) = 0$ and $I(w) = w$ if w is a nonempty word. Then, by (1.1),

$$\sum u \otimes P_u = \sum_{k \geq 1} \frac{(-1)^{k-1}}{k} \left(\sum_w w \otimes I(w)\right)^k$$

By lemma 2.3,

$$\left(\sum_w w \otimes I(w)\right)^k = \sum_w w \otimes (p_k \circ I^{\otimes k} \circ c_k(w))$$

This shows that

$$(2.4) \qquad \pi(u) = \sum_{k\geq 1} \frac{(-1)^{k-1}}{k} \, p_k \circ I^{\otimes k} \circ c_k(u)$$

As π is linear, this relation holds also for any polynomial P instead of u.
By lemma 2.4, we thus obtain that

$$\pi \circ f = f \circ \pi$$

for any algebra endomorphism f of Q<A> such that f(a) is a Lie polynomial for each letter a. Now let P be any Lie polynomial and f an algebra endomorphism of Q<A> such that f(a) = P for some fixed letter a. Then $\pi(P^q) = \pi \circ f(a^q) = f \circ \pi(a^q)$. This shows that it is enough to prove the above assertions for a. For this, let g be the algebra endomorphism of Q<A> such that g(a) = a and g(b) = 0 for the other

letters. Then

$$\pi(a) = \pi \circ g(a)$$

which shows that we just have to consider the one-letter case. But

$$\sum_{n\in N} a^n \otimes a^n = \exp (a \otimes a)$$

in $\mathcal{T}$, because the n-th shuffle power of a is $n!\, a^n$. Hence

$$\log (\sum_n a^n \otimes a^n) = a \otimes a$$

which shows that

$$\pi\,(a^q) = \delta_{q,1}\, a$$

as desired.

Now, let $a_1,\ldots,a_n$ be distinct letters and

$$\pi(a_1\ldots a_n) = \sum_{\sigma\in S_n} \alpha_\sigma \, a_{\sigma(1)}\ldots a_{\sigma(n)}$$

For $\sigma \in S_n$, let f_σ be the algebra endomorphism of Q<A> defined by

$$f_\sigma\,(a_i) = a_{\sigma(i)}$$

As π is a projection, we have

$$\sum_{\sigma\in S_n} \alpha_\sigma \, a_{\sigma(1)}\ldots a_{\sigma(n)} = \pi\,(a_1\ldots a_n)$$

$$= \pi \circ \pi\,(a_1\ldots a_n)$$

$$= \sum_\sigma \alpha_\sigma \, \pi\,(a_{\sigma(1)}\ldots a_{\sigma(n)})$$

$$= \sum_\sigma \alpha_\sigma \, \pi \circ f_\sigma \, (a_1 \dots a_n)$$

$$= \sum_\sigma \alpha_\sigma \, f_\sigma \circ \pi \, (a_1 \dots a_n)$$

by lemma 2.5.

Hence

$$\sum_\sigma \alpha_\sigma \, a_{\sigma(1)} \dots a_{\sigma(n)} = \sum_\sigma \alpha_\sigma \, f_\sigma \, (\sum_\tau \alpha_\tau \, a_{\tau(1)} \dots a_{\tau(n)})$$

$$= \sum_{\sigma,\tau} \alpha_\sigma \alpha_\tau \, a_{\sigma\tau(1)} \dots a_{\sigma\tau(n)}$$

which shows that

$$\sum_\sigma \alpha_\sigma \, \sigma = \sum_{\sigma,\tau} \alpha_\sigma \alpha_\tau \sigma\tau$$

is idempotent in $Q[S_n]$. □

<u>Remark</u> We have used, in the course of the proof, the fact that in the one-letter case, one has in $\mathcal{T}$: $\sum_{n \in N} a^n \otimes a^n = \exp (a \otimes a)$. Note that when $|A| \geq 2$, it is not true in $\mathcal{T}$ that

$$\sum_{w \in A^*} w \otimes w = \exp (\sum_{a \in A} a \otimes a)$$

This is only true in the one-letter case.

<u>Corollary 1.6</u> <u>The canonical projection</u> $\pi_1 : Q\langle A\rangle \to \mathcal{L}\langle A\rangle$ <u>is given, for any Lie polynomials</u> $P_1, \dots, P_n$, <u>by</u>

$$\pi_1(P_1 \dots P_n) = \sum_{\sigma \in S_n} \frac{(-1)^{d_\sigma - 1}}{d_\sigma} \binom{n}{d_\sigma}^{-1} P_{\sigma(1)} \dots P_{\sigma(n)}$$

$$= \sum_{\sigma \in S_n} \frac{(-1)^{d_\sigma - 1}}{d_\sigma \; n} \binom{n}{d_\sigma}^{-1} [P_{\sigma(1)} \dots P_{\sigma(n)}]$$

<u>where</u> $[Q_1 \dots Q_n]$ <u>denotes the bracketing (from left to right)</u> $[\dots[Q_1Q_2]\dots\dots Q_n]$.

The last formula has already been proved by Solomon [14]; he starts directly from the P. - B. - W theorem, and does not use the logarithm as here.

<u>Proof</u> The first formula is a consequence of theorem 1.1, theorem 2.1 and lemma 2.5. For the second, apply the formula of Dynkin - Specht - Wever, see e.g. [10] th. 2.3. □

3. The other projections

We have shown that the canonical projection $\pi_1 : Q\langle A\rangle \to \mathcal{L}\langle A\rangle$ is determined by the logarithm of $\sum_{w\in A^*} w \otimes w$; in fact in terms of the product of End $(Q\langle A\rangle)$ introduced in the remark after lemma 2.3, one has $\pi_1 = \log(\mathrm{id})$, where id is the identity of $Q\langle A\rangle$ (this seems to be a formal analogue of the exponential function in a Lie group). What about the other projections?

Theorem 3.1 The projection $\pi_q : Q\langle A\rangle \to Q\langle A\rangle$ is completely specified by the formula

$$(3.1) \qquad \frac{1}{q!}\left(\log\left(\sum_{w\in A+} w \otimes w\right)\right)^q = \sum_{u\in A^*} u \otimes \pi_q(u)$$

which holds in the algebra $\mathcal{T}$. Moreover, with $A = \{a_1,\dots,a_n\}$, one has

$$(3.2) \qquad \pi_q(a_1\dots a_n) = \sum_{\sigma\in S_n} \left(\sum_k \frac{s(k,q)}{k!} \binom{n-d_\sigma}{k-d_\sigma}\right) a_{\sigma(1)}\dots a_{\sigma(n)}$$

where $s(k,q)$ is the Stirling number of first kind. Viewed as elements of $Q[S_n]$, the n elements $\pi_q(a_1\dots a_n)$, $1\le q\le n$, are orthogonal idempotents of sum 1 (for $q = 0$ or $q > n$, $\pi_q(a_1\dots a_n) = 0$).

Formula 3.1 was already proved by Hain [6], in the context of graded Hopf algebras.

Recall that the Stirling numbers are defined by

$$x(x-1)(x-2)\dots(x-k+1) = \sum_q s(k,q)\, x^q, \text{ see [13].}$$

Unfortunately, I could not completely identify the coefficient of $a_{\sigma(1)}\dots a_{\sigma(n)}$ in (3.2). One obtains for each n, an n by n table of coefficients, depending on q and d_σ. I give the table for $n = 6$, and a description of the known parts of the table.

$q \backslash d_\sigma$	1	2	3	4	5	6
1	$\frac{1}{6}$	$\frac{-1}{30}$	$\frac{1}{60}$	$\frac{-1}{60}$	$\frac{1}{30}$	$\frac{-1}{6}$
2	$\frac{137}{360}$	$\frac{-13}{360}$	$\frac{1}{180}$	$\frac{1}{180}$	$\frac{-13}{360}$	$\frac{137}{360}$
3	$\frac{5}{16}$	$\frac{1}{48}$	$\frac{-1}{48}$	$\frac{1}{48}$	$\frac{-1}{48}$	$\frac{-5}{16}$
4	$\frac{17}{144}$	$\frac{5}{144}$	$\frac{-1}{144}$	$\frac{-1}{144}$	$\frac{5}{144}$	$\frac{17}{144}$
5	$\frac{1}{48}$	$\frac{1}{80}$	$\frac{1}{240}$	$\frac{-1}{240}$	$\frac{-1}{80}$	$\frac{-1}{48}$
6	$\frac{1}{720}$	$\frac{1}{720}$	$\frac{1}{720}$	$\frac{1}{720}$	$\frac{1}{720}$	$\frac{1}{720}$

First row: inverses of the binomial coefficients of order $n - 1$ multiplied by $\pm \frac{1}{n}$

First and last column: Stirling numbers $s(n,q)$ multiplied by $\pm \frac{1}{n!}$

Last row: $\frac{1}{n!}$

Sum of the first column: 1

Sum of the other columns: 0

<u>Proof of theorem 3.1</u>

Let the left-member of (3.1) be equal to $\Sigma\, u \otimes \alpha(u)$, for some linear endomorphism α of $Q<A>$. We want to show that $\alpha = \pi_q$. By lemma 2.3, we have

$$\alpha = \frac{1}{q!}\, P_q \circ \pi_1^{\otimes q} \circ c_q$$

From this, the first assertion will easily follow. Let P be a Lie polynomial. Then $c_q(P^q) = c_q(P)^q = (P\otimes 1\otimes \ldots \otimes 1 + \ldots + 1\otimes \ldots \otimes 1 \otimes P)^q$ by lemma 2.2: it is the sum of $q!\; P\otimes \ldots \otimes P$ and of terms $P_1 \otimes \ldots \otimes P_q$ where at least one P_i is equal to 1. Hence, as $\pi_1(1) = 0$, we obtain that $\pi_1^{\otimes q} \circ c_q\, (P^q) = q!\; P \otimes \ldots \otimes P$, by the definition of π_1. Thus $\alpha\, (P^q) = P^q$. Now, let $r \neq q$. Then $C_q(P^r)$ is a sum of terms $P_1 \otimes \ldots \otimes P_q$ where at least one P_i is equal to P^j for some $j \neq 1$: as $\pi_1\, (P^j) = 0$, we obtain that
$\alpha\, (P^r) = 0$. Thus $\alpha = \pi_q$, by definition of π_q.

It is well-known [13] that

$$\frac{1}{q!}\, (\log\, (1+x))^q = \sum_k \frac{s(k,q)}{k!}\, x^k$$

Thus we have by (3.1)

$$\Sigma\, u \otimes \pi_q(u) = \sum_k \frac{s(k,q)}{k!}\, \Big(\sum_{w \in A^*} w \otimes w\Big)^k$$

Hence, by (1.2), we obtain

$$\pi_q(u) = \sum_k \frac{s(k,q)}{k!} \sum_{u_1,\ldots,u_k \in A^+} (u_1 \circ \ldots \circ u_k, u)\, u_1 \ldots u_k$$

Now, this and lemma 1.2 imply formula (3.2).

It is obvious, from the definition of π_q, that $\pi_q(a_1 \ldots a_n) = 0$ if $q = 0$ or $q > n$. Evidently $id = \sum_q \pi_q$ which shows that

$$\sum_{1\leq q\leq n} \pi_q(a_1 \ldots a_n) = a_1 \ldots a_n,$$

and hence the last assertion follows. The fact that $\pi_q(a_1 \ldots a_n)$ is idempotent is a consequence, as in the proof of theorem 2.1, of the fact that π_q is a projection. The fact that these idempotents are orthogonal is similarly proved by using the relations

$$\pi_q \circ \pi_{q'} = 0$$

if $q \neq q'$. □

4. Representations of the symmetric group.

We have obtained n orthogonal idempotents of sum 1 of $Q[S_n]$, which are $\pi_q(a_1 \ldots a_n)$, $1\leq q\leq n$. We compute now the degrees of the associated representations of the symmetric group. A surprising fact is that Stirling numbers step in again.

Note that if $A = \{a_1 \ldots a_n\}$, then $Q[S_n]$ acts naturally on $Q\langle A\rangle$ by $\sigma . a_i = a_{\sigma(i)}$, extended in an algebra endomorphism of $Q\langle A\rangle$. In particular, let E be the subspace of $Q\langle A\rangle$ generated by the words $a_{\sigma(1)} \ldots a_{\sigma(n)}$, $\sigma \in S_n$. This is of course stable under the action of S_n, and the associated representation is the left regular representation of S_n.

Moreover, let $V^q = U^q \cap E$, where $U^q = \pi_q\ (Q\langle A\rangle)$, see section 2. Now, it is well-known, by multilinearization, that U^q is generated by the polynomials

$$(P_1, \ldots, P_q) = \sum_{\sigma \in S_q} P_{\sigma(1)} \ldots P_{\sigma(q)}$$

where $P_1, \ldots, P_q$ are Lie polynomials. This shows that one has

$$E = \bigoplus_q V^q$$

and that each V^q is stable under the action of S_n. It is easy to show that $V^q = 0$ if $q = 0$ or $q > n$. Hence

$$E = \bigoplus_{1\leq q\leq n} V^q$$

or equivalently

$$Q[S_n] = \bigoplus_{1\leq q\leq n} Q[S_n]\ \pi_q(a_1 \ldots a_n)$$

if $\pi_q(a_1 \ldots a_n)$, defined by (3.2), is viewed as an element of $Q[S_n]$. Note that we have also, by lemma 2.4

$$\pi_q(\sigma.P) = \sigma.\pi_q(P)$$

for any polynomial P (which implies in fact all the previous assertions). Now, we have the perhaps classical result.

Theorem 4.1 The dimension of V^q is $|s(n,q)|$.

Proof. We show that each permutation σ in S_n defines an element $[\sigma]$ of a basis of V^q, where q is the number of cycles of σ. As it is well-known ([13] p. 71) that the number of permutations in S_n with q cycles is $|s(n,q)|$, the result will follow.

First, we associate to σ a multilinear word: decompose σ in cycles

$$\sigma = (i_1,i_2,\ldots,i_u)(j_1,\ldots,j_v)(k_1,\ldots,k_w)\ldots$$

with $i_1 = \inf\{i_1,\ldots,i_u\} > j_1 = \inf\{j_1,\ldots,j_v\} > k_1 = \inf\{k_1,\ldots,k_w\}$ etc...

Then associate to σ the word

$$w = a_{i_1}a_{i_2}a_{i_u}\ a_{j_1}\ldots a_{j_v}\ a_{k_1}\ldots a_{k_w}\ldots$$

This is clearly a bijection. Moreover, the factorization of w

$$w = (a_{i_1}\ldots a_{i_u})(a_{j_1}\ldots a_{j_v})(a_{k_1}\ldots a_{k_w})\ldots$$

is just the decomposition of w into Lyndon words, see [8]. Denote by [u] the Lie polynomial which corresponds to a Lyndon word in the Lyndon basis of $\mathcal{L}\langle A\rangle$ (ibid.). If $w = u_1\ldots u_q$ is the decomposition of w into Lyndon words, then let

$$[\sigma] = [w] = \sum_{\sigma\in S_q}[u_{\sigma(1)}]\ldots[u_{\sigma(q)}]$$

Let B be a subset of A. Let M_B be the set of multilinear words which have exactly B as set of letters. Let E_B be the space generated by M_B and $V_B^q = U^q \cap E_B$. Then the space $V_B^1 = \mathcal{L}\langle A\rangle \cap E_B$ admits as a basis the set of

$$[u],\ u\in M_B,\ u \text{ Lyndon, see [8].}$$

Now, let $1\le q\le n$. By homogeneity, the space V^q is generated by the polynomials

$$(P_1,\ldots,P_q)$$

where $P_i \in V_{B_i}^1$ for some partition $A = \bigcup_{1\le i\le q} B_i$. This shows, by multilinearity, that V^q is generated by the polynomials

$$([u_1],\ldots,[u_q])$$

where $u_i \in M_{B_i}$ is a Lyndon word. But there are $|s(n,q)|$ polynomials of this type (by the above bijection); thus $n! = \sum_q \dim(V_q) \leq \sum_q |s(n,q)| = n!$. This shows that these polynomials form a basis of V^q, whose dimension is consequently $|s(n,q)|$. □

Example $A = \{a,b,c\}$

$$\left.\begin{array}{l}[abc] = [a,[b,c]] \\ [acb] = [[a,c],b]\end{array}\right\} V^1$$

$$\left.\begin{array}{l}[bac] = b[a,c] + [a,c]b \\ [bca] = [b,c]a + a[b,c] \\ [cab] = c[a,b] + [a,b]c\end{array}\right\} V^2$$

$$[cba] = cba + cab + bca + bac + acb + abc \} \; V^3$$

($[x,y]$ denotes $xy - yx$).

5. Conclusion

In the course of computing the coefficients of the series of Ree, we were lead to discover several striking facts. First, that the elements of the algebra of the symmetric group which appear, as noncommutative polynomials, are idempotents: this is a priori not obvious, due to the fact that they are defined by concatenation and shuffle of words, and not in term of the product of the symmetric group.

To explain this idempotence, we have shown that Ree's series may be interpreted as the canonical projection of the free associative algebra onto the free Lie algebra (any enveloping algebra would however work).

More precisely, in terms of the product of the endomorphism algebra defined by the Hopf algebra structure of Q<A>, this projection is the logarithm of the identity; or, the identity is the exponential of the projection, which seems to be a kind of analogue of the exponential in a Lie group.

Another surprising fact is that Stirling numbers intervene separately twice: once, in the coefficients of the idempotents and secondly as dimensions of the associated representations.

What should be done now is the exact identification of the coefficients in formula (3.2). Moreover, more information should be given about the representations of the symmetric group which were introduced here.

Let me give some more comments . As pointed out above, it is surprising that concatenation and shuffle of words have something to do with the composition of permutations. I give here two other illustrations of this. By the formula of Dynkin - Specht - Wever, one has for each homogeneous Lie polynomial P of degree n

$$[P] = n\, P$$

where the endomorphism $P \to [P]$ is defined for any word $a_1 \dots a_n$ ($a_i \in A$) by

$$[a_1 \dots a_n] = [\dots[a_1,a_2],a_3],\dots,a_n]$$

(bracketing from left to right). Interpreting $[a_1 \dots a_n]$ as an element e of $Q[S_n]$, this implies that this element satisfies

$$e^2 = n\, e$$

i.e. e/n is idempotent (this may be proved as in theorem 2.1).

Example

$$[[a_1,a_2],a_3] = a_1a_2a_3 - a_2a_1a_3 - a_3a_1a_2 + a_3a_2a_1$$

$$= 1 - (12) - (132) + (13)$$

It is easily shown, moreover that the element $[a_1 \dots a_n]$ of $Q[S_n]$ may be factorized as

$$[a_1 \dots a_n] = (1 - (12)(23)\dots(n-1,n))\dots\dots(1 - (12)(23))\ (1 - (12))$$

This gives a further connection between Lie brackets and composition of permutations.

Acknowledgements

The main part of the present work was done during a 6 months stay of the author at the University of Saarbrucken, in spring 1982, by invitation of Pr. G. Hotz, who is gratefully acknowledged.

Correspondence with J. Dixmier and conversations with D. Perrin and P. Leroux were also helpful.

Added in proof:

The representation of the symmetric group on V^1 is obtained by A. Joyal in a different way: it corresponds to the logarithm in the theory of species. Moreover, the methods of generating series of species (more precisely: the "séries indicatrices", see [A. Joyal, une théorie combinatoire des séries formelles, Advances in Maths. 42 (1981) 1-82]) allow him to give formulas for the computation of the multiplicities of the irreducible components of this representation. These where computed up to n = 12 by Nantel Bergeron at UQAM. It seems that, except for the trivial and alternating representations and a few exceptions, each irreducible representation appears in V^1 (personal communication).

References

[1] N. Bourbaki, Groupes et algèbres de Lie, chapitre 1, Hermann (1971).

[2] K.T. Chen, Integration of paths, geometric invariants and a generalized Baker-Haussdorff formula, Annals Maths 65 (1957) 163-178.

[3] J. Dixmier, Algèbres enveloppantes, Hermann (1974)

[4] M. Fliess, D. Normand-Cyrot, Algèbres de Lie nilpotentes, intégrales itérées de K.T. Chen et formule de Baker-Campbell-Hausdorff, Lect. Notes Maths 920 (1982) 257-265.

[5] M. Fliess, C. Reutenauer, Picard-Vessiot theory of bilinear systems, IEEE 23rd Congress on Decision and Control, Proc., 1153-1157 (1983).

[6] R.M. Hain, On the indecomposable elements of the bar construction, preprint (1985); see also: de Rham homotopy theory of complex algebraic varieties, manuscript (1984), appendix.

[7] G.P. Hochschild, Basic theory of algebraic groups and Lie algebras, Springer Verlag (1981).

[8] M. Lothaire, Combinatorics on words, Addison Wesley (1983)

[9] D. Perrin, G. Viennot, A note on shuffle algebras (1981), manuscript.

[10] R. Ree, Lie elements and an algebra associated with shuffles, Annals Maths 68 (1958) 210-220.

[11] C. Reutenauer, Point générique du plus petit groupe algébrique dont l'algèbre de Lie contient plusieurs matrices données, Comptes Rendus Acad. Sci. Paris 293 (1981) 577-580.

[12] C. Reutenauer, The shuffle algebra on the factors of a word is free, J. Combin. Theory 38 (1985) 48-57.

[13] J. Riordan, An introduction to combinatorial analysis, John Wiley 1967.

[14] L. Solomon, On the Poincaré-Birkhoff-Witt theorem, J. Combin. Theory (A) 4 (1968) 363-375.

A BAKER'S DOZEN OF CONJECTURES CONCERNING PLANE PARTITIONS

Richard P. Stanley*
Department of Mathematics
Massachusetts Institute of Technology
Cambridge, MA 02139

Many remarkable conjectures have been made recently concerning the explicit enumeration of certain classes of tableaux. Most of these are due to or arise from the work of W. Mills, D. Robbins, and H. Rumsey. Here we will survey the most prominent of these conjectures (omitting some rather technical refinements). We will for the most part not discuss the background of these conjectures and their connections with symmetric functions and representation theory. We will also for the most part ignore a host of known results which are very similar to many of the conjectures and which make the conjectures considerably more tantalizing. The reader should consult the references cited below for further information.

We begin with the necessary definitions. A plane partition π is an array $\pi = (\pi_{ij})_{i,j\geq 1}$ of nonnegative integers π_{ij} with finite sum $|\pi| = \Sigma\, \pi_{ij}$, which is weakly decreasing in rows and columns [10]. The nonzero π_{ij} are called the parts of π, and normally when writing examples only the parts are displayed. Such terminology as "number of rows of π" refers only to the parts of π. Thus, for example,

```
443211
43311
321
22
1
```

is a plane partition π with $|\pi| = 38$, and with 17 parts, 5 rows, and 6 columns. We now list some special classes of plane partitions.

column-strict: the parts strictly decrease in each column.

row-strict: the parts strictly decrease in each row.

symmetric: $\pi_{ij} = \pi_{ji}$ for all i,j.

cyclically symmetric: the i-th row of π, regarded as an ordinary partition, is conjugate (in the sense of [4, p. 21]) to the i-th column, for all i.

totally symmetric: symmetric and cyclically symmetric.

(r,s,t)-self-complementary: π has $\leq r$ rows, $\leq s$ columns, largest part $\leq t$, and $\pi_{ij} + \pi_{r-i+1,s-j+1} = t$ for all $1 \leq i \leq r$, $1 \leq j \leq s$.

*Partially supported by NSF Grant # 8104855-MCS

Example. Consider the three plane partitions

```
4431        4432        44321
3321        4331        4222
321         332         321
2           21
```

The first is cyclically but not totally symmetric, while the second is totally symmetric. Moreover, the third is (3,5,4) - self-complementary.

A Gelfand pattern (see [3]) is a triangular array

$$\begin{matrix} a_{11} & a_{12} & \cdots & a_{1n} \\ & a_{22} & \cdots & a_{2n} \\ & & \vdots & \\ & & & a_{nn} \end{matrix}$$

of nonnegative integers a_{ij} which weakly increase in rows and such that $a_{i-1,j-1} \leq a_{ij} \leq a_{i-1,j}$ for all $2 \leq i \leq j \leq n$. A Gelfand pattern is strict if the rows stricly increase. A strict Gelfand pattern with first row $1,2,\ldots,n$ is called a monotone triangle of length n. An $n \times n$ alternating sign matrix is an $n \times n$ matrix whose entries are $0, \pm 1$, whose row and column sums are all equal to 1, and such that the nonzero entries of every row and column alternate in sign. An element a_{ij} of a strict Gelfand pattern T is special if $2 \leq i \leq j \leq n$ and $a_{i-1,j-1} < a_{ij} < a_{i-1,j}$. Let $s(T)$ denote the number of special elements of T. There is a simple bijection [6] between monotone triangles T of length n and alternating sign matrices A of length n, for which $s(T)$ is the number of -1's in A. There is also a simple bijection (e.g., [2]) between Gelfand patterns with first row $\lambda_n \leq \lambda_{n-1} \leq \cdots \leq \lambda_1$ and column-strict plane partitions of shape $\lambda = (\lambda_1, \lambda_2, \ldots, \lambda_n)$ (i.e., λ_i parts in row i) and largest part $\leq n$.

Example. The seven monotone triangles T of length 3 are given by

```
123   123   123   123   123   123   123
12    12    13    13    13    23    23
1     2     1     2     3     2     3
```

All of them satisfy $s(T) = 0$ except the fourth, for which $s(T) = 1$.

A shifted plane partition is defined analogously to plane partition, except that the array (π_{ij}) is defined only for $1 \leq i \leq j$. Such terminology as "column-strict" and "number of rows" is carried over in an obvious way to shifted plane partitions. For example,

```
554331
 4322
  11
```

is a column-strict shifted plane partition with 3 rows and 6 columns.

Let μ be an integer. A column-strict shifted plane partition (CSSPP) is of <u>class μ</u> if the first entry of each row exceeds the row length by precisely 2μ. There is a simple bijection [8] between CSSPP's of class 1 with $\leq n$ columns and <u>descending plane partitions</u> (as defined by G. Andrews [1]) with largest part $\leq n+1$. There is also a simple bijection between CSSPP's of class 0 with $\leq n$ columns and cyclically symmetric plane partitions with largest part $\leq n$ (see [8]). A part π_{ij} of a CSSPP of class μ is <u>special</u> if $\mu < \pi_{ij} \leq j-i+\mu$, and we write $s(T)$ for the number of special parts of T.

<u>Example</u>. The seven CSSPP's of class 1 with ≤ 2 columns are given by

$$\phi \quad 3 \quad 41 \quad 42 \quad 43 \quad 44 \quad \begin{matrix} 44 \\ 3 \end{matrix}$$

All of these satisfy $s(T) = 0$ except the fifth, for which $s(T) = 1$.

We now are ready to list the conjectures (as of November, 1985), together with some related theorems.

<u>Theorem</u> (equivalent to [1, Thm. 7]). The number of CSSPP's of class 1 and $\leq n-1$ columns is equal to

$$A_n := \prod_{i=0}^{n-1} \frac{(3i+1)!}{(n+i)!} .$$

<u>Conjecture 1</u> [6]. The number of $n \times n$ alternating sign matrices is equal to A_n.

<u>Conjecture 2</u> [7, Conj. 1][11, Case 10]. The number of totally symmetric $(2n,2n,2n)$-self-complementary plane partitions is equal to A_n.

<u>Note</u>. One can give a bijection [7] between totally symmetric $(2n,2n,2n)$-self-complementary plane partitions and shifted plane partitions $\pi = (\pi_{ij})$ of shape $(n-1,n-2,\ldots,1)$ such that $n-1 \leq \pi_{ij} \leq n$ for all parts π_{ij} of π.

<u>Note</u>. It is not known whether the number of $n \times n$ alternating sign matrices is equal to the number of totally symmetric $(2n,2n,2n)$-self-complementary plane partitions.

<u>Conjecture 3</u> [6, Conj. 2]. The number of monotone triangles of length n with bottom entry $a_{nn} = r$ (equivalently, the number of $n \times n$ alternating sign matrices (α_{ij}) with $\alpha_{nr} = 1$) is equal to

$$\binom{2n-2}{n-1}^{-1}\binom{n+r-2}{n-1}\binom{2n-r-1}{n-1}A_{n-1}$$

Note. One easily deduces Conjecture 1 from Conjecture 3 .

Conjecture 4 [6, Conjs. 4 and 5]. Define $A_n(x) = \sum_T x^{s(T)}$, where T ranges over all monotone triangles of length n . Define $B_{2n+1}(x) = \sum_T x^{s(T)}$, where T ranges over all strict Gelfand patterns with first row 1,3,5,...,2n-1. Then there exist polynomials $B_{2n}(x)$ for which

$$A_n(x) = \begin{cases} B_n(x)B_{n+1}(x) & , n \text{ odd} \\ 2\ B_n(x)B_{n+1}(x) & , n \text{ even} . \end{cases}$$

Note. For a conjectured explicit value of $B_{2n+1}(1)$, see the note following Conjecture 9.

Note. Conjecture 1 is equivalent to the assertion $A_n(1) = A_n$. It is not difficult to show [6, Cor. on p. 358] that $A_n(2) = 2^{\binom{n}{2}}$. In fact, much more can be said concerning the weight $2^{s(T)}$ of a strict Gelfand pattern T , and there are strong connections with the theory of symmetric functions. For instance, if $\sigma_i(T)$ denotes the i-th row sum of T , then it can be shown that

$$\sum_T 2^{s(T)} x_1^{\sigma_1(T)-\sigma_2(T)} x_2^{\sigma_2(T)-\sigma_3(T)} \cdots x_n^{\sigma_n(T)}$$

$$= s_\lambda(x_1,\ldots,x_n) \prod_{1\le i<j\le n}(x_i+x_j) ,$$

where T ranges over all strict Gelfand patterns with first row $(\lambda_n,\lambda_{n-1}+1,\ldots,\lambda_1+n-1)$, and where s_λ denotes the Schur function (as defined, e.g., in [4] or [10]) corresponding to the partition $\lambda = (\lambda_1,\ldots,\lambda_n)$.

Conjecture 5 [6, Conj. 6]. $A_n(3) = 3^{t(n)}H_n$, where

$$t(n) = \begin{cases} m(m-1) & , n = 2m \\ m^2 & , n = 2m+1 , \end{cases}$$

amd where H_n is determined by the recurrence

$$H_0 = 1 , \quad \frac{H_{2n+1}}{H_{2n}} = \frac{\binom{3n}{n}}{\binom{2n}{n}} , \quad \frac{H_{2n}}{H_{2n-1}} = \frac{4}{3}\frac{\binom{3n}{n}}{\binom{2n}{n}} .$$

Conjecture 6 [8, Conj. in Sect. 4]. Define $Z_n(x,\mu) = \sum_T x^{s(T)}$, where T ranges over all CSSPP of class μ and rows of length $\leq n$. Then $Z_n(2,\mu)$ is determined by the recurrence $Z_1(2,\mu) = 2$,

$$\frac{Z_{2m}(2,\mu)}{Z_{2m-1}(2,\mu)} = 2^m \prod_{i=1}^{m} \frac{\mu+2m+2i-1}{m+i}$$

$$\frac{Z_{2m+1}(2,\mu)}{Z_{2m}(2,\mu)} = 2^{m+1} \prod_{i=1}^{m} \frac{\mu+2m+2i-1}{m+i} .$$

Note. A strengthening of Conjecture 1 is given by $Z_n(x,1) = A_n(x)$, where $A_n(x)$ is defined in Conjecture 4 (see [8, Sect. 4]).

Conjecture 7. (see [11, Case 4]). The number of totally symmetric plane partitions with largest part $\leq n$ is equal to

$$T_n = \prod_{1\leq i\leq j\leq k\leq n} \frac{i+j+k-1}{i+j+k-2} .$$

Note. It is not hard to show that the number of totally symmetric plane partitions with largest part $\leq n$ is also equal to

a) the number of row-strict shifted plane partitions with largest part $\leq n$,

b) the number of order ideals of the poset $L(3,n)$ of Ferrers diagrams fitting in a $3\times n$ rectangle, ordered by inclusion,

c) the sum of the minors of all orders (including the void minor equal to 1) of the matrix whose (i,j)-entry is $\binom{i}{j}$ for $0\leq i, j\leq n-1$.

Note. All quantities arising in connection with Conjecture 7 have natural q-analogues. The q-analogue of T_n is

$$T_n(q) = \prod_{1\leq i\leq j\leq k\leq n} \frac{1-q^{i+j+k-1}}{1-q^{i+j+k-2}}$$

The q-analogue of the number of totally symmetric plane partitions with largest part $\leq n$ is the polynomial $N'_G(B;q)$ defined in [11], where $B = B(n,n,n)$ and $G = S_3$. The q-analogue of (a) is just $\Sigma q^{|\pi|}$, summed over all π satisfying (a). The q-analogue of (b) is $\Sigma q^{|I|}$, summed over all order ideals I of $L(3,n)$. Finally, the q-analogue of

(c) corresponds to the matrix with (i,j)-entry $q^{i+1+\binom{j+1}{2}}\begin{bmatrix} i \\ j \end{bmatrix}$ $0 \leq i, j \leq n-1$. As in Conjecture 7, the last four quantities are known to be equal, and are conjectured to equal $T_n(q)$.

Conjecture 8 (D. Robbins, et al.; see [11, Case 9]). The number of cyclically symmetric (2n,2n,2n)-self-complementary plane partitions is equal to A_n^2.

Note. It is not known whether the number of cyclically symmetric (2n,2n,2n)-self-complementary plane partitions is the square of the number which are also symmetric (Conjecture 2). Perhaps there is a bijection which shows the equivalence of Conjectures 2 and 8 without proving either one.

Conjecture 9 (implicit in [8]). The number F_n of $n\times n$ alternating sign matrices which are invariant under a reflection about a vertical axis is given by the recurrence

$$F_1 = 1,\ F_{2n} = 0\ ,\ \frac{F_{2n+1}}{F_{2n-1}} = \frac{\binom{6n-2}{2n}}{2\binom{4n-1}{2n}}\ .$$

Note. It is easy to see that $F_{2n+1} = B_{2n+1}(1)$, as defined in Conjecture 4. Moreover, the number of strict Gelfand patterns (a_{ij}) with first row 1,3,...2n-1 which are "flip-symmetric", in the sense that $a_{ij} + a_{i,n+i-j} = 2n$ for all $1 \leq i \leq j \leq n$, is equal to P_{2n+1}, as defined in Conjecture 12.

Conjecture 10 [7, Conj. 5]. The number of $n\times n$ alternating sign matrices which are invariant under a 180° rotation is equal to the quantity H_n of Conjecture 5.

Note. It is not known whether Conjectures 5 and 10 are equivalent, i.e., whether $3^{-t(n)}A_n(3)$ is equal to the number of $n\times n$ alternating sign matrices invariant under a 180° rotation.

Conjecture 11 (D. Robbins; see [9, Sect. 3.5]). The number Q_n of $n\times n$ alternating sign matrices which are invariant under a 90° rotation is given by the recurrence

$$Q_1 = 1\ ,\ Q_{4n+2} = 0\ ,\ \frac{Q_{4n+3}}{Q_{4n+1}} = \frac{\binom{3n+1}{n}^2}{\binom{2n}{n}^2}$$

$$\frac{Q_{4n+5}}{Q_{4n+3}} = \frac{3\binom{3n+2}{n}^2}{\binom{2n+1}{n}^2}, \quad \frac{Q_{4n}}{Q_{4n-1}} = \frac{2\binom{3n-1}{n}}{\binom{2n}{n}}.$$

Conjecture 12 (W.H. Mills; see [9, Sect. 4.2]). The number P_n of $n \times n$ alternating sign matrices which are invariant under reflections in both a horizontal axis and a vertical axis is given by the recurrence $P_1 = 1$, $P_{2n} = 0$,

$$\frac{P_{4n+3}}{P_{4n+1}} = \frac{(3n+1)\binom{6n}{2n}}{(4n+1)\binom{4n}{2n}}, \quad \frac{P_{4n+1}}{P_{4n-1}} = \frac{(3n-1)\binom{6n-3}{2n-1}}{(4n-1)\binom{4n-2}{2n-1}}.$$

Conjecture 13 (D. Robbins; see [9, Sect. 3.7]). The number X_n of $n \times n$ alternating sign matrices which are invariant under reflections in both diagonals satisfies $X_1 = 1$,

$$\frac{X_{2n+1}}{X_{2n-1}} = \frac{\binom{3n}{n}}{\binom{2n-1}{n}}.$$

Note. There are no conjectures at present for the cardinalities of two additional symmetry classes of $n \times n$ alternating sign matrices, viz., those that are symmetric matrices (i.e., invariant under a reflection in the main diagonal), and those that are invariant under the full symmetry group of the square. Call these cardinalities S_n and K_n, respectively. Moreover, no conjecture is known for X_{2n} as defined by Conjecture 13.

Note. There are a total of ten symmetry classes of plane partitions with $\le r$ rows, $\le s$ columns, and largest part $\le t$ [11]. Seven of these classes have been successfully counted, while the remaining three correspond to Conjectures 2, 7, and 8.

Note. In [9] many of the above conjectures related to symmetry classes of alternating sign matrices are strengthened by considering various weights on the alternating sign matrices under consideration. There also appear some surprising connections between different symmetry classes (which follow from the conjectures themselves, but which perhaps can be proved independently). For instance, it follows from Conjecture 5 above that $H_{2n} = Z_n(1,0)A_n$ (a special case of

[9, Conj. 3.3.1]), and from Conjectures 1,5, and 11 above that $Q_{4n} = A_n^2 H_{2n}$, $Q_{4n+1} = A_n^2 H_{2n+1}$, $Q_{4n-1} = A_n^2 H_{2n-1}$ (a special case of [9, Conj. 3.5.1]).

We conclude with a table listing some of the values of the functions discussed above. Many of these values are taken from [9]. An entry marked * denotes a number of eight digits or more whose value we omit.

n	1	2	3	4	5	6	7	8
A_n	1	2	7	42	429	7436	218348	*
H_n	1	2	3	10	25	140	588	5544
T_n	2	5	16	66	352	2431	21760	252586
$Z_n(1,0)$	2	5	20	132	1452	26741	826540	*
F_{2n-1}	1	1	3	26	646	45885	9304650	*
Q_n	1	0	1	2	3	0	12	40
P_{2n-1}	1	1	1	2	6	33	286	4420
X_n	1	2	3	8	15	52	126	568
S_n	1	2	5	16	67	368	2630	24376
K_{2n-1}	1	1	1	2	4	13	46	248

Moreover: $Q_9 = 100$, $Q_{10} = 0$, $Q_{11} = 1225$, $Q_{12} = 6860$,

$X_9 = 1782$, $X_{10} = 10436$, $X_{11} = 42471$, $X_{12} = 323144$, $X_{13} = 1706562$, $X_{14} = 16866856$

$K_{17} = 1516$

$B_1(x) = B_2(x) = B_3(x) = 1$, $B_4(x) = 6+x$, $B_5(x) = 2+x$, $B_6(x) = 60+70x+12x^2+x^3$, $B_7(x) = 6+13x+6x^2+x^3$, $B_8(x) = 840+3080x + 3038x^2 + 1224x^3 + 195x^4 + 20x^5 + x^6$, $B_9(x) = 24 + 136x + 234x^2$ $176x^3 + 63x^4 + 12x^5 + x^6$

$Z_1(2,\mu) = 2$, $Z_2(2,\mu) = 2(\mu+3)$, $Z_3(2,\mu) = 4(\mu+3)^2$, $Z_4(2,\mu) = \frac{4}{3}(\mu+3)^2(\mu+5)(\mu+7)$, $Z_5(2,\mu) = \frac{8}{9}(\mu+3)^2(\mu+5)^2(\mu+7)^2$.

REFERENCES

1. G.E. Andrews, Macdonald's conjecture and descending plane partitions, in Combinatorics, Respresentation Theory, and Statistical Methods in Groups (T.V. Narayana, R.M. Mathsen, and J.G. Williams, eds.), Marcel Dekker, New York and Basel, 1980, 91-106.
2. L. Carlitz and R.P. Stanley, Branchings and partitions, Proc. Amer. Math. Soc. 53 (1975), 246-249.
3. I.M. Gelfand and M.L. Tseitlin, Finite-dimensional representations of the group of unimodular matrices (Russian), Dokl. Akad. Nauk. USSR 71 (1950), 825-828.
4. I.G. Macdonald, Symmetric Functions and Hall Polynomials, Oxford Univ. Press, London, 1979.
5. W.H. Mills, D.P. Robbins, and H. Rumsey, Jr., Proof of the Macdonald conjecture, Invent. math. 66 (1982), 73-87.
6. W.H. Mills, D.P. Robbins, and H. Rumsey, Jr., Alternating sign matrices and descending plane partitions, J. Combinatorial Theory (A) 34 (1983), 340-359.
7. W.H. Mills, D.P. Robbins, and H. Rumsey, Jr., Self-complementary totally symmetric plane partitions, to appear.
8. W.H. Mills, D.P. Robbins, and H. Rumsey, Jr., Enumeration of a symmetry class of plane partitions, to appear.
9. D.P. Robbins, Symmetry classes of alternating sign matrices, preprint.
10. R.P. Stanley, Theory and application of plane partitions, Parts 1 and 2, Studies in Applied Math. 50 (1971), 167-188, 259-279.
11. R.P. Stanley, Symmetries of plane partitions, J. Combinatorial Theory (A), to appear.

Combinatorics of Jacobi configurations I : Complete oriented matchings

Volker Strehl
Institut fuer Mathematische Maschinen und Datenverarbeitung I
Universitaet Erlangen-Nuernberg

D-8520 Erlangen, Fed.Rep.Germany

Abstract:

The first combinatorial model for the Jacobi polynomials has been introduced by Foata/Leroux (Proc.AMS 87(1983)). Here a second model - complete oriented matchings - is presented and the equivalence of both models is proved combinatorially. The new model allows rather simple derivations for a number of explicit expressions for generating polynomials for either kind of configurations - this fact is illustrated in special cases related to the Gegenbauer polynomials.

0 Introduction

The investigation of identities for special functions from a combinatorial point of view has attracted quite a number of authors in recent years. The reader is referred to A.Garsia's "statement of policy" in [GR] and to the survey article by D.Foata [F2]. Two particularly nice examples of combinatorial approaches to identities in classical analysis are

- the derivation of Mehler's formula for the Hermite polynomials, based on the "superposition" of involutions, by Foata [F1];
- the derivation of Jacobi's generating function for the Jacobi polynomials using an endofunction-model, by Foata and Leroux [FL].

The present article is related to both of these examples. Indeed, what is presented here is a new combinatorial model for Jacobi polynomials - complete oriented matchings - equivalent to the Foata-Leroux model in a nontrivial way, (the proof of equivalence is the major result of this article), and based on fixed point free involutions and the idea of superposition (=joint action) of involutions. This model has the advantage of being easier to handle in some situations, in particular when one considers special cases related to the Gegenbauer polynomials, and

of being very close to the methods used in [ST] for the generalized Hermite polynomials. The first of these aspects will be illustrated in the second half of this article.
The notion of "Jacobi configuration" in the title of this article refers to combinatorial configurations related to Jacobi polynomials: the endofunctions of Foata-Leroux, the complete oriented matchings introduced here, and possibly others. More methods and results about Jacobi configurations will be presented in subsequent articles (e.g. material related to generating functions, recurrence relations etc).
For a combinatorial treatment of the "basic" properties of the Jacobi polynomials the reader is referred to [LS], where a slight extension of the Foata-Leroux model is used.
Due to the lack of space the present article definitely suffers from a lack of illustrative examples. The reader should consult [FL] for graphical illustrations of Jacobi endofunctions, and he should try to visualize the proofs given here by drawing his own examples of complete oriented matchings.

1 Definitions and notation - Statement of the main result

Let S denote a finite set. The set JAC(S) of Jacobi endofunctions on S is the set of all pairs ((A,B),f) where (A,B) is an ordered bipartition of S(i.e. $A \cup B = S$) and $f:S \longrightarrow S$ is an endofunction such that its restrictions $f|A:A \longrightarrow S$ and $f|B:B \longrightarrow S$ are both injective. JAC(A,B) will denote the set of all Jacobi endofunctions using the bipartition (A,B). We will denote by cyc(f) (cyc(f|A) resp., cyc(f|B) resp.) the number of cycles of f ((f|A) resp., (f|B) resp.) , i.e. the number of type-m (type-a resp., type-b resp.) connected components in the terminology of [FL]. An element $x \in S$ is said to be regular (singular resp.) with respect to ((A,B),f) (or simply: w.r.t. f) if $|f^{-1}(f(x))|=2$ (=1 resp.). The number of regular points of f is $2*(|S|-|f(S)|) = 2*\mathrm{def}(f)$, where def(f) denotes the defect of f. The function f is reduced if it has no singular points - in this case |S| must be even. The notation $JAC_{red}(S)$ and $JAC_{red}(A,B)$ refers to reduced Jacobi endofunctions.

For any finite set S let #S denote a copy of S. This can be used to form the set <S> := $S \cup \#S$. The sign # will also be used to denote the obvious involution acting on <S>. By a complete matching on <S> we mean any fixed point free involution s of <S>; CM<S> will denote the set of all complete matchings on <S> . For any $s \in CM\langle S\rangle$ we will write its transpositions as (unordered) pairs {x,y}, where $x \neq y$ and $s(x)=y$.

A transposition (x,y) is said to be positive (negative resp., mixed resp.) if both (none resp., exactly one resp.) of x and y belongs to S; pt(s) (nt(s) resp., mt(s) resp.) denotes the number of positive (negative resp., mixed resp.) transpositions of s. Note that

$$pt(s) = nt(s) \quad \text{and} \quad pt(s)+nt(s)+mt(s) = |S| \ .$$

s is reduced if it has no mixed transpositions. The joint action on S of a complete matching $s \in CM<S>$ and # induces a set of connected components ("s-#-cycles", see [F1] for an interesting application of this fact), the cardinality of which is denoted by cyc(s).
Let now a,b be two distinct symbols ("colours"). The set $COM<S>$ of complete oriented matchings consists of all pairs (s,c) where

$$s \in CM<S> \ , \qquad c : <S> \longrightarrow \{a,b\} \ ,$$

which satisfy the usual compatibility condition:

$$s(x) = y \implies c(x) \neq c(y) \ .$$

For any bipartition (A,B) of S we let $COM<A,B>$ denote the set of those $(s,c) \in COM<S>$ such that $A = c^{-1}(\{a\})\cap S$, $B = c^{-1}(\{b\})\cap S$. The notation $COM_{red}<S>$ and $COM_{red}<A,B>$ will be used for reduced complete oriented matchings.

Apart from Jacobi endofunctions and complete oriented matchings the following classes of configurations will be used in the sequel:

BIJ(A,B) : the set of bijective mappings from A onto B , where it is understood that A,B are disjoint ,finite sets of equal cardinality;

$LAG_2(A,B)$: the set of two-colored Laguerre configurations, i.e. the set of constructs $(((A_1,A_2),B),f)$, where (A_1,A_2) is an ordered bipartition of A, A and B are disjoint finite sets, and f is an injective mapping from A into $A\cup B$.
(Thus ((A,B),f) is a Laguerre configuration in the sense of [FS] - see that article for a justification of the terminology).

We are now ready to state the main result of this article.

Theorem: For any pair (A,B) of disjoint, finite sets there is a bijection

$$F : JAC(A,B) \longrightarrow COM<A,B> : f \longmapsto (s_f, c_f)$$

which satisfies

$$cyc(f) = cyc(s_f) \quad \text{and} \quad def(f) = pt(s_f) \ .$$

In particular, $JAC_{red}(A,B)$ is mapped onto $COM_{red}<A,B>$.

As will be evident from the proof given below, connectivity is preserved by F in the following sense: $x,y \in A\cup B$ are in the same component w.r.t. f (i.e. there exist m,n>=0 s.th. $f^m(x) = f^n(y)$) if and only if they are in the same s_f-#-cycle (i.e. there exists r>=0 s.th. $(s_f\circ\#)^r(x)=y$ or $(s_f\circ\#)^r(x)=\#(y)$).

Illustration: Let $S = \{1,2,\ldots,7\}$; an element $s \in CM\langle S\rangle$, given by its set of transpositions $\{\{1,4\},\{2,\#2\},\{3,\#4\},\{\#1,\#3\},\{5,\#7\},\{6,\#5\},\{7,\#6\}\}$ may be visualized by

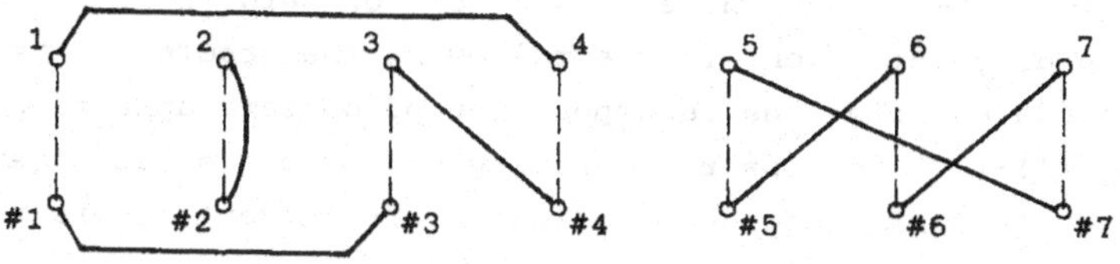

where the dashed lines indicate the action of the involution #.
One observes : $cyc(s) = 3$, $pt(s) = nt(s) = 1$, $mt(s) = 5$.

Let now $c:\langle S\rangle \longrightarrow \{a,b\}$ be given by
$c^{-1}(\{a\}) = \{1,2,3,6,\#3,\#6,\#7\}$, $c^{-1}(\{b\}) = \{4,5,7,\#1,\#2,\#4,\#5\}$.
Then c is compatible with s above, as is clear from the visual presentation:

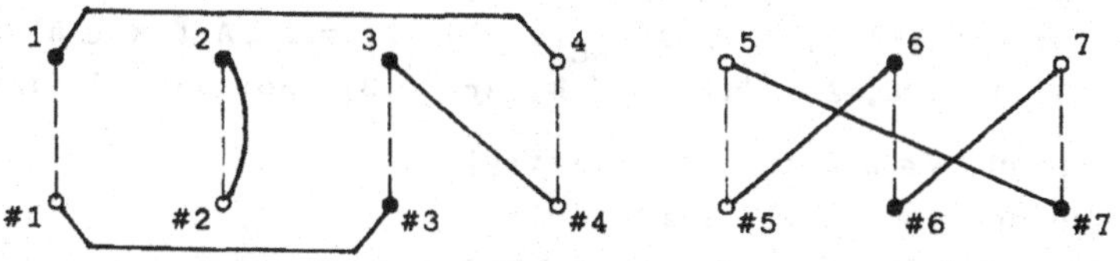

Indeed, (s,c) belongs to $COM\langle A,B\rangle$, where $A=\{1,2,3,6\}$ and $B=\{4,5,7\}$.

2 Proof of the theorem

The proof given in this section uses the idea of recursive descent, thus the bijection F, as announced in the statement of the theorem, is not given explicitely. It is possible to give an explicit description of F, but here we have decided to present the recursive method for two reasons:

- it is considerably more elegant;
- it introduces in a very natural way the important concept of 'order' (see the comment at the end of this section).

Our recursive proof is based on the following four facts:

(I) For any finite set S:

$$JAC(S) \cong \bigoplus \{ JAC_{red}(D) \times LAG_2(E,D) \ ; \ D \cup E = S, \ |D| \text{ even} \} \ .$$

(II) For any finite set S of even cardinality:

$$JAC_{red}(S) \cong \bigoplus \{ JAC(A) \times BIJ(A,B) \ ; \ A \cup B = S, \ |A| = |B| \} \ .$$

(III) For any finite set S :

$$COM\langle S\rangle \cong \bigoplus \{ COM_{red}\langle D\rangle \times LAG_2(E,D) \ ; \ D \cup E = S, \ |D| \text{ even} \} \ .$$

(IV) For any finite set S of even cardinality :

$$\mathrm{COM}_{red}\langle S\rangle = \bigoplus \{ \mathrm{COM}\langle A\rangle \times \mathrm{BIJ}(A,B) \; ; \; A\cup B=S, \; |A|=|B| \}$$

Here "..=.." is to be read as: "there is a bijection between .. and ..", the $\bigoplus$ indicates a disjoint union. In the following the proofs of these facts will be indicated, i.e. the decomposition of objects appearing on the l.h.s. of (I)-(IV) into pairs appearing on the r.h.s. is given. The reader may verify that these decompositions are in fact reversible.

Proof of (I) : For any $((A,B),f)\in \mathrm{JAC}(S)$ let D_f (E_f resp.) denote the set of regular (singular resp.) points of f. For $x\in D_f$ let $g(x):=f^m(x)$, where m is the least $i>=1$ such that $f^i(x)\in D_f$. For $x\in E_f$ let $h(x):=f(x)$. One checks that

$((A\cap D_f,B\cap D_f),g)\in \mathrm{JAC}_{red}(D_f)$ and $(((A\cap E_f,B\cap E_f)D_f),h)\in \mathrm{LAG}_2(E_f,D_f)$.

Proof of (II) : For any $((A,B),f)\in \mathrm{JAC}_{red}(S)$ let $A_f:=\{x\in A;f(x)\in A\}$ and $B_f:=\{x\in A;f(x)\in B\}$, i.e. $A_f=A\cap f^{-1}(A)$ and $B_f=A\cap f^{-1}(B)$. For $x\in A$ define

$g(x):=$ the unique $y\in A$ s.th. $f(f(x))=f(y)$,

$h(x):=$ the unique $y\in B$ s.th. $f(x)=f(y)$.

One checks that $((A_f,B_f),g)\in \mathrm{JAC}(A)$ and $h\in \mathrm{BIJ}(A,B)$.

Proof of (III) : For any $(s,c)\in \mathrm{COM}\langle S\rangle$ let D_s (E_s resp.) denote the elements of S belonging to positive (mixed resp.) transpositions of s. We then define $(t,d)\in \mathrm{COM}_{red}\langle D_s\rangle$ as follows:

for $x\in D_s$: $t(x):=s(x)$ and $d(x):=c(x)$;

for $x\in \#D_s$: let $x_s:=(\#\circ s)^m(x)$, where m is the least $i>=0$ such that $(\#\circ s)^i(x)$ belongs to a negative transposition of s; we now put $t(x):=$the unique $y\in \#D_s$ s.th. (x_s,y_s) is a negative transposition of s, $d(x):=c(x_s)$.

On the other hand, an element of $\mathrm{LAG}_2(E_s,D_s)$ is given by $((A\cap E_s,B\cap E_s),D_s,h)$, where $h(x):=\#\circ s(x)$ for $x\in E_s$.

Proof of (IV) : For any $(s,c)\in \mathrm{COM}_{red}\langle S\rangle$ let $A_c:=c^{-1}(\{\underline{a}\})\cap S$ and $B_c:=c^{-1}(\{\underline{b}\})\cap S$. An element $(t,d)\in \mathrm{COM}\langle A_c\rangle$ is then specified by the following requirement: transpositions of t are those pairs $\{x,y\}$ where

$x,y\in A_c$ and $\{\#x,\#y\}$ is an s-transposition,

$x\in A_c,\#y\in A_c$ and $\{\#x,\#s(y)\}$ is an s-transposition,

$\#x\in A_c,\#y\in A_c$ and $\{\#s(x),\#s(y)\}$ is an s-transposition .

An element $h\in \mathrm{BIJ}(A_c,B_c)$ is given by $h(x):=s(x)$ $(x\in A_c)$.

The proof of the theorem uses facts (I)-(IV) inductively. Let

$$\mathrm{JAC} := \bigoplus \{ \mathrm{JAC}(S) \ ; \ S \text{ a finite set} \} ,$$

and define JAC_{red}, COM, and COM_{red} similarly. From the proofs of (I) and (III) we have "reduction" mappings

$$R : \mathrm{JAC} \longrightarrow \mathrm{JAC}_{red} : ((A,B),f) \longmapsto ((A \cap D_f, B \cap E_f),g) ,$$

$$R : \mathrm{COM} \longrightarrow \mathrm{COM}_{red} : (s,c) \longmapsto (t,d) ,$$

and from the proofs of (II) and (IV) we have "contraction" mappings

$$C : \mathrm{JAC}_{red} \longrightarrow \mathrm{JAC} : ((A,B),f) \longmapsto ((A_f,B_f),g) ,$$

$$C : \mathrm{COM}_{red} \longrightarrow \mathrm{COM} : (s,c) \longmapsto (t,d) .$$

Repeated application of reduction-contraction pairs will make any configuration disappear eventually, since each contraction reduces the "size" by one half. This leads to the concept of "order" for both Jacobi endofunctions and complete oriented matchings:
The sets $\mathrm{JAC}^{(k)}$ ($\mathrm{JAC}^{(k)}_{red}$ resp.) of Jacobi endofunctions (reduced Jacobi-endofunctions resp.) of order k are defined inductively by:

$\mathrm{JAC}^{(0)}_{red} := \{\emptyset\}$, where $\emptyset$ denotes the empty function ;

$\mathrm{JAC}^{(0)} := R^{-1}(\mathrm{JAC}^{(0)}_{red})$, which is the set of all ((A,B),f) with f a permutation of $A \cup B$;

$\mathrm{JAC}^{(k)}_{red} := C^{-1}(\mathrm{JAC}^{(k-1)})$, for k>0 ;

$\mathrm{JAC}^{(k)} := R^{-1}(\mathrm{JAC}^{(k)}_{red})$, for k>0 .

Similar definitions apply for (reduced) complete oriented matchings:

$\mathrm{COM}^{(0)}_{red} := \{\emptyset\}$, where $\emptyset$ denotes the empty matching ;

$\mathrm{COM}^{(0)} := R^{-1}(\mathrm{COM}^{(0)}_{red})$, which is the set of all (s,c) s.th. s has mixed transpositions only ;

$\mathrm{COM}^{(k)}_{red} := C^{-1}(\mathrm{COM}^{(k-1)})$, for k>0 ;

$\mathrm{COM}^{(k)} := R^{-1}(\mathrm{COM}^{(k)}_{red})$, for k>0 .

Note that there is a local correspondence between $\mathrm{JAC}^{(0)}$ and $\mathrm{COM}^{(0)}$, where "local" means that this correspondence exists between $\mathrm{JAC}^{(0)}(S)$ and $\mathrm{COM}^{(0)}(S)$ for any finite set S :

$((A,B),f) \longmapsto (s,c)$, where the transpositions of s are given by the pairs $\{x,\#f(x)\}$ for $x \in S$; c is determined by $c^{-1}(\{\underline{a}\}) \cap S = A$, $c^{-1}(\{\underline{b}\}) \cap S = B$, and $c(x) \neq c(s(x))$ for $x \in S$.

This local correspondence can be lifted to a local correspondence between (reduced) configurations of order k, for any k>0, using (I)-(IV) inductively:

$$\begin{aligned}
JAC^{(k)}_{red}(S) &\cong \bigoplus\left\{ JAC^{(k-1)}(A) \times BIJ(A,B) \; ; \; A\cup B=S,\ |A|=|B| \right\} \\
&\cong \bigoplus\left\{ COM^{(k-1)}(A) \times BIJ(A,B) \; ; \; A\cup B=S,\ |A|=|B| \right\} \\
&\cong COM^{(k)}_{red}\langle S\rangle \quad , \\
JAC^{(k)}(S) &\cong \bigoplus\left\{ JAC^{(k)}_{red}(D) \times LAG_2(E,D) \; ; \; D\cup E=S,\ |D| \text{ even} \right\} \\
&\cong \bigoplus\left\{ COM^{(k)}_{red}(D) \times LAG_2(E,D) \; ; \; D\cup E=S,\ |D| \text{ even} \right\} \\
&\cong COM^{(k)}\langle S\rangle \quad .
\end{aligned}$$

This concludes the proof of the theorem.

Comment: The notion of 'order' introduced here for both Jacobi endofunctions and complete oriented matchings is closely related to the concept of 'register number' (of a binary tree) in computer science. See Francon's recent article [FR] for an exposé and further references. For the Jacobi endofunctions this observation is not too surprising if one looks at the visualization via 'Catalan trees' in [FL]. This interesting aspect will be pursued in more detail in a subsequent article. The reader is referred to M.Vauchaussade's thesis ([VA]) for an in depth treatment of 'order' phenomena in trees, paths, microbiological structures etc. .

3 Generating polynomials associated to complete (oriented) matchings

In this section two families $(Q_n^{(\gamma,\delta)}(X,Y))_{n>=0}$ and $(R_n^{(\gamma,\delta)}(X,Y))_{n>=0}$ of generating polynomials related to complete (oriented) matchings - hence to Jacobi endofunctions via the theorem - are introduced; their relation to the familiar Gegenbauer polynomials will be studied in the next section. In this article we do not aim at a complete treatment of analytic properties (generating functions, recurrence relations, differential formulas etc.) of these polynomials - this will be done elsewhere. The main purposes of this section are

- to demonstrate that the combinatorial model of (oriented) complete matchings leads to analytical results in a simple and elegant way;
- to provide the necessary information for the two applications given in the next section.

For a positive integer n let $[n]:=\{1,2,...,n\}$; we will write <n> instead of <[n]. Let γ,δ be variables. Each $s \in CM<n>$ will be weighted by

$$w(s) := \gamma^{cyc(s)} * \delta^{pt(s)} \quad .$$

For n>=1, we define the generating polynomial

$$Q_n^{(\gamma,\delta)}(x) := \sum \{ \; w(s)*x^{mt(s)} \quad ; \; s \in CM<n> \; \} \; .$$

As an illustration, the $s \in CM<7>$ shown at the end of sec.1 will contribute $\gamma^3 * \delta^1 * x^5$ to the polynomial $Q_7^{(\gamma,\delta)}(x)$.
It will be convenient to consider also the homogeneous polynomials

$$Q_n^{(\gamma,\delta)}(X,Y) := (X-Y)^n * Q_n^{(\gamma,\delta)}((X+Y)/(X-Y)) \; .$$

Each $(s,c) \in COM<n>$ will be given the weight

$$z(s,c) := w(s)*X^{|A_c|}*Y^{|B_c|} \quad ,$$

where $A_c = c^{-1}(\{a\}) \cap [n]$, $B_c = c^{-1}(\{b\}) \cap [n]$, as in the proof of the theorem.
We then define for n>=1:

$$R_n^{(\gamma,\delta)}(X,Y) := \sum \{ \; z(s,c) \; ; \; (s,c) \in COM<n> \; \} \; .$$

Again, as an illustration, the $(s,c) \in COM<7>$ shown at the end of sec.1 will contribute $\gamma^3 * \delta^1 * X^4 * Y^3$ to the polynomial $R_7^{(\gamma,\delta)}(X,Y)$.

<u>Proposition</u>: The polynomials $Q_n^{(\gamma,\delta)}(X,Y)$ and $R_n^{(\gamma,\delta)}(X,Y)$ are "inverse" to each other via the substitution

$$u(X,Y) := (X+Y)/2 + (XY)^{1/2} \quad ,$$

$$v(X,Y) := (X,Y)/2 - (XY)^{1/2} \quad ,$$

i.e. one has

$$R_n^{(\gamma,\delta)}(X,Y) = Q_n^{(\gamma,\delta)}(u(X,Y),v(X,Y)) \; ,$$

$$Q_n^{(\gamma,\delta)}(X,Y) = R_n^{(\gamma,\delta)}(u(X,Y),v(X,Y)) \; .$$

Proof: It suffices to prove the first one of the two foregoing identities since the (u,v)-substitution is involutive:

$$u(u(X,Y),v(X,Y))=X \; , \; v(u(X,Y),v(X,Y))=Y \; .$$

Substituting into Q_n we get

$$Q_n^{(\gamma,\delta)}(u(X,Y),v(X,Y)) =$$

$$= \sum \{ \; w(s)*[u(X,Y)+v(X,Y)]^{mt(s)} * [u(X,Y)-v(X,Y)]^{n-mt(s)} ; \; s \in CM<n> \; \}$$

$$= \sum \{ \; w(s)*[X+Y]^{mt(s)} * [2XY]^{pt(s)} * 2^{nt(s)} ; \; s \in COM<n> \; \} \quad .$$

But extending $s \in CM\langle n\rangle$ to $(s,c) \in COM\langle n\rangle$ can be done in 2^n ways, "orienting" each of the n transpositions independently. The contribution to z coming from each

mixed transposition is $(X+Y)*w(s)$,

positive transposition is $2XY*w(s)$,

negative transposition is $2*w(s)$,

so that the last summation can be written as

$$\sum \{ z(s,c) \ ; \ (s,c) \in COM\langle n\rangle \} ,$$

which proves the proposition.

The polynomials $R_n^{(\gamma,\delta)}(X,Y)$ can be written in a rather explicit form due to the following result.

Proposition: For any pair of disjoint, finite sets A,B with $|A|=a, |B|=b$

$$\sum \{ w(s) \ ; \ (s,c) \in COM\langle A,B\rangle \} = (\gamma)_{a+b} * {}_2F_1\left[\begin{matrix}-a & -b \\ (\gamma+1)/2\end{matrix} ; \delta\right] ,$$

where $(\gamma)_n$ denotes the rising factorial $\gamma*\gamma+1*...*\gamma+n-1$, and ${}_2F_1$ denotes the familiar hypergeometric series, which terminates in the case under consideration.

Proof: For convenience, let $A=\{1,2,...,a\}$ and $B=\{a+1,a+2,...a+b\}$. An arbitrary element $(s,c) \in COM\langle A,B\rangle$ can be constructed by the following stepwise procedure:

1) choose some k (s.th. $0<=k<=\min(a,b)$) and k-subsets $A' \subseteq A$, $B' \subseteq B$;
2) choose a bijection r between A' and B' (representing exactly the positive transpositions of the s under construction) ;
3) extend r to a complete matching s of $\langle a+b\rangle$ without adding any positive transpositions ;
4) introduce an orientation for each of the k negative transpositions of s .

We have $\binom{a}{k}\binom{b}{k}$ choices for step 1), k! choices for step 2), and 2^k choices for step 4). The crucial part is step 3), for which the reader is urged to verify for himself the following

Lemma: Choose any set of k transpositions in [n], where $0<=k<=\lfloor n/2\rfloor$. Summing $\gamma^{cyc(s)}$ over the set of all those s COM<n> which have exactly these k positive transpositions gives

$$(\gamma+2k)_{n-2k}*(\gamma/2)_k*2^k \quad .$$

(Note that for k=0 we are back to the familiar counting of permutations classified by their number of cycles).

Modulo this lemma the proof of our proposition is now complete:

$$\sum \{ w(s) \ ; \ (s,c) \in COM\langle A,B\rangle \} =$$

$$= \sum_k \delta^k * \binom{a}{k}\binom{b}{k} * (\gamma+2k)_{a+b-2k} * (\gamma/2)_k * 2^k * 2^k =$$

$$= \sum_k \frac{(-a)_k * (-b)_k * (\gamma)_{a+b}}{k! * ((\gamma+1)/2)_k} * \delta^k .$$

<u>Corollary</u>: a) $R_n^{(\gamma,\delta)}(X,Y) = (\gamma)_n \sum_k \binom{n}{k} \ {}_2F_1\left[\begin{matrix} -k \ -(n-k) \\ (\ +1)/2 \end{matrix} ; \delta\right] * X^k Y^{n-k}$,

b) $R_n^{(\gamma,1)}(X,Y) = \dfrac{(\gamma)_n}{((\gamma+1)/2)_n} \sum_k \binom{n}{k} \left(\frac{\gamma+1}{2}+k\right)_{n-k} \left(\frac{\gamma+1}{2}+n-k\right)_k X^{n-k} Y^k$.

Part a) is obvious from the previous proposition and the definition of R_n; part b) follows from a) by the well known evaluation of ${}_2F_1$ for unit argument, see e.g.[RA],§32. It is not necessary, however, to make use of this classical fact. One may prove b) directly from the combinatorial model; the rest of this section is devoted to a sketch of such a proof.

Note that part b) of the corollary is equivalent to

$$\sum \{ \gamma^{cyc(s)} ; (s,c) \in COM\langle A,B\rangle \} = (\gamma)_{a+b} * \frac{\left(\frac{\gamma+1}{2} + a\right)_b * \left(\frac{\gamma+1}{2} + b\right)_a}{\left(\frac{\gamma+1}{2}\right)_{a+b}}$$

for disjoint, finite sets A,B with |A|=a, |B|=b. Let us write l(a,b) (r(a,b) resp.) for the l.h.s. (r.h.s resp.) of this identity; both are polynomials in the variable γ. This is not obvious for r(a,b), but it is easily veryfied that r(a,b) can be written as

$$r(a,b) = \prod \{ \gamma+i \ ; \ i \in I(a,b) \} ,$$

where I(a,b) contains all the even numbers i such that 0<=i<a+b and all the odd numbers j such that 2*min(a,b)<j<a+b or 2*max(a,b)<j<2*(a+b). This presentation leads to:

$$r(a+b,0) = (\gamma)_{a+b} ,$$

$$(\gamma+2a+1)*r(a+1,b-1) = (\gamma+2b-1)*r(a,b) ,$$

which can be shown to be equivalent to the recursion

$$r(a,0) = (\gamma)_a \ , \ r(0,b) = (\gamma)_b ,$$

$$r(a,b) = a*r(a-1,b) + (\gamma+2a+b-1)*r(a,b-1) .$$

Thus it suffices to verify the same recursion for the polynomials l(a,b). The boundary conditions $l(a,0)=(\gamma)_a$ and $l(0,b)=(\gamma)_b$ are satisfied since in this case we are simply counting cycles of permutations (cf. the argument in sec.2). For the recursion identity we will employ a mapping

$$COM\langle A,B\rangle \longrightarrow COM\langle A,B'\rangle \cup \bigoplus_{x\in A} COM\langle A\backslash\{x\},B'\cup\{x\}\rangle \ : \ (s,c) \longmapsto (s',c')$$

where, for convenience, we may assume that

A={1,2,...,a}, B={a+1,a+2,...,a+b}, B'=B\{a+b} with a>0 and b>0.

Several cases have to be considered:

1) if s(a+b) = #(a+b), then (s',c') is nothing but the restriction of (s,c) to <A∪B'>;

2) if s(a+b) ≠ #(a+b), let s(a+b)=i and s(#(a+b))=j. Then {i,j} will be a transposition of s' and the other a+b-2 transpositions of s' are those of s which are contained in <A∪B'>. c' will coincide with c on the set <A∪B'>\{j} and c'(j):=b.

   The following subcases occur:

   1) c(#(a+b))=a ;

   2) c(#(a+b))=b and j ∈ #(A∪B') ;

   3) c(#(a+b))=b and j ∈ A .

Note that (s',c') belongs to COM<A,B'> in cases 1, 2.1, 2.2, and to COM<A\{j},B'∪{j}> in case 2.3, where j varies over A. Note further that cyc(s')=cyc(s) in case 2, whereas cyc(s')=cyc(s)-1 in case 1, and finally, that the following multiplicities occur: 1 in cases 1 and 2.3, a+b-1 in case 2.1, and a in case 2.2. Putting all this together gives

$$\begin{aligned} l(a,b) &= \gamma * l(a,b-1) + (a+b-1)*l(a,b-1) + a*l(a,b-1) + a*l(a-1,b) \\ &= a*l(a-1,b) + (\gamma+2a+b-1)*l(a,b-1) \ , \end{aligned}$$

as desired.

4 Two applications:
Gegenbauer polynomials and an identity of Tricomi's

In this section we will show how the polynomials $Q_n^{(\gamma,\delta)}(X,Y)$ and $R_n^{(\gamma,\delta)}(X,Y)$ introduced in the previous section are related to Jacobi polynomials, and more specifically to Gegenbauer polynomials. Using the homogeneous version of the Jacobi polynomials as introduced in [FL] :

$$\mathcal{P}_n^{(\alpha,\beta)}(X,Y) = \sum_{i+j=n} \binom{n}{i} * (1+\alpha+j)_i * (1+\beta+i)_j * X^i Y^j \ ,$$

which are related to the Jacobi polynomials $P_n^{(\alpha,\beta)}(x)$ in their standard notation (cf. e.g. [ER], [RA], [SM]) simply by

$$\mathcal{P}_n^{(\alpha,\beta)}(X,Y) = n!*(X-Y)^n * P_n^{(\alpha,\beta)}((X+Y)/(X-Y)) \ ,$$

$$n!*P_n^{(\alpha,\beta)}(x) = \mathcal{P}_n^{(\alpha,\beta)}((x+1)/2,(x-1)/2) \ ,$$

we find from part b) of the corollary that

$$R_n^{(\gamma,1)}(X,Y) = \frac{(\gamma)_n}{((\gamma+1)/2)_n} * \mathcal{P}_n^{((\gamma-1)/2,(\gamma-1)/2)}(X,Y) \ .$$

Thus we may write

$$R_n^{(\gamma,1)}(X,Y) = \mathcal{L}_n^{\gamma/2}(X,Y) ,$$

where $\mathcal{L}_n^{\lambda}(X,Y)$ denotes the homogeneous Gegenbauer polynomials:

$$\mathcal{L}_n^{\lambda}(X,Y) = n!*(X-Y)^n*C_n^{\lambda}((X+Y)/(X-Y)) .$$

Here $C_n^{\lambda}(x)$ refers to the standard notation for Gegenbauer polynomials see e.g. the references cited above.

Refering now back to our theorem in sec.1 and the work of Foata/Leroux in [FL], we can state:

Corollary: The homogeneous Gegenbauer polynomials $\mathcal{L}_n^{\gamma/2}(X,Y)$ are related to Jacobi endofunctions JAC in two different ways:

$$\sum \{ \gamma^{cyc(f)}*X^{|A|}Y^{|B|} ; ((A,B),f) \in JAC[n] \} =$$

$$= R_n^{(\gamma,1)}(X,Y) = \mathcal{L}_n^{\gamma/2}(X,Y) =$$

$$= \sum \{ (\tfrac{\gamma+1}{2})^{cyc(f|A)+cyc(f|B)} *X^{|A|}Y^{|B|} ; ((A,B),f) \in JAC[n] \} .$$

Thus two essentially different valuations on the "species" (in the sense of [JO]) of Jacobi endofunctions - one of them counting all cycles, the other one counting only type-a and type-b cycles - lead to the same generating polynomial.

As to the polynomials $Q_n^{(\gamma,\delta)}(X,Y)$, we will now use their nonhomogeneous version $Q_n^{(\gamma,\delta)}(x)$, as introduced at the beginning of section 3. The combinatorial definition given there leads to the

Lemma: The polynomials $Q_n^{(\gamma,\delta)}(x)$ are determined by the recurrence

$$Q_0^{(\gamma,\delta)}(x) = 1 ,$$

$$Q_{n+1}^{(\gamma,\delta)}(x) = [(2n+\gamma)*x + (\delta-x^2)(d/dx)]\, Q_n^{(\gamma,\delta)}(x) .$$

Proof(Sketch): We consider the mapping CM<n+1> $\longrightarrow$ CM<n>:s $\longmapsto$ s' underlying the mapping (s,c) $\longmapsto$ (s',c') in the proof of part b) of the corollary (end of section 3). We have no colors a,b here, but we have to take care of the number of positive (counted via δ) and mixed (counted via x) transpositions. Note that

pt(s')=pt(s) and mt(s')=mt(s)-1, unless $i \in [n]$ and $j \in \#[n]$.

In this latter case we have

pt(s')=pt(s)-1 and mt(s')=mt(s)+1.

This is exactly what the correcting term $(\delta-x^2)(d/dx)$ takes into account, the derivative introducing mt(s') as a factor (= number of possibilities for selecting a mixed transposition in $s' \in$ CM<n> when going backwards to some preimage $s \in$ CM<n+1>).

Corollary: The polynomials $Q_n^{(\gamma,\delta)}(x)$ are given by the Rodriguez type formula: $Q_n^{(\gamma,\delta)}(x) = (\delta-x^2)^{n+(\gamma/2)}*(\frac{d}{dx})^n(\delta-x^2)^{-(\gamma/2)}$.

This follows from the lemma by a simple induction.

We are now ready to re-read the inverse relations from the first proposition in sec.3; replacing X by (x+1)/2 and Y by (x-1)/2 we find:

Corollary: The polynomials $Q_n^{(\gamma,\delta)}(x)$ are related to the Gegenbauer polynomials $C_n^{\lambda}(x)$ via

$$Q_n^{(\gamma,\delta)}(x) = n!*(x^2-\delta)^{n/2}*C_n^{\gamma/2}(x/(x^2-\delta)^{1/2}) \quad .$$

For the proof we make use of

$$Q_n^{(\gamma,\delta)}(x) = \delta^{n/2}*Q_n^{(\gamma,1)}(x/\delta^{1/2})$$

- which follows from the definition of these polynomials - and

$$R_n^{(\gamma,1)}(x) = n!*C_n^{\gamma/2}(x).$$

This last result goes back to Hermite([HE]) in the case $\gamma=\delta=1$, and to Tricomi([TR]) in the case $\delta=1$, see also [ER],,vol.II,sec.10.9,p.178, and [SM],sec.8.1,p.409. A different combinatorial approach to Tricomi's identity has been presented by Dumont ([DU]).

5 References

[DU] Dumont,D. Etude combinatoire d'une suite de polynômes associés aux polynômes ultrasphériques, Publ.Elektr.Fak.Univ.Beograd, Ser.Mat.Fiz.(1979),116-125.

[ER] Erdelyi et al.(eds.) Higher Transcendental Functions, 3 vols. McGraw-Hill, 1953, (reprint: Krieger Publ.Comp. 1981).

[F1] Foata,D. A combinatorial proof of the Mehler formula, J.Comb.Theory,ser.A,24(1978),250-259.

[F2] Foata,D. Combinatoire des identités sur les polynômes orthogonaux, Proc.Int.Congr.Math.(Warsaw, 16-24 Aug.1983), to appear.

[FL] Foata,D. and P.Leroux Polynômes de Jacobi, interprétation combinatoire et fonction génératrice, Proc.Amer.Math.Soc.87(1983), 47-53.

[FS] Foata,D. and V.Strehl Combinatorics of Laguerre Polynomials, Proc. Waterloo Silver Jubilee Conf.,in: Enumeration and Design (D.M.Jackson and S.A.Vanstone, eds.), Academic Press, Toronto 1984.

[FR] Françon,J. Sur le nombre de registres nécessaires à l'évaluation d'une expression arithmétique, R.A.I.R.O. Informatique Théorique 18(1984), 355-364.

[GR] Garsia,A. and J.Remmel A combinatorial interpretation of q-derangement numbers and q-Laguerre numbers, Europ.J.Combinatorics 1(1980), 47-59.

[HE] Hermite,C. Sur l'intégrale $\int_{-1}^{+1} \frac{dx}{(a-x)\sqrt{1-x^2}}$
Ann.Mat.Pura Appl.3(1869),p.89.

[JO] Joyal,A. Une théorie combinatoire des séries formelles, Adv.in Math.42(1981),1-82.

[LS] Leroux,P. and V.Strehl Jacobi Polynomials: Combinatorics of the Basic Identities, Discrete Mathematics 57(1985),167-187.

[RA] Rainville,E. Special Functions, Chelsea, Bronx, N.Y.1960.

[RI] Riordan,J. An introduction to combinatorial analysis, Wiley, N.Y. 1958.

[SM] Srivastava,H. and H.Manocha A treatise on generating functions, Ellis Horwood Series in Mathematics and its Applications, J.Wiley, 1984.

[ST] Strehl,V. Polynômes d'Hermite généralisés et identités de Szegö - une version combinatoire, Proc.Symp. E.N.Laguerre sur les Polynômes Orthogonaux, Bar-le-Duc, 1984.

[TR] Tricomi,F. Sul comportamento asintotico dei polinomi di Laguerre, Ann.Math.Pura Appl.28(1949),263-289.

[VA] Vauchaussade de Chaumont,M. Nombres de Strahler des arbres, langages algébriques et dénombrement de structures secondaires en biologie moléculaire, Thèse 3me cycle, Université de Bordeaux, 1985.

ABOUT THE INEQUALITIES OF ERDÖS AND MOSER ON THE LARGEST TRANSITIVE SUBTOURNAMENT OF A TOURNAMENT*

Claudette Tabib
Collège Edouard-Montpetit
945 Chemin Chambly, Longueuil,
Québec, Canada, J4H 3M6

1. INTRODUCTION

By using the properties of homogeneous tournaments, the upper bound (for the values $n \leq 78$) and the lower bound in the estimate of Erdös and Moser, on the largest transitive subtournament of a tournament, are improved. Two homogeneous tournaments of order 27 are shown to be non isomorphic. In fact, one of them does not contain any transitive subtournament of order 6, whereas the other does. The links that exist between homogeneity and transitive subtournaments lead us to find such a subtournament T_7 of order 7 in two homogeneous tournaments of order 31. Several other T_7's are found to exist in all tournaments of order 29, obtained from either one or the other of the two tournaments of order 31, by deleting any pair of distinct vertices. Further, existence of, not only one, but at least three non isomorphic tournaments of order 31 which contain no T_8 is shown. These tournaments are also homogeneous.

A *tournament* $T = (V(T), A(T))$ is a directed finite graph without loops, in which each pair of distinct vertices x and y is joined by exactly one of the arcs (x, y) or (y, x). A *homogeneous tournament* is a tournament with at least one cycle such that every arc lies on the same number of 3-cycles. Recall that the order of a homogeneous tournament is congruent to -1 modulo 4 [4]. Let T be a rotational tournament of order 2n + 1. Label its vertices $u_0, u_1, \ldots, u_{2n}$ in such a way that

$$(u_i, u_j) \in A(T) \Longrightarrow (u_{i+1}, u_{j+1}) \in A(T),$$

for every pair of indices i, j; suppose that

$$u_p = u_q \Longleftrightarrow p \equiv q \pmod{2n+1}.$$

The *characteristic* $X(T) = (x_1, x_2, \ldots, x_n)$ of T is defined as follows:

$$x_i = \begin{cases} 1, \text{ if } (u_0, u_i) \in A(T), \\ 0, \text{ if } (u_i, u_0) \in A(T), \end{cases}$$

for every $i = 1, 2, \ldots, n$. See Moon [5] or Reid and Beineke [6] for any undefined terms.

---

* Research supported by a FCAR Grant (ACSAIR).

DEFINITION 1. Let $v = v(n)$ be the largest integer such that every tournament T_n of order n contains a transitive subtournament T_v of order v.

As every tournament T_n, $n \geq 4$, contains at least one transitive subtournament T_3, but not every tournament T_n is itself transitive, the following question arises: What is the value of $v(n)$, for each $n > 1$?

Erdös and Moser [1] (see also [9]) gave the following bounds for $v(n)$:

$$[\log_2 n] + 1 \leq v(n) \leq [2 \log_2 n] + 1.$$

2. IMPROVEMENT OF THE ESTIMATE OF ERDÖS AND MOSER

Denote by $\tau(T)$ the number of the 3-cycles in a tournament T. From [3, Theorem 2], it follows that

$$\tau(H) = \frac{1}{3}k(2k - 1)(4k - 1),$$

if H is a homogeneous tournament of order $4k - 1$.

THEOREM 1. *Let H be a homogeneous tournament of order $4k - 1$ and T be a transitive subtournament of H of order v, $v > 1$. Then*

$$(1)\quad k \geq \frac{1}{3}(w + 2)(w + 1), \quad \textit{if } v = 2w + 1,$$

and

$$(2)\quad k \geq \frac{1}{6}(2w + 1)(w + 1), \quad \textit{if } v = 2w.$$

PROOF. Consider a transitive subtournament T of order v of a homogeneous tournament H of order $4k - 1$. As

$$\tau(H) - \tau(H\backslash T) = v(2k - 1)k - \binom{v}{2}k$$
$$= \frac{1}{2}vk(4k - v - 1),$$

then, by the above remark, we obtain that

$$\tau(H\backslash T) = \frac{1}{3}k(2k - 1)(4k - 1) - \frac{1}{2}vk(4k - v - 1)$$
$$= \frac{1}{6}k(16k^2 - 12k + 2 - 12kv + 3v^2 + 3v).$$

Denote by $m = 4k - v - 1$ the order of $H\backslash T$; thus

$$m = 2(2k - w - 1) \equiv 0 \pmod 2 \quad , \text{ if } v = 2w + 1,$$

and

$$m = 2(2k - w - 1) + 1 \equiv 1 \pmod 2, \text{ if } v = 2w.$$

It follows by [5, p. 9] that

$$\tau(H\backslash T) \le \begin{cases} \frac{1}{24}[(4k - 2w - 2)^3 - 4(4k - 2w - 2)], & \text{if } v = 2w + 1, \\ \frac{1}{24}[(4k - 2w - 1)^3 - (4k - 2w - 1)], & \text{if } v = 2w. \end{cases}$$

Therefore

$$\tau(H\backslash T) \le \begin{cases} \frac{1}{3}(8k^3 - 12k^2w + 6kw^2 + 12kw - 12k^2 + 4k - w^3 - 3w^2 - 2w), & \text{if } v = 2w + 1 \\ \frac{1}{6}(16k^3 - 24k^2w + 12kw^2 + 12kw - 12k^2 + 2k - 2w^3 - 3w^2 - w), & \text{if } v = 2w. \end{cases}$$

But as, from what has been obtained above,

$$\tau(H\backslash T) = \begin{cases} \frac{1}{3}(8k^3 - 12k^2w + 6kw^2 + 9kw - 12k^2 + 4k), & \text{if } v = 2w + 1, \\ \frac{1}{6}(16k^3 - 24k^2w + 12kw^2 + 6kw - 12k^2 + 2k), & \text{if } v = 2w, \end{cases}$$

then

$$\begin{cases} 9kw \le 12kw - w^3 - 3w^2 - 2w, & \text{if } v = 2w + 1, \\ 6kw \le 12kw - 2w^3 - 3w^2 - w, & \text{if } v = 2w. \end{cases}$$

It follows that

$$\begin{cases} k \ge \frac{1}{3}(w + 2)(w + 1), & \text{if } v = 2w + 1, \\ k \ge \frac{1}{6}(2w + 1)(w + 1), & \text{if } v = 2w. \end{cases}$$

□

<u>THEOREM 2</u>. $\nu(n) \le \left[-\frac{3}{2} + \sqrt{3n + \frac{13}{4}}\right]$.

PROOF. Consider a homogeneous tournament H of order $n = 4k - 1$. If T is a transitive subtournament of H of order v, then

$$k \ge \begin{cases} \frac{1}{3}\left(\frac{v - 1}{2} + 2\right)\left(\frac{v - 1}{2} + 1\right), & \text{if } v \text{ is odd}, \\ \frac{1}{6}(v + 1)\left(\frac{v}{2} + 1\right), & \text{if } v \text{ is even}, \end{cases}$$

by Theorem 1. Therefore

$$k \ge \frac{1}{12}(v + 2)(v + 1).$$

It follows that

$$3n + 3 \ge v^2 + 3v + 2,$$

which implies, by Definition 1, that

$$v(n) \leq \left[-\frac{3}{2} + \sqrt{3n + \frac{13}{4}}\right]. \qquad \square$$

Theorem 2 gives us a better estimate for the upper bound of $v(n)$ than that of Erdös and Moser, for the values $n \leq 78$. Figure 1 allows to compare the values of

$$f(n) = [2 \log_2 n] + 1$$

with those of

$$g(n) = \left[-\frac{3}{2} + \sqrt{3n + \frac{13}{4}}\right]$$

for $n = 2, 3, \ldots, 100$.

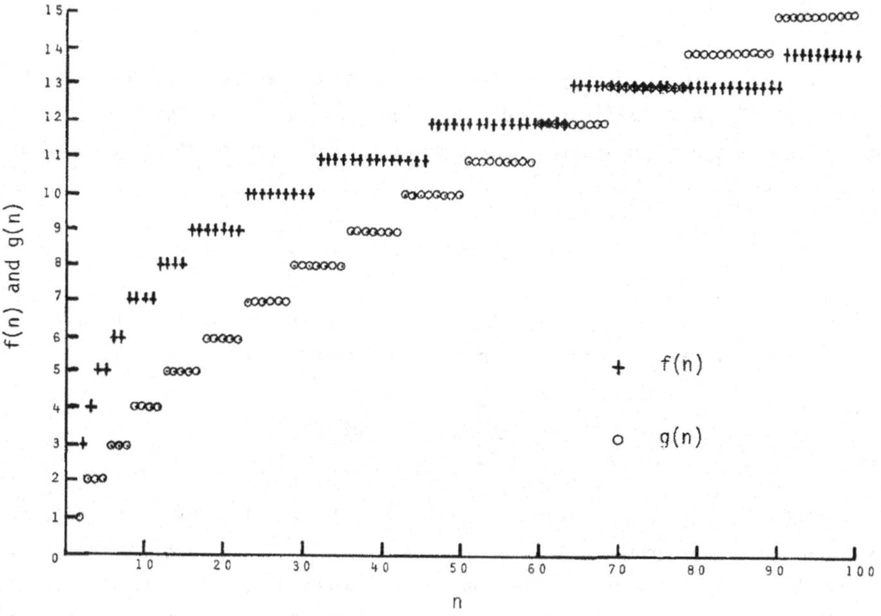

Figure 1. The upper bounds $f(n)$ and $g(n)$ of $v(n)$.

We are seeking to determine the exact value of $v(n)$, $n > 1$. We will present, in the following, examples for some small values of n:

a) $v(2) = v(3) = 2.$

b) Consider the rotational tournament of order 7 whose characteristic equals (1, 1, 0). This tournament contains no transitive subtournament of order 4. As $v(n) \geq 3$ for $n \geq 4$, it follows that
$$v(4) = \ldots = v(7) = 3.$$

c) Likewise, the rotational tournament of order 13, the characteristic of which is (1, 1, 1, 0, 1, 1), contains no transitive subtournament of order 5. But, $v(n) \geq 4$ for $n \geq 8$, by the inequalities of Erdös and Moser, then

$$v(8) = \ldots = v(13) = 4.$$

For all these values of n, the lower bound $[\log_2 n] + 1$ of $v(n)$ is thus attained. Erdös and Moser [2] conjectured that

$$v(n) = [\log_2 n] + 1$$

for all $n > 1$.

THEOREM 3. *Every tournament of order 14 contains at least one transitive subtournament of order 5.*

Because limitation on space precludes the presentation of the proof of this theorem in this paper, the interested reader may find it in [10, Chapter 5]. The method used in the proof is simple, new and entirely different from that employed by Reid and Parker [8].

COROLLARY 4. *$v(n) \geq 5$, for $n \geq 14$.*

Thus, the conjecture of Erdös and Moser is false for $n = 14$. The next result allows us to obtain a best estimate for the lower bound of $v(n)$.

COROLLARY 5. *$v(n) \geq \left[\log_2 \frac{n}{14}\right] + 5$, $n \geq 14$.*

PROOF. Let T be a tournament of order 7×2^k, $k \geq 1$. There certainly exist a vertex u and a subtournament S in T, of order $7 \times 2^{k-1}$, each vertex of which is dominated by u. By induction, if we suppose that S contains a transitive subtournament with at least $k + 3$ vertices, then these vertices together with u generate a transitive subtournament with at least $k + 4$ vertices. It follows that T contains at least one transitive subtournament of order $k + 4$. □

Corollary 5 implies that $v(n) \geq k + 4$ for $n = 7 \times 2^k$, $k \geq 1$. As $[\log_2 n] + 1 = k + 3$, then $v(n) \neq [\log_2 n] + 1$. Consequently, the conjecture of Erdös and Moser is false, not only for $n = 14$, but for an infinite number of values of n.

3. THE HOMOGENEOUS TOURNAMENTS OF ORDER 27

DEFINITION 2. Let $N = N(m)$ be the smallest integer such that every tournament T_N of order N contains a transitive subtournament T_m of order m.

Actually, N(3) = 4, N(4) = 8 and N(5) = 14. By Corollary 5, we have for n = 28, in particular, that every tournament of order 28 contains at least one T_6. We will obtain that N(6) = 28.

THEOREM 6. *The following matrix* M_1 *is the dominance matrix of a homogeneous tournament of order 27 which contains no transitive subtournament of order 6.*

0	1	1	0	0	1	1	1	0	0	0	0	0	0	0	0	1	1	1	1	0	1	0	1	0	1	1
0	0	0	1	0	0	0	1	1	1	1	0	1	1	0	0	0	1	1	1	0	0	0	1	1	0	1
0	1	0	1	1	1	0	0	0	0	1	1	0	0	0	0	0	0	1	0	1	0	1	1	1	1	1
1	0	0	0	0	0	1	0	1	0	1	1	0	0	0	1	0	1	0	1	1	1	1	1	0	0	1
1	1	0	1	0	1	0	0	0	1	0	0	0	0	1	1	1	1	0	1	0	0	1	0	1	0	1
0	1	0	1	0	0	1	0	1	0	0	1	1	1	0	1	1	1	1	0	0	0	1	0	0	1	0
0	1	1	0	1	0	0	1	0	0	1	0	1	1	0	1	1	0	0	1	1	0	1	0	0	0	1
0	0	1	1	1	1	0	0	1	1	0	1	0	1	0	0	1	1	0	0	1	1	0	0	0	0	1
1	0	1	0	1	0	1	0	0	1	0	0	0	1	0	0	0	1	1	1	1	0	1	0	1	1	0
1	0	1	1	0	1	1	0	0	0	1	0	1	1	1	0	0	0	1	0	0	1	1	0	0	0	1
1	0	0	0	1	1	0	1	1	0	0	0	1	0	1	1	0	1	1	0	1	0	0	0	0	1	1
1	1	0	0	1	0	1	0	1	1	1	0	0	1	1	0	1	0	0	0	0	0	0	1	0	1	1
1	0	1	1	1	0	0	1	1	0	0	1	0	0	1	0	1	0	1	1	0	0	1	1	0	0	0
1	0	1	1	1	0	0	0	0	0	1	0	1	0	0	1	1	1	0	0	0	1	0	1	1	1	0
1	1	1	1	0	1	1	1	1	0	0	0	0	1	0	1	0	0	0	0	1	0	0	1	1	0	0
1	1	1	0	0	0	0	1	1	1	0	1	1	0	0	0	0	0	0	0	0	1	1	0	1	1	1
0	1	1	1	0	0	0	0	1	1	1	0	0	0	1	1	0	0	1	1	1	1	0	0	0	1	0
0	0	1	0	0	0	1	0	0	1	0	1	1	0	1	1	1	0	1	0	1	0	0	1	1	0	1
0	0	0	1	1	0	1	1	0	0	0	1	0	1	1	1	0	0	0	1	0	1	0	0	1	1	1
0	0	1	0	0	1	0	1	0	1	1	1	0	1	1	1	0	1	0	0	0	0	1	1	0	1	0
1	1	0	0	1	1	0	0	0	1	0	1	1	1	0	1	0	0	1	1	0	1	0	1	0	0	0
0	1	1	0	1	1	1	0	1	0	1	1	1	0	1	0	0	1	0	1	0	0	0	0	1	0	0
1	1	0	0	0	0	0	1	0	0	1	1	0	1	1	0	1	1	1	0	1	1	0	0	1	0	0
0	0	0	0	1	1	1	1	1	1	1	0	0	0	0	1	1	0	1	0	0	1	1	0	1	0	0
1	0	0	1	0	1	1	1	0	1	1	1	1	0	0	0	1	0	0	1	1	0	0	0	0	1	0
0	1	0	1	1	0	1	1	0	1	0	0	1	0	1	0	0	1	0	0	1	1	1	1	0	0	0
0	0	0	0	0	1	0	0	1	0	0	0	1	1	1	0	1	0	0	1	1	1	1	1	1	1	0

M_1

	1	2	3	4	5	6	7	8	9	10	11	12	13	14	15	16	17	18	19	20	21	22	23	24	25	26	27
1	0	1	1	1	1	1	1	1	1	1	1	1	1	1	0	0	0	0	0	0	0	0	0	0	0	0	0
2	0	0	0	0	0	1	0	0	1	1	0	1	1	1	1	1	0	1	0	0	0	1	1	1	0	1	0
3	0	1	0	0	0	0	1	0	0	1	1	0	1	1	0	1	1	0	1	0	0	0	1	1	1	0	1
4	0	1	1	0	0	0	0	1	0	0	1	1	0	1	1	0	1	1	0	1	0	0	0	1	1	1	0
5	0	1	1	1	0	0	0	0	1	0	0	1	1	0	0	1	0	1	1	0	1	0	0	0	1	1	1
6	0	0	1	1	1	0	0	0	0	1	0	0	1	1	1	0	1	0	1	1	0	1	0	0	0	1	1
7	0	1	0	1	1	1	0	0	0	0	1	0	0	1	1	1	0	1	0	1	1	0	1	0	0	0	1
8	0	1	1	0	1	1	1	0	0	0	0	1	0	0	1	1	1	0	1	0	1	1	0	1	0	0	0
9	0	0	1	1	0	1	1	1	0	0	0	0	1	0	0	1	1	1	0	1	0	1	1	0	1	0	0
10	0	0	0	1	1	0	1	1	1	0	0	0	0	1	0	0	1	1	1	0	1	0	1	1	0	1	0
11	0	1	0	0	1	1	0	1	1	1	0	0	0	0	0	0	0	1	1	1	0	1	0	1	1	0	1
12	0	0	1	0	0	1	1	0	1	1	1	0	0	0	1	0	0	0	1	1	1	0	1	0	1	1	0
13	0	0	0	1	0	0	1	1	0	1	1	1	0	0	0	1	0	0	0	1	1	1	0	1	0	1	1
14	0	0	0	0	1	0	0	1	1	0	1	1	1	0	1	0	1	0	0	0	1	1	1	0	1	0	1
15	1	0	1	0	1	0	0	0	1	1	1	0	1	0	0	1	1	1	0	1	1	0	0	1	0	0	0
16	1	0	0	1	0	1	0	0	0	1	1	1	0	1	0	0	1	1	1	0	1	1	0	0	1	0	0
17	1	1	0	0	1	0	1	0	0	0	1	1	1	0	0	0	0	1	1	1	0	1	1	0	0	1	0
18	1	0	1	0	0	1	0	1	0	0	0	1	1	1	0	0	0	0	1	1	1	0	1	1	0	0	1
19	1	1	0	1	0	0	1	0	1	0	0	0	1	1	1	0	0	0	0	1	1	1	0	1	1	0	0
20	1	1	1	0	1	0	0	1	0	1	0	0	0	1	0	1	0	0	0	0	1	1	1	0	1	1	0
21	1	1	1	1	0	1	0	0	1	0	1	0	0	0	0	0	1	0	0	0	0	1	1	1	0	1	1
22	1	0	1	1	1	0	1	0	0	1	0	1	0	0	1	0	0	1	0	0	0	0	1	1	1	0	1
23	1	0	0	1	1	1	0	1	0	0	1	0	1	0	1	1	0	0	1	0	0	0	0	1	1	1	0
24	1	0	0	0	1	1	1	0	1	0	0	1	0	1	0	1	1	0	0	1	0	0	0	0	1	1	1
25	1	1	0	0	0	1	1	1	0	1	0	0	1	0	1	0	1	1	0	0	1	0	0	0	0	1	1
26	1	0	1	0	0	0	1	1	1	0	1	0	0	1	1	1	0	1	1	0	0	1	0	0	0	0	1
27	1	1	0	1	0	0	0	1	1	1	0	1	0	0	1	1	1	0	1	1	0	0	1	0	0	0	0

M_2

A computer has been used to analyse all the possibilities and no T_6 has been found in this tournament.

COROLLARY 7. $\nu(14) = \ldots = \nu(27) = 5$.

Denote by $H_{27}^{(1)}$ the tournament defined above. Let $H_{27}^{(2)}$ be the homogeneous tournament of order 27 whose dominance matrix is M_2 [10, p. 194].

THEOREM 8. *There exists a homogeneous tournament of order 27 which contains at least one transitive subtournament of order 6.*

PROOF. Consider the subtournament of the homogeneous tournament $H_{27}^{(2)}$ induced by the set

$$\{u_{15}, u_5, u_9, u_3, u_{13}, u_{16}\}$$

of vertices of $H_{27}^{(2)}$. This subtournament of $H_{27}^{(2)}$ is transitive of order 6. □

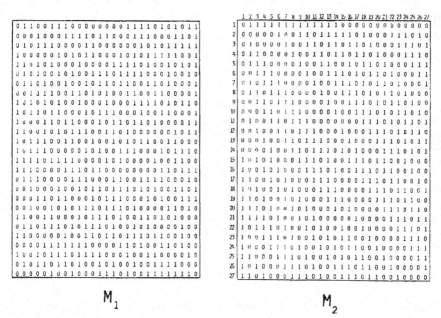

COROLLARY 9. *There exists a pair of non isomorphic homogeneous tournaments of order 27.*

PROOF. The tournaments $H_{27}^{(1)}$ and $H_{27}^{(2)}$ are obviously non isomorphic. □

REMARK. The existence of a homogeneous tournament of order p^n, where p is an odd prime number and n a positive integer, is well known. In fact, there always exists a unique field with p^n elements (*Galois Field*), because any two finite fields with the same number of elements are isomorphic. This field is a quotient field of $Z_p[x]$, where every element can be expressed as a polynomial of degree k, $k \le n - 1$, with coefficients in Z_p and x is a root of an irreducible polynomial of degree n over Z_p. We determine the directions in the tournament in the following way: for distinct vertices g_i and g_j, (g_i, g_j) is an arc if and only if $(g_j - g_i)(x)$ is a *quadratic residue* modulo the irreducible polynomial P(x); that is, iff

$$\lfloor f(x) \rfloor^2 \equiv [(g_j - g_i)(x)\rfloor (\text{mod } P(x))$$

has solutions. This tournament, of order p^n, is homogeneous. The tournament of order 3^3 obtained in this way contains also no T_6 (there is a note, to this effect, presented by one of the authors of [8]).

4. HOMOGENEITY AND TRANSITIVE SUBTOURNAMENTS

REMARK. The unique [2] tournament of order 7, which contains no T_4, is homogeneous. Reid and Parker [8] have shown also the uniqueness of the tournament of order 13 which contains no T_5; this tournament is nearly homogeneous (the definition and a characterization of such tournaments are given in [11]). In section 3, we have found that it is a homogeneous tournament of order 27 which contains no T_6.

This remark leads us thus to consider the homogeneous tournaments of order 31. By Corollary 5 and [5, p. 16], we have:

$$6 \le v(28) \le \ldots \le v(31) \le 7.$$

Let RH_{31} be the rotational homogeneous tournament of order 31 whose characteristic equals

$$(1, 1, 0, 1, 1, 0, 1, 1, 1, 1, 0, 0, 0, 1, 0).$$

THEOREM 10. *The homogeneous tournament RH_{31} contains a transitive subtournament of order 7.*

PROOF. Let $\{u_0, u_1, \ldots, u_{30}\}$ be a labelling of the vertices of RH_{31} which is associated with the characteristic mentioned above. The subtournament of RH_{31}, induced by the set

$$\{u_0, u_1, u_8, u_5, u_2, u_9, u_{10}\}$$

of vertices of RH_{31}, is transitive of order 7. □

As RH_{31} is rotational, there is obviously several other copies of T_7 in it. Further, every tournament obtained from RH_{31} by deleting any vertex contains also a T_7.

Consider now the homogeneous tournament of order 31, obtained beginning with a homogeneous tournament of order 15 (which is also obtained starting from the unique homogeneous tournament of order 7), whose dominance matrix is the following:

	1	2	3	4	5	6	7	8	9	10	11	12	13	14	15	16	17	18	19	20	21	22	23	24	25	26	27	28	29	30	31
1	0	1	0	0	1	1	0	0	1	1	0	0	1	1	0	0	1	1	0	0	1	1	0	0	1	1	0	0	1	1	0
2	0	0	1	0	0	1	1	0	0	1	1	0	0	1	1	0	0	1	1	0	0	1	1	0	0	1	1	0	0	1	1
3	1	0	0	0	1	0	1	0	1	0	1	0	1	0	1	0	1	0	1	0	1	0	1	0	1	0	1	0	1	0	1
4	1	1	1	0	0	0	0	0	1	1	1	1	0	0	0	0	1	1	1	1	0	0	0	0	1	1	1	1	0	0	0
5	0	1	0	1	0	0	1	0	0	1	0	1	1	0	1	0	0	1	0	1	1	0	1	0	0	1	0	1	1	0	1
6	0	0	1	1	1	0	0	0	0	0	1	1	1	1	0	0	0	0	1	1	1	1	0	0	0	0	1	1	1	1	0
7	1	0	0	1	0	1	0	0	1	0	0	1	0	1	1	0	1	0	0	1	0	1	1	0	1	0	0	1	0	1	1
8	1	1	1	1	1	1	1	0	0	0	0	0	0	0	0	0	1	1	1	1	1	1	1	0	0	0	0	0	0	0	0
9	0	1	0	0	1	1	0	1	0	0	1	1	0	0	1	0	0	1	0	0	1	1	0	1	1	0	1	1	0	0	1
10	0	0	1	0	0	1	1	1	1	0	0	1	1	0	0	0	0	0	1	0	0	1	1	1	1	1	0	1	1	0	0
11	1	0	0	0	1	0	1	1	0	1	0	1	0	1	0	0	1	0	0	0	1	0	1	1	0	1	1	1	0	1	0
12	1	1	1	0	0	0	0	1	0	0	0	0	1	1	1	0	1	1	1	0	0	0	0	1	0	0	0	1	1	1	1
13	0	1	0	1	0	0	1	1	1	0	1	0	0	1	0	0	0	1	0	1	0	0	1	1	1	0	1	0	1	1	0
14	0	0	1	1	1	0	0	1	1	1	0	0	0	0	1	0	0	0	1	1	1	0	0	1	1	1	0	0	0	1	1
15	1	0	0	1	0	1	0	1	0	1	1	0	1	0	0	0	1	0	0	1	0	1	0	1	0	1	1	0	1	0	1
16	1	1	1	1	1	1	1	1	1	1	1	1	1	1	1	0	0	0	0	0	0	0	0	0	0	0	0	0	0	0	0
17	0	1	0	0	1	1	0	0	1	1	0	0	1	1	0	1	0	0	1	1	0	0	1	1	0	0	1	1	0	0	1
18	0	0	1	0	0	1	1	0	0	1	1	0	0	1	1	1	1	0	0	1	1	0	0	1	1	0	0	1	1	0	0
19	1	0	0	0	1	0	1	0	1	0	1	0	1	0	1	1	0	1	0	1	0	1	0	1	0	1	0	1	0	1	0
20	1	1	1	0	0	0	0	0	1	1	1	1	0	0	0	1	0	0	0	0	1	1	1	1	0	0	0	0	1	1	1
21	0	1	0	1	0	0	1	0	0	1	0	1	1	0	1	1	1	0	1	0	0	1	0	1	1	0	1	0	0	1	0
22	0	0	1	1	1	0	0	0	0	0	1	1	1	1	0	1	1	1	0	0	0	0	1	1	1	1	0	0	0	0	1
23	1	0	0	1	0	1	0	0	1	0	0	1	0	1	1	1	0	1	1	0	1	0	0	1	0	1	1	0	1	0	0
24	1	1	1	1	1	1	1	0	0	0	0	0	0	0	0	1	0	0	0	0	0	0	0	0	1	1	1	1	1	1	1
25	0	1	0	0	1	1	0	1	0	0	1	1	0	0	1	1	1	0	1	1	0	0	1	0	0	1	0	0	1	1	0
26	0	0	1	0	0	1	1	1	1	0	0	1	1	0	0	1	1	1	0	1	1	0	0	0	0	0	1	0	0	1	1
27	1	0	0	0	1	0	1	1	0	1	0	1	0	1	0	1	0	1	1	1	0	1	0	0	1	0	0	0	1	0	1
28	1	1	1	0	0	0	0	1	0	0	0	0	1	1	1	1	0	0	0	1	1	1	1	0	1	1	1	0	0	0	0
29	0	1	0	1	0	0	1	1	1	0	1	0	0	1	0	1	1	0	1	0	1	1	0	0	0	1	0	1	0	0	1
30	0	0	1	1	1	0	0	1	1	1	0	0	0	0	1	1	1	1	0	0	0	1	1	0	0	0	1	1	1	0	0
31	1	0	0	1	0	1	0	1	0	1	1	0	1	0	0	1	0	1	1	0	1	0	1	0	1	0	0	1	0	1	0

Denote this tournament by $Ext[Ext(H_7)]$.

THEOREM 11. *The homogeneous tournament* $Ext[Ext(H_7)]$ *of order 31 contains a transitive subtournament of order 7.*

PROOF. The set of vertices

$S_1 = \{u_{24}, u_1, u_{25}, u_5, u_{29}, u_2, u_{26}\}$

induces a subtournament of $Ext[Ext(H_7)]$ which is transitive. □

REMARK. $Ext[Ext(H_7)]$ contains additional T_7's. In fact, the sets

$S_2 = \{u_{28}, u_8, u_{20}, u_1, u_{21}, u_2, u_{22}\}$

and

$S_3 = \{u_{24}, u_6, u_{30}, u_4, u_{28}, u_3, u_{27}\}$

induce two other transitive subtournaments of $Ext[Ext(H_7)]$.

As u_5, u_{24}, u_{25}, u_{26} and u_{29} do not belong to S_2 and u_1 and u_2 are not elements of S_3, it follows immediately from the last remark that:

THEOREM 12. *Every tournament of order 30, obtained from either RH_{31} or $Ext[Ext(H_7)]$ by deleting any vertex, contains a transitive subtournament of order 7.*

THEOREM 13. *Every tournament of order 29, obtained from either RH_{31} or $Ext[Ext(H_7)]$ by deleting any pair of distinct vertices, contains a transitive subtournament of order 7.*

PROOF. By Theorem 10 (or 11), the homogeneous tournament RH_{31} (or $Ext[Ext(H_7)]$) contains a T_7. The tournament obtained from RH_{31} (or $Ext[Ext(H_7)]$), by deleting any pair of distinct vertices (x, y), contains also such a T_7, provided that x and y do not belong to a set of vertices which induce a T_7 in RH_{31} (or $Ext[Ext(H_7)]$). This is actually the case for RH_{31}, because this tournament is rotational.

In $Ext[Ext(H_7)]$, the sets of vertices S_1, S_2, S_3 (defined above) and

$S_4 = \{u_{20}, u_2, u_{30}, u_{10}, u_{22}, u_3, u_{23}\}$,

$S_5 = \{u_{20}, u_1, u_{29}, u_9, u_{21}, u_2, u_{22}\}$

and

$S_6 = \{u_{31}, u_8, u_{23}, u_4, u_{19}, u_1, u_{18}\}$

induce transitive subtournaments of order 7 in $Ext[Ext(H_7)]$. Let (u_i, u_j) be any pair of distinct vertices of $Ext[Ext(H_7)]$. As u_i and u_j do not belong to one of the sets $S_1, S_2, \ldots, S_6$, then the tournament of order 29, obtained by deleting the vertices u_i and u_j, contains a T_7. □

From these last results we conclude that not only both the homogeneous tournaments RH_{31} and Ext[Ext (H_7)] or order 31, but also all the tournaments of order 29, which are obtained from either one or the other of these tournaments by deleting any pair of distinct vertices, contain a T_7. This may be useful for future work in order to determine the exact value of $v(n)$ for $n \geq 28$.

As $v(31) \leq 7$, there exists at least one tournament of order 31 which contains no T_8. In the next section, we will see that such a tournament is also homogeneous. Moreover, we will find that there exist, not only one, but at least three non isomorphic homogeneous tournaments of order 31 which contain no T_8.

5. EXISTENCE OF A TRIPLE OF NON ISOMORPHIC TOURNAMENTS OF ORDER 31 WHICH CONTAIN NO T_8

Let $T = (V(T), A(T))$ be a tournament. Denote by $P(u)$ the subtournament of T induced by $\{v \in V(T): (u, v) \in A(T)\}$, and by $Q(u)$ the subtournament of T induced by $\{w \in V(T): (w, u) \in A(T)\}$. The number of vertices in $P(u) \cap P(v)$ is denoted by $\Gamma(u, v)$, whereas $\tau(u, v)$ denotes the number of 3-cycles which contain the arc (u, v).

Recall the following properties of a homogeneous tournament T or order $4k - 1$, $k \geq 1$, [7 and 4]:

The order of $P(u)$ equals $2k - 1$, for every vertex u of T, (1)

$\Gamma(u, v) = k - 1$, for every pair of distinct vertices u and v of T, (2)

$\tau(u, v) = k$, for every arc(u, v) of T. (3)

THEOREM 14. *No homogeneous tournament of order 31 contains a transitive subtournament of order 8.*

PROOF. Suppose that there exists a homogeneous tournament H of order 31 which contains a transitive subtournament T of order 8. Let $u_0, u_1, \ldots, u_7$ be the vertices of T, such that $|V[P(u_i) \cap T]| = i$, for every $i = 0, 1, \ldots, 7$. Then $u_0, u_1, \ldots, u_5$ are vertices of $P(u_6) \cap P(u_7)$. But $\Gamma(u_6, u_7) = 7$ implies the existence of exactly one vertex of $P(u_6) \cap P(u_7)$ which does not belong to $\{u_0, \ldots, u_5\}$. Denote this vertex by v.

Let $M = P(u_7) \cap Q(u_6)$, $R = Q(u_7) \cap Q(u_6)$ and $S = Q(u_7) \cap P(u_6)$. By (1) and (2), we have

$$|V(M)| = |V(R)| = |V(S)| - 1 = 7. \qquad (4)$$

As $\Gamma(u_6, u_0) = 8$, by (3), then

$$|V[P(u_0) \cap Q(u_6)]| = |V[P(u_0) \cap (M \cup R)]| = 8.$$

Thus

$$|V[P(u_0) \cap M]| + |V[P(u_0) \cap R]| = 8. \tag{5}$$

As the order of $P(u_0)$ equals 15, then

$$|V[P(u_0) \cap Q(u_7)]| = 15 - \Gamma(u_0, u_7) = 8$$

and, consequently,

$$|V[P(u_0) \cap R]| + |V[P(u_0) \cap S]| = 8. \tag{6}$$

Equations (5) and (6) imply that

$$|V[P(u_0) \cap M]| = |V[P(u_0) \cap S]|. \tag{7}$$

But $u_1, u_2, \ldots, u_6$ are vertices of $Q(u_0) \cap P(u_7)$; then

$$|V[P(u_0) \cap M]| = \begin{cases} 6, & \text{if } (u_0, v) \in A(H), \\ 7, & \text{if } (v, u_0) \in A(H). \end{cases} \tag{8}$$

Consider the arc (u_6, u_5). By (3), we have

$$|V[P(u_5) \cap Q(u_6)]| = |V[P(u_5) \cap M]| + |V[P(u_5) \cap R]| = 8.$$

As the order of $P(u_5)$ equals 15 and $\Gamma(u_5, u_7) = 7$, then

$$|V[P(u_5) \cap R]| + |V[P(u_5) \cap S]| = 8.$$

From the last two equations, we obtain

$$|V[P(u_5) \cap M]| = |V[P(u_5) \cap S]|. \tag{9}$$

We will show now that $(v, u_5) \in A(H)$. If this were not true, then u_0, u_1, u_2, u_3, u_4 and v would be vertices of $P(u_5) \cap P(u_7)$. As $\Gamma(u_5, u_7) = 7$, it would follow that $|V[P(u_5) \cap M]| = 1$ and, by (9), that $|V[P(u_5) \cap S]| = 1$. Equation (4) would imply that

$$|V[Q(u_5) \cap M]| = |V[Q(u_5) \cap S]| - 1 = 6.$$

But,

$$|V[P(u_0) \cap M]| = |V[P(u_0) \cap S]| \geq 6,$$

by (7) and (8). Therefore,

$$|V[P(u_0) \cap Q(u_5) \cap M]| \geq 5$$

and

$$|V[P(u_0) \cap Q(u_5) \cap S]| \geq 5,$$

which would imply that $\tau(u_5, u_0) \geq 10$. As the latter statement is in contradiction with (3), it follows that $(v, u_5) \in A(H)$.

As $\Gamma(u_5, u_7) = 7$, then $|V[P(u_5) \cap M]| = 2$. From (4) and (9), we obtain

$$|V[Q(u_5) \cap M]| = |V[Q(u_5) \cap S]| - 1 = 5.$$

Consequently,

$$|V[P(u_0) \cap Q(u_5) \cap M]| \geq 4, \quad \text{if } (u_0, v) \in A(H),$$

$$|V[P(u_0) \cap Q(u_5) \cap M]| = 5, \quad \text{if } (v, u_0) \in A(H),$$

and

$$|V[P(u_0) \cap Q(u_5) \cap S]| \geq \begin{cases} 4, & \text{if } (u_0, v) \in A(H), \\ 5, & \text{if } (v, u_0) \in A(H). \end{cases}$$

As $\tau(u_5, u_0) = 8$, by (3), then $(u_0, v) \in A(H)$. But (v, u_5) is also an arc of H. Therefore, v is a vertex of $P(u_0) \cap Q(u_5)$. It follows that

$$\tau(u_5, u_0) \geq 4 + 4 + 1,$$

which is in contradiction with (3). Thus, the initial assumption is false. As a consequence, every homogeneous tournament H of order 31 contains no transitive subtournament of order 8. This completes the proof of the theorem. □

COROLLARY 15. *There exist at least three non isomorphic homogeneous tournaments of the same order 31 which contain no transitive subtournament of order 8.*

PROOF. Denote by H_1 and H_2 the two non isomorphic homogeneous tournaments of order 15 [13, Theorem 2.9]. Let $H_{31}^{(1)} = \text{Ext}(H_1)$ and $H_{31}^{(2)} = \text{Ext}(H_2)$ be the homogeneous extensions of H_1 and H_2, respectively. The tournaments $H_{31}^{(1)}$ and $H_{31}^{(2)}$ are not isomorphic, by [14]. From [13, Section 4], it follows that the rotational homogeneous tournament $H_{31}^{(3)} = RH_{31}$, already defined, is neither isomorphic to $H_{31}^{(1)}$ nor to $H_{31}^{(2)}$. The above theorem implies that the non isomorphic tournaments

$$H_{31}^{(1)}, \quad H_{31}^{(2)} \quad \text{and} \quad H_{31}^{(3)}$$

contain no transitive subtournament of order 8. □

ACKNOWLEDGMENTS

The financial support provided by a Quebec FCAR Grant (ACSAIR) is gratefully acknowledged. Mr. Jean-Luc Meloche is thanked for his assistance in performing the computer analysis. The use of the computer facilities of the Collège Edouard-Montpetit is appreciated.

REFERENCES

[1] P. ERDÖS and L. MOSER, *A problem on tournaments*, Can. Math. Bull. 7 (1964), 351-356.

[2] P. ERDÖS and L. MOSER, *On the representation of directed graphs as unions of orderings*, Publ. Math. Inst. Hungar. Acad. Sci. 9 (1964), 125-132.

[3] A. KOTZIG, *Cycles in a complete graph oriented in equilibrium*, Mat. - Fyz. Časopis 16 (1966), 175-182.

[4] A. KOTZIG, *Sur les tournois avec des 3-cycles régulièrement placés*, Mat. Časopis 19 (1969), 126-134.

[5] J.W. MOON, *Topics on Tournaments*, Holt, Rinehart and Winston, 1968.

[6] K.B. REID and L.W. BEINEKE, *Tournaments*, Chapter 7 in "Selected Topics in Graph Theory", edited by Beineke and Wilson, Academic Press, 1978.

[7] K.B. REID and E. BROWN, *Doubly regular tournaments are equivalent to skew Hadamard matrices*, Jour. Combin. Theory 12 (1972), 332-338.

[8] K.B. REID and E.T. PARKER, *Disproof of a conjecture of Erdös and Moser on tournaments*, Jour. Combin. Theory 9 (1970), 225-238.

[9] R. STEARNS, *The voting problem*, Am. Math. Monthly 66 (1959), 761-763.

[10] C. TABIB, *La théorie des tournois réguliers pourvus de symétries supplémentaires*, Ph.D. Thesis, Université de Montréal, 1978.

[11] C. TABIB, *Caractérisation des tournois presqu'homogènes*, Ann. of Discrete Math. 8 (1980), 77-82.

[12] C. TABIB, *The number of 4-cycles in regular tournaments*, Utilitas Math. 22 (1982), 315-322.

[13] C. TABIB, *Pluralité des tournois homogènes de même ordre*, Ann. sc. math. Québec 8 (1984), no 1, 81-95.

[14] C. TABIB, *Construction of an infinity of pairs of homogeneous tournaments*, Proceedings of the Tenth British Combinatorial Conference, Ars Combinatoria 20A/20B (1986), 7 p. In press.

HEAPS OF PIECES, I :
BASIC DEFINITIONS AND COMBINATORIAL LEMMAS

Gérard Xavier VIENNOT
Université de Bordeaux I
U.E.R. de Mathématiques et d'Informatique
Talence Cedex France

Abstract. We introduce the combinatorial notion of heaps of pieces, which gives a geometric interpretation of the Cartier-Foata's commutation monoid. This theory unifies and simplifies many other works in Combinatorics : bijective proofs in matrix algebra (MacMahon Master theorem, inversion matrix formula, Jacobi identity, Cayley-Hamilton theorem), combinatorial theory for general (formal) orthogonal polynomials, reciprocal of Rogers-Ramanujan identities, graph theory (matching and chromatic polynomials). Heaps may bring new light on classical subjects as poset theory. They are related to other fields as Theoretical Computer Science (parallelism) and Statistical Physics (directed animals problem, lattice gas model with hard-core interactions). Complete proofs and definitions are given in sections 2, 3,4,5. Other sections give a summary of possible applications of heaps.

1. Introduction

Following some work of Foata [24] on combinatorial properties of rearrangements of sequences, Cartier and Foata [9] introduced in 1969 the monoids generated by an alphabet A with relations ab = ba, for all pairs of letters a,b of A such that $(a,b) \in C$, where C is a fixed subset of $A \times A$. The basic properties of these monoids, especially the so-called *flow monoid* and *rearrangement monoid*, appear nowadays to be a classical model in combinatorics (see for example the corresponding chapters of the books of Lallement [39] or Lothaire [40]). These monoids are sometimes called *free partially abelian monoids*. For short, we propose to call them *commutation monoids*.

This model has been used to prove combinatorially (i.e. with bijections) some classical formulae of matrix algebra : the celebrated MacMahon Master theorem in Cartier-Foata [9], the inversion matrix formula in Foata [26] and the Jacobi identity in Foata [27]. More recently, Gessel [30] has shown how to deduce, from the commutation monoid model, Stanley's relation between *chromatic polynomials* and *acyclic orientations* of graphs. Very recently, a new active area of research has grown up in Theoretical Computer Science, using commutation monoids as an algebraic and combinatorial model for *parallelism* problems and concurrency access to data bases, see §10 below.

In this paper, we introduce another model : the notion of *heaps of pieces*. This model will appear to be equivalent to the commutation monoid model. At the beginning, the reader may have certain doubts about the interest of presenting this new version of the commutation monoid with heaps of pieces. These doubts will probably be reinforced after reading the abstract definitions 2.1,2.4, 2.5 and 2.7 below where the heaps model seems more complicated than the commutation monoid.

Once the reader has overpassed these abstract preliminaries, heaps give a powerful "geometric" visualization of the commutation monoids. Many basic lemmas and bijections become really simple. The heaps model appears to be related to other domains, as for example Statistical Physics, although the relationship with commutation monoids was not obvious. Using the heaps model, we have solved combinatorially some open questions about the *directed animals* model introduced by physicists in 1982 (see a survey in Viennot [47]).

Now we give with an example an intuitive introduction to the notion of heaps. Suppose we have an 8×8 chessboard and some dimers. Each dimer is a piece of wood which can cover two consecutive cells of the chessboard. Suppose we put the dimers, one by one, on the chessboard. Each time, one choose a "geographical position" for the next dimer (i.e. two consecutive cells of the chessboard). Then the dimer is put vertically above this position and lowered until it touches the chessboard, covering the two cells of the chosen geographical position, or until it touches one (or two) other dimers previously placed. Placing the new dimer under other dimers is not allowed. In other words, it must be possible to remove the dimers one by one, whithout moving the other dimers, as in the game called "Mikado". What is seen on the chessboard is the visualization of the mathematical notion of *heaps of dimers*, (see Fig. 1).

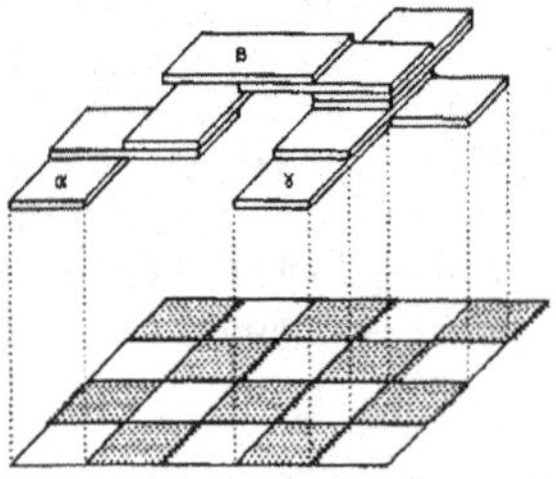

Fig.1. Heap of dimers
(or equivalence class of a
commutation monoid)

When we consider such heaps of dimers, we thus forget some informations about the exact order of placements of the dimers. For example, between the two dimers α and γ of figure 1, one cannot tell which one was placed first. Nevertheless, looking this same figure, one can say that the dimer α was put before the dimer β. In other words, we define the relation $\alpha \leqslant \beta$ iff it is impossible to remove the dimer α from the heap whithout removing the dimer β. The relation $\leqslant$ is a partial order relation. A heap will be a *poset* (partially order set) satisfying certain axioms relating the order relation $\leqslant$, called "*to be above*", and another relation called *concurrency relation*. Here this relation is defined on the set of "geographical positions" for dimers (there are $2 \times 8 \times 7 = 112$ such positions).Two positions are concurrent iff they have one (or two) cells in common.

If we take as alphabet the set A of the 112 possible "geographical positions" for dimers on the chessboard, a word w of letters in A is an encoding of the placements of the dimers of the heap (remembering the order of placements). Forgetting this exact order corresponds to consider the word w up to the commutations ab = ba , where the geographical positions a and b are disjoint (i.e. not in concurrence). The heap of dimers is exactly the geometric visualization of the equivalence class of the word w in the corresponding commutation monoid.

This paper is the first of a series devoted to the theory of heaps and its various applications. It contains two parts. In sections 2,3,4,5 we give the basic definitions and lemmas of the theory, with complete proofs. Sections 6,7,8,9,10 present a summary of the other papers [15],[17],[50],[51] and related works.

Heaps are defined in §2, together with the *heap monoid* $H(P,\mathcal{C})$ related to a set of basic pieces P equipped with a concurrency relation.

In §3, we show the equivalence between the heap monoids and the *commutation monoids*.

In §4, we show that every heap monoid can be realized with a concurrency relation analogous to the one described above with dimers. Basic pieces are subsets of a set, each of these subsets being equipped with a certain combinatorial structure. We also show that every poset can be "realized" as a heap of pieces. This section is just a preliminary step of a promising area of research : studying *posets theory* with the heaps point of view, in particular *realizations* of family of posets as a family of heaps $H(P,\mathcal{C})$.

Basic lemmas about *heaps generating functions* are given in §5: inversion lemma (in fact the equivalent of the Möbius function of the commutation monoid, defined by Cartier, Foata [9]), heaps with given maximal pieces and the logarithmic property "log(heap)=pyramid". A *pyramid* is a heap having only one maximal piece (as in figure 1).

After the work of Cartier, Foata [9] and Foata [26],[27] giving combinatorial proof of classical *matrix algebra* theorems, and also works of Jackson [36], Straubing [46] and simplifications of Zeilberger [51], we present in Dulucq, Viennot [18] an ultimate step, unifying all these bijections as simple consequences of a few basic properties of heaps. A summary is given in §6.

After Flajolet [22], the author has proposed in [47] (survey in [48]) a combinatorial theory of formal *orthogonal polynomials* with weighted paths. Some part of this theory can be simplified and reinterpreted with heaps of pieces. This is summarized in §7. Closely related is a property of Andrews [2] about the "reciprocal" of the famous Rogers-Ramanujan identities. Andrew's interpretation can also be deduced from heaps basic properties.

In §8 we give some relations between heaps and *graph theory* : *chromatic polynomials* and acyclic orientations of graphs from Gessel [30] and a summary of Desainte-Catherine, Viennot [15] relating heaps and *matching polynomials* of graphs. Godsil's tree-like paths [31] fit very well with the heaps model.

In §9 we present a brief summary of Viennot [50],[51] giving two applications of heaps theory in *statistical physics* : the combinatorial solution of the *directed animal* problem and a combinatorial interpretation of the *density of a gas* with hard-core interactions.

In §10, we give a flavor of the connections with *parallelism* problems in *Theoretical Computer Science*.

2. Basic terminology for heaps

Let P be a set equipped with a symmetric and a reflexive binary relation $\mathscr{C}$ (i.e. $a \mathscr{C} b \Leftrightarrow b \mathscr{C} a$ and $a \mathscr{C} a$ for every $a,b \in P$). The elements of P are called *basic pieces*. The relation $\mathscr{C}$ is called the *concurrency relation*.

Definition 2.1. A *labeled heap* with pieces in P is a triple $(E,\leqslant,\varepsilon)$ where $(E,\leqslant)$ is a finite *poset* (i.e., partially ordered set) with order relation denoted by $\leqslant$ and ε is a map $\varepsilon : E \to P$ satisfying the two following conditions

(i) for every $\alpha,\beta \in E$ such that $\varepsilon(\alpha) \; \mathscr{C} \; \varepsilon(\beta)$, then α and β are comparable (i.e $\alpha \leqslant \beta$ or $\beta \leqslant \alpha$).

(ii) for every $\alpha,\beta \in E$ such that $\alpha < \beta$ and β covers α (i.e. $\alpha \leqslant \gamma \leqslant \beta \Rightarrow \gamma = \alpha$ or $\gamma = \beta$) then $\varepsilon(\alpha) \; \mathscr{C} \; \varepsilon(\beta)$.

The elements of E will be called *pieces*. When $\alpha \leqslant \beta$, we will say that the piece β is *above* the piece α.

Remark that P is not necessarily finite but it is important to set down that all the heaps we consider in this theory are *finite*.

Example 2.2. Let $B = [0,8] \times [0,8]$. A *cell* (or *elementary square*) is the set of points (x,y) of B such that $i<x<i+1$, $j<y<j+1$ for certain i, j of $[0,7]$. The set P of basic pieces is the set of subsets of B formed by the union of two cells joined by an edge. The concurrency relation $\mathscr{C}$ is defined by $a \mathscr{C} b$ iff $a \cap b \neq \emptyset$. A heap E with pieces in P was visualized on Fig.1.

Here the map ε is the projection associating to each dimer of the heap its "geographical position", i.e. an element of P (see the heuristic introduction in §1).

Example 2.3. Let $P = \mathbb{Z}$ be the set of integers. The concurrency relation $\mathscr{C}$ is defined by : $i\mathscr{C}j$ iff $|j-i|\leq 1$ for $i,j\in P$. The poset E is defined on Fig.2 by its *Hasse diagram*, (i.e. an edge goes upward from α to β iff β covers α). The map ε is defined on Fig.2 by: for $\alpha\in E$ lying on the vertical line $x=i$, then $\varepsilon(\alpha)=i$.

The reader will verify that the axioms (i) and (ii) are satisfied.

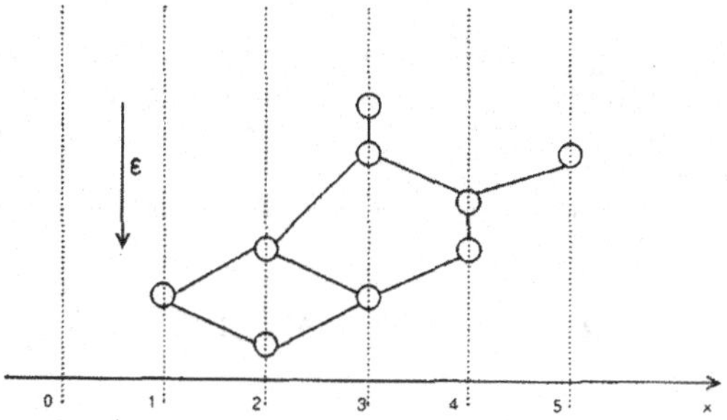

Fig.2. A heap of pieces.

Equivalent definitions for heaps

a) Conditions (i) and (ii) can be replaced by (i) and (ii') where (ii') is the following condition

(ii') for every $\alpha,\beta\in E$ with $\alpha\leq\beta$, there exists a sequence $\alpha=\alpha_1\leq\ldots\leq\alpha_k=\beta$ of pieces of E such that $\varepsilon(\alpha_i)\ \mathscr{C}\ \varepsilon(\alpha_{i+1})$ for every i, $1\leq i<k$.

b) A second formulation is (i) and (ii") where (ii") is the following condition

(ii") the order relation $\leq$ is the transitive closure of the relation $\mathscr{C}'$ defined by : for $\alpha,\beta\in E$, $\alpha\ \mathscr{C}'\ \beta$ iff $\alpha\leq\beta$ and $\varepsilon(\alpha)\ \mathscr{C}\ \varepsilon(\beta)$.

c) A third formulation is the following. Let ε: $E \rightarrow P$ be a map of the set E in P. Let $G(E,\varepsilon,\mathscr{C})$ be the graph which vertices are the elements of E and with an edge between α and β iff $\varepsilon(\alpha)\ \mathscr{C}\ \varepsilon(\beta)$. Then defining an order relation $\leq$ on E such that $(E,\leq,\varepsilon)$ satisfies conditions (i) and (ii), is nothing but defining an *acyclic orientation* of the graph $G(E,\varepsilon,\mathscr{C})$ (i.e. an orientation of each edge such that the graph does not contain cycles).

Subheap. Let $(E,\leq,\varepsilon)$ be a heap and F a subset of E. Let ε' be the restriction of ε to F. Let $\mathscr{R}$ be the relation defined on F by $\alpha\ \mathscr{R}\ \beta$ iff $\alpha\leq\beta$ and $\varepsilon(\alpha)\ \mathscr{C}\ \varepsilon(\beta)$. Let $\leq'$ be the transitive closure of $\mathscr{R}$. Then $(F,\leq',\varepsilon')$ is called a *subheap*.

Definition 2.4. Let $(E,\leq,\varepsilon)$ and $(E',\leq',\varepsilon')$ be two heaps of pieces in P with the same concurrency relation $\mathscr{C}$. We say that they are *isomorphic* iff there exists a bijection $\varphi:E \rightarrow E'$ which is a poset isomorphism (i.e. $\alpha\leq\beta$ in E iff $\varphi(\alpha)\leq'\varphi(\beta)$ in E'), and such that $\varepsilon=\varepsilon'\circ\varphi$.

Definition 2.5. A *heap* of pieces (in French : *empilement de pièces*) in P with concurrency relation $\mathscr{C}$ is a labeled heap (definition 2.1) defined up to a heap isomorphism (or equivalence class, for isomorphism, of labeled heaps).

In the following, a heap will be denoted by one of its representative $(E,\leq,\varepsilon)$ or E for short. We will sometimes use the notation $E = (E,\leq,\varepsilon)$. We will again say that the elements of E are *pieces*, and call the order relation $\leq$ as "to be above". The set of all finite heaps with pieces in P and concurrency relation $\mathscr{C}$ is denoted by $H(P,\mathscr{C})$.

Lemma 2.6. Any automorphism φ of a labeled heap $(E,\leq,\varepsilon)$ is trivial (i.e. is the identity map of E).

Proof. For any basic piece $a \in P$, $\varepsilon^{-1}(a)$ is a finite chain of E (from (i) and the reflexivity of $\mathscr{C}$). The relation $\varepsilon = \varepsilon \circ \varphi$ implies that φ preserves this chain. As it is a poset automorphism, we deduce that φ is the identity map.

Let a_n be the number of heaps of $H(P,\mathscr{C})$ having n pieces. From lemma 2.6, we deduce that the number b_n of labeled heaps, with set of labels any set of n elements as for example $E = \{1,2,\ldots,n\}$, is $b_n = n!a_n$. When enumerating heaps (resp. labeled heaps) we will use the ordinary (resp. exponential) generating function $\sum_{n\geq 0} a_n t^n$ (resp. $\sum_{n\geq 0} b_n t^n/n!$).

In fact, labeled heaps are an example of Joyal's *species* [38]. Heaps are the corresponding *type of species*.

Definition 2.7. Let E and F be two heaps of $H(P,\mathscr{C})$. the *product* $H = E \odot F$ (or *superposition* of F over E) is the heap H defined by the following : if $E = (E,\leq_E,\varepsilon)$, $F = (F,\leq_F,\varepsilon')$, $H = (H,\leq_H,\varepsilon'')$, then

(i) $H = E+F$ (disjoint union of E and F)

(ii) ε'' is the unique map $\varepsilon'' : H \to P$ which restriction to E (resp. F) is ε (resp. ε').

(iii) the order relation $\leq_H$ is the transitive closure of the following relation $\mathscr{R}$ for $\alpha,\beta \in H$, $\alpha \mathscr{R} \beta$ iff

- $\alpha,\beta \in E$ and $\alpha \leq_E \beta$

or - $\alpha,\beta \in F$ and $\alpha \leq_F \beta$

or - $\alpha \in E, \beta \in F$ and $\varepsilon(\alpha) \mathscr{C} \varepsilon'(\beta)$.

Remark that E and F are subheaps of $E \odot F$.

Such a definition is compatible with isomorphisms and thus is well defined on the set $H(P,\mathscr{C})$ of heaps.

This product of heaps is associative and $H(P,\mathscr{C})$ is a *monoid*, called the *heap monoid*, which neutral element is the *empty heap* denoted by $\emptyset$.

An element α of P will be identified with a heap reduced to a single piece. The heap $E\circledcirc\alpha$ is said to be obtained by *adding* (or *putting*) the (basic) piece α *above* the heap E. Any heap E is a product (in general in several different ways) of its (basic) pieces. Remark that for any two basic pieces.

(1) $\alpha,\beta\in P$, $\alpha\circledcirc\beta=\beta\circledcirc\alpha$ iff $\alpha \not\mathscr{C} \beta$ (i.e. α and β are not in concurrency).

The product of heaps is a left and right *simplifiable* product, that is : $E\circledcirc F=E\circledcirc F' \Rightarrow F=F'$ and $E\circledcirc F=E'\circledcirc F \Rightarrow E=E'$. If $E\circledcirc\alpha=F$, for $\alpha\in P$ and $E,F\in H(P,\mathscr{C})$, we will say that E is obtained by *deleting* the piece α from the top of the heap F.

Definition 2.8. A *trivial* heap is a heap such that the order relation $\leqslant$ is trivial, that is no pieces are above another.

We will denote by $T(P,\mathscr{C})$ the set of trivial heaps of $H(P,\mathscr{C})$. If the concurrency relation $\mathscr{C}$ is "empty", every heap is trivial and the heap monoid $H(P,\mathscr{C})$ is isomorphic to the free commutative monoid generated by P.

Lemma 2.9. <u>Any heap</u> $E \in H(P,\mathscr{C})$ <u>can be written in a unique way as a product of trivial heaps</u> $E = T_1\circledcirc\ldots\circledcirc T_p$ <u>satisfying the condition</u> :

(2) <u>for any</u> $1\leqslant j<p$, <u>any pieces of</u> T_{j+1} <u>is above a piece of</u> T_j .

It suffices to take T_1 as the subheap formed by the minimum elements of E. Then one can write $E = T_1\circledcirc E_1$. Repeating recursively this factorization, we get the unique factorization satisfying (2).

This factorization can be characterized in another way. It is the unique factorization into a product of trivial heaps, each factor having maximum cardinality.

3. The Cartier-Foata commutation monoid

Let A be a set and A^* be the *free monoid* generated by A, that is the set of words $u=a_1a_2\ldots a_p$ with *letters* a_i in the set A (called *alphabet*), together with the multiplicative law *concatenation* of two words : $u=a_1..a_p$ and $v=b_1..b_q$, $uv=a_1..a_pb_1..b_q$. The *empty word* is denoted by e.

Let C be a symmetric and antireflexive relation on A (i.e. $a\not C a$ for every $a\in A$).

Definition 3.1. The *commutation monoid* $Co(A,C)$ is the quotient of the free monoid by the congruence $\equiv_C$ generated by the (commutation) relations :

(3) for every $a,b \in A$ with $a\ C\ b$, then $ab \equiv_C ba$.

The words u and v are equivalent iff one can transform u into v by a sequence of transpositions of two consecutive letters a and b such that a C b. The monoids $Co(A,C)$, introduced by Cartier and Foata [9], are also called *free partially abelian monoids.*

We suppose that the alphabet A is the set P of basic pieces equipped with the concurrency relation $\mathscr{C}$. Let $C = \bar{\mathscr{C}}$ be the complementary relation (i.e. $a\, C\, b$ iff $a \not\mathscr{C} b$). We are going to show that the heap monoid $H(P,\mathscr{C})$ is a commutation monoid isomorphic to $Co(P,C)$.

We define the map $\varphi : P^* \to H(P,\mathscr{C})$ by the relation

(4) for $w = \alpha_1\alpha_2\ldots\alpha_n \in P^*$, $\varphi(w) = \alpha_1 \odot \alpha_2 \odot \ldots \odot \alpha_n \in H(P,\mathscr{C})$.

In other words φ is the unique morphism (of monoids) such that for $\alpha \in P$, $\varphi(\alpha)$ is the heap identified with the basic piece α.

Let $(E,\leqslant)$ be a poset having n elements. A *natural labeling* of the poset $(E,\leqslant)$ is a bijection $f : E \to [n]=\{1,2,\ldots,n\}$ such that

(5) for every $\alpha,\beta \in E$, $\alpha \leqslant \beta \Rightarrow f(\alpha) \leqslant f(\beta)$.

Another equivalent definition is the so-called *linear extension* of a poset.

Lemma 3.2. Let $(E,\leqslant,\varepsilon)$ be a heap of $H(P,\mathscr{C})$. For $u = \alpha_1\ldots\alpha_n \in \varphi^{-1}(E)$, let $\lambda(u) = f$ be the labeling $f:E\to[n]$ defined by $\varepsilon(f^{-1}(i)) = \alpha_i$. The map λ is a bijection between the set of words $\varphi^{-1}(E)$ and the set $\mathscr{L}(E)$ of natural labelings of E.

Proof. a) From the definition 2.7 of the product of heaps, the heap $E = \alpha_1\odot\alpha_2\odot\ldots\odot\alpha_n$ is obtained by adding vertices $s_1,s_2,\ldots s_n$ to the empty heap, with the map ε defined by $\varepsilon(s_i) = \alpha_i$, and the order relation $\leqslant$ defined by

(6) $\leqslant$ is the transitive closure of the relation $\mathscr{R}$ defined by $s_i \,\mathscr{R}\, s_j$ iff $i \leqslant j$ and $\alpha_i \,\mathscr{C}\, \alpha_j$.

Thus the map $f=\lambda(\alpha_1\ldots\alpha_n)$ defined by $f(s_i) = i$ is a natural labeling of E.

b) Conversely, let $f: E \to [n]$ be a natural labeling of E. Let $t_i=f^{-1}(i) \in E$ $(1\leqslant i\leqslant n)$ and $\beta_i=\varepsilon(t_i)$. Let F be the heap $F=\varphi(\beta_1\ldots\beta_n)$. We can identify the vertices of the two heaps E and F. We show that these heaps are isomorphic. If $s \underset{E}{\leqslant} t$, then from heap axiom (ii'), there exists a sequence $t_{i_1} = s \underset{E}{\leqslant} \ldots \underset{E}{\leqslant} t_{i_k} = t$ of vertices of E such that for $j, 1\leqslant j<k$, $t_{i_j} \,\mathscr{C}\, t_{i_{j+1}}$. As the map $t_i \to i$ is a natural labeling of E, then $i_1 \leqslant \ldots \leqslant i_k$. From the definition of the product $\beta_1\odot\beta_2\odot \ldots \odot\beta_n$, we deduce $t_{i_j} \underset{F}{\leqslant} t_{i_{j+1}}$ and thus $s \underset{F}{\leqslant} t$. The heaps E and F are isomorphic and $\beta_1\odot\ldots\odot\beta_n \in \varphi^{-1}(E)$.

Combining a) and b) the map λ is a surjection from $\varphi^{-1}(E)$ onto $\mathscr{L}(E)$. As it is obviously an injection, the lemma is proved.

□

Lemma 3.3. <u>For every heap</u> $E \in H(P,\mathscr{C})$, <u>the set of words</u> $\varphi^{-1}(E)$ <u>is an equivalence class for the commutation relation</u> $\equiv_C$.

Proof. a) If α and β are two basic pieces not in concurrency, the two heaps $\alpha\odot\beta$ and $\beta\odot\alpha$ are trivial (definition 2.8.) . Thus $\alpha\odot\beta=\beta\odot\alpha$. For $u,v \in P^*$, we deduce that $u \equiv_C v$ implies $\varphi(u) = \varphi(v)$.

b) Conversely let $u=\alpha_1...\alpha_n$ and $v=\beta_1...\beta_n$ be two words such that $\varphi(\alpha_1...\alpha_n)=\varphi(\beta_1...\beta_n)$ is the heap E. As in the proof a) of lemma 3.2, let $s_1,...,s_n$ be the vertices of $E = (E,\leq,\varepsilon)$ with $\varepsilon(s_i)=\alpha_i$. From (6), the vertex s_i of E is minimal (for $\leq$) iff no pieces α_j, $1\leq j<i$ are in concurrency with α_i, that is α_i commutes with all the letters located at its left in the word $u = \alpha_1..\alpha_n$. We can write $u \equiv_C u_1u_1'$, where u_1 is the word containing all the letters (commuting two by two) of u corresponding to minimal elements of E.

Similarly, we can write $v \equiv_C v_1v_1'$, where v_1 is the word containing all the letters (commuting two by two) of v corresponding to minimal elements of E.

Thus $u_1 \equiv_C v_1$ and $\varphi(u_1')=\varphi(v_1')$ is the subheap E_1 obtained from E by deleting all its minimal elements (see lemma 2.9).

By a recurrence on the common length of the words u and v, we deduce that u and v are equivalent modulo $\equiv_C$.

□

Combining lemmas 3.2 and 3.3, we deduce

Proposition 3.4. <u>Let</u> $H(P,\mathscr{C})$ <u>be a heap monoid with pieces in</u> P <u>and concurrency relation</u> $\mathscr{C}$. <u>Let</u> C <u>be the complementary relation of</u> $\mathscr{C}$. <u>The morphism of monoid</u> φ: $P^* \rightarrow H(P,\mathscr{C})$ <u>defined by</u> (4) <u>induces an isomorphism</u> $\bar{\varphi}$ <u>between the monoid</u> $H(P,\mathscr{C})$ <u>and the commutation monoid</u> Co(P,C).

It may be useful to restate the definition of this isomorphism $\bar{\varphi}$:Co(P,C) $\rightarrow H(P,\mathscr{C})$, together with its main properties coming from the proof of lemmas 3.2 and 3.3.

a) Let $\bar{u}$ be an element of Co(P,C).Choose any representative $u=\alpha_1\alpha_2...\alpha_n$ of this class of words. Then the heap $\varphi(u)=\alpha_1\odot\alpha_2\odot...\odot\alpha_n$ is independent of the choice of $u\in\bar{u}$ and will be denoted by $\bar{\varphi}(\bar{u})$.

b) Conversely, if $E = (E,\leq,\varepsilon)$ is a heap of $H(P,\mathscr{C})$, taking any natural labeling $f: E \rightarrow [n]$ of the poset $(E,\leq)$, we define a word $u=u_1...u_n$ with $u_i=\varepsilon(f^{-1}(i))$.Let $\bar{u}$ be the equivalence class of u for $\equiv_C$. Then the map $E \rightarrow \bar{u}$ is the reverse bijection of the bijection $\bar{\varphi}$:Co(P,C) $\rightarrow H(P,\mathscr{C})$.

c) Let $\bar{u}$ be a commutative class of Co(P,C) and $u=\alpha_1...\alpha_n$ be one of its representants. We define a poset $([n],\leq)$ in the following way. The vertices are the integers 1,2,..,n. The order relation $\leq$ is defined by the following relation

(7) $i\leq j$ iff there exists a sequence $1\leq i_1=i<...<i_k=j\leq n$ such that the letters α_{i_j} and $\alpha_{i_{j+1}}$ do not commute (for $1\leq j<k$).

Defining the map ε: $[n] \to P$ by $\varepsilon(i)=\alpha_i$, we have now a labeled heap $E(u) = ([n],\leq,\varepsilon)$ which is a representant of the heap $\bar{\varphi}(\bar{u})$. This labeling f of the vertices of $\bar{\varphi}(\bar{u})$ by the integers 1,2,...,n is a natural labeling. The map $u \to f$ is a bijection between the words of the equivalence class $\bar{u}$ and the natural labelings (or linear extensions) of the poset underlying the heap $\bar{\varphi}(\bar{u})$.

Remark that the order relation defined by (7) is the transitive closure of the relation defined by Cori and Métivier from the directed graph denoted by $\Gamma(u)$ in [12]. Also, to give the labeled heap $E(u) = ([n],\leq,\varepsilon)$ is equivalent to give the so-called *"dependency graph"* of the word u introduced by Perrin in [43].

If we restate lemma 2.9 in terms of commutation monoids, we get the classical property (see Cartier, Foata [9]) :

Corollary 3.5. Let u be a word of P^* and Co(P,C) be a commutation monoid. Then $\bar{u}$ can be written in a unique way $\bar{u}=\bar{u}_1..\bar{u}_p$ where each u_j is a block of letters commuting two by two, and for each pair of consecutive blocks $u_j u_{j+1}$, any letter of u_{j+1} does not commute with at least a letter of u_j.

This unique factorization is called the *normal form* of u in [43] and V-factorization in Cartier-Foata [9]. In fact this corollary also comes from part b) of the proof of lemma 3.3.

4. Graphs, Heaps and Posets

Let P and B be two sets and let $\pi: P \to \mathcal{P}(B)$ be a map from P into the set of non empty subsets of B. We define the concurrency relation $\mathscr{C}$ by the relation.

(8) for $a,b \in P$, $a \,\mathscr{C}\, b$ iff $\pi(a) \cap \pi(b) \neq \emptyset$.

In this fundamental example of heaps, the heap monoid $H(P,\mathscr{C})$ will also be denoted by $H(P,\pi,B)$. The set B is called the *basis*. The subset $\pi(a)$ is called the *support* of the basic piece $a \in P$.

Let $E = (E,\leq,\varepsilon)$ be a heap of $H(P,\pi,B)$ and $\alpha \in E$ be a piece of E. The subset $\pi\circ\varepsilon(\alpha)$ will also be called the *support* of the piece α. We say that two pieces $\alpha,\beta \in E$ (resp. basic pieces $a,b \in P$) are disjoint if their support are *disjoint*. In the contrary, that is $\varepsilon(\alpha) \,\mathscr{C}\, \varepsilon(\beta)$ (resp. $a \,\mathscr{C}\, b$) they are said to be *intersecting*. Two heaps E and F are said to be *intersecting* iff one piece of E intersects one piece of F.

Example 4.1. Let $B = \mathbb{Z}$ and P be the set of *dimers*, that is the set of subsets of the form $\{i,i+1\}$, $i \in \mathbb{Z}$. We define π as the restriction to P of the identity map of $\mathcal{P}(B)$. The heap displayed on Fig.3 is" isomorphic" to the heap of Fig.2. (here the term "isomorphic" would be an extension of definition 2.4 to the case of two heaps with different set of basic pieces, see below just before remark 4.4).

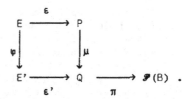

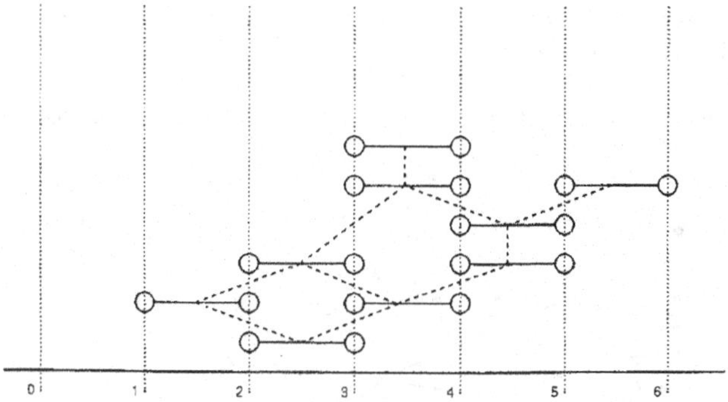

Fig.3. A heap of dimers on $\mathbb{Z}$.

Example 4.2. Let $B = [0,8] \times [0,8]$ and P as in example 2.2. Let π be the restriction to P of the identity map of $\mathscr{P}(B)$. Heaps of $H(P,\pi,B)$ were considered in example 2.2 and are visualized on Fig.1 of the introduction.

In fact, any heap monoid $H(P,\mathscr{C})$ can be identified with a heap monoid $H(Q,\pi,B)$. For that, we need the following definition.

Definition 4.3. Let $\mathscr{C}$ be a concurrency relation (i.e. symmetric and reflexive) on the set P. The *concurrency graph* is the graph $G(\mathscr{C})$ with vertices in P and with edges $\{a,b\} \in A$ iff $a \,\mathscr{C}\, b$ and $a \neq b$.

Let $B = P \cup A$. For each $a \in P$, we define the subset $\mu(a)$ of B by

$$\mu(a) = \{a\} \cup \{\{a,b\} \in A\}. \qquad (9)$$

The map μ is a bijection between P and $\mu(P)=Q$. Now $a \,\mathscr{C}\, b$ iff $\mu(a) \cap \mu(b) \neq \emptyset$. Let π be the restriction to Q of the identity map of $\mathscr{P}(B)$.

Any heap $(E,\leq,\varepsilon) \in H(P,\mathscr{C})$ is "*isomorphic*" to a heap $(E',\leq',\varepsilon')$ of $H(Q,\pi,B)$, i.e. there exists a poset isomorphism $\varphi: E \to E'$ and a bijection $\mu: P \to Q$ "*preserving*" the concurrency relations of P and Q, such that the following diagram is commutative

$$\begin{array}{ccccc} E & \xrightarrow{\ \varepsilon\ } & P & & \\ {\scriptstyle\varphi}\downarrow & & \downarrow{\scriptstyle\mu} & & \\ E' & \xrightarrow[\ \varepsilon'\ ]{} & Q & \xrightarrow[\ \pi\ ]{} & \mathscr{P}(B) \end{array} .$$

Remark 4.4. The construction of the map μ defined by (9) is related to the so-called *line graph* (or *median graph*) of the concurrency graph $G(\mathscr{C})$.

If the graph $G(\mathscr{C})$ is represented by points of $\mathbb{R}^d$ joined by segments, then one can represent $\mu(a)$ as the set of points formed by the vertex a and the middle of the edges containing this vertex a. The pieces look like *starfishes* (see Fig.4).

The monoid $H(Q,\pi,B)$ constructed above from P and $\mathscr{C}$ will be called a *starfish monoid*.

Proposition 4.5. Every heap monoid is isomorphic to a starfish monoid.

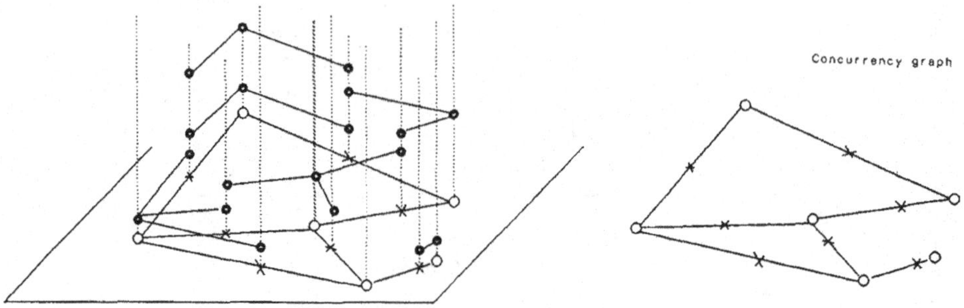

Fig.4. A heap of starfishes.

Proposition 4.6. Every poset $(E,\leq)$ can be represented as the underlying poset of a heap $(E,\leq,\varepsilon)$.

For the proof of this proposition, we need the following definitions.

Let $(E,\leq)$ be a poset and $\mathscr{F} = (C_i)_{i\in I}$ be a family of chains. We say that $\mathscr{F}$ *strongly covers* E iff, for every pair α,β of elements of E such that β covers α (i.e. α and β are connected by an edge in the Hasse diagram of E), then there exists a chain C_i of $\mathscr{F}$ containing both α and β.

Let $\mathscr{F} = (C_i)_{i\in I}$ be such a family (always exists). We take as basis the set B=I. The basic pieces are the subsets of I and π is the identity of $P = \mathscr{P}(I)$. We define the map $\varepsilon : E \longrightarrow P$ by $\varepsilon(\alpha)=\{i\in I, \alpha\in C_i\}$. The concurrency relation $\mathscr{C}$ is defined by (8).

The triple $(E,\leq,\varepsilon)$ satisfies condition (i) of definition 2.1 of labeled heap : if $\alpha,\beta\in E$, $\varepsilon(\alpha)\mathscr{C}\,\varepsilon(\beta)$ implies that α and β belongs to a same chain C_i of $\mathscr{F}$. Condition (ii) of definition 2.1 follows from the strongly covering property. □

It would be interesting to represent some known families of posets as families of heaps $H(P,\pi,B)$. Is it possible to give a poset characterization of the posets underlying heaps of a given heaps monoid Many questions arise about *representations* of posets with heaps. Here we will mainly be interested in heaps as a tool for combinatorial enumeration and combinatorial interpretation of classical results or identities. Nevertheless we will mention the following property.

Let $E = (E,\leq,\varepsilon)$ be a heap of $H(P,\pi,B)$. For $x\in B$, the *fiber* of E over x is the set defined by

(10) $$F_x(E) = \{\alpha \in E,\ x \in \pi\circ\varepsilon(\alpha)\}\ .$$

Such fibers are chains. The family $\{F_x(E)\}_{x\in B}$ strongly covers the poset $(E,\leq)$.

The minimum cardinality of the basis set B such that the poset $(E,\leq)$ is realized as a heap of $H(P,\pi,B)$ is the minimum number of chains strongly covering E. This number is not less that the minimum number of chains covering E. This last number is more classical in poset theory and, from Dilworth's theorem is known to be equal to the maximum cardinality of *antichains* of E (set of elements two by two incomparable).

The reader may ask the interest of introducing the map $P \xrightarrow{\pi} \mathcal{P}(B)$ instead of simply introducing the basic pieces as subsets of B. We will need basic pieces where a combinatorial structure is defined on their support. An important example will be heaps of *cycles*. Here P is the set of all cycles of B (in the sense of cycle of permutation : that is a circular permutation on a subset of B). The map π associates to a cycle its underlying set of vertices. An example is displayed on Fig.5. The order $\leq$ between the cycles is defined by the fibers (corresponding to the vertical lines).

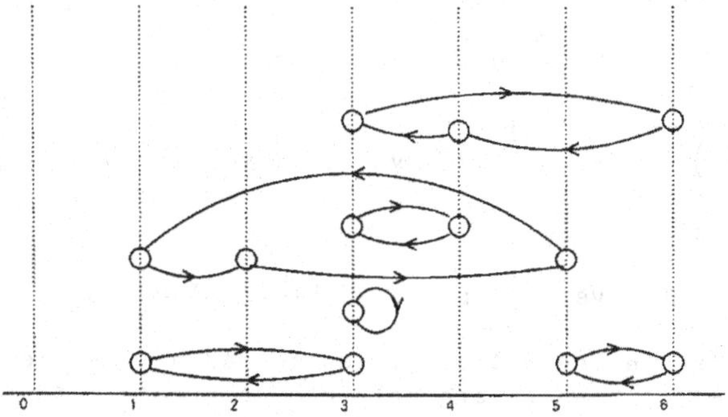

Fig.5. A heap of cycles on $\mathbb{N}$.

5. Heap Generating functions

Let $H(P,\mathscr{C})$ be a heap monoid with basic pieces in P and concurrency relation $\mathscr{C}$. Let $\mathbb{K}[[X]]$ be the algebra of formal power series with variables in a set X (not necessarily finite) and with coefficients in the commutative ring $\mathbb{K}$. We define a *valuation* (or *weight* function) as a map $v : P \to \mathbb{K}[[X]]$ which associates to every basic piece $\alpha\in P$ a power series $v(\alpha)$ *having no constant term*. This last condition is necessary for the summability of heaps generating functions. In general, most examples of the theory will be such that $v(\alpha)$ is a monomial in variables from X.

The *valuation* (or *weight*) $v(E)$ of a heap $E=(E,\leq,\varepsilon)$ is the product of the valuations of its pieces $v(E) = \prod_{\alpha\in E} v(\varepsilon(\alpha))$.

In all this work, we suppose that the valuation v satisfies the following condition

(11) for every monomial μ in the variables X, there exists a finite number of heaps E of $H(P,\mathscr{C})$ such that the coefficient of μ in the series $v(E)$ is $\neq 0$.

This condition, which is satisfied when the set P of basic pieces is finite, implies the summability of the heaps generating function $\sum_{E\in H(P,\mathscr{C})} v(E)$.

Proposition 5.1. (Inversion lemma) _Let_ $H(P,\mathscr{C})$ _be a heap monoid with a valuation_ v _satisfying_ (11). _The generating function of the weighted heaps of_ $H(P,\mathscr{C})$ _is given by_

$$\sum_{E\in H(P,\mathscr{C})} v(E) = \frac{1}{\sum_{F\in T(P,\mathscr{C})} (-1)^{|F|} v(F)}, \tag{12}$$

where $T(P,\mathscr{C})$ _denotes the set of trivial heaps_ (definition 2.8).

The identity (12) is equivalent to the identity

$$\sum_{(E,F)} (-1)^{|F|} v(E)\, v(F) = 1, \tag{13}$$

where the summation is over all pairs (E,F) of $HT=H(P,\mathscr{C}) \times T(P,\mathscr{C})$.

Let M(E,F) be the set of pieces formed by the pieces of F and the maximal pieces of E which are not in concurrence with pieces of F. Let L be a non-empty trivial heap. In the summation (13), we select only pairs (E,F) such that M(E,F) = L. We can write $L=L_1\odot L_2$ with $E=E_1\odot L_1$ and $F=L_2$ (see Fig.6). Thus, we have the identity

$$\sum_{\substack{(E,F)\in HT\\ M(E,F)=L}} (-1)^{|F|} v(E)\,v(F) = v(L) \sum_{E_1} v(E_1) \left[\sum_{\substack{L_1,L_2\in T(P,\mathscr{C})\\ L=L_1\odot L_2}} (-1)^{|L_2|}\right],$$

where the first summation of the right hand-side is over heaps $E_1\in H(P,\mathscr{C})$ such that all their maximal pieces are in concurrence with at least a piece of L. The second summation of the right-hand side is 0.

Thus, the only non-vanishing term in (13) is the pair corresponding to $M(E,F) = \emptyset$, that is $E=\emptyset$, $F=\emptyset$. Its weight is 1.

□

Remark 5.2. If the reader prefers a proof with bijections, it would be possible to define a sign-reversing involution on the set HT. The idea is simply to "transfer" a piece α of F on the top of E, or vice-versa (see Fig.6). We totally order the set P of basic pieces. For a pair $(E,F)\in HT$, $(E,F)\neq(\emptyset,\emptyset)$, we take the smallest piece α of M(E,F). If α is a piece of E, then $E=E_1\odot\alpha$ and we define $\Psi(E,F)=(E_1,\alpha\odot F)$. If α is a piece of F, then $F=F_1\odot\alpha$ $(=\alpha\odot F_1)$ and we define $\Psi(E,F)=(E\odot\alpha,F_1)$. The map $\Psi(E,F)\longrightarrow(E',F')$ is an involution such that

$$(-1)^{F}\, v(E)\,v(F) = -(-1)^{F'}\, v(E')\,v(F').$$

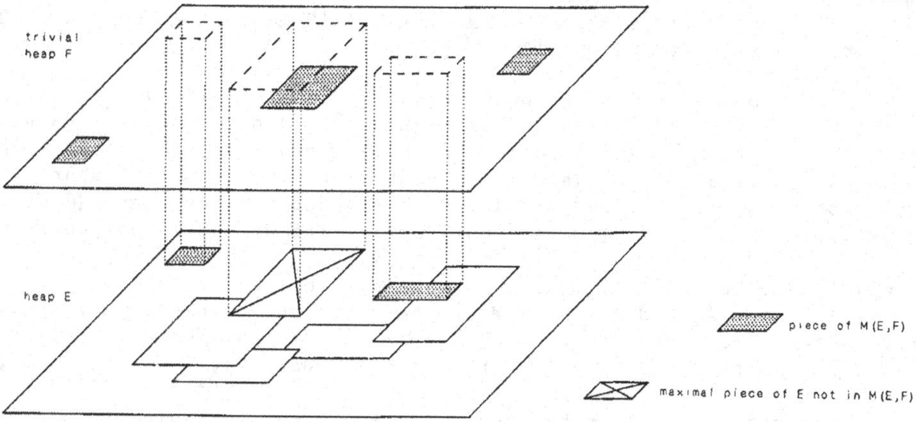

The concurrency relation $\mathscr{C}$ is the intersection relation

Fig.6. Proof of proposition 5.1.

In term of commutation monoid, relation (12) is nothing but expressing the *Möbius function* of that monoid (see theorem 2.4 of Cartier, Foata [9]). Möbius inversion of posets is a classical chapter of Combinatorics which has been popularized by Rota [42]. Content, Lemay, Leroux [11] present a synthesis of Rota and Cartier-Foata's Möbius inversion. We give the following extension, which generalizes a proposition of Desainte-Catherine [14], [15].

Proposition 5.3. Let $H(P,\mathscr{C})$ be a heap monoid with a valuation v satisfying (11). Let M be a set of basic pieces of P. The generating function of weighted heaps of $H(P,\mathscr{C})$ such that the maximal pieces are in M is given by.

$$(14)\qquad \sum_{\substack{E\in H(P,\mathscr{C})\\ \textbf{Maximal pieces } M}} v(E) = \frac{N}{D}, \qquad \text{with}$$

$$D=\sum_{F\in T(P,\mathscr{C})} (-1)^{|F|}\, v(F) \qquad \text{and} \qquad N=\sum_{F\in T(P\setminus M,\mathscr{C})} (-1)^{|D|}\, v(F).$$

where $T(P,\mathscr{C})$ denotes the set of trivial heaps (definition 2.8).

It would be possible to define a sign-reversing involution as in remark 5.2 transfering a piece α from E to F, by taking M(E,F) as the set of minimal pieces of E, not in concurrence with any pieces of F, together with the set of pieces of F which are in M or in concurrence with at least one piece of E. The pairs corresponding to $M(E,F) = \emptyset$ are the pairs involved in the summation for N.

A simpler proof, as suggested by A.Joyal, is to apply proposition 5.1 and the following lemma.

Lemma 5.4. Let $M \subset P$ and $E \in H(P,\mathscr{C})$. Then the heap E has a unique factorization $E = E_1 \circ E_2$ where E_1 is a heap with maximal pieces in M and E_2 is a heap with pieces not in M.

The power serie D appearing as the denominator of generating function (12) and (14) plays an important role in heaps theory. We will call this power serie the *exclusion power serie* for the pieces P and concurrency relation $\mathscr{C}$ (and the valuation v), or for short the *exclusion serie* of the heap monoid. We will denote it by D or $D(P,\mathscr{C})$ or $D(P,\mathscr{C},v)$. If P is finite, D is a polynomial, the *exclusion polynomial*.

Example 5.5. Let $A = (a_{ij})$ be an n×n matrix. Let $B = [n]$, P be the set of cycles on B and $\pi : P \to \mathscr{P}(B)$ the map associating to a cycle its underlying set. The concurrency relation is defined by (8). The valuation of the cycle $\gamma=(x_1 \ldots x_m)$ is the product $\lambda^m a_{x_1 x_2} \ldots a_{x_{m-1} x_m} a_{x_m x_1}$

The letters λ and a_{ij} can be considered as formal variables in X. Then it is almost the definition of the determinant to say that $D(P,\mathscr{C}) = \det(I-\lambda A)$.

Thus the *characteristic polynomial* of the matrix A can be considered as the reciprocal of an exclusion polynomial.

Example 5.6. Let $B = \mathbb{N}$ be the basis and P be the set of monomers $\{i\}$, $i \geq 0$ and dimers $\{i,i+1\}$, $i \geq 0$, with concurrency relation the intersection relation as in §4.

Let $\{b_k\}_{k\geq 0}$ and $\{\lambda_k\}_{k\geq 1}$ be two sequences of the ring $\mathbb{K}$. The monomer $\{i\}$, $i \geq 0$ is weighted $b_i x$. The dimer $\{i-1,i\}$, $i \geq 1$, is weighted $\lambda_i x^2$.

If we restrict the basis to be $B_n = [0,n-1]$, with pieces in B_n, let $P_n(x)$ be the corresponding exclusion polynomial. These polynomials satisfy the three-terms linear recurrence relation

(15) $P_{n+1}(x) = (x-b_n)\, P_n(x) - \lambda_n\, P_{n-1}(x)$, with $P_0(x)=1$, $P_1(x)=x-b_0$.

From Favard's theorem the sequence $\{P_n(x)\}_{n\geq 0}$ is a sequence of formal *orthogonal polynomials* and conversely,any orthogonal polynomials are obtained this way. (see for example Chihara [10], Viennot [47]).

Example 5.7. Let $G = (V,A)$ be a graph with vertices in V and edges in A. Let $P = A$ with concurrency relation $\mathscr{C}$ be the intersection relation. The weight of an edge is x^2. Then the *matching polynomial* of the graph G (see for example [20],[32],[35]) is the reciprocal of the corresponding exclusion polynomial.

Example 5.8. Let G = (V,A) be a graph with vertices in V and edges in A. The set of basic pieces is P=V with concurrency relation $\mathscr{C}$ defined by : a $\mathscr{C}$ a and a $\mathscr{C}$ b iff $\{a,b\} \in A$ (i.e. G is the concurrency graph of $\mathscr{C}$). Each vertex is weighted x. We propose to call the corresponding exclusion polynomial the *independency polynomial* of the graph G. It is less classical than the matching polynomial, but it appears in some statistical mechanics models (see below §9,b). In fact, up to a change of variable, the matching polynomial of the graph G is the independency polynomial of its so-called *line graph* .

We interpret below the logarithm of the generating function of weighted heaps. We suppose that the ring $\mathbb{K}$ is the field $\mathbb{Q}$ of rational numbers. We need the following definition.

Definition 5.9. A *pyramid* is a heap having a unique maximal piece.

Proposition 5.10. Let $H(P,\mathscr{C})$ be a heap monoid with a valuation v satisfying (11). Then

(16) $$\log\left[\sum_{E\in H(P,\mathscr{C})} v(E)\right] = \sum_{F} \frac{v(F)}{|F|} ,$$

where the second summation is over all pyramids of $H(P,\mathscr{C})$.

Condition (11) implies the summability of both sides of identity (16).

Here we work with exponential generating function, that is labeled heaps :

$$\sum_{E\in H(P,\mathscr{C})} v(E) = \sum_{n\geq 0}\left[\sum_{|E|=n} n!\,\frac{v(E)}{n!}\right] .$$

This generating function is the exponential generating function for labeled (by 1,2,...,n) weighted heaps. We decompose such heaps E into pyramids in the following way .

We select the piece α_1 of E with minimal label (i.e. 1). Let E_1 be the subheap formed by all the pieces below α_1 (order ideal). E_1 is a pyramid and in fact E can be factorized $E=E_1 \circ E_1'$. Another way to define E_1 is to say that there exists a unique factorization $E=E_1\circ E_1'$ such that E_1 is a pyramid with maximal piece α_1. We select the piece α_2 of E_1' with minimal label and get a factorization $E=E_1\circ E_2\circ E_2'$. Recursively we have a factorization of the heap E into a product of labeled pyramids with the property

(17) the piece with minimal label is the maximal piece of the pyramid.

Conversely, from the set $\{E_1,...,E_k\}$ of such pyramids, one can reconstruct E by taking their product in the increasing order of the label of their maximal element.

In the context of species of Joyal [38], or of Foata's "*composé partitionnel*" [25], a heap is an "*assembly*" of labeled pyramids satisfying (17). Their exponentiel generating function is the right-hand side of (16). The proposition comes from standard result on assembly of species or on "composé partitionnel". □

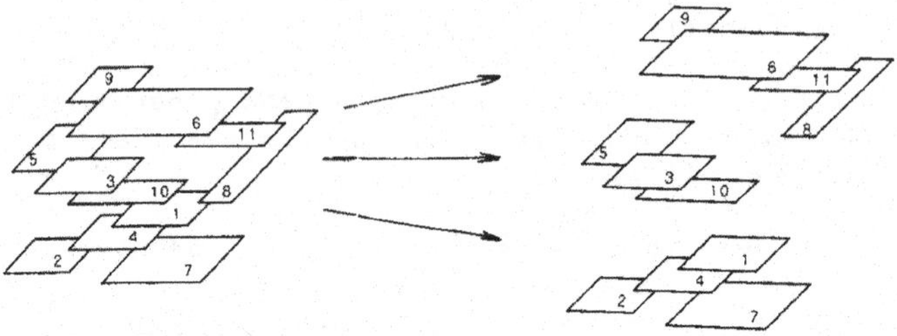

heap assembly of pyramids

Fig.7. log(heap) = pyramid.

In the next sections, we give a summary of the possible applications of heaps theory to enumerative and interpretative combinatorics. This will be done in details in the papers [15],[18],[50],[51].

6. Flow monoid and combinatorial proofs in linear algebra

(summary of [18])

We use the notations of §4. The basis is B. The set of basic pieces is P = B × B. The projection $\pi : P \longrightarrow \mathcal{P}(B)$ is defined by

(18) for any $(s,t) \in P = B \times B$, $\pi(s,t) = s$.

Definition 6.1. The *flow monoid* is the heap monoid $F(B)=H(P,\pi,B)$ defined by (18).

This corresponds to the flow monoid introduced by Cartier, Foata [9] : the edges (s,t) and (s',t') commute iff $s \neq s'$.

The heaps of $H(P,\pi,B)$ are called *flows*. Such a flow E is defined by its fibers. The fiber $F_s(E)$ over $s \in B$ is isomorphic to a word of B_s^* with $B_s = \{s\} \times B$. In fact there is no order relation between two elements of distinct fibers and the flow monoid is isomorphic to a direct product of free monoids $H(P,\pi,B) \simeq \prod_{s \in B} B_s^*$ (see Fig.8).

Definition 6.2. A *rearrangement* is a flow $E=(E,\leq,\varepsilon)$ of $H(P,\pi,B)$ such that for every $s \in B$, the fiber $F_s(E)$ defined by (10) satisfies

(19) $|F_s(E)| = |\{\alpha \in E,\ \varepsilon(\alpha) = (t,s)\}|$.

In other words the number of edges (t,s) of E coming in s is the same as the number of edges (s,t) starting from s.

The rearrangements form a submonoid R(B) of the flow monoid F(B).

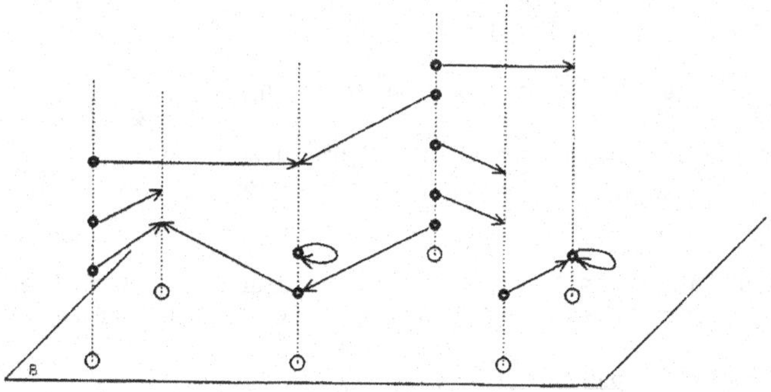

Fig.8. A flow.

The two following propositions are typical examples of bijections transforming a heap into another heap of bigger pieces obtained by "gluing" the small pieces (or conversely "breaking" pieces into smaller pieces).

A *path* ω of B is any sequence $\omega=(s_0,s_1,\ldots,s_n)$ of points of B. We consider the heap monoid SPCy(B) $= H(Q,\pi,B)$ which pieces Q are *cycles* (Cy) on B or *self-avoiding* paths (SP) on B (i.e. no two vertices appear twice in ω) and the projection π is the map associating to a piece its underlying set of vertices of B. A path ω can be identified with the flow $(s_0,s_1)\circledcirc(s_1,s_2)\circledcirc \ldots \circledcirc(s_{n-1},s_n)$, (product of heaps).

A cycle $\gamma = (s_1,\ldots,s_n)$ (see the definition at the end of §4 and also see example 5.5) can be identified with the rearrangement $(s_1,s_2)\circledcirc\ldots\circledcirc(s_{n-1},s_n)\circledcirc(s_n,s_1)$. The submonoid Cy(B) of SPCy(B) is formed by heaps of cycles.

Proposition 6.3. Let $u,v \in B$. There exists a bijection between paths ω of B going from u to v and pyramids E of SPCy(B) such that all pieces are cycles of B, except the maximal piece, which is a self-avoiding path η going from u to v. This bijection is such that the number of edges (s,t) in ω (or elementary steps) in the same as the number of edges (s,t) contained in the cycles and the paths η of the pyramid E.

This bijection is particularly useful for the enumeration of certain families of heaps (see below the directed animal problem).

Proposition 6.4. There exists an isomorphism of monoids Ψ:Cy(B)$\longrightarrow$R(B) between the heap monoid of cycles and the heap monoid of rearrangements. Moreover, for any $s,t\in B$, Ψ preserves the number of edges (s,t) in each heap.

Each bijection of propositions 6.3 and 6.4 is obtained by "breaking" the heap of cycles (and self-avoiding path) into its elementary components : the edges (s,t) considered as elements of the flow monoid.

Combining the above propositions with the propositions of §5 gives combinatorial proofs of classical identities in linear algebra (see [9],[26],[27],[36],[46],[52]).

Let $A = (a_{ij})$ be an $n \times n$ matrix and $X = \begin{bmatrix} x_1 \\ . \\ x_n \end{bmatrix}, Y = \begin{bmatrix} y_1 \\ . \\ y_n \end{bmatrix} = AX$.

Paths and cycles are weighted as in example 5.5 by

$v(\omega) = v(s_0,s_1) \ldots v(s_{n-1},s_n)$ and $v(i,j) = a_{ij}$.

Corollary 6.5. (MacMahon Master theorem). <u>The coefficient of</u> $x_1^{\alpha_1} \ldots x_n^{\alpha_n}$ <u>in the formal serie</u> $1/\det(I-AX)$ <u>is the same as the coefficient of</u> $x_1^{\alpha_1} \ldots x^{\alpha_n}$ <u>in the polynomial</u> $y_1^{\alpha_1} \ldots y_n^{\alpha_n}$.

This is a combination of proposition 5.1, example 5.5 and proposition 6.4.

Corollary 6.6. (Inversion matrix formula) <u>The term</u> (i,j) <u>of the inverse matrix</u> $(I-A)^{-1}$ <u>is</u> $N_{ij}/\det(I-A)$ <u>where</u> N_{ij} <u>is the term</u> (j,i) <u>of the adjoint matrix (cofactor)</u>.

This is a combination of proposition 5.3, example 5.5 (together with a companion formula for the cofactor) and proposition 6.3.

Corollary 6.7. (Jacobi identity)

$$\frac{1}{\det(I-A)} = \exp(\mathrm{Tr}(\log(I-A)^{-1})). \tag{20}$$

This identity comes from a combination of proposition 5.1, example 5.5 and proposition 5.3 (in a slightly more general version).

Also, Cayley-Hamilton theorem can be obtained by using a slightly more general form of the identity (14) of proposition 5.3.

7. Orthogonal polynomials

Any sequence $\{P_n(x)\}_{n \geqslant 0}$ of (formal) orthogonal polynomials appears as the sequence of reciprocal of the exclusion polynomials of weighted monomers and dimers on the segment [0,n-1] (see example 5.6).

A combinatorial theory of classical properties valid for any sequences of orthogonal polynomials has been made by Viennot [47], following work of Flajolet [22]. This combinatorial theory is written in terms of certain weighted paths (called *Dyck* and *Motzkin* paths). Some of the bijective proofs can be simplified by using heaps terminology. The Dyck (resp. Motzkin) paths are transformed (by proposition 6.3) into pyramids of dimers (resp. monomers and dimers) on $B = \mathbb{N}$.

The main property is the following. Let $P_n(x)$ be the sequence of polynomials defined by the recurrence (15) and $H(P,\pi,B)$ be the heap monoid of monomers and dimers on $B = \mathbb{N}$, weighted by the sequences $\{b_k\}_{k\geq 0}$ and $\{\lambda_k\}_{k\geq 1}$ of elements of the ring $\mathbb{K}$ as in example 5.6.

Let μ_n be the sequence defined by

(21) $$\mu_n = \sum_{E} v(F),$$

where the summation is over all weighted pyramids of m monomers and d dimers such that n = m+2d and such that the maximal piece contains the value 0 (i.e. this maximal piece is either {0} or {0,1}), see Fig.9.)

Let f be the unique linear functional $f:\mathbb{K}[x] \rightarrow \mathbb{K}$ such that $f(x^n) = \mu_n$ $(n\geq 0)$. Suppose that $\lambda_k \neq 0$ $(k\geq 1)$ and that $\mathbb{K}$ has no zero divisors.

Proposition 7.1 - The polynomials $P_n(x)$ defined by the three-terms linear recurrence (15) are orthogonal with respect to the sequence of moments μ_n defined by (21), that is :

(22) $f(P_k P_l) = 0$ if $k \neq l$ and $f(P_k^2) \neq 0$, for every $k,l\geq 0$.

The proof follows from the same generalization of identity (14) of proposition 5.3 mentioned at the end of §6 about a bijective proof of Cayley-Hamilton theorem.

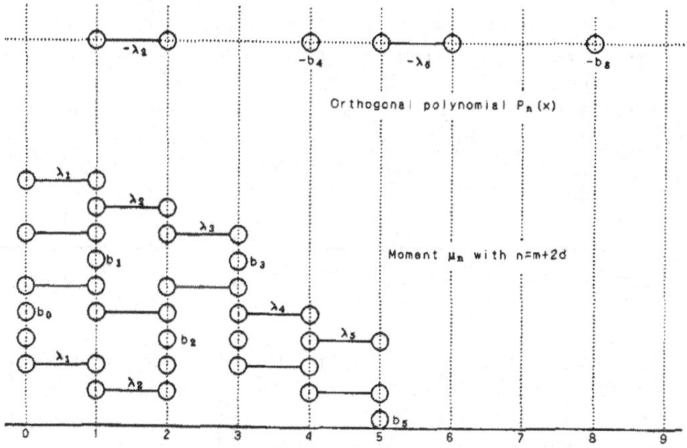

Fig.9. Orthogonal polynomials and moments interpretated as exclusion polynomials and pyramids of monomers-dimers.

Many other properties of general (i.e. formal) orthogonal polynomials can be deduced from heaps basic lemmas. In particular the Jacobi continued fraction expansion (corresponding to Flajolet's theorem about weighted Motzkin paths) here becomes a simple consequence of a decomposition lemma about the pyramids interpretating μ_n into other pyramids. This decomposition is the analog, for ordinary generating functions, of the decomposition given in the proof of proposition 5.10 with exponential generating functions.

Corolallary 7.2. With the above notation (21) ,

$$(23)\qquad \sum_{n\geq 0} \mu_n t^n = \cfrac{1}{1-b_0 t - \cfrac{\lambda_1 t^2}{1-b_1 t - \cfrac{\lambda_2 t^2}{\cfrac{\cdots\cdots\cdots}{1-b_k t - \cfrac{\lambda_{k+1} t^2}{\cdots\cdots\cdots}}}}}$$

The convergents of the continued fraction (23) are nothing but the generating functions of the pyramids interpretating μ_n and bounded on the segment $[0,n]$. Thus, applying proposition 5.3, these convergents are

$$(24)\qquad \delta P_n^*(t) \ / \ P_{n+1}^*(t),$$

where $P_{n+1}^*(t)$ is the reciprocal $t^{n+1}P_{n+1}(1/t)$ of $P_{n+1}(t)$ and $\delta P_n(t)$ is the exclusion polynomial for heaps of monomers and dimers on $[0,n]$ not containing the value 0, that is the nth orthogonal polynomial corresponding to the "shifted" valuations $b'_k = b_{k+1}$, $\lambda'_k = \lambda_{k+1}$.

If we take $b_k=0$ and $\lambda_k = -q^k$, then we get the exclusion power serie $D(P,\mathscr{C})$. We are in the case of an infinite set of pieces and (11) is satisfied. Taking the basis $B = \mathbb{N}$, the exclusion power serie $D(q)$ is the left hand-side of the famous (first) Rogers-Ramanujan identity (see for example Andrews [1]) :

$$(25)\qquad 1 + \sum_{n\geq 1} \frac{q^{n^2}}{(1-q)(1-q^2)\dots(1-q^n)} = \prod_{n\geq 0} \frac{1}{(1-q^{5n+1})(1-q^{5n+4})} .$$

The left hand side of the second Rogers-Ramanujan identity

$$(26)\qquad 1 + \sum_{n\geq 1} \frac{q^{n^2+n}}{(1-q)(1-q^2)\dots(1-q^n)} = \prod_{n\geq 0} \frac{1}{(1-q^{5n+2})(1-q^{5n+3})} ,$$

can be interpretated as the exclusion power serie $N(q)$ for weighted trivial heaps of dimers (with valuation $\lambda_k = -q^k$) not containing 0. Proposition 5.1 and 5.3 gives interpretations of the generating functions $1/D(q)$ and $N(q)/D(q)$, respectively in terms of heaps and moments pyramids. We can easily deduce Andrews's interpretations [2] with *quasi-partitions*.

8. Heaps and algebraic graph theory

a) Matching polynomials of graphs (summary of [15])

Let G be a graph. The *matching* polynomial of G is the reciprocal of the exclusion polynomial D(G;x) defined in example 5.7 : pieces are dimers on G (i.e. edges) weighted by x^2. Several work has been done on these polynomials, in relation with physics and chemistry, see for example [20],[31],[32],[35].

Proposition 5.1 and 5.3 give combinatorial interpretation of the coefficients of the power series $1/D(G;x)$ and $D(G\setminus M;x)/D(G;x)$ where $G\setminus M$ denotes the graph obtained by deleting from G the set of edges M (resp. set of vertices M).

If M is the set of edges containing a vertex s, then $D(G\setminus M;x)/D(G;x)$ is the generating function for the so-called *tree-like* paths introduced by Godsil [31] in order to give a nice proof of the fact that the roots of matching polynomials are real numbers (Heilmann, Lieb [35]). In Desainte-Catherine, Viennot [15] we deduce bijectively Godsil's result and give some generalizations.

Remark that tree-like paths correspond exactly (via the bijection of proposition 6.3) to pyramids with the restriction that all the cycles have length 2.Such cycles can be identified with dimers of G

b) Chromatic polynomials and acyclic orientations of graphs (from Gessel [30])

Let G be a finite graph with n vertices, $\gamma(G;x)$ be the *chromatic polynomial* of G and $\alpha(G)$ be the number of *acyclic orientations* of G. In [45] Stanley has proved the following identity.

$$\gamma(G;-1) = (-1)^n \alpha(G). \tag{27}$$

Gessel [30] has given a nice proof of this identity, using the commutation monoid. Here we just sketch the idea of his proof, translated in terms of heaps.

Let G = (S,A) with set of vertices S (resp. edges A). Let $\mathscr{C}$ be the concurrency relation such that G is its concurrency graph (see §4). Let E be a heap of $H(S,\mathscr{C})$. For $k \geq 1$, we denote by $\beta_k(E)$ the number of factorizations of E in the form $E = T_1 \circledcirc \ldots \circledcirc T_k$ where each T_i is a non-empty trivial heap (remark that condition (2) of lemma 2.9 is not necessarily satisfied). Let v be a valuation on the heap monoid $H(S,\mathscr{C})$ as in §5. A heap E is called *linear* (resp. *covering*) iff each basic piece appears at most (resp. at least) once in E. A linear and covering heap E is a product (in $H(S,\mathscr{C})$) of all the basic pieces S. Let $LC(S,\mathscr{C})$ be the set of such heaps. We have the relation

$$\gamma(G;x) = \sum_{k\geq 0} \frac{1}{k!} \left[\sum_{E\in LC(S,\mathscr{C})} \beta_k(E) \right] x(x-1)\ldots(x-k+1), \tag{28}$$

which leads us to introduce the *complete chromatic power serie* of the graph G.

(29)
$$\Gamma(G;x) = \sum_{k\geq 0} \frac{1}{k!} \left[\sum_{E\in H(S,\mathscr{C})} \beta_k(E)\, v(E) \right] x(x-1)\ldots(x-k+1).$$

We have

(30)
$$\sum_{E,k\geq 0} \beta_k(E)\ v(E)\ t^k = \left[1 - t \sum_{F} v(F) \right]^{-1},$$

where the first summation is over all heaps $E\in H(S,\mathscr{C})$ and the second is restricted to non-empty trivial heaps. From relation (12) of proposition 5.1, we deduce (a bijective proof would also be possible)

(31)
$$\Gamma(G;-1) = (-1)^n \sum_{E\in H(S,\mathscr{C})} v(E)\ .$$

The restriction to linear and covering heaps gives (27).

9. Heaps and Statistical Physics

a) The directed animal problem (summary of Viennot [50])

In 1982, physicists have introduced and studied the following problem. A *directed animal* is a set A of points of $\mathbb{N}\times\mathbb{N}$ such that $(0,0)\in A$ and any point (x,y) of A can be reached by a path going from (0,0) to (x,y), with vertices in A, and using elementary steps North or East. The point (0,0) is called the *source point* and North-East is called *priviligied direction.* The size of the animal A is described by its *width* and *length* (i.e. size of the smallest rectangle containing A with edges parallel or perpendicular to the priviligied direction). Let a_n be the number of directed animals with n points. Considering these animals equidistributed, let ℓ_n (resp. L_n) be the average width (resp. length).

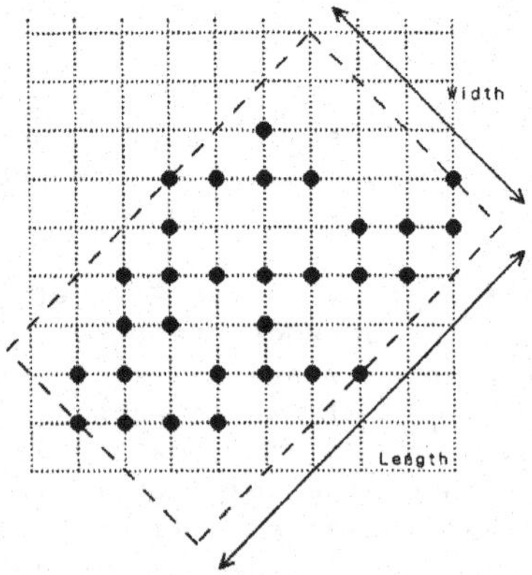

Fig.10. A directed animal, one source point, square lattice.

Physicists expect the following asymptotic behaviour

(32) $$a_n \sim \mu^n n^{-\theta}, \qquad \ell_n \sim n^{\nu_\perp}, \qquad L_n \sim n^{\nu_{/\!/}}.$$

The constants θ, $\nu_\perp$ and $\nu_{/\!/}$ are called *critical exponents*. Such numbers are of particular importance in the models for *phase transitions* and *critical phenomena*.

A surprising fact is that very simple exact formulae exist for a_n and ℓ_n from which one get immediately $\mu=3$, $\theta = \nu_\perp = 1/2$. After many other works (for a survey see [49]) , physics solutions are given by Dhar [16], [17] and Hakim, Nadal [34] following Nadal, Derrida, Vannimenus [41] .

A complete combinatorial solution (for a_n and ℓ_n) can be given by using heaps basic properties and a bijection between directed animals (with one source point) and certain pyramids of dimers on $\mathbf{Z}$. This is done in [50], where some conjectures of Dhar [16] are proved. A survey of the directed animal model, with both physics and combinatorial solutions, and relationship with other problems and models, is given in Viennot [49] . The case of directed animals on a *triangular lattice* is easier. A "brute force" bijection between directed animals and certain paths has been given by Gouyou-Beauchamps, Viennot [33] . This bijection is the same as the one obtained using heaps.

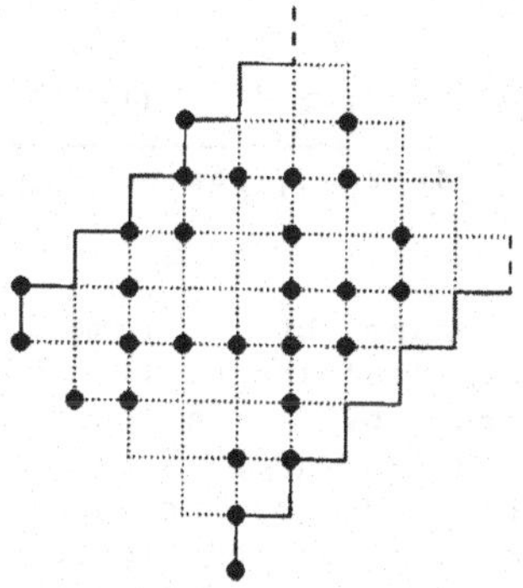

Fig.11. Directed animals on a bounded strip

In the physics solution, Nadal, Derrida, Vannimenus [41], Hakim, Nadal [34] consider directed animals on a *bounded strip* : several source points are now possible (see Fig. 11). The borders may be identified (*circular strip*). Using transition matrices acting on a space of spins, they give a formula for the number of such animals with given source points. This formula is easily obtained from the generating function of such animals, which is a rational serie $N(t)/D(t)$. The polynomials $N(t)$ and $D(t)$ can be deduced from §7 and proposition 5.3, using Tchebycheff polynomials first kind (circular strip) and second kind (bounded strip).

The problem of the existence and determination of the exponent $\nu_{/\!/}$ is still open. It is conjectured [41] to be 9/11.

b) **Combinatorial interpretation of the density of a gas with hard-core interactions** (summary of Viennot [51])

Here we use propositions 5.1 and 5.10 in the case of the independency polynomials of example 5.8. The heaps model put some light on the so-called "*thermodynamic limit*" of the independency polynomials. We obtain a combinatorial interpretation of the *partition function* $Z(t)$ (on an infinite lattice) and of the *density*

(33) $$\rho(t) = t\ d/dt \log Z(t).$$

In fact,

(34) $$-\rho(-t) = \sum_{n \geqslant 0} a_n t^n,$$

where the a_n are positive integers enumerating certain pyramids.

Using statistical mechanics techniques, Baxter has recently solved [6] the famous *hard hexagon* model. This model has a phase transition for the "*activity*" $t_c = (11+5\sqrt{5})/2$. For $0<t<t_c$, the partition fonction $Z(t)$ is given by the following system of equations

Let $R_I(q)$ (resp. $R_{II}(q)$) be the left hand-side of the first (resp. second) Rogers-Ramanujan identity (25) (resp. (26)). The partition fonction $Z(t)$ is obtained by eliminating q between the two following equations

(35) $$t = -q \left[\frac{R_{II}(q)}{R_I(q)}\right]^5,$$

(36) $$Z = \prod_{n \geqslant 0} \frac{(1-q^{6n+2})(1-q^{6n+3})^2(1-q^{6n+4})(1-q^{5+1})^2(1-q^{5n+4})^2(1-q^{5n})^2}{(1-q^{6n+1})(1-q^{6n+5})(1-q^{6n})^2(1-q^{5n+2})^3(1-q^{5n+3})^3}.$$

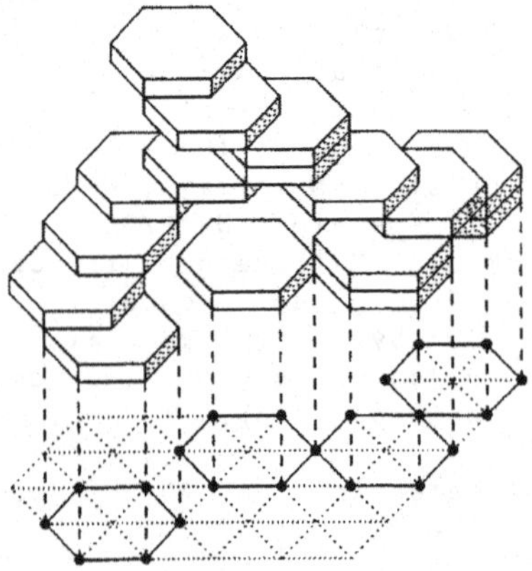

Fig.12. A pyramid of hexagons interpretating the density of the gas in the hard hexagon model.

From heaps basic lemmas, we deduce that the coefficient a_n defined by equations (33), (34), (35) and (36) is the number of pyramids of hexagons on a triangular lattice, formed with n hexagons, as shown on Fig.12. It would be of great interest to prove directly the equations enumerating such pyramids of hexagons, without using Baxter's solution (which has nothing to do with heaps of pieces). Also the combinatorics of heaps proves the equivalence between directed animals problems and hard-core gas model, as shown recently by Dhar [17] using physics arguments.

10. Heaps and parallelism in Computer Science

Finally, commutation monoids have recently appeared in Theoretical Computer Science as a model for *parallelism* and *concurrent access to databases*. This is an active and promising area of research and a meeting on this subject took place in Paris, March 1985, and another is planned in 1986.

A database is a set of objects called *entities*. A *transaction* is any sequence of atomic actions operating on the entities. Several transactions can access concurrently to the same database. An action is identified with a letter and a transaction with a word. Commutations are defined on these letters, describing the possible concurrency access to the database. The model can be developped from an *algebraic* point of view (*rational* and *recognizable* languages in this monoid,...) in analogy with the *free monoid* case (see for example [7],[12],[13],[19],[23],[42]). Another direction introduced by Françon [28],[29] and Arques et al.[4],[5] is *combinatorial*. This direction, following some ideas of Papadimitriou ([42] and related papers) allows the comparison of the performances of concurrency control algorithms with the computation of the cost of *serialization* of an execution or with the determination of Françon's *parallelism ratio*, frequency of *deadlocked executions*,etc... These problems can be reduced to the asymptotic enumeration of certain sets of commutation classes, or enumeration of words in these classes.

The heap monoid model may bring other ideas about these questions. First of all, replacing the alphabet (which letters are a coding of the atomic actions) by a set of basic pieces P, equipped with a basis B and a projection map π: P $\rightarrow \mathcal{P}$(B) can be closer to concurrency considerations. For example one can consider the basis B to be the set of entities. An atomic action, symbolized by the basic piece $\alpha\in$P, will operates on the subset $\pi(\alpha)\subset$B of entities. This atomic action, exactly as the basic piece, is a certain "structure" on the support $\pi(\alpha)$. The concurrency relation of §4 will corresponds to atomic actions having access to common entities. Under this model, two actions will commute iff they operate on disjoint subsets of entities.

Another idea is to use heaps of pieces as a new data structure in Computer Science. This structure appears as a generalization of the *binary tree* structure (which is a poset and thus can be *"realized"* as a heap) and the structure formed by several independant *stacks*. One can implement the heap data structure by its fibers, or by defining *links* between a piece α and the pieces β which are covered by α. This data structure would be of particular advantage in parallel algorithms.

In conclusion, one of the interest of the heaps formulation is to relate some problems coming from completely different fields, as for example the determination of the critical exponents of the directed animals problem in Statistical Physics and the computation of Françon's ratio of parallelism. Both are equivalent to asymptotic enumeration of certains heaps. Some problems are equivalent to enumerate the number of words in a commutation class (as for example *"aggregats"* problems in Statistical Physics). From §3, this is equivalent to enumerate linear extensions of a given poset. This problem is well-known in poset theory and explicit formulae exist only in certain particular cases (standard Young tableaux, standard shifted Young tableaux, trees,...). Of course, the difficulty of the problem remains the same in both points of view, but the spatial intuition, the powerful basic heaps lemmas and the connections made between different domains may be useful.

Acknowledgements

I address my special thanks to all the organizers of this wonderful colloque UQAM 85 in Montréal, especially to Pierre Leroux for his warm encouragements in writing and finishing this paper, to the referee for his nice work and to André Joyal for fruitful discussions.

A preliminary version of the heap theory, containing some parts of this paper together with papers [15],[18] has been written by Jacques Bourret in his UQAM "Mémoire de maîtrise" from lectures and manuscripts of the author, with Serge Dulucq and Myriam Desainte-Catherine.

This paper is edited with "MULTITEXT".Special thanks for Maylis Delest and Michel Pallard in the help for the drawing of the figures with "GED&ON". These text and figures editors are being developped in the Mathematics and Computer Science department of Univ. Bordeaux I.

References

[1] G.ANDREWS, The Theory of partitions, Encyclopedia of Maths. and its applications, G.C.Rota ed., vol 2, Addison Wesley, Reading , 1976

[2] G.ANDREWS, The Rogers-Ramanujan reciprocal and Minc's partition function, Pacific J.Math, 95 (1981) 251-256.

[3] G.ANDREWS, The hard-hexagon model and Rogers-Ramanujan identities, Proc.Nat.Acad.Sci.U.S.A, 78 (1981) 5290-5292.

[4] D.ARQUES, J.FRANCON, M.T.GUICHET and P.GUICHET , Comparison of Algorithms controlling concurrent access to a data base : A combinatorial approach, Research rep. n°38, Mulhouse Univ. to appear in Proc.ICALP 1986.

[5] D.ARQUES, P.GUICHET, Asymptotic behaviour and comparison of algorithms controlling concurrent access to a data base, Research report n° 40, Mulhouse Univ.,Jan.1986.

[6] R.J.BAXTER, Exactly solved models in statistical mechanics, Academic Press, New-York, 1982.

[7] A.BERTONI,G.MAURI and N.SABADINI, Equivalence and membership problems for regular trace languages, Proc. of the 9th ICALP 1982, Lecture Notes in Comp.Sci. n°140(1982) 61-71, Springer-Verlag.

[8] J.BOURRET, Applications algébriques des empilements de pièces et de la méthode involutive en Combinatoire, Mémoire de Maîtrise, UQAM, Montréal, Mai 1985.

[9] P.CARTIER and D.FOATA, Problèmes combinatoires de commutation et réarrangements, Lecture Notes in Maths. n° 85, Springer-Verlag, New-York/Berlin, 1969.

[10] T.S.CHIHARA, An introduction to orthogonal polynomials, Gordon and Breach, New-York, 1978.
[11] M.CONTENT, F.LEMAY and P.LEROUX, Catégories de Möbius et fonctorialités : un cadre général pour l'inversion de Möbius, J.of Combinatorial Th.A,28 (1980) 169-190.
[12] R.CORI and Y.METIVIER, Rational subsets of some partially abelian monoids, Theor.Comp.Sci.35 (1985) 179-189.
[13] R.CORI and D.PERRIN, Sur la reconnaissabilité dans les monoïdes partiellement commutatifs libres, R.A.I.R.O Info.Th. 19 (1985) 21-32.
[14] M.DESAINTE-CATHERINE, Couplages et Pfaffiens en Combinatoire Physique et Informatique, thèse 3ème cycle, Université Bordeaux I, 1983.
[15] M.DESAINTE-CATHERINE and G.X.VIENNOT, Heaps of pieces, III : Matching polynomials of graphs, in preparation.
[16] D.DHAR, equivalence of the two-dimensional directed-site animal problem to Baxter's hard-square lattice-gas model, Phys.Rev.Lett.,49(1982) 959-962.
[17] D.DHAR, Exact solution of a directed site-animals enumeration problem in three dimensions,J.Phys.A 15(1982)1849-1859.
[18] S.DULUCQ and G.X.VIENNOT, Heaps of pieces, II : the Cartier- Foata flow monoid revisited and combinatorial proofs in matrix algebra, in preparation.
[19] C.DUBOC, Some properties of commutation in free partially commutative monoids, Inform.Proc.Let.20 (1985) 1-4.
[20] E.J.FARREL, An introduction to matching polynomials, J. Combinatorial Theory B, 27 (1979) 75-86.
[21] E.J. FARRELL, A survey of the unifying effects of F-polynomials in "Combinatorics and Graph Theory", Proc. 4th Yugoslav Seminar on Graph Theory, Novi Sad, (1983),137-150.
[22] P.FLAJOLET, Combinatorial aspects of continued fractions, Discrete Math., 32 (1980) 125-161.
[23] M.P.FLE and G.ROUCAIROL, Maximal serializability of iterated transactions, ACM SIGACT SIGOPS, (1982) 194-200.
[24] D.FOATA, Etude algébrique de certains problèmes d'Analyse Combinatoire et du calcul des Probabilités, Publ. Inst.Stat.Univ.Paris, 14 (1965) 81-24.
[25] D.FOATA, "La série génératrice exponentielle dans les problèmes d'énumérations", Les Presses de l'Univ.de Montréal,1974
[26] D.FOATA, A non-commutative version of the matrix inversion formula, Adv. in Maths., 31 (1979) 330-349.
[27] D.FOATA, A combinatorial proof of Jacobi's identity, Annals of Discrete Math., 6 (1980), 125-135.
[28] J.FRANÇON , Une approche quantitative de l'exclusion mutuelle to appear in R.A.I.R.O.
[29] J.FRANÇON, Sérialisabilité, commutation, mélange et tableaux de Young, Research report n° 27, Mulhouse Univ.,1985.
[30] I.GESSEL, Personnal communication.
[31] C.D.GODSIL, Matchings and walks in graphs, J. of Graph Th., 5 (1981) 285-291.
[32] C.D.GODSIL and I.GUTMAN, On the theory of the matching polynomial, J. of Graph Th., 5 (1981) 137-144.

[33] D.GOUYOU-BEAUCHAMPS and G.X.VIENNOT , Equivalence of the 2d directed animals problem to a 1d path problem, preprint 1985, submitted to Adv. in Maths.

[34] V.HAKIM and J.P.NADAL, exact results for 2d directed animals on a strip of finite width, J.Phys. A : Math. Gen.; 16 (1983) L 213-218.

[35] O.T.HEILMANN and E.H.LIEB, Monomers and dimers, Phys. Rev. Lett., 24 (1970) 1412-1414.

[36] D.M.JACKSON, The Combinatorial interpretation of the Jacobi identity from Lie algebras, J.Combinatorial Th.A,23 (1977) 233-257.

[37] S.N.JONI and G.C.ROTA, Coalgebras and bialgebras in Combinatorics, studies in Applied Maths, 61(1979)93-134.

[38] A.JOYAL, Une théorie combinatoire des séries formelles, Advances in Maths., 42 (1981) 1-82.

[39] G.LALLEMENT, Semigroups and Combinatorial applications, John Wiley, New-York, 1979

[40] M.LOTHAIRE, Combinatorics on words, Encyclopedia Maths. and its applications, G.C.Rota ed., vol 2, Addison- Wesley, Reading, 1983.

[41] J.P.NADAL, B.DERRIDA and J.VANNIMENUS, Directed lattice animals in 2 dimensions : numerical and exact results, J.Physique, 43 (1982) 1561.

[42] C.H.PAPADIMITRIOU, Concurrency control by Locking, SIAM J. on Comp.,12 (1983) 215-226.

[43] D.PERRIN, Words over a partially commutative monoid,in "Combinatorial algorithms on words", eds.A.Apostolics and Z.Galil, NATO ASI Series, vol F12, Springer- Verlag, Berlin Heidelberg, 1985, 329-340.

[44] G.C.ROTA, On the foundations of combinatorial theory I.Theory of Möbius functions, Z.Wahrsch., 2 (1964)340-368.

[45] R.STANLEY, Acyclic orientations of graphs, Discrete Maths. 5(1973) 171-178.

[46] H.STRAUBING, A combinatorial proof of the Cayley-Hamilton theorem, Discrete Math.43 (1983) 273-279.

[47] G.X.VIENNOT, Une théorie combinatoire des polynômes orthogonaux généraux, lecture Notes, Université du Québec à Montréal, Dpt. de Maths. 1984, 215p.

[48] G.X.VIENNOT, A combinatorial theory for general orthogonal polynomials with extensions and applications, in "Polynômes orthogonaux and Applications", Proc. Bar-le-duc 1984, eds C.Brezinski, A.Draux A.P.Magnus P.Maroni and A.Ronveaux, Lecture Notes in Maths. n° 1171, 139-157, Springer-Verlag, Berlin, 1985.

[49] G.X.VIENNOT, Problèmes combinatoires posés par la physique statistique, Séminaire N.BOURBAKI, exposé n° 626 36^{è} année, in Astérisque n°121-122 (1985) 225-246, SMF.

[50] G.X.VIENNOT, Heaps of pieces, IV : Combinatorial solution of the directed animal problem, in preparation.

[51] G.X.VIENNOT, Heaps of pieces, V : Combinatorial interpretation of the density of a gas with hard-core interactions, in preparation.

[52] D.ZEILBERGER, A combinatorial approach to matrix algebra, Discrete Maths, 56 (1985) 61-72.

THE CALCULUS OF VIRTUAL SPECIES AND $\mathbb{K}$-SPECIES.

by

Yeong-Nan Yeh

Université du Québec à Montréal*

Introduction

In [3], Joyal introduces the category of species together with several operations such as $+, \cdot, \times, \circ$ and $'$. In [4], he states the substitution rule for virtual species. In this paper, we develop a method for proving the correctness of this rule; we also further study and extend some aspects of the theory of virtual species. In particular, we will

(1) Show that the ring of virtual species (resp d-species) is a unique factorization domain (**UFD**).

(2) Give a relation between $\times$ and $\circ$.

(3) Extend all the identities involving $+, \cdot, \times, \circ, '$, 0 and 1 to the setting of virtual species and, more generally, $\mathbb{K}$-species.

(4) Give some $\mathbb{K}$-species which are analogues of the logarithm and trigonometric functions.

I am grateful to Professor Stephen Schanuel, my thesis advisor, for his encouragement and valuable discussions throughout this work. I would like to thank P. Bouchard, A. Joyal, J. Labelle, G. Labelle, P. Leroux and D. Rawlings for their useful suggestions.

---

* Stagiaire post-doctoral. Avec l'aide financière des subventions d'équipes FCAR (Québec EQ - 1608) et CRSNG (Canada A5660).

Chapter I : Background

§ I.1. Algebra

In this paper, "ring" always means commutative ring with 1.

Definition I.1.1. $(\mathbb{K}, 0, 1, +, \cdot)$ is a **half-ring** iff $(\mathbb{K}, +)$ and $(\mathbb{K}, \cdot)$ are commutative monoids and the two "distributive" laws: (1) $(a+b)c = ac + bc$; (2) $0c = 0$ hold in $\mathbb{K}$.

If $\mathbb{K}$ is a half-ring and $(\mathbb{M}, \cdot)$ is a monoid with the property that for each $m \in \mathbb{M}$, there are only finitely many pairs (m_1, m_2) such that $m = m_1 m_2$, then the set of all functions $f: \mathbb{M} \to \mathbb{K}$, denoted $\mathbb{K}[[\mathbb{M}]]$, gets a half-ring structure with pointwise addition and multiplication by convolution:

$$(f\cdot g)(m) \quad = \quad \sum_{m=m_1\cdot m_2} f(m_1)\, g(m_2).$$

Obviously, $\mathbb{K}[[\mathbb{M}]]$ is a ring iff $\mathbb{K}$ is a ring. The map $\mathbb{M} \to \mathbb{K}[[\mathbb{M}]]$ sending each m to its characteristic function is an embedding of monoids if $\mathbb{K} \neq 0$, and it is customary to identify $\mathbb{M}$ with its image, and to write $\sum_{m \in \mathbb{M}} f(m)\, m$ instead of f, when this is convenient.

Let K, H be two groups of permutations of the finite sets F, E respectively. The **wreath product** $K \wr H$ is defined to be the group of permutations t of the set $F \times E$ which are of the form $t(f,e) = (\alpha(e)(f), h(e))$ where α is a function: $E \to K$ and $h \in H$. Thus t is determined by an element of H and a function α. So $|K \wr H| = |K|^{|E|} . |H|$. If G is a group of permutations of a set D, then $(K \wr H) \wr G = K \wr (H \wr G)$.

Example I.1.2. $2^{\S} \wr 2^{\S} = \mathbf{D}_4$ where $\mathbf{D}_4$ is the dihedral group of order 8.

Definition I.1.3. ([13]) Let $H \subset E_1^{\S} \times E_2^{\S} \times \cdots \times E_r^{\S}$ and $K_i \subset F_i^{\S}$ for $1 \leq i \leq r$. The **wreath product** $(K_1, K_2, \ldots, K_r) \wr H$ is defined to be the group of permutations t of the set $F_1 \times E_1 + F_2 \times E_2 + \cdots + F_r \times E_r$, which are of the form: For $1 \leq i \leq r$, $t(f_i, e_i) = (\varphi_i(e_i)(f_i), h(e_i))$ where φ_i is a function $E_i \to K_i$ and $h \in H$. Thus t is determined by an element of H and functions φ_i where $1 \leq i \leq r$.

So

$$|(K_1, K_2, \ldots, K_r) \wr H| \;=\; |K_1|^{|E_1|} . |K_2|^{|E_2|} \ldots |K_r|^{|E_r|} . |H| .$$

Given a finite set E, a partition π of E is a family E_i of non-empty subsets of E such that $E_i \cap E_j = \emptyset$ if $i \neq j$ and $\cup E_i = E$. Two partitions are equal iff they have the same elements. Let $P[E]$ denote the set of all partitions of E. Let $\Sigma\, \underline{E}$ denote the disjoint union of $E_1, E_2, \ldots, E_d$ where $\underline{E} = (E_1, E_2, \ldots, E_d) \in \mathbf{B}^d$; We write $\Sigma\, \underline{E} = E_1 + E_2 + \cdots + E_d$.

§ I.2. Commutative algebra.

Definition I.2.1. ([13]). Let $\mathbb{R}$ be a ring. The **length** of any element r in $\mathbb{R}$, $\ell(r)$, is defined by: (a) $\ell(0) = \infty$; (b) $\ell(r) = 0$ if r is a unit; (c) otherwise, $\ell(r) = \sup\{k \mid r = x_1 \cdot x_2 \cdots x_k$ with x_i non-zero and non-unit $\}$.

Definition I.2.2. ([13]). Let $\mathbb{R}$ and $\mathbb{S}$ be two rings. A ring homomorphism $f: \mathbb{R} \to \mathbb{S}$ is called **local** if $f(r)$ unit in $\mathbb{S}$ implies r unit in $\mathbb{R}$ and is called **unit-surjective** if s unit in $\mathbb{S}$ implies $\exists r \in \mathbb{R}$ with $f(r) = s$.

Let $(\mathbb{R}_n)_{n\in\mathbb{N}}$ be a sequence of **UFD**'s and $(\alpha_n)_{n\in\mathbb{N}}$ be a sequence of local, unit-surjecive ring homomorphisms where $\alpha_n: \mathbb{R}_{n+1} \to \mathbb{R}_n$, and let $\langle \mathbb{R}, (\varphi_n)_{n\in\mathbb{N}} \rangle$ be the inverse limit of $\langle (\mathbb{R}_n)_{n\in\mathbb{N}}, (\alpha_n)_{n\in\mathbb{N}} \rangle$ where φ_n is the canonical homomorphism from $\mathbb{R}$ to $\mathbb{R}_n$. In fact φ_n is a local unit-surjective ring homomorphism. We often write r_n instead of $\varphi_n(r)$ for all $r \in \mathbb{R}$.

Proposition I.2.3. The inverse limit $\mathbb{R}$ of a sequence $\mathbb{R}_n$ of **UFD**'s and local, unit-surjective homomorphisms is an **UFD**.

Proof. Every non-zero and non-unit element r in $\mathbb{R}$ can be factored into a finite product of irreducible elements since $\ell(r) \le \ell(r_n)$ $\forall n$. If $r \in \mathbb{R}$ and $\ell(r) = 1$ then $\lim_{n\to\infty} \ell(r_n) = 1$. It can be proved that every irreductible element in $\mathbb{R}$ is a prime. So $\mathbb{R}$ is an **UFD**.

Proposition I.2.4. Let $(\mathbb{M}, \cdot)$ be a free commutative monoid and $\mathbb{R}$ be an **UFD** then $\mathbb{R}[\mathbb{M}]$ and $\mathbb{R}[[\mathbb{M}]]$ are **UFD**'s.

Chapter II: The concepts of species and $\mathbb{K}$-species

§ II.1. Group Sets

If X is a finite set, a permutation of X is a bijective map $g: X \to X$. Under the operation of composition, the set of all permutations of X forms a group $X\,\text{¥}$. We have $|X\text{¥}| = |X|!$, where we use $|\ |$ to denote cardinality. If G is a subgroup of $X\text{¥}$, then we shall say that the pair (G,X) is a **group-set**. A subset Y of X is called a **G-invariant** subset if $g(Y) \subset Y$ for any $g \in G$. Let (G,X) be a group-set, U be a finite set containing X, and Y be a G-invariant subset of X. For any $g \in G$, the **extension of** g **to** U, g^U, is defined by: $g^U(u) = g(u)$ if $u \in X$; $g^U(u) = u$ otherwise. The **restriction of** g **to** Y, g_Y, is defined by: $g_Y(y) = g(y)$ if $y \in Y$. We denote $G^U = \{g^U \mid g \in G\}$ and $G_Y = \{g_Y \mid g \in G\}$.

Under the operation of composition, G^U and G_Y form groups and (G^U, U), (G_Y,Y) are group-sets. Since Y is a G-invariant subset of X, then $X - Y$ is a G-invariant subset of X and $(G_{X-Y}, X-Y)$ is a group-set.

Definition II.1.1 ([13]). Let (G,X) and (H,Y) be two group-sets. (H,Y) is called a **reducing group-set** of (G,X) if it satisfies the following conditions:

(a) Y is a G-invariant subset of X; (b) $H = G_Y$; (c) $H^X \subset G$.

Definition II.1.2 ([13]). Let (H,Y) and (K,Z) be two group-sets, then

(a) For any $h \in H$ and $k \in K$, let $h * k \in (Y+Z)!$ be defined by: $(h * k)(u) = h(u)$ if $u \in Y$; $(h*k)(u) = k(u)$ if $u \in Z$.
(b) Let $H * K$ denote the subgroup $\{h * k | h \in H, k \in K\}$ of $(Y + Z)!$.
(c) The group-set $(H * K, Y + Z)$ is called **external product** of the two group-sets (H,Y) and (K,Z) and is denoted : $(H * K, Y + Z) = (H,Y) * (K,Z)$.

From definition I.2, we find the group $H*K$ is the direct product of H^{Y+Z} and K^{Y+Z}. It is easy to check that the external product, $*$, satisfies the associative law.

Lemma II.1.3. If (G_Y,Y) is a reducing group-set of (G,X), then

(a) $(G_{X-Y}, X-Y)$ is a reducing group-set of (G,X); (b) $(G,X) = (G_Y,Y) * (G_{X-Y}, X-Y)$.

Lemma II.1.4. Let (G_Y,Y), (G_Z,Z) be two reducing group-sets of (G,X), then so is $(G_{Y\cap Z}, Y\cap Z)$.

Lemma II.1.5. If (H,Y) is a reducing group-set of (G,X) and (K,Z) is a reducing group-set of (H,Y), then (K,Z) is reducing group-set of (G,X).

Definition II.1.6. ([13]) A group-set (G,X) is called an **atomic group-set** if $X \neq \emptyset$ and (G,X) has no non-empty proper reducing group-set.

Proposition II.1.7. Every group-set (G,X) can be decomposed uniquely into an external product of atomic group-sets.

Let (G,X) and (H,Y) be two group-sets. We write $(G,X) \sim (H,Y)$ if there exists a bijection $f: Y \to X$ such that $f^{-1}Gf = H$. It is easy to prove that $\sim$ is an equivalence relation. Let $\mathcal{G}$ be the set of equivalence classes of group - sets. We have:

Proposition II.1.8. $(\mathcal{G}, \cdot)$ is a free monoid.

§ II.2. Species

Let ***Sets*** be the category of (small) sets and maps and **B** be the category of finite sets and bijections.

Definition II.2.1 ([3]). A **species** is a functor $S: \mathbf{B} \rightarrow$ ***Sets***, and a **morphism** τ from species S to species T is a natural transformation from functor S to functor T.

If there is an isomorphism τ from species S to species T, then we write $S \approx T$. (and use the notation $S = T$ when we work "up to an isomorphism"). In what follows, the symbol S will be used sometimes to represent a species, and some other times to represent it's isomorphism class. The usage at a particular point in the text should be clear from the context. For any $E \in \mathbf{B}$ and any species S we write S[E] for the image of E under S. Every element in S[E] is called an S - structure on E.

The reader is referred to [3] (or [5]) for the definitions of the sum $S + T$, product $S \cdot T$, cartesian product $S \times T$, derivative S', and substitution $S \circ T$ (if $T[\emptyset] = \emptyset$), of two species S and T. They are summarized as follows:

Definition II.2.2 ([3]). For any $E \in \mathbf{B}$,

(a) $(S + T)[E] = S[E] + T[E]$ (b) $(S \cdot T)[E] = \sum_{E = E_1 + E_2} S[E_1] \times T[E_2]$.

(c) $(S \times T)[E] = S[E] \times T[E]$ (d) $S'[E] = S[E + 1]$

(d) $(S \circ T)[E] = \sum_{\pi \in P[E]} S[\pi] \times \prod_{C \in \pi} T[C]$

where P[E] is the set of all partitions of E.

A species S is called a **subspecies** of the species U if $S[E] \subset U[E]$ for all finite sets E and the inclusion is a natural transformation. It is obvious that if S is a subspecies of U then there exists a unique species T such that $U = S + T$.

Example II.2.3. The zero species, 0, is defined by: $0[E] = \emptyset$ for any finite set E. 0 is the unit element for addition.

Example II.2.4. $1 = \mathbf{B}(\emptyset,-)$, so the species 1 satisfies $1[E] = \emptyset$ for any non-empty finite set E and $1[\emptyset] = \{*\}$; i.e. there is a unique 1-structure on the empty set. 1 is the unit element for multipliction.

Example II.2.5. $X = \mathbf{B}(\{*\},-)$, so $X[E] = \{*\}$ if $|E| = 1$; $X[E] = \emptyset$ if $|E| \neq 1$.

Example II.2.6. For $n \in \mathbb{N}$, write $\mathbf{n} = \{1, 2, ..., n\}$. We have $X^n = \mathbf{B}(\mathbf{n},-) = X \cdot X \cdots X$. More generally, let $H \subset \mathbf{n}!$; then we use X^n/H to denote the species $\mathbf{B}(\mathbf{n},-)/H$, i.e. $X^n/H[E]$ = the set of all "left cosets" of H in $\mathbf{B}(\mathbf{n}, E)$ where $\mathbf{B}(\mathbf{n}, E)$ is the set of all bijections from $\mathbf{n}$ to E ($\mathbf{B}(\mathbf{n}, E)$ is not a group). In fact $X^n/H[\mathbf{n}]$ = the set of all left cosets of H in $\mathbf{n}!$.

Example II.2.7. The exponential species $e^X = \mathbf{B}(-, \{*\})$ is defined by: $e^X[E] = \{*\}$ for any finite set E, i.e. there is a unique e^X-structure on any finite set. We have:

$$e^X = \sum_{n \geq 0} X^n/\mathbf{n}! \;.$$

A species U is called a **molecule** if $U \neq 0$, and $U = S + T$ implies either $S = 0$ or $T = 0$. Every species is a (possibly infinite) sum of its molecular subspecies. The molecules are of the type :

$$X^n/H \qquad \text{where } H \text{ is a subgroup of } \mathbf{n}! \;.$$

It is easy to prove that $X^n/H = X^m/K$ iff $n = m$ and H, K are conjugate in $\mathbf{n}!$. Let $\mathcal{M}$ denote the set of isomorphism classes of all molecular species and $\mathcal{M}^*$ denote the set of isomorphism classes of all non-constant molecular species.

Proposition II.2.8 ([13]). Let $n, m \in \mathbb{N}$, $H \subset \mathbf{n}!$ and $K \subset \mathbf{m}!$, then:

(1) $X^n/H \cdot X^m/K = X^{n+m}/(H*K)$ where " $*$ " is the external product.

(2) $X^n/H \times X^m/K = \begin{cases} \sum_L |L|\,|A_L|\; X^n/L & \text{if } n = m;\text{ where } A_L = \{g \in \mathbf{n}! \mid gHg^{-1} \cap K = L\}. \\ 0 & \text{otherwise,} \end{cases}$

(3) $X^n/H \circ X^m/K = X^{mn}/(K \wr H)$ where " $\wr$ " is the wreath product.

(4) $(X^n/H)' = \sum_{e \in O_{n,H}} X^n/(H \cap (\mathbf{n}-\{e\})!)$ where $O_{n,H}$ denotes a complete set of representatives for the orbits of H in $\mathbf{n}$.

By propositions II.1.8 and II.2.8, we have

Proposition II.2.9. $(\mathcal{M}, \cdot)$ is a free commutative monoid.

Definition II.2.10. ([13]). A species S is called **finitary** if $S[E]$ is finite for all $E \in \mathbf{B}$. A finitary species S is called **strictly finite** if $\exists\, n > 0$ such that $S[E] = \emptyset$ for all $E \in \mathbf{B}$ with $|E| > n$.

The set of all finitary species (resp strictly finite species) forms a half-ring which is isomorphic to $\mathbb{N}[[\mathfrak{M}]]$ (resp $\mathbb{N}[\mathfrak{M}]$). The universal ring **V** (resp **SV**) containing this is called **the ring of virtual species** (or **$\mathbb{Z}$-species**). Every element in **V** can be represented as $S - T$ where S and T are two species. The ring **V** (resp **SV**) is isomorphic to $\mathbb{Z}[[\mathfrak{M}]]$ (resp $\mathbb{Z}[\mathfrak{M}]$). From propositions II.2.4 and II.2.9, we have

Theorem II.2.11. These two rings $\mathbb{Z}[[\mathfrak{M}]]$ and $\mathbb{Z}[\mathfrak{M}]$ are **UFD's.**

There are many identities involving $+, \cdot, \times, \circ, ', 0$ and 1 ([3],[5],[13]). Let S, T and U be species, then

(i) $(S + T) \circ U = (S \circ U) + (T \circ U)$;
(ii) $(S \cdot T) \circ U = (S \circ U) \cdot (T \circ U)$;
(iii) $(S \circ T) \circ U = S \circ (T \circ U)$;
(iv) $(S + T)' = S' + T'$;
(v) $(S \cdot T)' = S' \cdot T + S \cdot T'$;
(vi) $(S \times T)' = S' \times T'$;
(vii) $(S \circ T)' = (S' \circ T) \cdot T'$
... etc.

One objective is to extend all these identities to the setting of $\mathbb{K}$-species. This is done in chapter three.

§ II.3. d-species.

Definition II.3.1 ([3]). Let d be an integer > 0. A **d-species** is a functor $S: \mathbf{B}^d \longrightarrow$ ***Sets***, and a **morphism** τ from d-species S to d-species T is a natural transformation τ from functor S to functor T.

Let S, T be d-species and $T_1, T_2, \ldots, T_d$ be r-species (where $d, r \in \mathbb{N}$). The <u>sum</u> $S + T$, <u>product</u> $S \cdot T$, <u>cartesian product</u> $S \times T$, <u>partial derivatives</u> $(\partial S/\partial X_i)$, $1 \le i \le d$, and <u>substitution</u> $S \circ (T_1, T_2, \ldots, T_d)$ are defined as follows :

Definition II.3.2 ([3]). For any $\underline{E} = (E_1, E_2, \ldots, E_d) \in \mathbf{B}^d$ and $\underline{A} = (A_1, \ldots, A_r) \in \mathbf{B}^r$, define

(a) $(S + T)[\underline{E}] = S[\underline{E}] + T[\underline{E}]$
(b) $(S \cdot T)[\underline{E}] = \sum_{\underline{E} = \underline{D} + \underline{F}} S[\underline{D}] \times T[\underline{F}]$

where $\underline{E} = \underline{D} + \underline{F}$ means $E_i = D_i + F_i$ for $1 \le i \le d$,

(c) $(S \times T)[\underline{E}] = S[\underline{E}] \times T[\underline{E}]$
(d) $(\partial S/\partial X_i)[\underline{E}] = S[\underline{E} + \underline{e}_i]$

where $\underline{e}_i = (F_1, F_2, \ldots, F_d)$ with $F_i = \{*\}$ and $F_j = \emptyset$ if $i \ne j$, $1 \le i, j \le d$,

(d) $S \circ (T_1, \ldots, T_d)[\underline{A}] = \sum_{\pi \in P[\underline{A}]} \sum_{f: \pi \to \mathbf{d}} S[(f^{-1}(1), \ldots, f^{-1}(d))] \times \prod_{C \in \pi} T_{f(C)}[C \cap A_1, \ldots, C \cap A_r]$

where $P[\underline{A}]$ denotes the set of all partitions of $A_1 + \cdots + A_d$.

Example II.3.3. $X_i = \mathbf{B}^d(\underline{e}_i,-)$ where $\underline{e}_i = (F_1,F_2,\dots,F_d)$ with $F_i = \{*\}$ and $F_j = \emptyset$ if $i \neq j$. For $\underline{E} = (E_1,E_2,\dots,E_d) \in \mathbf{B}^d$, $X_i[\underline{E}] = \{*\}$ if $E_i = \underline{e}_i$; $X_i[\underline{E}] = \emptyset$ otherwise.

Example II.3.4. $X_1^{n_1} \cdot X_2^{n_2} \cdots X_d^{n_d} = \mathbf{B}^d(\underline{n},-)$ where $\underline{n} = (n_1,n_2,\dots,n_d)$. and $\mathbf{B}^d(\underline{n},\underline{E})$ is the set of all $(f_1,f_2,\dots,f_d)$ where all f_i are bijections from $\mathbf{n}_i$ to E_i. Note that $\mathbf{B}^d(\underline{n},\underline{E})$ is empty unless $|E_i| = n_i$ for all i. More generally, let $H \subset \mathbf{n}_1! \times \mathbf{n}_2! \times \cdots \times \mathbf{n}_d!$, then $(X_1^{n_1} \cdot X_2^{n_2} \cdots X_d^{n_d}/H)[\underline{E}]$ is the set of all "left cosets" of H in $\mathbf{B}^d(\underline{n},\underline{E})$; we often use $X_1^{n_1} \cdot X_2^{n_2} \cdots X_d^{n_d}/H$ to denote the d-species $\mathbf{B}^d(\underline{n},-)/H$.

As in the single variable case, every d-species is uniquely a (possibly infinite) sum of its molecular d-subspecies. The molecular d-species are of the type:

$$X_1^{n_1} \cdot X_2^{n_2} \cdots X_d^{n_d}/H \qquad \text{where} \quad H \subset \mathbf{n}_1! \times \mathbf{n}_2! \times \cdots \times \mathbf{n}_d!.$$

Let $\mathfrak{M}_d$ be the set of all isomorphism classes of molecular d-species.

Definition II.3.5 ([13]). Let $n_i, m_i \in \mathbb{N}$ for $1 \le i \le d$, $H \subset \mathbf{n}_1! \times \cdots \times \mathbf{n}_d!$, $K \subset \mathbf{m}_1! \times \cdots \times \mathbf{m}_d!$. For any $h = (h_1, \dots, h_d) \in H$, $k = (k_1, \dots, k_d) \in K$ (h_i and k_i are the restriction of h, k to $\mathbf{n}_i$, $\mathbf{m}_i$ respectively for $1 \le i \le d$) and $u = (u_1,u_2,\dots,u_d) \in (\mathbf{n}_1+\mathbf{m}_1) \times \dots \times (\mathbf{n}_d+\mathbf{m}_d)$, we define: $(h *_d k)(u) = (g_1(u_1), g_2(u_2), \dots, g_d(u_d))$ where $g_i(u_i) = h_i(u_i)$ if $u_i \in \mathbf{n}_i$; $g_i(u_i) = k_i(u_i)$ if $u_i \in \mathbf{m}_i$ for $1 \le i \le d$, and $H *_d K = \{h *_d k \mid h \in H \text{ and } k \in K\}$.

From the above definition , we have

$$(X_1^{n_1} \cdots X_d^{n_d}/H) \cdot (X_1^{m_1} \cdots X_d^{m_d}/K) = X_1^{n_1+m_1} \cdots X_d^{n_d+m_d}/(H *_d K).$$

Lemma II.3.6. $(\mathfrak{M}_d, \cdot)$ is a free commutative monoid.

Theorem II.3.7. The ring of virtual finitary species $\mathbf{Z}[[\mathfrak{M}_d]]$ and the ring of virtual strictly finite species $\mathbf{Z}[\mathfrak{M}_d]$ are **UFD**'s.

Proposition II.3.8 ([13]). Let $n_i, m_i \in \mathbb{N}$, $K_i \subset \mathbf{m}_i!$ for $1 \le i \le d$, $H \subset \mathbf{n}_1! \times \cdots \times \mathbf{n}_d!$, and $K \subset \mathbf{m}_1! \times \cdots \times \mathbf{m}_d!$, then

$$(X_1^{n_1} \cdots X_d^{n_d}/H) \circ (X_1^{m_1}/K_1, \cdots, X_d^{m_d}/K_d) = (X_1^{n_1 m_1 + \cdots + n_d m_d}/(K_1, \dots, K_d) \wr H)$$

$$(X_1^{n_1} \cdots X_d^{n_d}/H) \times (X_1^{m_1} \cdots X_d^{m_d}/K) = \begin{cases} \sum_L |L| \cdot |A_{d,L}| \cdot (X_1^{n_1} \cdots X_d^{n_d}/L) & \text{if } m_i = n_i \text{ for } 1 \le i \le d; \\ 0 & \text{if } m_i \neq n_i \text{ for some } i. \end{cases}$$

where $A_{d,L} = \{ g \in \mathbf{n}_1! \times \mathbf{n}_2! \times \cdots \times \mathbf{n}_d! \mid gHg^{-1} \cap K = L \}$.

Just as in the case of one variable, there are many identities involving the operations $+, \cdot, \times, \circ$ and $'$ in d-species ([3]). We can also extend those identities of d-variable species to the setting of d-variable $\mathbb{K}$-species.

§ II.4. $\mathbb{K}$-species.

Let $\mathbb{K}$ be a half-ring. We can extend the operations $+, \cdot, \times$ and $'$ to the set

$$\mathbb{K}[[\mathfrak{M}]] = \{ \textstyle\sum_{T \in \mathfrak{M}} a_T T \mid a_T \in \mathbb{K} \}$$

as follows:

$$\begin{array}{llcll}
\text{(a)} & (\sum_{T \in \mathfrak{M}} a_T T) + (\sum_{T \in \mathfrak{M}} b_T T) & = & \sum_{T \in \mathfrak{M}} & (a_T + b_T) T \\
\text{(b)} & (\sum_{T \in \mathfrak{M}} a_T T) \cdot (\sum_{S \in \mathfrak{M}} b_S S) & = & \sum_{T,S \in \mathfrak{M}} & (a_T \cdot b_S)(T \cdot S) \\
\text{(c)} & (\sum_{T \in \mathfrak{M}} a_T T) \times (\sum_{S \in \mathfrak{M}} b_S S) & = & \sum_{T,S \in \mathfrak{M}} & (a_T \cdot b_S)(T \times S) \\
\text{(d)} & (\sum_{T \in \mathfrak{M}} a_T T)' & = & \sum_{T \in \mathfrak{M}} & a_T T'.
\end{array}$$

Of course, the terms must be collected on the right sides of (b), (c), (d). It is possible to do so because: given a molecular species M, there are only finitely many pairs of molecular species (S,T) such that $M = S \cdot T$, finitely many pairs of molecular species (U,V) such that M is a subspecies of $U \times V$, and finitely many molecular species W such that M is a subspecies of W'.

Let σ be the unique half-ring homomorphism: $\mathbb{N} \to \mathbb{K}$; then σ induces a half-ring homomorphism $\hat{\sigma}: \mathbb{N}[[\mathfrak{M}]] \to \mathbb{K}[[\mathfrak{M}]]$. The homomorphism preserves $+, \cdot, \times$ and $'$. We hope to extend the concept of substitution, $\circ$, to $\mathbb{K}[[\mathfrak{M}]]$ in such a way that $\hat{\sigma}$ preserves $\circ$ and all the identities involving $+, \cdot, \times, \circ, '$ continue to hold.

Unfortunately it cannot succeed for all half-rings. For example:

(1) Let $\mathbb{K} = \mathbb{F}_2$, then $(X^2/2!) \circ (X+X) = (X^2/2!) \circ (0) = 0$, but $(X^2/2!) \circ (X+X) = (X^2/2!) + X^2 + (X^2/2!) = X^2$. This is a contradiction.

(2) Let $\mathbb{K} = \mathbb{Z}[i]$. Let $(X^2/2!) \circ (iX) = aX^2 + b(X^2/2!)$ since $\deg((X^2/2!) \circ (iX)) = 2$. (Here we are assuming a bit more about the extended substitution, namely that degrees multiply under substitution of $\mathbb{K}$-species of the form scalar times molecule.) More detailed computations show that $(a,b) = (i,(-1-i)/2)$ or $(i,(-1+i)/2)$. This is a contradiction since $b \notin \mathbb{Z}[i]$.

For examples above, we want $\binom{i}{2} = (-1-i)/2 \in \mathbb{K}$ if $i \in \mathbb{K}$. This suggests that some special half-rings, "binomial half-rings", will satisfy our desire.

Definition II.4.1 ([13]). A half-ring $\mathbb{K}$ is called a **binomial half-ring** if

(a) there exists a $\mathbb{Q}$-algebra $\mathbb{L}$ containing $\mathbb{K}$, and
(b) for every $a \in \mathbb{K}$ and $i \in \mathbb{N}$, $\binom{a}{i} = a(a-1)(a-2)\cdots(a-i+1)/i! \in \mathbb{K}$.

For example $\mathbb{N}$, $\mathbb{Z}$, $\mathbb{Q}$, $\mathbb{R}$, $\mathbb{C}$, $\mathbb{Q}[i]$ and $\mathbb{N} + \mathbb{Q}\varepsilon$ ($\varepsilon^2 = 0$) are all binomial half-rings, but $\mathbb{F}_p$, p prime, and $\mathbb{Z}[i]$ are not binomial half-rings.

Definition II.4.2 ([13]). Let $\mathbb{K}$ be a binomial half-ring. A **$\mathbb{K}$ - species** is an element S of $\mathbb{K}[[\mathcal{M}]]$, i.e. a formal linear combination of the molecular species with coefficients in $\mathbb{K}$.

The concepts of species (resp. virtual species) and $\mathbb{N}$-species (resp. $\mathbb{Z}$-species) coincide.

Chapter III: The calculus of $\mathbb{K}$-species

§ III.1. Extension of substitution to $\mathbb{K}$-species.

In this section, $\mathbb{K}$ is a given binomial half-ring. We will define the operation $\circ$ for $\mathbb{K}$-species and prove that the identities in chapter II involving $\circ$ continue to hold.

Proposition III.1.1. Let T_1 and T_2 be two species, then $e^{T_1 + T_2} = e^{T_1} \cdot e^{T_2}$.

Notation III.1.2. a) Let $\mathbb{L}$ be a $\mathbb{Q}$-algebra, $a \in \mathbb{L}$ and $r_1, r_2, \ldots, r_n \in \mathbb{N}$. We write

$$\binom{a}{r_1, r_2, \ldots, r_n} = a(a-1)\cdots(a-\Sigma r+1) / r_1!\, r_2! \cdots r_n!$$

where Σr means $r_1 + r_2 + \cdots + r_n$.

b) Let $(p_j)_{j\in J}$ be a family of formal variables. We denote by $\mathbb{N}[\binom{p_j}{i}]$ the sub half-ring of $\mathbb{Q}[(p_j)_{j\in J}]$ generated by the polynomials $\binom{p_j}{i}$, $j \in J$, $i \in \mathbb{N}$.

Remark III.1.3. If $f((p_j)_{j\in J}) \in \mathbb{N}[\binom{p_j}{i}]$ and $(a_j)_{j\in J}$ is an arbitrary family of elements of the binomial half-ring $\mathbb{K}$, then $f((a_j)_{j\in J}) \in \mathbb{K}$.
We also have

$$\binom{a}{r_1, r_2, \ldots, r_n} = \binom{\Sigma r}{r_1, r_2, \ldots, r_n}\binom{a}{\Sigma r} \in \mathbb{N}[\binom{a}{i}].$$

Corollary III.1.4. For all $n \in \mathbb{N}$,

$$e^{nX} = (e^X)^n = \sum_{k \geq 0} \sum_{r_1+2r_2+\ldots+kr_k = k} \binom{n}{r_1, r_2, \ldots, r_k} (X/1!)^{r_1} \cdot (X^2/2!)^{r_2} \cdots (X^k/k!)^{r_k}$$

$$= \sum_{M \in \mathfrak{M}} g_M(n)\, M$$

where all r_i are non-negative integers and all $g_M(p) \in \mathbb{N}[(^p_i)]$.

Proposition III.1.5. $S \times e^{nX} = S \circ (nX)$ for all $n \in \mathbb{N}$.

Proof. It is easy to show that for any $E \in \mathbf{B}$,

$$(S \circ (nX))[E] = S[E] \times \mathbf{n}^E$$

In particular for $S = e^X$, this gives $e^{nX}[E] = \mathbf{n}^E$. Substituting this back into the above equality gives

$$(S \circ (nX))[E] = S[E] \times e^{nX}[E].$$

Naturality in E is easily verified, so the proof is completed. □

Lemma III.1.6. $(\sum_{A \in \mathfrak{M}_d} a_A A) \cdot (\sum_{A \in \mathfrak{M}_d} b_A A) = \sum_{A \in \mathfrak{M}_d} c_A A$, where $c_A = \sum_{A_1 \cdot A_2 = A} a_{A_1} b_{A_2}$ is a finite sum.

Lemma III.1.7. $(\sum_{A \in \mathfrak{M}_d} a_A A) \times (\sum_{A \in \mathfrak{M}_d} b_A A) = \sum_{A \in \mathfrak{M}_d} c_A A$, where $c_A = \sum_{A_1, A_2} n_{A, A_1, A_2} a_{A_1} b_{A_2}$ is a finite sum, with $n_{A, A_1, A_2} \in \mathbb{N}$ defined by $A_1 \times A_2 = \sum_{A \in \mathfrak{M}_d} n_{A, A_1, A_2} A$.

Proposition III.1.8. Let S be a species and $n \in \mathbb{N}$, then $S(nX) = \sum_{M \in \mathfrak{M}} f_M(n) M$ for some $f_M(p) \in \mathbb{N}[(^p_i)]$.

Now, we can extend proposition III.1.8 to $\mathbb{K}$ - species:

Definition III.1.9. Let $\mathbb{K}$ be a binomial half-ring, $a \in \mathbb{K}$ and S be a $\mathbb{K}$-species. Then $S(aX) = \sum_{M \in \mathfrak{M}} f_M(a)\, M$ with $f_M(p)$ defined in proposition III.1.8.

Tables 4 and 5 give $S(-X)$ and $S(nX)$ for molecular species of small degree.

Lemma III.1.10. $X_1^{n_1} \cdots X_d^{n_d}/H \circ (X_1^{m_1}/K_1, \ldots, X_d^{m_d}/K_d) = X^{m_1 n_1 + m_2 n_2 + \cdots + m_d n_d}/((K_1, K_2, \cdots, K_d) \wr H)$ where $n_i, m_i \in \mathbb{N}$, $K_i \subset \mathbf{m}_i!$ for $1 \leq i \leq d$, and $H \subset \mathbf{n}_1! \times \mathbf{n}_2! \times \cdots \times \mathbf{n}_d!$.

Corollary III.1.11. Let $T_1, T_2, \ldots, T_d$ be d-species, then $e^{T_1 + T_2 + \ldots + T_d} = e^{T_1} \cdot e^{T_2} \ldots e^{T_d}$.

Lemma III.1.12. $e^X \circ (n_1 X_1 + \cdots + n_d X_d) = \sum_{A \in \mathfrak{M}_d} f_A(n_1, \ldots, n_d)\, A$

where $f_A(p_1,\dots,p_d) \in \mathbb{N}[\binom{p_i}{j}]_{1\le i\le d}$

Lemma III.1.13. For any $n_1,n_2,\dots,n_d \in \mathbb{N}$ and d-species S, we have:

$$S \circ (n_1X_1, n_2X_2,\dots,n_dX_d) = S \times (e^X \circ (n_1X_1+\dots+ n_dX_d)).$$

Lemma III.1.14. Let $M_1,\dots,M_d \in \mathfrak{M}^*$, then $(\sum_{A\in\mathfrak{M}_d} a_A A)\circ(M_1,\dots,M_d) = \sum_{B\in\mathfrak{M}} c_B B$ where $c_B = \sum (a_A \mid A \in \mathfrak{M}_d,\ A \circ (M_1, M_2, \dots, M_d) = B)$, a finite sum.

Lemma III.1.15. Let T be species, $A_j \in \mathfrak{M}^*$ for $1 \le j \le d$. Then we have $T\circ(n_1A_1+\dots+n_dA_d) = \sum_{B\in\mathfrak{M}} f_B(n_1,n_2,\dots,n_d)\, B$, where $f_B(p_1,p_2,\dots,p_d) \in \mathbb{N}[\binom{p_j}{i}]_{1\le j\le d}$ for all $B \in \mathfrak{M}$.

Remark III.1.16. Let S be a species and $S \circ (\sum_{A\in\mathfrak{M}} n_A A) = \sum_{B\in\mathfrak{M}} f_B((n_A)_{A\in\mathfrak{M}})\, B$ where f_B depends only on S and on the n_A's with $\deg A \le \deg B$. So we have:

Proposition III.1.17. Let T be a species, then

$$T \circ (\sum_{A\in\mathfrak{M}^*} n_A A) = \sum_{B\in\mathfrak{M}} f_B((n_A)_{A\in\mathfrak{M}})\, B \quad \text{where} \quad f_B((p_A)_{A\in\mathfrak{M}}) \in \mathbb{N}[\binom{p_A}{j}]_{A\in\mathfrak{M}}.$$

Definition III.1.18 ([13]). Let $\mathbb{K}$ be a binomial half-ring and S, T be two $\mathbb{K}$-species with $T = \sum_{A\in\mathfrak{M}^*} n_A A$ for $n_A \in \mathbb{K}$. The **substitution** of T in S, $S\circ T$, is defined by

$$\sum_{B\in\mathfrak{M}} f_B((n_A)_{A\in\mathfrak{M}})\, B \quad \text{with } f_B((p_A)_{A\in\mathfrak{M}}) \text{ given in proposition III.1.17.}$$

If S is a $\mathbb{K}$-species, then $S = \sum_{n\in\mathbb{N}} \sum_H a_{nH} \cdot X^n/H$ where $a_{nH} \in \mathbb{K}$ and H ranges over representatives for the conjugacy classes of subgroups of $n\mathfrak{F}$. S_n denotes the n-th term of the outer sum. (If S is an actual species, then $S_n[E] = S[E]$ if $|E| = n$; $S_n[E] = \emptyset$ if $|E| \ne n$.)

Theorem III.1.19. Let S, T and U be $\mathbb{K}$-species with $T_0 = U_0 = 0$, then

$$(S \circ T) \circ U = S \circ (T \circ U)$$

Proof. Let $S = \sum_{A\in\mathfrak{M}} s_A A$, $T = \sum_{B\in\mathfrak{M}} t_B B$ and $U = \sum_{C\in\mathfrak{M}} u_C C$. We have

$$(S \circ T) \circ U = \sum_{M\in\mathfrak{M}} f_M((s_A,t_B,u_C)_{A,B,C\in\mathfrak{M}})M, \quad S \circ (T \circ U) = \sum_{M\in\mathfrak{M}} g_M((s_A,t_B,u_C)_{A,B,C\in\mathfrak{M}})M$$

where $f_M((p_A,q_B,r_C)_{A,B,C\in\mathfrak{M}}),\ g_M((p_A,q_B,r_C)_{A,B,C\in\mathfrak{M}}) \in \mathbb{N}[\binom{p_A}{i},\binom{q_B}{j},\binom{r_C}{k}]_{A,B,C\in\mathfrak{M}}$.

By associativity of substitution for actual species, f_M and g_M agree when natural number are substituted for p_A, q_B and r_C, and hence they agree when arbitrary elements of $\mathbb{K}$ are substituted. □

Similar arguments prove all the identites involving $+$, $\cdot$, $\times$, $\circ$, $'$, 0 and 1. Substitution for several-variable $\mathbb{K}$-species is defined in the analogous way and identities from actual species can be lifted to these by arguments similar to the one variable case.

§ III.2. The $\mathbb{K}$-species SIN, COS, and LG.

The trigonometric functions, $\cos x$ and $\sin x$ have properties such as $(\sin x)' = \cos x$, $(\cos x)' = -\sin x$ and $\sin^2 x + \cos^2 x = 1$. Here we try to find some special $\mathbb{K}$-species which have similar properties.

In fact, we can't find any $\mathbb{Z}$-species (=virtual species) which have the properties above. Suppose S and C are two $\mathbb{Z}$-species with $S_0 = 0$, $C_0 = 1$ such that $S' = C$, $C' = -S$ and $S \cdot S + C \cdot C = 1$.

Let $S = a_1X + a_2X^2 + a_3X^2/2! + \cdots$ and $C = 1 + b_1X + b_2X^2 + b_3X^2/2! + \cdots$. We have:

(i) $a_1 + (2a_2 + a_3)X + \ldots = 1 + b_1X + \cdots$, since $S' = C$;
(ii) $b_1 + (2b_2 + b_3)X + \ldots = -(a_1X + \cdots)$, since $C' = -S$;
(iii) $1 + 2b_1X + (a_1^2 + b_1^2 + 2b_2)X^2 + 2b_3X^2/2! + \cdots = 1$, since $S \cdot S + C \cdot C = 1$.

Comparing the coefficients of each molecular species on both sides, we have: $a_1 = 1$, $b_1 = 0$, and $a_1^2 + b_1^2 + 2b_2 = 1 + 2b_2 = 0$. This is a contradiction since $b_2 \notin \mathbb{Z}$.

Definition III.2.1. Let $\mathbb{K}$ be a binomial ring containing $\mathbb{Q}$, then

$$\mathbf{COS}\, X = 1/2\,(e^{iX} + e^{-iX}) \qquad \text{and} \qquad \mathbf{SIN}\, X = -i/2\,(e^{iX} + e^{-iX}).$$

Of course, in this definition, e^{iX} and e^{-iX} are both computed by substituting $\mathbf{i}$ for n in corollary III.1.4. Let the ring homomorphism $\sigma\colon \mathbb{Q}[\mathbf{i}] \to \mathbb{Q}[\mathbf{i}]$ be defined by: $a + b\mathbf{i} \mapsto a - b\mathbf{i}$. The induced homomorphism $\hat{\sigma}\colon \mathbb{Q}[\mathbf{i}][[\mathfrak{M}]] \to \mathbb{Q}[\mathbf{i}][[\mathfrak{M}]]$ fixes species **SIN** and species **COS**. So **SIN**, **COS** $\in \mathbb{Q}[[\mathfrak{M}]]$.

Proposition III.2.2. Let S be a $\mathbb{Z}$-species with $S_0 = 0$. If $e^X \circ S = 1$ then $S = 0$.

Proof. $1 = e^X \circ S = \sum_{n \geq 0} \sum_{r_1 + 2r_2 + \cdots + nr_n = n} ((X^{r_1}/r_1!) \circ (S_1)) \cdot ((X^{r_2}/r_2!) \circ (S_2)) \cdots ((X^{r_n}/r_n!) \circ (S_n))$ where $r_i \geq 0$ for all i. Comparing terms of degree n on both sides gives:

$$0 = \sum_{r_1 + 2r_2 + \cdots + nr_n = n} ((X^{r_1}/r_1!) \circ (S_1)) \cdot ((X^{r_2}/r_2!) \circ (S_2)) \cdots ((X^{r_n}/r_n!) \circ (S_n))$$

The n-th equation has highest term S_n (from $r_1 = \ldots = r_{n-1} = 0$, $r_n = 1$) and all lower terms involve only $S_1, S_2, \ldots, S_{n-1}$. The system of equations can be solved recursively. We have $S_n = 0$ for $n \geq 1$. i.e. $S = 0$. □

Corollary III.2.3. Let S, T be two $\mathbb{Z}$-species such that $S_0 = T_0 = 0$ and $e^X \circ S = e^X \circ T$ then $S = T$.

Definition III.2.4. The species **LG** $= \sum_{k\geq 0} S_k$ is recursively defined by

$$S_0 = 0, \qquad S_1 = -X$$

and

$$\sum_{r_1+2r_2+\cdots+nr_n=n} ((X^{r_1}/r_1!)\circ(S_1))\cdot((X^{r_2}/r_2!)\circ(S_2))\cdots((X^{r_n}/r_n!)\circ(S_n)) = 0, \quad n \geq 2.$$

Proposition III.2.5. $e^X \circ \mathbf{LG}\, X = 1 - X$.

Let $\mathbf{V}_i = \{T \in \mathbb{Z}[[\mathfrak{M}]] \mid T_0 = i\}$ then $T \mapsto e^T$ $(= e^X \circ T)$ gives a group homomorphism exp: $(\mathbf{V}_0, +) \to (\mathbf{V}_1, \cdot)$. From the propositions III.1.1, III.2.2 and III.2.5, we know that exp is a group isomorphism and that its inverse log is given by: $T \mapsto \mathbf{LG}(1 - T)$ (log is not a species).

Proposition III.2.6. $\mathbf{LG}(1 - S\cdot T) = \mathbf{LG}(1 - S) + \mathbf{LG}(1 - T)$ for any $S, T \in \mathbf{V}_1$.

NOTATION FOR TABLES

$\mathbf{n}!$ = The group of all permutations on **n**; A_n = The group of all even permutations on **n**; C_n= The cyclic subgroup of $\mathbf{n}!$ generated by (12...n); D_n= The dihedral group of order 2n; A·B = The direct product of group A and group B; K_4 = {id,(12)(34),(13)(24),(14)(23)}; H = {id, (12)(34)}; L = {id,(123),(132),(12)(45),(13)(45),(23)(45)} = $A_5 \cap$ Stabilizer {4,5}; T = The normalizer of C_5 = The affine group $\{ax + b \mid a,b \in \mathbb{F}_5,\ a \neq 0\}$ = {id, (12345), (13524), (14253), (15432), (2354), (25)(34), (2453), (1534), (13)(45), (1435), (1452), (15)(24), (1254), (1523), (12)(35), (1325), (1243), (14)(23), (1342)}.

The cartesian product between molecular species of degree ≤ 3

$X \times X = X$			
$X^2 \times X^2 = 2X^2$	$X^2 \times X^2/2! = X^2$	$X^2/2! \times X^2/2! = X^2/2!$	
$X^3 \times X^3 = 6X^3$	$X^3 \times X^3/2! = 3X^3$	$X^3 \times X^3/A_3 = 2X^3$	$X^3 \times X^3/3! = X^3$
$X^3/2! \times X^3/2! = X^3/2! + X^3$	$X^3/2! \times X^3/A_3 = X^3$	$X^3/2! \times X^3/3! = X^3/2!$	
$X^3/A_3 \times X^3/A_3 = 2X^3/A_3$	$X^3/A_3 \times X^3/3! = X^3/A_3$	$X^3/3! \times X^3/3! = X^3/3!$	

Table 1

The cartesian product between molecular species of degree 4

× cartesian product	$\frac{X^4}{4\nabla}$	$\frac{X^4}{A_4}$	$\frac{X^4}{D_4}$	$\frac{X^4}{3\nabla}$	$\frac{X^4}{2\nabla 2\nabla}$	$\frac{X^4}{K_4}$	$\frac{X^4}{C_4}$	$\frac{X^4}{A_3}$	$\frac{X^4}{2\nabla}$	$\frac{X^4}{H}$	X^4
$\frac{X^4}{4\nabla}$	$\frac{X^4}{4\nabla}$	$\frac{X^4}{A_4}$	$\frac{X^4}{D_4}$	$\frac{X^4}{3\nabla}$	$\frac{X^4}{2\nabla 2\nabla}$	$\frac{X^4}{K_4}$	$\frac{X^4}{C_4}$	$\frac{X^4}{A_3}$	$\frac{X^4}{2\nabla}$	$\frac{X^4}{H}$	X^4
$\frac{X^4}{A_4}$	$\frac{X^4}{A_4}$	$2\frac{X^4}{A_4}$	$\frac{X^4}{K_4}$	$\frac{X^4}{A_3}$	$\frac{X^4}{H}$	$2\frac{X^4}{K_4}$	$\frac{X^4}{H}$	$2\frac{X^4}{A_3}$	X^4	$2\frac{X^4}{H}$	$2X^4$
$\frac{X^4}{D_4}$	$\frac{X^4}{D_4}$	$\frac{X^4}{K_4}$	$\frac{X^4}{D_4}+\frac{X^4}{K_4}$	$\frac{X^4}{2\nabla}$	$\frac{X^4}{2\nabla 2\nabla}+\frac{X^4}{H}$	$3\frac{X^4}{K_4}$	$\frac{X^4}{C_4}+\frac{X^4}{H}$	X^4	$\frac{X^4}{2\nabla}+X^4$	$3\frac{X^4}{H}$	$3X^4$
$\frac{X^4}{3\nabla}$	$\frac{X^4}{3\nabla}$	$\frac{X^4}{A_3}$	$\frac{X^4}{2\nabla}$	$\frac{X^4}{3\nabla}+\frac{X^4}{2\nabla}$	$2\frac{X^4}{2\nabla}$	X^4	X^4	$\frac{X^4}{A_3}+X^4$	$2\frac{X^4}{2\nabla}+X^4$	$2X^4$	$4X^4$
$\frac{X^4}{2\nabla 2\nabla}$	$\frac{X^4}{2\nabla 2\nabla}$	$\frac{X^4}{H}$	$\frac{X^4}{2\nabla 2\nabla}+\frac{X^4}{H}$	$2\frac{X^4}{2\nabla}$	$2\frac{X^4}{2\nabla 2\nabla}+X^4$	$3\frac{X^4}{H}$	$X^4+\frac{X^4}{H}$	$2X^4$	$2\frac{X^4}{2\nabla}+2X^4$	$2\frac{X^4}{H}+2X^4$	$6X^4$
$\frac{X^4}{K_4}$	$\frac{X^4}{K_4}$	$2\frac{X^4}{K_4}$	$3\frac{X^4}{K_4}$	X^4	$3\frac{X^4}{H}$	$6\frac{X^4}{K_4}$	$3\frac{X^4}{H}$	$2X^4$	$3X^4$	$6\frac{X^4}{H}$	$6X^4$
$\frac{X^4}{C_4}$	$\frac{X^4}{C_4}$	$\frac{X^4}{H}$	$\frac{X^4}{C_4}+\frac{X^4}{H}$	X^4	$X^4+\frac{X^4}{H}$	$3\frac{X^4}{H}$	$2\frac{X^4}{C_4}+X^4$	$2X^4$	$3X^4$	$2X^4+2\frac{X^4}{H}$	$6X^4$
$\frac{X^4}{A_3}$	$\frac{X^4}{A_3}$	$2\frac{X^4}{A_3}$	X^4	$\frac{X^4}{A_3}+X^4$	$2X^4$	$2X^4$	$2X^4$	$2\frac{X^4}{A_3}+2X^4$	$4X^4$	$4X^4$	$8X^4$
$\frac{X^4}{2\nabla}$	$\frac{X^4}{2\nabla}$	X^4	$\frac{X^4}{2\nabla}+X^4$	$2\frac{X^4}{2\nabla}+X^4$	$2\frac{X^4}{2\nabla}+2X^4$	$3X^4$	$3X^4$	$4X^4$	$2\frac{X^4}{2\nabla}+5X^4$	$6X^4$	$12X^4$
$\frac{X^4}{H}$	$\frac{X^4}{H}$	$2\frac{X^4}{H}$	$3\frac{X^4}{H}$	$2X^4$	$2\frac{X^4}{H}+2X^4$	$6\frac{X^4}{H}$	$2X^4+2\frac{X^4}{H}$	$4X^4$	$6X^4$	$4\frac{X^4}{H}+4X^4$	$12X^4$
X^4	X^4	$2X^4$	$3X^4$	$4X^4$	$6X^4$	$6X^4$	$6X^4$	$8X^4$	$12X^4$	$12X^4$	$24X^4$

Table 2

The derivative of molecular species of degree ≤ 5

Molecular	Derivative	Molecular	Derivative
1	0	X^5	$5X^4$
X	1	X^5/H	$X^4/H + 2X^4$
X^2	$2X$	$X^5/2\underset{\circ}{\nabla}$	$3X^4/2\underset{\circ}{\nabla} + X^4$
$X^2/2\underset{\circ}{\nabla}$	X	X^5/A_3	$2X^4/A_3 + X^4$
X^3	$3X^2$	X^5/C_4	$X^4/C_4 + X^4$
$X^3/2\underset{\circ}{\nabla}$	$X^2/2\underset{\circ}{\nabla} + X^2$	X^5/K_4	$X^4/K_4 + X^4$
X^3/A_3	X^2	$X^5/2\underset{\circ}{\nabla}\cdot 2\underset{\circ}{\nabla}$	$X^4/2\underset{\circ}{\nabla}\cdot 2\underset{\circ}{\nabla} + 2X^4/2\underset{\circ}{\nabla}$
$X^3/3\underset{\circ}{\nabla}$	$X^2/2\underset{\circ}{\nabla}$	X^5/C_5	X^4
X^4	$4X^3$	X^5/L	$X^4/A_3 + X^4/H$
X^4/H	$2X^3$	$X^5/A_3\cdot 2\underset{\circ}{\nabla}$	$X^4/A_3 + X^4/2\underset{\circ}{\nabla}$
$X^4/2\underset{\circ}{\nabla}$	$2X^3/2\underset{\circ}{\nabla} + X^3$	$X^5/3\underset{\circ}{\nabla}$	$2X^4/3\underset{\circ}{\nabla} + X^4/2\underset{\circ}{\nabla}$
X^4/A_3	$X^3/A_3 + X^3$	X^5/D_4	$X^4/D_4 + X^4/2\underset{\circ}{\nabla}$
X^4/C_4	X^3	X^5/D_5	X^4/H
X^4/K_4	X^3	$X^5/2\underset{\circ}{\nabla}\cdot 3\underset{\circ}{\nabla}$	$X^4/2\underset{\circ}{\nabla}\cdot 2\underset{\circ}{\nabla} + X^4/3\underset{\circ}{\nabla}$
$X^4/2\underset{\circ}{\nabla}\cdot 2\underset{\circ}{\nabla}$	$2X^3/2\underset{\circ}{\nabla}$	X^5/A_4	$X^4/A_4 + X^4/A_3$
$X^4/3\underset{\circ}{\nabla}$	$X^3/3\underset{\circ}{\nabla} + X^3/2\underset{\circ}{\nabla}$	X^5/T	X^4/C_4
X^4/D_4	$X^3/2\underset{\circ}{\nabla}$	$X^5/4\underset{\circ}{\nabla}$	$X^4/3\underset{\circ}{\nabla} + X^4/4\underset{\circ}{\nabla}$
X^4/A_4	X^3/A_3	X^5/A_5	X^4/A_4
$X^4/4\underset{\circ}{\nabla}$	$X^3/3\underset{\circ}{\nabla}$	$X^5/5\underset{\circ}{\nabla}$	$X^4/4\underset{\circ}{\nabla}$

Table 3

The substitution of −X in molecular species of degree ≤ 5

$1\circ(-X)=1$	$X\circ(-X)=-X$
$X^2\circ(-X)=X^2$	$X^2/\mathbf{2}\underset{\circ}{\nabla}\circ(-X)=X^2-X^2/\mathbf{2}\underset{\circ}{\nabla}$
$X^3\circ(-X)=-X^3$	$X^3/\mathbf{2}\underset{\circ}{\nabla}\circ(-X)=X^3/\mathbf{2}\underset{\circ}{\nabla}-X^3$
$X^3/A_3\circ(-X)=-X^3/A_3$	$X^3/\mathbf{3}\underset{\circ}{\nabla}\circ(-X)=2X^3/\mathbf{2}\underset{\circ}{\nabla}-X^3-X^3/\mathbf{3}\underset{\circ}{\nabla}$
$X^4\circ(-X)=X^4$	$X^4/H\circ(-X)=X^4/H$
$X^4/\mathbf{2}\underset{\circ}{\nabla}\circ(-X)=X^4-X^4/\mathbf{2}\underset{\circ}{\nabla}$	$X^4/A_3\circ(-X)=X^4/A_3$
$X^4/C_4\circ(-X)=X^4/H-X^4/C_4$	$X^4/K_4\circ(-X)=3X^4/H-X^4-X^4/K_4$
$X^4/\mathbf{2}\underset{\circ}{\nabla}\cdot\mathbf{2}\underset{\circ}{\nabla}\circ(-X)=X^4/\mathbf{2}\underset{\circ}{\nabla}\cdot\mathbf{2}\underset{\circ}{\nabla}+X^4-2X^4/\mathbf{2}\underset{\circ}{\nabla}$	$X^4/\mathbf{3}\underset{\circ}{\nabla}\circ(-X)=X^4/\mathbf{3}\underset{\circ}{\nabla}+X^4-2X^4/\mathbf{2}\underset{\circ}{\nabla}$
$X^4/D_4\circ(-X)=X^4/\mathbf{2}\underset{\circ}{\nabla}\cdot\mathbf{2}\underset{\circ}{\nabla}+X^4/H-X^4/\mathbf{2}\underset{\circ}{\nabla}-X^4/D_4$	$X^4/A_4\circ(-X)=2X^4/A_3+X^4/H-X^4-X^4/A_4$
$X^4/\mathbf{4}\underset{\circ}{\nabla}\circ(-X)=X^4/\mathbf{2}\underset{\circ}{\nabla}\cdot\mathbf{2}\underset{\circ}{\nabla}+2X^4/\mathbf{3}\underset{\circ}{\nabla}+X^4-3X^4/\mathbf{2}\underset{\circ}{\nabla}-X^4/\mathbf{4}\underset{\circ}{\nabla}$	
$X^5\circ(-X)=-X^5$	$X^5/H\circ(-X)=-X^5/H$
$X^5/\mathbf{2}\underset{\circ}{\nabla}\circ(-X)=X^5/\mathbf{2}\underset{\circ}{\nabla}-X^5$	$X^5/A_3\circ(-X)=-X^5/A_3$
$X^5/C_4\circ(-X)=X^5/C_4-X^5/H$	$X^5/K_4\circ(-X)=X^5/K_4+X^5-3X^5/H$
$X^5/\mathbf{2}\underset{\circ}{\nabla}\cdot\mathbf{2}\underset{\circ}{\nabla}\circ(-X)=2X^5/\mathbf{2}\underset{\circ}{\nabla}-X^5-X^5/\mathbf{2}\underset{\circ}{\nabla}\cdot\mathbf{2}\underset{\circ}{\nabla}$	$X^5/C_5\circ(-X)=-X^5/C_5$
$X^5/L\circ(-X)=X^5/L+X^5-2X^5/H-X^5/A_3$	$X^5/A_3\cdot\mathbf{2}\underset{\circ}{\nabla}\circ(-X)=X^5/\mathbf{2}\underset{\circ}{\nabla}\cdot A_3-X^5/A_3$
$X^5/\mathbf{3}\underset{\circ}{\nabla}\circ(-X)=2X^5/\mathbf{2}\underset{\circ}{\nabla}-X^5-X^5/\mathbf{3}\underset{\circ}{\nabla}$	$X^5/D_4\circ(-X)=X^5/\mathbf{2}\underset{\circ}{\nabla}+X^5/D_4-X^5/\mathbf{2}\underset{\circ}{\nabla}\cdot\mathbf{2}\underset{\circ}{\nabla}-X^5/H$
$X^5/D_5\circ(-X)=-X^5/D_5$	$X^5/T\circ(-X)=2X^5/C_4-X^5/T-X^5/H$
$X^5/\mathbf{2}\underset{\circ}{\nabla}\cdot\mathbf{3}\underset{\circ}{\nabla}\circ(-X)=3X^5/\mathbf{2}\underset{\circ}{\nabla}+X^5/\mathbf{2}\underset{\circ}{\nabla}\cdot\mathbf{3}\underset{\circ}{\nabla}-X^5/\mathbf{3}\underset{\circ}{\nabla}-X^5-2X^5/\mathbf{2}\underset{\circ}{\nabla}\cdot\mathbf{2}\underset{\circ}{\nabla}$	
$X^5/A_4\circ(-X)=X^5+X^5/A_4-2X^5/A_3-X^5/H$	
$X^5/\mathbf{4}\underset{\circ}{\nabla}\circ(-X)=3X^5/\mathbf{2}\underset{\circ}{\nabla}+X^5/\mathbf{4}\underset{\circ}{\nabla}-2X^5/\mathbf{3}\underset{\circ}{\nabla}-X^5-X^5/\mathbf{2}\underset{\circ}{\nabla}\cdot\mathbf{2}\underset{\circ}{\nabla}$	
$X^5/A_5\circ(-X)=2X^5+2X^5/A_4+2X^5/L-3X^5/A_3-3X^5/H-X^5/A_5$	
$X^5/\mathbf{5}\underset{\circ}{\nabla}\circ(-X)=2X^5/\mathbf{2}\underset{\circ}{\nabla}\cdot\mathbf{3}\underset{\circ}{\nabla}+2X^5/\mathbf{4}\underset{\circ}{\nabla}+4X^5/\mathbf{2}\underset{\circ}{\nabla}-3X^5/\mathbf{3}\underset{\circ}{\nabla}-X^5/\mathbf{5}\underset{\circ}{\nabla}-X^5-3X^5/\mathbf{2}\underset{\circ}{\nabla}$	

Table 4

The substitution of nX in molecular species of degree ≤ 4

$$1\circ(nX) = 1$$

$$X\circ(nX) = \binom{n}{1} X$$

$$X^2\circ(nX) = \left(\binom{n}{1} + 2\binom{n}{2}\right)X^2 = \binom{n}{1}^2 X^2$$

$$X^2/2\mathbb{Z}\circ(nX) = \binom{n}{2} X^2 + \binom{n}{1} X^2/2\mathbb{Z}$$

$$X^3\circ(nX) = \left(\binom{n}{1} + 6\binom{n}{2} + 6\binom{n}{3}\right) X^3 = \binom{n}{1}^3 X^3$$

$$X^3/2\mathbb{Z}\circ(nX) = \left(\binom{n}{1} + 2\binom{n}{2}\right)X^3/2\mathbb{Z} + \left(2\binom{n}{2} + 3\binom{n}{3}\right) X^3$$

$$X^3/A_3\circ(nX) = \binom{n}{1} X^3/A_3 + \left(2\binom{n}{2} + 2\binom{n}{3}\right) X^3$$

$$X^3/3\mathbb{Z}\circ(nX) = 2\binom{n}{2} X^3/2\mathbb{Z} + \binom{n}{3} X^3 + \binom{n}{1} X^3/3\mathbb{Z}$$

$$X^4\circ(nX) = \left(\binom{n}{1} + 14\binom{n}{2} + 36\binom{n}{3} + 24\binom{n}{4}\right) X^4 = \binom{n}{1}^4 X^4$$

$$X^4/H\circ(nX) = \left(\binom{n}{1} + 2\binom{n}{2}\right) X^4/H + \left(6\binom{n}{2} + 18\binom{n}{3} + 12\binom{n}{4}\right) X^4$$

$$X^4/2\mathbb{Z}\circ(nX) = \left(4\binom{n}{2} + 15\binom{n}{3} + 12\binom{n}{4}\right) X^4 + \left(\binom{n}{1} + 6\binom{n}{2} + 6\binom{n}{3}\right) X^4/2\mathbb{Z}$$

$$X^4/A_3\circ(nX) = \left(4\binom{n}{2} + 12\binom{n}{3} + 8\binom{n}{4}\right) X^4 + \left(\binom{n}{1} + 2\binom{n}{2}\right) X^4/A_3$$

$$X^4/C_4\circ(nX) = \binom{n}{2} X^4/H + \binom{n}{1} X^4/C_4 + \left(3\binom{n}{2} + 9\binom{n}{3} + 6\binom{n}{4}\right) X^4$$

$$X^4/K_4\circ(nX) = 3\binom{n}{2} X^4/H + \left(2\binom{n}{2} + 9\binom{n}{3} + 6\binom{n}{4}\right) X^4 + \binom{n}{1} X^4/K_4$$

$$X^4/2\mathbb{Z}\cdot 2\mathbb{Z}\circ(nX) = \left(\binom{n}{1} + 2\binom{n}{2}\right) X^4/2\mathbb{Z}\cdot 2\mathbb{Z} + \left(\binom{n}{2} + 6\binom{n}{3} + 6\binom{n}{4}\right) X^4 + \left(4\binom{n}{2} + 6\binom{n}{3}\right) X^4/2\mathbb{Z}$$

$$X^4/3\mathbb{Z}\circ(nX) = \left(\binom{n}{1} + 2\binom{n}{2}\right) X^4/3\mathbb{Z} + \left(3\binom{n}{3} + 4\binom{n}{4}\right) X^4 + \left(4\binom{n}{2} + 6\binom{n}{3}\right) X^4/2\mathbb{Z}$$

$$X^4/D_4\circ(nX) = \binom{n}{2}X^4/2\mathbb{Z}\cdot 2\mathbb{Z} + \binom{n}{2}X^4/H + \left(2\binom{n}{2} + 3\binom{n}{3}\right)X^4/2\mathbb{Z} + \binom{n}{1}X^4/D_4 + \left(3\binom{n}{3} + 3\binom{n}{4}\right)X^4$$

$$X^4/A_4\circ(nX) = 2\binom{n}{2} X^4/A_3 + \binom{n}{2} X^4/H + \left(3\binom{n}{3} + 2\binom{n}{4}\right) X^4 + \binom{n}{1} X^4/A_4$$

$$X^4/4\mathbb{Z}\circ(nX) = \binom{n}{2} X^4/2\mathbb{Z}\cdot 2\mathbb{Z} + 2\binom{n}{2} X^4/3\mathbb{Z} + \binom{n}{4} X^4 + 3\binom{n}{3} X^4/2\mathbb{Z} + \binom{n}{1} X^4/4\mathbb{Z}$$

Table 5

REFERENCES

[1] N. Bourbaki, "Algebra". Paris: Hermann, 1942-1962.

[2] N. Bourbaki, "Algèbre Commutative". Paris: Herman, 1962.

[3] A. Joyal, "Une théorie combinatoire des séries formelles". Adv. in Math. Vol. 42, No. 1, pp. 1-82, Octobre 1981.

[4] A. Joyal,"Règle des signes en algèbre combinatoire", C.R. Acad. Sci., Soc. Roy., Canada, Vol. VII, No. 5, October 1985, pp. 285-290.

[5] J. Labelle, "Applications diverses de la théorie combinatoire des espèces de structures". Ann. Sc. Math. du Québec, Vol. 7, no. 1, pp. 59-94, 1983.

[6] J. Labelle, "Quelques espèces sur les ensembles de petite cardinalité". Ann. Sc. Math. du Québec, vol. 9, no. 1, pp. 31-58, 1985.

[7] G. Labelle, "Une nouvelle démonstration combinatoire des formules d'inversion de Lagrange. Adv. in Math. Vol. 42, pp. 217-247, 1981.

[8] G. Labelle, "On Combinatorial Differential Equations ". J. of Math. Analysis and Applications. Acad. Press. (to appear).

[9] P. Leroux and G. Viennot. "Combinatorial resolution of systems of differential equations ,I. Ordinary differential equations ". (This volume).

[10] S. Maclane, "Categories for the working Mathematician". New York: Springer-Verlag, 1971.

[11] D.S. Passman, "Permutation Group", New York: Benjamin, 1969.

[12] B.R. Tennison, "Sheaf Theory", New York: Cambridge University Press, 1975.

[13] Y.N Yeh, "On the combinatorial species of Joyal", Ph.D. Thesis. State University of NY at Buffalo, 1985.

Toward a Combinatorial Proof of the

Jacobian Conjecture?

Doron Zeilberger

Department of Mathematics, Drexel University, Philadelphia, PA 19104

Dominique Foata taught us how to do algebra and special functions combinatorially. Now André Joyal and his diciples teach us how to do calculus combinatorially. The first part of this paper will describe a new approach to combinatorial calculus which was highly inspired by the Québec philosophy and that correspond essentially to Joyal's linear species. However there is a slight conceptual twist in that in my approach the coefficents are indeterminates while in the Québec approach they count things.

Intuitively a function is just a <u>line of dots</u>. Informally, introduce the infinite set $\mathcal{F}$

$$\mathcal{F} = \{\phi, \cdot, \cdot\cdot, \cdot\cdot\cdot, \ldots, \overbrace{\cdot\cdot \cdots \cdot\cdot}^{n \text{ dots}}, \ldots\}$$

and put on the following weight:

$$\mathrm{weight}(\underbrace{\cdot\cdot \cdots \cdot\cdot}_{n \text{ dots}}) = f_n x^n$$

where $\{f_n\}$ and x are commuting indeterminates. Then the generic "function" (really formal power series) $f = \sum_{n=0}^{\infty} f_n x^n$ is the weight of $\mathcal{F}$: $f = \mathrm{weight}\ (\mathcal{F})$.

We will soon consider rows of dots with some dots circled. The weight of a row of dots with n dots, k of which are circled, is defined to be $f_n x^{n-k}$. The derived set $\mathcal{F}'$ of $\mathcal{F}$ is the set of all possible lines of dots with exactly one dot circled:

$$\mathcal{F}' = \{ \odot ; \odot \cdot , \cdot \odot ; \odot \cdot\cdot , \cdot \odot \cdot , \cdot\cdot \odot ; \ldots \}$$

then it is readily seen that

$$f' = \mathrm{weight}(\mathcal{F}') .$$

Similarly, for any fixed k, let $\mathfrak{F}^{(k)}$ be the set of all lines of dots with exactly k dots circled, then $\frac{D^k f}{k!} = \text{weight}(\mathfrak{F}^{(k)})$.

Product of two functions: Let $\mathfrak{G}$ be the same as $\mathfrak{F}$ only endowed with the weight $g_n x^n$ rather than $f_n x^n$ and let

$$\mathfrak{F} \times \mathfrak{G} = \left\{ \begin{matrix} \overbrace{\cdot\;\cdot\;\cdot\;\cdot}^{n_1} \\ \\ \underbrace{\cdot\;\cdot\;\cdot\;\cdot\;\cdot}_{n_2} \end{matrix} \;,\; n_1 \geq 0 \;,\; n_2 \geq 0 \right\}$$

with the weight

$$\text{weight}\left\{ \begin{matrix} \overbrace{\cdot\;\cdot\;\cdot\;\cdot}^{n_1} \\ \\ \underbrace{\cdot\;\cdot\;\cdot\;\cdot\;\cdot}_{n_2} \end{matrix} \right\} = f_{n_1} g_{n_2} x^{n_1 + n_2}$$

clearly $fg = \text{weight}(\mathfrak{F} \times \mathfrak{G})$.

Now the fun begins! $(fg)' = \text{weight}\left[(\mathfrak{F} \times \mathfrak{G})'\right]$, but $(\mathfrak{F} \times \mathfrak{G})'$ is the set of two rows of dots with one dot circled. Since that dot may be either on the top row $\left[\mathfrak{F}' \times \mathfrak{G}\right]$ or on the bottom row $\left[\mathfrak{F} \times \mathfrak{G}'\right]$ we have

$$\left[\mathfrak{F} \times \mathfrak{G}\right]' = \mathfrak{F}' \times \mathfrak{G} \cup \mathfrak{F} \times \mathfrak{G}' \quad .$$

Taking weights gives the good old product rule

$$(fg)' = f'g + fg' \quad .$$

Similarly, we can prove Leibnitz's rule

$$\underset{\substack{\uparrow \\ \text{n dots circled}}}{\frac{D^n}{n!}(fg)} = \sum_{k=0}^{n} \underset{\substack{\uparrow \\ \text{k in top}}}{\frac{(D^k f)}{k!}} \; \underset{\substack{\uparrow \\ \text{n-k in bottom}}}{\frac{(D^{n-k} g)}{(n-k)!}} \quad .$$

Composition: $f(g)$ is the weight enumerator of creatures of the form

(1) 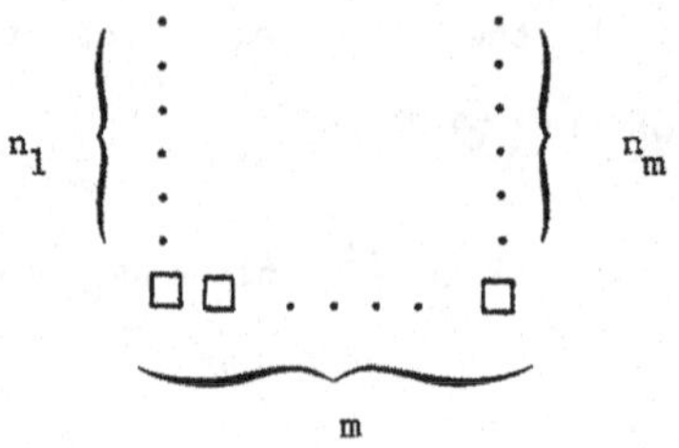

whose weight is $f_m g_{n_1} g_{n_2} \cdots g_{n_m} x^{n_1+\cdots+n_m}$.

Chain Rule: $[f(g)]' = f'(g)g'$

Proof: The act of differentiating is a decision: What dot to circle? This decision can be broken up into two decisions:

i. In what column to circle? i.e., to choose a □ , the weight of all these is $f'(g)$.

ii. In the chosen column, where to put the circle? i.e., g' .

Thus $[f(g)]' = f'(g) \cdot g'$.

Exercise: Prove Faa'di Bruno's formula

$$\frac{[f(g)]^{(n)}}{n!} = \sum_k \sum_{\substack{k_1+\cdots+k_n = k \\ k_1+2k_2+\cdots+nk_n = n}} \frac{k!}{k_1!\cdots k_n!} \frac{f^{(k)}}{k!} \left(\frac{g^{(1)}}{1!}\right)^{k_1} \cdots \left(\frac{g^{(n)}}{n!}\right)^{k_n}$$

Hint: For every way of circling n dots in (1) let there be k_i columns with i dots $(i=1,\ldots,n)$ then there are $k = k_1+\cdots+k_n$ columns that have at least one circled dot in them.

The Joni formula will be crucial for our approach to the Jacobian conjecture. Let's first state and prove it for $n=1$, in which case it is just one of the versions of Lagrange inversion.

Theorem (Lagrange): Let $U(x)$ be any formal power series and let $F(x)$ be a formal power series of the form $F(x) = x + f_2x^2 + \cdots$ then

$$U(x) = \sum_{p \in N} \frac{D^p}{p!} (U(F)F'(X-F)^p) .$$

<u>Proof</u>: By linearity it is enough to prove, for every m, that

$$x^m = \sum_{p \in N} \frac{D^p}{p!} (F^m F'(x-F)^p) .$$

Put $F = x-H$ where $H = -f_2x^2 - f_3x^3 - \ldots$ then we have to prove, for every m

$$(1) \qquad x^m = \sum_{p \in N} \frac{D^p}{p!} ((x-H)^m H^p (x-H)') .$$

$\frac{D^p}{p!}$ corresponds to circling p dots. The right hand side enumerates creatures of the following kind.

Left Section	Middle Section	Right Section
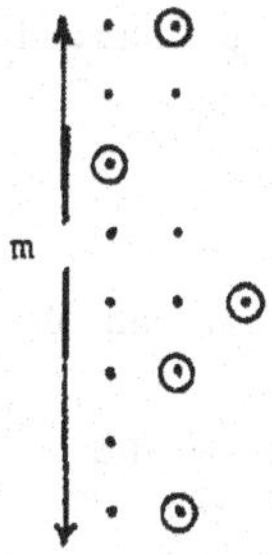 $(x-H)^m$	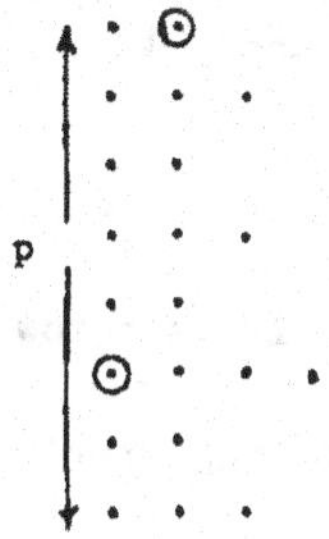 H^p	 $(x-H)'$
m rows some with one dot and some with more than one dot. the contribution to the weight from a row of more than one dot is the negative of what it is in the middle section.	p rows of dots ($p \geq 0$) all with at least two dots.	one row either with one dot or with more than one dot. But one dot is circled by a special circle and possibly other circles.

ALTOGETHER YOU HAVE $p (p \geq 0)$ DOTS CIRCLED.

I.E., # OF ROWS IN MIDDLE SECTION = # OF DOTS CIRCLED.

We need a sign reversing involution whose sole survivor is a creature whose weight is x^m .

Here it is:

Look at the left section from top to bottom. Every row is either

i. a circled dot,

ii. a row of at least two dots,

iii. an uncircled dot.

Look for the first row that is not of kind iii.

Case I: It is of kind i , i.e., it is a circled dot (whose weight is just 1)

DELETE THAT CIRCLED DOT AND MOVE TO THAT PLACE THE TOP ROW OF THE MIDDLE SECTION.

(This reduces both the # of circles and the # of rows in the middle section by 1 , i.e., $p \longleftarrow p-1$. The weight is preserved except for a factor of -1 .)

Case II: It is of kind ii , i.e., it is a row with more than one dot.

TRANSFER THAT ROW TO THE TOP OF THE MIDDLE SECTION AND IN THE VACANCY OBTAINED FROM THAT REMOVAL PUT A CIRCLED DOT.

(Here $p \longleftarrow p+1$, weight $\longleftarrow$ - weight.)

This is an involution whenever it can be applied. Case I and Case II are inverses of each other.

Examples:

Case I

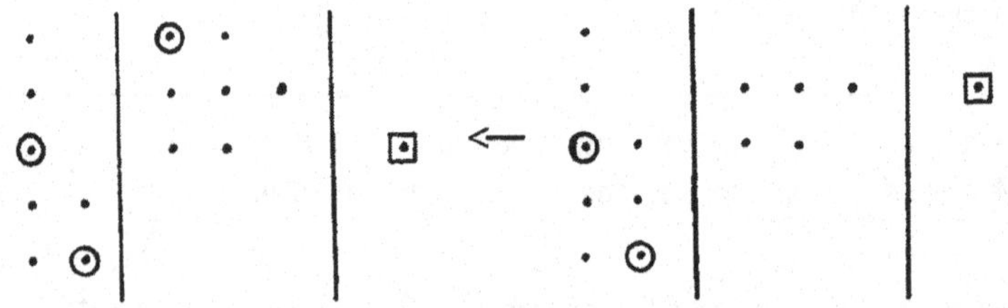

Case II 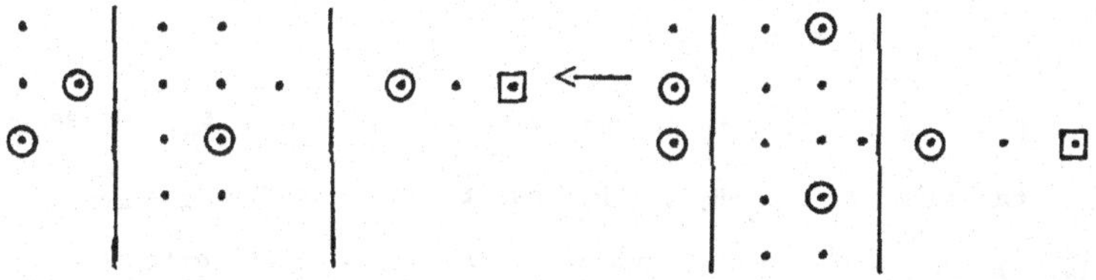

DO RIGHT NOW

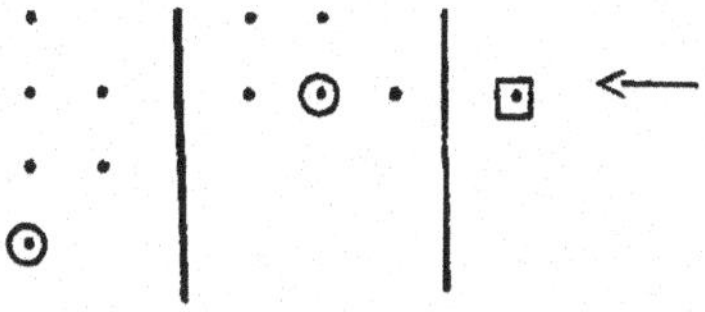

Who are the survivors of this involution?

Those in which all the left section is of kind iii, i.e., uncircled rows:

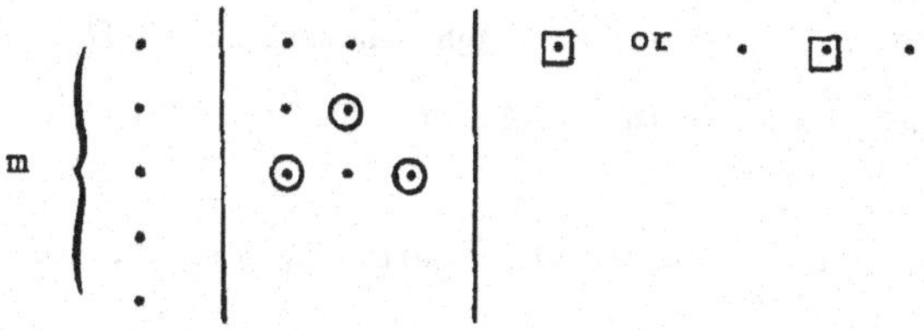

The total weight of these guys is $x^m \sum_p \frac{D^p}{p!}(H^p(1-H')) = x^m\left(\sum_p \frac{D^p}{p!} H^p - \sum \frac{D^p}{p!}\left(\frac{DH^{p+1}}{p+1}\right)\right) = x^m\left(\sum_p \frac{D^p}{p!} H^p - \sum \frac{D^{p+1}}{(p+1)!} H^{p+1}\right)$

telescoping

$\equiv x^m \cdot 1 = x^m$.

Note: This last algebraic step can also be made purely combinatorial.

The general Joni Formula ([2], [1]) is

$$\text{(J)} \quad U(X) = \sum_{p \in N^n} \frac{D^p}{p!} (U(F)j(F)H^p)$$

Here $p! = p_1! \dots p_n!$

$X = (X_1,\dots,X_n)$ U is a f.p.s. in n variables. $F = (F_1,\dots,F_n)$ We assume that F_i has the form $F_i = X_i - H_i$, H_i has terms of degree ≥ 2, $j(F) = \det(D_iF_j) = \det(\delta_{ij} - D_iH_j)$. The proof of $n=1$ can be extended to this and will appear in [5].

COMBINATORIAL ADVANCED CALCULUS

For the sake of simplicity, $n=2$.

$$\mathcal{F} = \{\underbrace{xx\dots x}_{m}\ \underbrace{yy\dots y}_{n}\ ;\ m \geq 0\ \ n \geq 0\}$$

$$\text{weight}(\underbrace{xx\dots x}_{m}\ \underbrace{yy\dots y}_{n}) = f_{mn}x^m y^n$$

f_{mn}, x, y are commuting indeterminates. D_1F is the weight enumerator of $\mathcal{F}^{(1,0)}$: $\mathcal{F}$ with one x circled. Similarly D_2F is the weight of $\mathcal{F}^{(0,1)}$.

$\frac{D_1^{p_1}}{p_1!}\frac{D_2^{p_2}}{p_2!}F$ is the weight enumerator of elements of $\mathcal{F}$ with p_1 x's circled and p_2 y's circled.

x ⓧ x ⓧ ... x y ⓨ ... ⓨ ⓨ y

↑ p_1 circles ↑ p_2 circles

Composition: $F(G_1,G_2)$ is the weight enumerator of creatures of the form

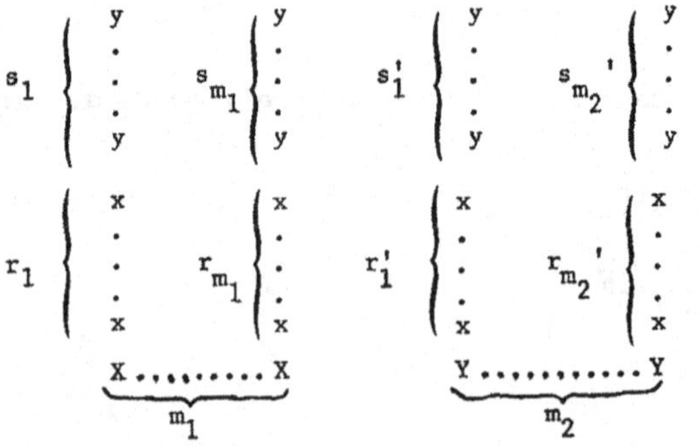

The weight of this is

$$f_{m_1m_2}\, g^{(1)}_{r_1s_1}\, x^{r_1}y^{s_1} \cdots g^{(1)}_{r_{m_1}s_{m_1}}\, x^{r_{m_1}}y^{s_{m_1}}$$

$$\cdot\, g^{(2)}_{r'_1s'_1}\, x^{r'_1}y^{s'_1} \cdots g^{(2)}_{r'_{m_2}s'_{m_2}}\, x^{r'_{m_2}}y^{s'_{m_2}}$$

<u>Exercise</u>: PROVE THE CHAIN RULE

$$D_1F(G_1,G_2) = \frac{\partial F}{\partial G_1}\frac{\partial G_1}{\partial x} + \frac{\partial F}{\partial G_2}\frac{\partial G_2}{\partial x}$$

$$D_2F(G_1,G_2) = \frac{\partial F}{\partial G_1}\frac{\partial G_1}{\partial y} + \frac{\partial F}{\partial G_2}\frac{\partial G_2}{\partial y}$$

<u>The Jacobian Conjecture</u>

Posed by Keller in 1941. (Several false proofs narrated with glee in Bass-Connel-Wright[1]).

> Let $F_1(X_1,\ldots,X_n),\ldots,F_n(X_1,\ldots,X_n)$ be <u>polynomials</u> such that
>
> $$J(F) = \det(D_iF_j) \equiv 1$$
>
> then the"inverse transformation" $(G_1,\ldots,G_n)$ (that exists because the Jacobian is not zero at the origin) are <u>actually</u> polynomials.

"One of the outstanding open problems in combinatorics", Zeilberger

The Joni formula (J) with $U(X) = G_i(X)$ gives

(2) $$G_i(X) = \sum_{p\in N^n} \frac{D^p}{p!}(X_iH^p)\,, \quad H = (X_1 - F_1, \ldots, X_n - F_n)$$

and with $U(X) \equiv 1$

$$(3)\qquad 1 = \sum_{p\, \in\, N^n} \frac{D^p}{p!}\,(H^p) \quad .$$

For the sake of simplicity take $n=2$ (The Jacobian Conjecture is even open for $n=2$.) and we have

$$G_1(x,y) = \sum_{p_1,p_2} \frac{D_1^{p_1}\, D_2^{p_2}}{p_1\,!\; p_2\,!}\left(x\, H_1^{p_1}\, H_2^{p_2} \right)$$

$$G_2(x,y) = \sum_{p_1,p_2} \frac{D_1^{p_1}\, D_2^{p_2}}{p_1\,!\; p_2\,!}\left(y\, H_1^{p_1}\, H_2^{p_2} \right)$$

Write

$$g_1^{(m,n)} = \text{coeff. of } x^m y^n \text{ in } G_1$$

$$g_2^{(m,n)} = \;=\;\;=\;\;=\;\;= G_2 \quad .$$

The Jacobian Conjecture will be proved if we can show that for $m+n \gg 0$, $g_1^{(m,n)} = 0$ and $g_2^{(m,n)} = 0$.

Let $\overline{\mathscr{S}}\,(m,n)$ be the set of creatures of the form, for some p_1 and p_2

$$p_1 \left\{ \begin{array}{l} x \;\; \textcircled{x} \;\; \ldots \;\; y \;\; y \\ \cdot \\ \cdot \\ \cdot \\ x \;\; x \;\; \textcircled{x} \;\; \ldots \;\; x \;\; y \ldots y \end{array} \right. \quad ; \quad \left. \begin{array}{l} x \;\; \textcircled{x} \;\; \ldots \;\; x \;\; y \ldots y \\ \cdot \\ \cdot \\ \cdot \\ x \;\; x \;\; y \ldots y \end{array} \right\} p_2$$

and altogether p_1 x's are circled and p_2 y's are circled and all rows have at least length 2.

That is,

# of rows at left = # of x's circled ,

# of rows at right = # of y's circled.

Furthermore, let the number of uncircled x's be m and the number of uncircled y's be n and the weight of a row with r x's and s y's be $h_{r,s}^{(1)}$. on the left and $h_{r,s}^{(2)}$ on the right. Finally, the weight of such a creature is the product of the weights of all the individual rows. It is readily seen that, by (3),

$$1 \equiv \sum \frac{D_1^{p_1} \; D_2^{p_2}}{p_1 ! \; p_2 !} \left(H_1^{p_1} \; H_2^{p_2} \right)$$

so the coefficient of $x^m y^n$ in the r.h.s. is the weight of $\overline{\mathscr{G}}(m,n)$. Thus for $m+n > 0$,

$$\text{weight } (\overline{\mathscr{G}}(m,n)) = 0 .$$

Now from (2) it follows that

$$G_1(x,y) = x + \sum_{p_1, p_2} \frac{D_1^{p_1-1}}{(p_1-1)!} \; \frac{D_2^{p_2}}{p_2 !} \left(H_1^{p_1} \; H_2^{p_2} \right) ,$$

and it is readily seen that $g_1^{(m,n)}$, the coefficient of $x^m y^n$ in G_1 is the weight of the set $\mathscr{G}(m,n)$ that has exactly the same form as $\overline{\mathscr{G}}(m,n)$ only with an ever so slightly different condition:

# of rows at left = # of x's circled + 1 .

$\mathscr{G}(m,n)$ and $\overline{\mathscr{G}}(m,n)$ are very similar sets and it is very frustrating that it is so easy to show that weight$(\overline{\mathscr{G}}(m,n)) \equiv 0$ but so hard to show that the weight of $\mathscr{G}(m,n)$ is zero.

<u>But there is hope</u>!

You can find lots and lots (in fact too many) subsets of $\mathscr{G}(m,n)$ that are "isomorphic" to some $\overline{\mathscr{G}}(r,s)$ with $r \leq m$, $s \leq m$ and thus lots of subsets whose weight is zero, e.g.,

$$\begin{array}{llll|lll} x & x & \textcircled{y} & y & \textcircled{x} & y & \\ x & y & \textcircled{y} & & x & y & y \end{array} ,$$

an element of $\mathscr{G}(4,5)$ equals, in weight, to $\left[x \;\; y \;\; \textcircled{y} \,|\, x \;\; y \;\; x \right] \cdot A$ where A is the element $x \;\; x \;\; \textcircled{y} \;\; y \,|\, \textcircled{x} \;\; y$ that belongs to $\overline{\mathscr{G}}(2,2)$. For sufficiently large $m+n$ it can be shown that <u>every</u>

element of $\mathcal{G}(m,n)$ belongs to some "zero set". Thus there is a huge family of subsets of $\mathcal{G}(m,n)$ s.t.

$$\mathcal{G}(m,n) = \bigcup_{A \in \mathcal{A}} A$$

and weight(A) = 0 .

Unfortunately this is not a disjoint union so we <u>cannot</u> conclude that weight($\mathcal{G}(m,n)$) = 0 .

My hope is that one day somebody (maybe you!) will find a subfamily $\mathcal{A}'$ such that it is still true that

$$\mathcal{G}(m,n) = \bigcup_{A \in \mathcal{A}'} A$$

but $\mathcal{A}'$ is closed under intersection, then it would follow from inclusion exclusion that

$$\text{weight}(\mathcal{G}(m,n)) = \Sigma\,\text{weight}(A) - \Sigma\,\text{weight}(A \cap B) + \dots = 0 \quad .$$

□

References

1. H. Bass, E. H. Connell and D. Wright, The Jacobian Conjecture: reduction of degree and formal expansion of the inverse, Bulletin of the A.M.S. 7 (1982) 287-330.

2. S. A. Joni, Lagrange inversion in higher dimensions and umbral operators, Linear and Multilinear Algebra 6 (1978) 111-121.

3. A. Joyal, Une théorie combinatoire des séries formelles, Adv. in Math. 42 No. 1, 1981.

4. G. Labelle, Une nouvelle démonstration combinatoire des formules d'inversion de Lagrange, Adv. in Math. 42 (1981) 217-247.

5. D. Zeilberger, A combinatorial proof of Joni's multidimensional Lagrange inversion formula, in preparation.

SÉANCE DE PROBLEMES

Ashok K. Agarwal, Pennsylvania State University, University Park, PA 16802.

See Agarwal's paper in this volume. The problems given at the problem session correspond to the cases $k = -2$ (now solved) and $k = 1$ (still open) in that paper.

◇◇◇◇◇◇◇

Richard Askey, University of Wisconsin, Madison, WI 53706, U.S.A.

A. Selberg (1944) proved that

$$\int_0^1 \cdots \int_0^1 \left[\prod_{1 \le i < j \le n} (t_i - t_j)^2\right]^z \prod_{1 \le i \le n} t_i^{x-1}(1 - t_i)^{y-1} dt_i$$

$$= \prod_{1 \le j \le n} \frac{\Gamma(x + (j-1)z)\,\Gamma(y + (j-1)z)\,\Gamma(jz + 1)}{\Gamma(x + y + (n+j-2)z)\ \Gamma(z+1)}$$

This is equivalent to finding the constant term in

$$\prod_{1 \le i \le n} (1-t_i)^a (1-t_i^{-1})^a (1-t_i^2)^b (1-t_i^{-2})^b$$

$$\cdot \prod_{1 \le i < j \le n} (1-t_i t_j)^c (1-t_i^{-1} t_j^{-1})^c (1-t_i t_j^{-1})^c (1-t_i^{-1} t_j)^c$$

There are many conjectured identities related to Selberg's integral. Three are as follows:

Prove that the constant terms in the following expressions are as given.

$$\text{C.T. } (1-x)^a (1-x^{-1})^a (1-y)^a (1-y^{-1})^a (1-xy)^a (1-x^{-1}y^{-1})^a$$
$$\cdot (1-xy^{-1})^b (1-x^{-1}y)^b (1-x^2y)^b (1-x^{-2}y^{-1})^b (1-xy^2)^b (1-x^{-1}y^{-2})^b$$

$$= \frac{(3a + 3b)!\,(3b)!\,(2a)!\,(2b)!}{(2a + 3b)!\,(a + 2b)!\,(a+b)!\,a!\,b!\,b!}$$

$$\text{C.T. } (x; q)_a (qx^{-1}; q)_a (y; q)_a (qy^{-1}; q)_a (xy; q)_a (qx^{-1}y^{-1}; q)_a$$
$$\cdot (x^{-1}y; q^3)_b (q^3xy^{-1}; q^3)_b (x^2y; q^3)_b (q^3x^{-2}y^{-1}; q^3)_b (xy^2; q^3)_b (q^3x^{-1}y^{-2}; q^3)_b$$

$$= \frac{(q;q)_{3a+3b}(q;q)_{3b}(q;q)_{2a}(q^3;q^3)_{a+3b}(q^3;q^3)_{2b}(q^3;q^3)_a}{(q;q)_{2a+3b}(q;q)_{a+3b}(q;q)_a^2(q^3;q^3)_{a+2b}(q^3;q^3)_{a+b}(q^3;q^3)_b^2}$$

$$\text{C.T. } (x;q)_a (qx^{-1};q)_a (y;q)_a (qy^{-1};q)_a (xy;q)_a (qx^{-1}y^{-1};q)_a \cdot (x^{-1}y;q)_b (qxy^{-1};q)_b (x^2y;q)_b (qx^{-2}y^{-1};q)_b (xy^2;q)_b (qx^{-1}y^{-2};q)_b$$

$$= \frac{(q;q)_{3a+3b} (q;q)_{3b} (q;q)_{2a} (q;q)_{2b}}{(q;q)_{2a+3b} (q;q)_{a+2b} (q;q)_{a+b} (q;q)_a (q;q)_b^2}$$

Here $(a;q)_n = (1-a)(1-aq)\dots(1-aq^{n-1})$.

"I will pay 25$ U.S. for the first proof of any one of these". For further conjectures of this type see the following two texts.

References

[1] I. G. Macdonald, *Some Conjectures for Root Systems*. SIAM J. Math. Anal., vol. 13, no. 6, 1982, pp. 988-1007.

[2] W. G. Morris, II, *Constant Term Identities for finite Affine Root Systems: Conjectures and Theorems*, Ph.D. Thesis, U. of Wisconsin-Madison, Jan. 1982.

◇◇◇◇◇◇◇

Omer Egecioglu, U. of California, Santa-Barbara CA 93106, U.S.A.

A theorem of Specht circa 1940 states the following: Let R be a real-closed field (for all practical purposes R = $\mathbb{C}$). Denote by * conjugate transpose. Consider the words in x, x* with homorphisms into $\mathfrak{M}_n$ (n × n matrices) defined by

$$\varphi_A x = A, \quad \varphi_B x = B, \quad \varphi_A x^* = A^*, \quad \varphi_B x^* = B^*$$

and extend. Suppose $\operatorname{tr}(\varphi_A \omega) = \operatorname{tr}(\varphi_B \omega)$ for all words ω in x and x*. Then A and B are unitary equivalent. This needs a combinatorial proof at least for the case * = transpose and unitary = orthogonal.

Reference Kaplansky, *Linear Algebra and Geometry*. Chelsea, N.Y. 1974. (p.71)

◇◇◇◇◇◇◇

G. Kreweras, Institut de statistique de l'Université de Paris, Un. Paris VI, F-75005 Paris France.

Toute permutation "alternée down-up" $\omega = x_1 x_2 x_3 x_4 \dots x_n$ possède une "forme dépliée" $\omega^* = x_2 x_4 x_6 \dots x_5 x_3 x_1 = y_1 y_2 \dots y_n$ $(y_1 = x_2,\ y_2 = x_4, \dots, y_{n-1} = x_3,\ y_n = x_1)$. ω^* peut présenter des "décroissances" (si $y_i > y_{i+1}$) et des "anticipations" (si $i < j$ et

$y_i = y_j+1$). Par exemple, si $\omega = 5371624$, alors $\omega^* = 3124675$ possède 2 décroissances ($3\searrow 1$ et $7\searrow 5$) et 2 anticipations (3 avant 2 et 6 avant 5). Soit $f_n(p,q)$ = nombre de permutations alternées down-up ω de $\{1, 2, ..., n\}$ telles que ω^* présente p décroissances et q anticipations.

Conjecture: $f_n(p,q) = f_n(q,p)$

"Pour l'instant, je sais seulement calculer (péniblement) $f_n(1,q)$ et $f_n(q,1)$ et montrer qu'ils ont tous deux pour expression $\sum_h \binom{h-1}{q-1}\binom{n-2h+1}{q+1}$."

◇◇◇◇◇◇◇

Gilbert Labelle, Université du Québec à Montréal, Montréal (Qué.) Canada, H3C 3P8.

1. Let $\mathbf{U}$ be a finite set and let $\beta: \mathbf{U} \xrightarrow{\sim} \mathbf{U}$ be a permutation of $\mathbf{U}$ of type $\beta_1, \beta_2, \beta_3, ...$ (where β_k is the number of cycles of length k in β). It is known (see [1]) that the number of rooted trees on $\mathbf{U}$ which are invariant under β is given by the formula

$$\beta_1{}^{\beta_1 - 1} \prod_{k\geq 2} \{(\mathrm{fix}(\beta^k))^{\beta_k} - k\beta_k (\mathrm{fix}(\beta^k))^{\beta_k - 1}\}$$

if $\beta_1 \geq 1$ and by 0 if $\beta_1 = 0$. Here $\mathrm{fix}(\beta^k)$ denotes the number of fixed points of the kth iterate of β. The problem is to find a direct combinatorial proof of the formula.
Hint: The same problem for endofunctions corresponds to the formula

$$\beta_1{}^{\beta_1} (\mathrm{fix}(\beta^2))^{\beta_2} (\mathrm{fix}(\beta^3))^{\beta_3} \cdots$$

Reference: [1] G. Labelle, *Some new computational methods in the theory of species*, this volume.

Editors' note: Doron Zeilberger proposed a nice proof of the formula using the matrix tree theorem. The problem to find a direct "bijective" proof of the formula is still open.

2. Let $F = F(X)$ be a (virtual) species. A. Joyal [1] has shown that the virtual species $G = G(X)$ given by the formula

$$G = E_1 F - E_2 F' + E_3 F'' - E_4 F''' + \ldots$$

(where E_n denotes the species of all n-points discrete graphs) is an integral of F in the sense that

$$G'(X) = F(X) \quad \text{and} \quad G[\emptyset] = \emptyset$$

The problem is to find an analogous formula for multisorted species. More specifically, given k (virtual) k-sorted species $F_i = F_i(X_1, \ldots, X_k)$, $i=1, \ldots, k$, with $\partial F_i/\partial X_j = \partial F_j/\partial X_i$, $1 \le i, j \le k$, find, in a canonical manner, a virtual k-sorted species $G = G(X_1, \ldots, X_k)$ such that

$$\partial G/\partial X_i = F_i(X_1, \ldots, X_k),\ i=1, \ldots, k,\ G[\emptyset, \ldots, \emptyset] = \emptyset.$$

References

[1] A. Joyal. *Calcul intégral combinatoire et homologie des groupes symétriques.* C. R. Math. Rep. Acad. Sci. Canada, vol. VII, no. 6, déc. 1985.

[2] G. Labelle. *On Combinatorial Differential Equations*. J. Math. Anal. and Appl. (to appear).

Editors' notes: This problem has been solved by Yeong Nan Yeh (to appear in the "Annales des Sciences Mathématiques du Québec").

◇◇◇◇◇◇◇

Simon Plouffe, Université du Québec à Montréal, Montréal (Qué.), Canada H3C 3P8.

Trouver une preuve combinatoire de la formule asymptotique

$$p(n) \sim (4n\sqrt{3})^{-1} \exp(\pi \sqrt{2n/3})$$

pour le nombre p(n) de partages arithmétiques de l'entier n.

Référence, G. H. Hardy, S. Ramanujan, *Asymptotic formulae in combinatory analysis*, Proc. London Math. Soc. (2), 17, (1918), 75-115.

Note des rédacteurs: "Hopeless" selon Richard Askey.

◇◇◇◇◇◇◇

Richard P. Stanley, Massachusetts Institute of Technology, Cambridge, Mass. 02139.

See Stanley's paper in this volume. Some of the conjectures contained in that paper were presented at the problem session.

◇◇◇◇◇◇◇

Loys Thimonier, U. Amiens, F-80039, Amiens, France.

Let $A = \{a_1, \ldots, a_m\}$ be an alphabet and $P = \{p_1, \ldots, p_m\}$ be a probability distribution associated with its letters. At each time $t = 1, 2, \ldots, n, \ldots,$ there is a drawing with replacement of a letter of A, up to the obtainment of a palindrome w (event $\mathcal{E}$); [recall that $w = a_{i_1} \ldots a_{i_k}$ is a palindrome if $w = w^t = a_{i_k} \ldots a_{i_1}$ and $|w| \geq 2$]. No prefix of w is a palindrome, because w is the first obtained palindrome. By considering mutually exclusive events,

$$\Pr(\mathcal{E}) = \sum_{n\geq 2} \Big(\sum_{\alpha_1 + \ldots + \alpha_m = n} C_{\alpha_1, \ldots, \alpha_m} \, p_1^{\alpha_1} \ldots p_m^{\alpha_m} \Big)$$

where $C_{\alpha_1, \ldots, \alpha_m}$ is the number of prefix-free palindromes of length n of commutative image $a_1^{\alpha_1} a_2^{\alpha_2} \ldots a_m^{\alpha_m}$.

The problem is to find "closed expressions" for f and g defined by

$$\Pr(\mathcal{E}) = f(p_1, \ldots, p_m) \ , \qquad E(X|\mathcal{E}) = g(p_1, \ldots, p_m)$$

where the random variable X is equal to the first time n at which event $\mathcal{E}$ is obtained. Begin with m = 3, because the case m = 2 is well-known (see Feller 1968):

$$\Pr(\mathcal{E}) = \sum_{n\geq 2} (p_1 p_2^{n-2} p_1 + p_2 p_1^{n-2} p_2) = 1 \ , \qquad E(X|\mathcal{E}) = 3.$$

◇◇◇◇◇◇◇

Gérard Viennot, Université de Bordeaux I, F-33405, Talence, France.

1. Donner une preuve bijective de la formule suivante:

$$p_{2n+8} = (2n+11)\, 4^n - 4(2n+1)\binom{2n}{n}$$

où p_{2n} est le nombre de polyominos convexes de périmètre 2n.

Un <u>polyomino convexe</u> est une union connexe de "carrés élémentaires" dont l'intérieur est connexe et telle que l'intersection avec toute droite verticale ou horizontale est un segment connexe. Par exemple:

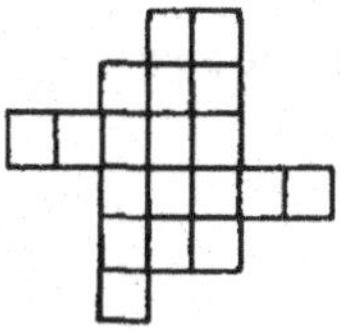

Référence: M.-P. Delest & G. Viennot, *Algebraic languages and polyominoes enumeration*. Theoretical Computer Science 34 (1984), 169-206. North-Holland.

Prix offert: 10 bouteilles de vin, Domaine des Mattes, 1981.

2. (Avec Mireille Vauchaussade de Chaumont)

Soit A un arbre planaire (enraciné, non étiquetté) énuméré par les nombres de Catalan. Un filament de A est une suite maximale de sommets $(s_1, \ldots, s_k)$ telle que pour $i = 1, \ldots, k-1$, s_{i+1} est fils unique de s_i, et s_k est une feuille de A. On dénote par R(A), l'arbre obtenu de A par émondage, c'est-à-dire en enlevant ses filaments. Le plus petit entier n tel que $R^n(A) = \{\bullet\}$ s'appelle l'ordre de A. Par exemple:

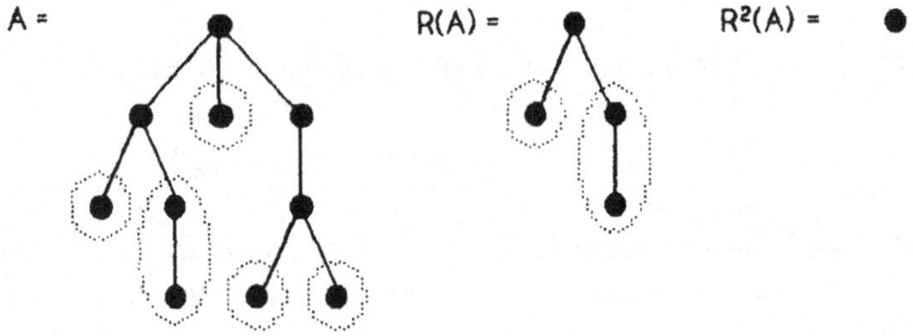

et A est d'ordre 2.

Montrer bijectivement que la distribution du paramètre ordre pour les arbres planaires ayant n + 1 sommets est la même que celle du paramètre ordre restreint aux arbres binaires complets ayant 2n + 1 sommets. Un arbre binaire complet est un arbre planaire tel que tout sommet possède soit zéro fils, soit deux fils.

Référence: M. Vauchaussade de Chaumont & G. Viennot, *Enumeration of RNAs by complexity*. Proc. Intern. Conf. of Medicine and Biology, Bari, Italie, 1983. Lecture Notes in Biomathematics, 1985.

Prix offert: 10 bouteilles de vin.

Note des rédacteurs. Ce problème a été résolu par Doron Zeilberger (article soumis au Europ J. Comb.) et, indépendamment, par Rodney Canfield et Ed Bender conjointement.

◊◊◊◊◊◊◊

Doron Zeilberger, Drexel University, Philadelphia, PA 19104, USA.

1. (50.00$) Find a nice combinatorial proof of the Askey-Gasper positivity result for sums of Jacobi polynomials:

$$\sum_{k\geq 0} P_k^{(\alpha,0)}(x) \geq 0\ , \quad \text{for } \alpha \geq -2\ ,\ -1 < x < 1\ ,$$

or of clausen's formula:

$$ {}_3F_2\left[\begin{matrix} 2a\ , 2b\ , a+b \\ 2a+2b\ , a+b+\frac{1}{2} \end{matrix}\ ; x\right] = \left\{ {}_2F_1\left[\begin{matrix} a\ , b \\ a+b+\frac{1}{2} \end{matrix}\ ; x\right]\right\}^2 , \qquad |x|<1\ .$$

These results are essential in de Brange's proof of the Bieberbach conjecture.

Références: R. Askey & G. Gasper, *Positive Jacobi polynomial sums*. II. Amer. J. Math. 98 (1976), 709-737.

G. Gasper, *A short proof of an inequality used by de Branges in his proof of the Bieberbach, Robertson, and Milin conjectures*. Preprint.

2. (25.00$) Give a Foata-style (bijective) proof of

$$\sum_w q^{Z(w)} = \sum_w q^{\text{inv}\,(w)}$$

Références. D. Foata, *On the Netto inversion number of a sequence*. Proc. Amer. Math. Soc. 19 (1968), 236-240.

D. Zeilberger & D. M. Bressoud, *A proof of Andrew's q-Dyson conjecture*. Discrete Math. 54 (1985), 201-224.

3. (40.00$). Give an analytical proof of the Rogers-Ramanujan identities without using the Jacobi triple product identity, preferably without using minus signs. Hopefully this will simplify the combinatorial proof of the R.-R. identities.

Editors'note: This was done by George Andrews, but using Watson's transformation ! Zeilberger paid 20.00$ ...

◇◇◇◇◇◇◇